MULTIVARIABLE CALCULUS

The Jones & Bartlett Learning Series in Mathematics

Geometry

Geometry with an Introduction to Cosmic Topology
Hitchman (978-0-7637-5457-0) © 2009

Euclidean and Transformational Geometry: A Deductive Inquiry
Libeskind (978-0-7637-4366-6) © 2008

A Gateway to Modern Geometry: The Poincaré Half-Plane, Second Edition
Stahl (978-0-7637-5381-8) © 2008

Understanding Modern Mathematics
Stahl (978-0-7637-3401-5) © 2007

Lebesgue Integration on Euclidean Space, Revised Edition
Jones (978-0-7637-1708-7) © 2001

Precalculus

Essentials of Precalculus with Calculus Previews, Fifth Edition
Zill/Dewar (978-1-4496-1497-3) © 2012

Algebra and Trigonometry, Third Edition
Zill/Dewar (978-0-7637-5461-7) © 2012

College Algebra, Third Edition
Zill/Dewar (978-1-4496-0602-2) © 2012

Trigonometry, Third Edition
Zill/Dewar (978-1-4496-0604-6) © 2012

Precalculus: A Functional Approach to Graphing and Problem Solving, Sixth Edition
Smith (978-0-7637-5177-7) © 2012

Precalculus with Calculus Previews (Expanded Volume), Fourth Edition
Zill/Dewar (978-0-7637-6631-3) © 2010

Calculus

Single Variable Calculus: Early Transcendentals, Fourth Edition
Zill/Wright (978-0-7637-4965-1) © 2011

Multivariable Calculus, Fourth Edition
Zill/Wright (978-0-7637-4966-8) © 2011

Calculus: Early Transcendentals, Fourth Edition
Zill/Wright (978-0-7637-5995-7) © 2011

Multivariable Calculus
Damiano/Freije (978-0-7637-8247-4) © 2011

Calculus: The Language of Change
Cohen/Henle (978-0-7637-2947-9) © 2005

Applied Calculus for Scientists and Engineers
Blume (978-0-7637-2877-9) © 2005

Calculus: Labs for Mathematica
O'Connor (978-0-7637-3425-1) © 2005

Calculus: Labs for MATLAB®
O'Connor (978-0-7637-3426-8) © 2005

Linear Algebra

Linear Algebra: Theory and Applications, Second Edition
Cheney/Kincaid (978-1-4496-1352-5) © 2012

Linear Algebra with Applications, Seventh Edition
Williams (978-0-7637-8248-1) © 2011

Linear Algebra with Applications, Alternate Seventh Edition
Williams (978-0-7637-8249-8) © 2011

Advanced Engineering Mathematics

A Journey into Partial Differential Equations
Bray (978-0-7637-7256-7) © 2012

Advanced Engineering Mathematics, Fourth Edition
Zill/Wright (978-0-7637-7966-5) © 2011

An Elementary Course in Partial Differential Equations, Second Edition
Amaranath (978-0-7637-6244-5) © 2009

Complex Analysis

Complex Analysis for Mathematics and Engineering, Sixth Edition
Mathews/Howell (978-1-4496-0445-5) © 2012

A First Course in Complex Analysis with Applications, Second Edition
Zill/Shanahan (978-0-7637-5772-4) © 2009

Classical Complex Analysis
Hahn (978-0-8672-0494-0) © 1996

Real Analysis

Elements of Real Analysis
Denlinger (978-0-7637-7947-4) © 2011

An Introduction to Analysis, Second Edition
Bilodeau/Thie/Keough (978-0-7637-7492-9) © 2010

Basic Real Analysis
Howland (978-0-7637-7318-2) © 2010

Closer and Closer: Introducing Real Analysis
Schumacher (978-0-7637-3593-7) © 2008

The Way of Analysis, Revised Edition
Strichartz (978-0-7637-1497-0) © 2000

Topology

Foundations of Topology, Second Edition
Patty (978-0-7637-4234-8) © 2009

Discrete Mathematics and Logic

Essentials of Discrete Mathematics, Second Edition
Hunter (978-1-4496-0442-4) © 2012

Discrete Structures, Logic, and Computability, Third Edition
Hein (978-0-7637-7206-2) © 2010

Logic, Sets, and Recursion, Second Edition
Causey (978-0-7637-3784-9) © 2006

The Jones & Bartlett Learning Series in Mathematics

For more information on this series and its titles, please visit us online at http://www.jblearning.com. Qualified instructors, contact your Publisher's Representative at 1-800-832-0034 or info@jblearning.com to request review copies for course consideration.

The Jones & Bartlett Learning International Series in Mathematics

For more information on this series and its titles, please visit us online at http://www.jblearning.com. Qualified instructors, contact your Publisher's Representative at 1-800-832-0034 or info@jblearning.com to request review copies for course consideration.

MULTIVARIABLE CALCULUS

DAVID B. DAMIANO
College of the Holy Cross

MARGARET N. FREIJE
College of the Holy Cross

JONES & BARTLETT
LEARNING

World Headquarters

Jones & Bartlett Learning
40 Tall Pine Drive
Sudbury, MA 01776
978-443-5000
info@jblearning.com
www.jblearning.com

Jones & Bartlett Learning
Canada
6339 Ormindale Way
Mississauga, Ontario L5V 1J2
Canada

Jones & Bartlett Learning
International
Barb House, Barb Mews
London W6 7PA
United Kingdom

Jones & Bartlett Learning books and products are available through most bookstores and online booksellers. To contact Jones & Bartlett Learning directly, call 800-832-0034, fax 978-443-8000, or visit our website, www.jblearning.com.

Production Credits:
Publisher: Cathleen Sether
Senior Acquisitions Editor: Timothy Anderson
Associate Editor: Melissa Potter
Production Director: Amy Rose
Associate Production Editor: Tiffany Sliter
Assistant Photo Researcher: Rebecca Ritter
Senior Marketing Manager: Andrea DeFronzo
V.P., Manufacturing and Inventory Control: Therese Connell
Cover and Title Page Design: Kristin E. Parker
Composition: Northeast Compositors, Inc.
Cover and Title Page Image: © Comstock Images/age fotostock
Printing and Binding: Malloy, Inc.
Cover Printing: Malloy, Inc.

Library of Congress Cataloging-in-Publication Data
Damiano, David B.
  Multivariable calculus / David B. Damiano and Margaret N. Freije.
    p. cm.
  Includes index.
  ISBN-13: 978-0-7637-8247-4 (casebound)
  ISBN-10: 0-7637-8247-5 (casebound)
  1. Calculus—Textbooks. I. Freije, Margaret N. II. Title.
  QA303.2.D36 2012
  515—dc22
                                        2010043010

6048
Printed in the United States of America
15 14 13 12 11   10 9 8 7 6 5 4 3 2 1

To our students

Contents

Preface

This text provides a complete set of teaching materials for a course in multivariable calculus that incorporates recent curricular and pedagogical developments in the teaching of calculus. The intended audience is mathematics, science, and engineering majors who have had the equivalent of a full year of single variable calculus but who have not had a course in linear algebra. Typically, these students are in their second or third semester of college. On a curricular level, the driving force behind this text is the desire to bridge the gap between mathematical concepts and their use in real-world applications outside of mathematics. Consequently, the ideas of multivariable calculus are presented in a context that is informed by their nonmathematical applications. Pedagogically, the text incorporates collaborative learning strategies and encourages the sophisticated use of technology.

Each chapter of the text begins with a collaborative learning exercise that is designed to engage students as active participants in the development of their understanding of the material. The collaborative exercise introduces an application of the material in the chapter or a key concept for the chapter. Each is designed to be used collaboratively by students working in groups of two to four. While some of the questions in the exercises require numerical answers, most of the questions involve an analysis of a problem or application that requires a prose answer. Thus an emphasis is placed on the development of the student's ability to communicate mathematically, both orally and in writing.

An additional set of collaborative learning exercises is available online from Jones & Bartlett Learning. Many of these exercises introduce new concepts and are to be used before the students read the corresponding section of the text. There are also several other exercises that are designed primarily to use the graphical and computational capabilities of the software package Maple™ to reinforce and explore topics that have been covered in class or to investigate an extended modeling application. Although the text can be used without the collaborative exercises, it becomes a more effective teaching tool when some collaborative work is introduced. The online instructor's manual contains summaries of the collaborative exercises and suggestions for their use in tandem with the text.

Content, Theory, and Method

The central theme of the text—the connections between mathematics and the sciences—is present throughout. The discussion of a topic is initiated and guided by a consideration of its uses outside of mathematics. As the mathematical ideas are developed, the text returns to their application in the sciences in extended examples and exercises. Furthermore, selected applications appear in the collaborative learning exercises to be used as in-class discussions or as the basis for extended modeling exercises. Applied topics are chosen both from the physical sciences and the life sciences and include traditional applications from mechanics as well as, for example, recently developed applications from physiology. A particular goal of this material is to provide sufficient information for students to appreciate the need to interpret a mathematical answer to a scientific question in its original context and to begin to carry out this interpretation themselves.

An effort has been made to present the material clearly and at a level of sophistication appropriate for the audience. It is important that, by the end of the third semester of calculus, students appreciate the need for precise mathematical statements and the need for justifying or proving these statements. Thus, from the beginning, the text uses set notation and the language and symbolism of functions, and definitions, propositions, and theorems are stated using precise mathematical language. The text also provides proofs of many important results in multivariable calculus, including, for example, the major theorems of line and surface integration, but it does not include concepts or proofs that require analytic capabilities beyond those of third-semester students, for example, the implicit and inverse function theorems. Furthermore, since the intended audience, in general, will not have had any formal experience with linear algebra, the focus is on $\mathbb{R}^2$ and $\mathbb{R}^3$ rather than $\mathbb{R}^n$ and vector spaces. Consequently, the algebra of linear transformations is not employed in the text. For example, differentiability of real-valued functions is phrased in terms of the existence of a linear approximation l to f, rather than the existence of a linear transformation that is called the derivative. Similarly, integration in nonrectangular coordinates is presented via Riemann sums rather than as an application of change of coordinates and the Jacobian of a coordinate transformation.

Vector fields are introduced early in the text, in Chapter 2, and then used in a variety of models, including the Lotka–Volterra predator–prey model and a phase plane model of a simple pendulum. By considering flow lines of a vector field and a graphical classification of the critical points of a vector field, we are able to make maximum use of the gradient vector field in discussing critical points of differentiable functions of two or three variables, the chain rule, and constrained optimization problems. Thus vector fields are familiar objects by the time we move to the study of line and flux integrals.

In addition to incorporating collaborative learning models, the text uses other developments in mathematical pedagogy. In particular, it emphasizes that functions can be represented numerically and graphically (in several ways) rather than only as a symbolic expression. This is particularly important for this text because most scientific applications begin with a collection of data that is represented numerically or graphically or a qualitative analysis that gives rise to differential equations. Indeed, it is a rare problem in the sciences that admits a closed-form symbolic solution. Thus the text contains numerical and graphical examples from outside of mathematics that cannot be expressed symbolically. Of course, some representations are better suited to thinking about particular mathematical ideas than others. For example, contour plots are useful when thinking about rates of change of functions, whereas data sets and density plots are better suited to discussions of total accumulation and Riemann sums.

The extensive end-of-section exercises include standard symbolic manipulations, simple proofs and verifications, problems that require a computer algebra system, and questions that require prose explanations. The exercises range in difficulty from simple calculations to more involved questions that do not follow templates provided in the text. The exercises both reinforce and extend the textual material. They also provide ample opportunities to continue the discussion of the applications of mathematical ideas to the sciences by asking questions about the particular examples that appear in the text, by varying data and models, and by presenting entirely new examples of the same type. Beyond simply asking students to carry out additional calculations, these problems ask students to explain and interpret their results in light of the phenomena being analyzed.

To illustrate the variety of ways that applications are used in the text, we offer the following brief summary of selected examples from Chapter 3, *Differentiation of Real-Valued Functions*. We begin the chapter with a discussion of an altitude function. In the opening collaborative exercise, students are asked to analyze a contour plot of a mountain in New Hampshire. In particular, they are asked to consider points that are local maxima and saddle points and to describe the contours around these points. They are also asked to consider a path along the contour plot and describe the corresponding path along the mountain. In this way, they are introduced to some of the key ideas of the chapter, critical points and rates of change, in a familiar context.

In Section 3.1, where real-valued functions are encountered formally for the first time, the first example focuses on how the language of functions might be used to describe quantities that depend on latitude and longitude, to describe physical laws, to describe quantities that depend on time and one or more other quantities, or to describe surfaces in $\mathbb{R}^3$. Subsequent examples in the same section explore these and other applications in more detail and connect basic mathematical questions we might ask about functions to questions that arise in the scientific context. Thus Example 3.5 discusses the motion of a vibrating string and, in particular, how we might describe traveling waves and standing waves with functions of time and position, and Example 3.6 examines implicitly defined surfaces by considering the isobars of sound pressure of an acoustic monopole. Later in the chapter, as the various notions of differentiation are introduced, they are used to carry out more sophisticated analyses of the applied situations. For instance, Example 3.11

computes the directional derivatives of the rms pressure of a monopole acoustic source; Example 3.13 gives a physical interpretation of the partial derivatives of the model of the vibrating string; Example 3.20 considers the question of continuity for the potential energy function of a system consisting of two molecules; and Example 3.22 considers the question of differentiability for a model of the intracellular concentration of calcium. Beyond simple questions of finding extreme values of physical quantities, the examples and accompanying exercises explore deeper questions, including, for example, why we might want the potential energy function of a molecular system to be discontinuous, or what possible physiological explanations there are for intracellular calcium concentrations to be nondifferentiable in time.

It may, of course, be necessary to employ a range of mathematical concepts and techniques to develop a complete understanding of a physical system. In order to emphasize this point, many physical systems appear in more than one chapter. For example, the acoustic systems that are introduced in Chapter 3 appear again in Chapter 7 when we consider absorption, reverberation, and total acoustic power. Systems of charged particles are first encountered in Chapter 1, *Euclidean Space and Vectors*, in the construction of the vector representing the force on a charged particle. They appear again Chapter 3, *Differentiation of Real-Valued Functions*, where we are concerned with the potential energy function of a system of two charged particles; in Chapter 4, *Critical Points and Optimization*, where we consider the extreme values of the potential energy of a system subject to a constraint; in Chapter 5, *Integration*, where we compute the potential energy of a planar charge distribution; and in Chapter 6, *Integration on Curves*, and Chapter 7, *Integration on Surfaces*, where we consider conservative forces and compute the total flux of an electric field in the plane and in space.

The Instructor's Materials

- The online instructor's manual contains commentaries on each chapter and each collaborative exercise and can be found at http://go.jblearning.com/damiano. The commentaries summarize the content of each chapter and suggest possible uses of the collaborative exercises in the classroom, in the laboratory, or outside of class in extended assignments. In addition, there is a daily schedule for each chapter that suggests how to coordinate textual material and collaborative exercises in order to take full advantage of the design of the project. This includes days for lectures on material that is most effectively presented in a lecture. It indicates what the students should have seen in the text and in class, either in discussion or lecture, prior to using a particular discussion. It also gives a summary of material that should be included in an introduction to the discussion and suggestions for summarizing or for follow-ups of the discussion.
- Solutions to the text's exercises are available for qualified instructors.
- PowerPoint® lecture slides are organized by chapter.
- WebAssign™, developed by instructors for instructors, is a premier independent online teaching and learning environment, guiding several million students through

their academic careers since 1997. With WebAssign, instructors can create and distribute algorithmic assignments using questions specific to this textbook. Instructors can also grade, record, and analyze student responses and performance instantly; offer more practice exercises, quizzes, and homework; and upload additional resources to share and communicate with their students seamlessly such as the PowerPoint slides supplied by Jones & Bartlett Learning. For more detailed information, please visit www.webassign.net.

■ As an added convenience this complete textbook is now available in eBook format for purchase by the student through WebAssign.

Acknowledgments

The authors would like to thank William Barker (Bowdoin College), Harvey Keynes (University of Minnesota), Richard Maher (Loyola University Chicago), Harriet Pollatsek (Mount Holyoke College), and Richard Wicklin (SAS) for their helpful reviews of this text. We would also like to thank the team at Jones & Bartlett Learning, including Tiffany Sliter, and Tim Anderson, our editor, for their work in bringing this project to print. John Little, our colleague, generously offered comments on the final manuscript and suggested several new exercises, for which we are very appreciative. We would also like to acknowledge the valuable comments of many of our students during the development of this text.

We gratefully acknowledge the support of our colleagues at College of the Holy Cross as well as the patience and support of our families throughout the project. Finally, David Damiano would like to thank Philip Auron for never failing to ask about the status of the text.

Euclidean Space and Vectors

This chapter introduces two themes that run through this text: first, the development of a mathematical language to extend the ideas of one-variable calculus to the calculus of several variables, and second, the investigation of ways that mathematics is applied in the sciences. While many of the concepts can be developed quite generally, we will work in two and three variables exclusively. That is, we will focus on the plane and space.

We begin by developing the Cartesian coordinate system in Section 1.1, where we use set notation, the formal language of mathematics, to describe subsets of the plane and space. However, we will also want to be able to describe sets in words and, where possible, work with plots of sets. Sections 1.2 and 1.3 introduce vectors and vector algebra. Vectors provide another way to describe subsets of the plane and space, and they also allow us to express important geometric ideas in a simple manner. In later chapters, we will use the language of vectors to express the ideas of the calculus of several variables in ways that closely parallel the terminology of one-variable calculus. In Section 1.4, we will use the Cartesian coordinate system and vectors to describe lines and planes.

We will also spend time in this chapter introducing examples of the types of problems from the sciences that will appear later in the text. We will use mathematical language to describe a variety of physical and biological phenomena. Later, we will be interested in using multivariable calculus to obtain a deeper understanding of these phenomena. We will also go in the other direction and use the applications to motivate the development of mathematical ideas and particular mathematical examples.

A Collaborative Exercise—Robotics

Two problems in robotics—the study, design, and use of robot systems for manufacturing industrial robots—make use of basic ideas about coordinate systems and subsets of Euclidean space. These are the forward-position kinematics problem and the inverse-position kinematics problem. We will need some terminology in order to explain these problems.

For our purposes, a robot will be an "arm" consisting of some number of segments or **links** of known length attached at joints. Here we consider a simple arm consisting of two links, with one segment extending from the origin O to P of length 1 and the second extending from P to Q, also of length 1. (See Figure 1.1.) The **effector** end of the arm is Q, the end that is free to move subject to the constraints of the links and joints. The **working envelope** of the arm is the region of the plane or space that can be reached by the effector.

The **forward-position kinematics problem** is to determine the location of the effector within the working envelope given the angles of the joints. The **inverse-position kinematics problem** is to determine the position of the joints given the position of the effector within the working envelope. We will consider only two types of joints, **hinges** and **sliders**. As the names indicate, a hinge joint allows rotation about an axis (think of a door hinge), and a slider joint allows one segment to slide along another without rotation. Since both have only one direction of motion, we say each has **one degree of freedom**. Also, since radian measure will be assumed throughout the text, we will omit explicit reference to radians.

To start, let us consider the case where $\overline{OP}$ is fixed as shown and the joint at P is a hinge joint free to move through an angle of 2π. The working envelope of the robotic arm is then a circle of radius 1 centered at the point $(0, 1)$. If, for instance, $\angle OPQ = \pi/3$, the coordinates of the effector (the point Q) are $(0 + \sin(\pi/3), 1 - \cos(\pi/3)) = (\sqrt{3}/2, 1/2)$. (Why?) This answers the forward-position kinematics problem. Additionally, this is the only position of the segments that will result in the effector being located at $(\sqrt{3}/2, 1/2)$. This answers the inverse-position kinematics problem.

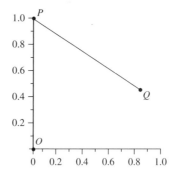

Figure 1.1   A robot with two segments or links of length 1.

Here are several questions about two different sets of hinges for this arm.

1. Suppose that in addition to the hinge joint at P, there is a hinge joint located at O that is also free to move through an angle of 2π.

   a. Sketch the working envelope of this robot.
   b. If the segment $\overline{OP}$ makes an angle of $2\pi/3$ with the positive x-axis (measured counterclockwise) and $\angle OPQ = \pi/3$, what are the coordinates of the location of the effector?
   c. Are there any other positions of the links that will result in the same position of the effector?
   d. If the location of the effector is known, how many different positions of the links will result in that effector location? (*Hint:* Your answer should depend on the position of the effector.)

2. Suppose that O is a slider joint free to move from $x = 0$ to $x = 2$ on the x-axis, $\overline{OP}$ is fixed at a right angle to the x-axis (as shown in Figure 1.1), and there is hinge joint at P free to move through an angle of 2π.

   a. Sketch the working envelope of this robot.
   b. If O is located at $(1, 0)$ and $\angle OPQ = \pi/3$, what are the coordinates of the location of the effector?
   c. If the location of the effector is known, how many different positions of the links will result in that effector location? (*Hint:* Your answer should depend on the position of the effector.)
   d. Based on your answer to (c) and on your sketch of the envelope from (a), color subsets of the envelope for which the inverse-position kinematics problem has the same number of solutions.

■ 1.1 The Cartesian Coordinate System

Our focus in this section will be on developing the basic terminology that we need to describe subsets of the plane and of space. From a mathematical point of view, the plane and space and their subsets are abstract objects that have no physical reality. However, we often visualize or imagine them in physical terms, identifying points in the plane or space with physical locations and subsets with physical objects. Conversely, we also use subsets of the plane or space to represent idealized versions of physical objects, thus providing a way to apply the tools of mathematics to analyze the object. We might call this the first example of using mathematics to "model" phenomena in the real world, and we will regularly exploit this correspondence when we plot mathematical objects. Let us begin by reviewing the Cartesian coordinate system in the plane.

In previous mathematics courses, you have seen that the coordinate plane can be represented by the set of ordered pairs, (x, y), of real numbers. We will represent the real numbers by the symbol $\mathbb{R}$ and the coordinate plane by the symbol $\mathbb{R}^2$. Using set notation, we write

$$\mathbb{R}^2 = \{(x, y) : x,\ y \in \mathbb{R}\}.$$

This should be read as "The coordinate plane is the set of all ordered pairs (x, y) such that x and y are real numbers." When we want to refer to a particular point in the plane without giving numerical values for the coordinates x and y, we will use the notation (x_0, y_0). The ordered pair (x_0, y_0) can be read either as "x-zero, y-zero" or "x-naught, y-naught." Generally, when we use a subscripted variable, like x_0, it will refer to a quantity that is constant or fixed.

The *x*-***axis*** is the line $\{(x, y) : y = 0\}$, and the *y*-***axis*** is the line $\{(x, y) : x = 0\}$. The point with coordinates $(0, 0)$ is called the ***origin***. It is the intersection of the *x*- and *y*-axes. When plotting points and subsets of the plane, we will represent the *x*-axis by a horizontal line and the *y*-axis by a vertical line, so that the axes are perpendicular and they intersect at the origin. The ***positive x-axis extends to the right of the origin***, and the ***positive y-axis extends up from the origin***, so that ***moving in a counterclockwise direction from the positive x-axis, we first encounter the positive y-axis.*** (See Figure 1.2(a).)

This scheme extends directly to a representation of coordinate space, which we denote by $\mathbb{R}^3$. ***Coordinate space*** will be represented by the set of ***ordered triples*** (x, y, z) of real numbers. In set notation,

$$\mathbb{R}^3 = \{(x, y, z) : x,\ y,\ z \in \mathbb{R}\}.$$

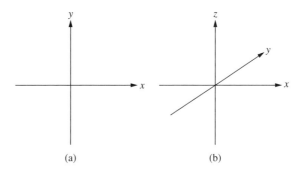

(a)   (b)

Figure 1.2   The coordinate axes in the plane (a) and in space (b). The arrows indicate the positive directions on the axes. Note the description of the relationship between the positive directions in the text. In (a) we imagine the coordinate plane as the plane of the page. In (b) we imagine the *x*- and *z*-axes as lying in the page and the *y*-axis as pointing into the page.

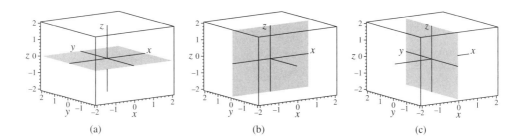

Figure 1.3   The coordinate axes in space showing (a) the xy-plane, (b) the xz-plane, and (c) the yz-plane. Notice that the plot displays also each contain a "bounding box." This is a device for lending perspective to a plot. Notice that the hash marks for the coordinate directions are shown on the edges of each box.

We will use (x, y, z) to denote a point in space, and we will represent a particular point P by (x_0, y_0, z_0). The point with coordinates $(0, 0, 0)$ is also called the ***origin***. The ***x-axis*** is the line $\{(x, y, z) : y = 0, \ z = 0\}$, the ***y-axis*** is the line $\{(x, y, z) : x = 0, \ z = 0\}$, and the ***z-axis*** is the line $\{(x, y, z) : x = 0, \ y = 0\}$. When plotting in space, ***we will represent the coordinate axes by three mutually perpendicular lines that intersect in a point, the origin.*** The direction of the axes is determined by the ***right-hand rule***. If you place your right hand around the z-axis at the origin so that your fingers curl in the direction from the positive x-axis to the positive y-axis, your thumb should point in the direction of the positive z-axis. (See Figure 1.2(b).) Each pair of coordinate axes is contained in a ***coordinate plane***. For example, the x- and y-axes determine the xy-***plane***. In set notation, we have

$$xy\text{-plane: } \{(x, y, z) : z = 0\},$$
$$xz\text{-plane: } \{(x, y, z) : y = 0\},$$
$$yz\text{-plane: } \{(x, y, z) : x = 0\}.$$

(See Figure 1.3.)

   Throughout this text, we will encounter examples of physical phenomena that can be modeled using multivariable calculus. In each case, it will be possible to use Cartesian coordinates to represent quantities that describe the phenomenon. Here is a preview of several examples that appear in later chapters.

Example 1.1   **Uses of Coordinate Systems**

A. Position. We will often use coordinate systems in the plane or space to represent the positions of objects. In situations like this, we will usually position the coordinate system so that our subsequent calculations are as simple as possible. Thus we might locate or center objects at the origin. In the collaborative exercise, the base of the robotic arm was first positioned at the origin. Later in this chapter and in subsequent chapters, coordinates will be used, for example, to locate one or more charged particles.

B. Shape. The shapes of objects will be described in set notation using coordinates in the plane or space. We have already described the links of robots and their working envelopes using coordinates. Later in the text, for example, we will consider the shape of an electrical conductor.

C. Modeling Populations. When studying interacting populations, the size of each population is given by a coordinate in the plane or space. There are many such examples. We will consider a predator–prey model for two species, one of which preys on the other. When considering the spread of epidemics, the populations are the people in different categories of exposure to the disease, for example, those susceptible to the disease, those infected by the disease, and those who have recovered from the disease. When considering the human immune system, the populations are cell and pathogen populations and are measured by the number of cells or particles per unit volume of blood.

D. Topographical Models. Topographical features provide a rich collection of examples of objects in space that we will use to motivate the development of mathematical concepts. We will identify the latitude, longitude, and altitude of a location on the surface of the earth with the coordinates of a point in coordinate space.

Subsets of $\mathbb{R}^2$ and $\mathbb{R}^3$

To describe subsets of the plane or space in mathematical terms, we will specify one or more conditions on the coordinates of the points of the subset. The conditions will be in the form of equalities or inequalities involving the coordinates, and we will often use set notation to produce a compact symbolic description of the subset. Since the Cartesian coordinate system employs real numbers for its coordinates, we will only be interested in the real numbers that satisfy these equations. For example, since the solutions of the equation $x^2 + 1 = 0$, $\sqrt{-1}$ and $-\sqrt{-1}$, are not real numbers, the relevant information for us is that there are no real solutions to the equation. Consequently, rather than explicitly writing in words (or symbols) "all real numbers x" that satisfy a condition, we will usually write "all x" that satisfy the condition.

Here are a few simple examples in $\mathbb{R}^2$.

Example 1.2	**Subsets of $\mathbb{R}^2$**

A. Each coordinate axis divides the plane into **half-planes.** For example, the **right half-plane** is given by $\{(x, y) : x \geq 0\}$. Together, the coordinate axes divide the plane into four **quadrants.** For example, the first quadrant is given by $\{(x, y) : x \geq 0, \ y \geq 0\}$.

B. A *horizontal line passing through the point* $P = (x_0, y_0)$ is parallel to the x-axis and can be written $\{(x, y) : y = y_0\}$. Notice that every point on this line has the same y-coordinate as P. A *vertical line passing through the point* P is parallel to the y-axis and can be written $\{(x, y) : x = x_0\}$. Every point on this line has the same x-coordinate as P.

C. A *circle* is the set of all points in the plane equidistant from a fixed point. For example, the circle of radius r centered at $P = (x_0, y_0)$ is the set $\{(x, y) : (x - x_0)^2 + (y - y_0)^2 = r^2\}$. To write out this formula, we used the fact that the distance between two points $P = (x_0, y_0)$ and $Q = (x_1, y_1)$ in the plane is given by the *distance formula*

$$d(P, Q) = \sqrt{(x_1 - x_0)^2 + (y_1 - y_0)^2}.$$

The subsets of $\mathbb{R}^2$ we described in Example 1.2 have counterparts in $\mathbb{R}^3$.

Example 1.3   **Subsets of** $\mathbb{R}^3$

A. Each coordinate plane divides $\mathbb{R}^3$ into two *half-spaces*. For example, the *upper half-space* is the set $\{(x, y, z) : z \geq 0\}$. Together, the coordinate planes divide space into eight regions, called *octants*. The octant given by $\{(x, y, z) : x \geq 0, \ y \geq 0, \ z \geq 0\}$ is called the *first octant*.

B. Through every $P \in \mathbb{R}^3$ there are *lines parallel to the coordinate axes* and *planes parallel to the coordinate planes*. For example, the *line parallel to the z-axis through the point* $P = (x_0, y_0, z_0)$ can be described by

$$\{(x, y, z) : x = x_0, \ y = y_0\}.$$

Points in this line have the same x- and y-coordinates as P. The *plane parallel to the yz-plane and passing through* P can be described by

$$\{(x, y, z) : x = x_0\}.$$

Points in this plane have the same x-coordinate as P.

C. A *solid cube* in the first octant with edges of length 1 and a vertex at the origin is given by

$$\{(x, y, z) : 0 \leq x \leq 1, \ 0 \leq y \leq 1, \ 0 \leq z \leq 1\}.$$

Notice the faces of the cube are contained in planes parallel to the coordinate planes.

A solid cube centered at the origin with edges of length 1 and faces parallel to the coordinate planes is given by

$$\{(x, y, z) : |x| \le \tfrac{1}{2}, \ |y| \le \tfrac{1}{2}, \ |z| \le \tfrac{1}{2}\}.$$

Before we can consider the analogue of Example 1.2C in space, we need a symbolic expression for the distance between two points in space that extends the formula for the distance between points in the plane. Suppose that $P = (x_0, y_0, z_0)$ and $Q = (x_1, y_1, z_1)$. The distance from P to Q is the length of the line segment from P to Q, which we can determine by two applications of the Pythagorean theorem. (See Figure 1.4.) Let $R = (x_1, y_1, z_0)$, so that R lies in the vertical line passing through Q and in the horizontal plane passing through P. Let $S = (x_1, y_0, z_0)$, so that S lies in the horizontal line through P parallel to the x-axis and in the horizontal line through R parallel to the y-axis. Then the line segments PS and SR meet in a right angle, so that we can apply the Pythagorean theorem to find that $|PR| = \sqrt{|PS|^2 + |SR|^2}$, where, for example, $|PR|$ denotes the length of the line segment PR. Also, the line segments PR and RQ meet in a right angle, so that $|PQ| = \sqrt{|PR|^2 + |RQ|^2}$. Combining these two equations, we see that

$$|PQ| = \sqrt{|PS|^2 + |SR|^2 + |RQ|^2}.$$

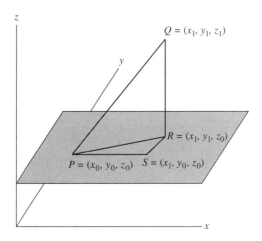

Figure 1.4   The right triangles necessary to calculate the distance from P to Q in space. The shaded plane is parallel to the xy-plane.

Since $|PS| = x_1 - x_0$, $|SR| = y_1 - y_0$, and $|RQ| = z_1 - z_0$, the **distance from** $P = (x_0, y_0, z_0)$ **to** $Q = (x_1, y_1, z_1)$ is given by

$$|PQ| = \sqrt{(x_1 - x_0)^2 + (y_1 - y_0)^2 + (z_1 - z_0)^2}.$$

Notice that if P is the origin, $P = (0, 0, 0)$, then $|PQ| = \sqrt{x_1^2 + y_1^2 + z_1^2}$ is the **distance from Q to the origin.**

Now let us return to the analogue of Example 1.2C in space. A **sphere of radius** r, $r > 0$, **centered at** $P = (x_0, y_0, z_0)$ is the set of all points in space distance r from P. That is, it is the set

$$\{(x, y, z) : (x - x_0)^2 + (y - y_0)^2 + (z - z_0)^2 = r^2\}.$$

The Method of Slices

To analyze more complicated subsets of space, it is often helpful to employ the **method of slices**. The idea is to look at the intersections of a set $\mathcal{S}$ in $\mathbb{R}^3$ with planes parallel to the coordinate planes and use the intersections to understand the shape or geometry of $\mathcal{S}$. We will refer to the process of intersecting $\mathcal{S}$ with a plane $\mathcal{P}$ as **slicing $\mathcal{S}$ by $\mathcal{P}$**. The intersection will be called a **slice** of $\mathcal{S}$.

We can visualize the slicing process by plotting the set $\mathcal{S}$ and the plane $\mathcal{P}$ on the same set of coordinate axes in space. (See Figure 1.5.) The slice is the set of points where the plots of $\mathcal{S}$ and $\mathcal{P}$ intersect, which we denote by $\mathcal{S} \cap \mathcal{P}$. Using coordinates, planes parallel to a coordinate plane are obtained by holding one variable constant. For example, planes parallel to the xy-plane are obtained by holding z constant, which we do by setting

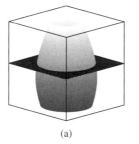   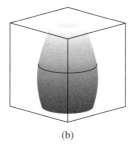   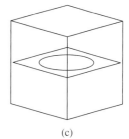

(a)   (b)   (c)

Figure 1.5   (a) A plot of a set $\mathcal{S}$ and a horizontal plane $\mathcal{P}$ on the same set of coordinate axes. The slice of $\mathcal{S}$ by $\mathcal{P}$ is the intersection of the two plots. (b) A plot of the set $\mathcal{S}$ and the slice of $\mathcal{S}$ by the same plane $\mathcal{P}$. The slice is the curve encircling $\mathcal{S}$. (c) A plot of $\mathcal{P}$ showing only the slice of $\mathcal{S}$.

$z = z_0$. The slice of a surface by this plane is obtained by substituting z_0 for z in the equation for the surface. For instance, if the surface $\mathcal{S}$ is defined by the equation

$$3x^2 - y^2 + z^2 = 1,$$

then the slice by the horizontal plane with z-coordinate $z = 2$, satisfies the equation

$$3x^2 - y^2 + 2^2 = 1.$$

To demonstrate how slicing works, we will look at surfaces in space defined by equations of the form

$$Ax^2 + By^2 + Cz^2 = r^2,$$

where A, B, C, and $r > 0$ are constants. These are examples of what are known as **quadric surfaces**. Quadric surfaces are defined by a single polynomial equation of degree 2, that is, with the largest exponent equal to 2. For example, we have already seen that the equation $x^2 + y^2 + z^2 = R^2$ defines a sphere of radius R, which is the set of points distance R from the origin.

Before we analyze the slices of quadric surfaces, it will be useful to review curves defined by equations of degree 2 in two variables. An **ellipse** is described by an equation of the form

$$\frac{x^2}{a^2} + \frac{y^2}{b^2} = 1,$$

where a and b are positive constants. An ellipse with an equation of this form is centered at the origin with axes of length $2a$ and $2b$ parallel to the coordinate axes. If $a = b = r$, we can rewrite the equation in the form

$$x^2 + y^2 = r^2,$$

which is the equation of a **circle of radius r centered at the origin.** The equation

$$\frac{x^2}{a^2} - \frac{y^2}{b^2} = 1$$

defines a **hyperbola** whose branches open to the left and to the right and intersect the x-axis at the points $(-a, 0)$ and $(a, 0)$. The equation

$$-\frac{x^2}{a^2} + \frac{y^2}{b^2} = 1$$

defines a **hyperbola** whose branches open up and down and intersect the y-axis at the points $(0, b)$ and $(0, -b)$.

In the next example, we will investigate the slices of three quadric surfaces. The accompanying plots show the sets $\mathcal{S}$ and their intersections with several planes. The planes themselves are not shown.

Example 1.4 **Quadric Surfaces**

A. An Ellipsoid. Consider the set defined by

$$\mathcal{S} = \{(x, y, z) : \ x^2 + y^2/4 + z^2/9 = 1\}.$$

A horizontal slice is the intersection of $\mathcal{S}$ with a plane $\mathcal{P}$ parallel to the xy-plane, $\mathcal{P} = \{(x, y, z) : \ z = z_0\}$. The intersection is given by

$$\mathcal{S} \cap \mathcal{P} = \{(x, y, z) : \ x^2 + y^2/4 = 1 - z^2/9, \ z = z_0\}.$$

If $|z_0| > 3$, $1 - z_0^2/9 < 0$ so that $x^2 + y^2/4 = 1 - z_0^2/9$ has no solutions and the intersection is empty. If $|z_0| = 3$, the equation becomes $x^2 + y^2/4 = 0$, which has one solution, $x = 0$ and $y = 0$, so that the intersection consists of the single point $(0, 0, 3)$ when $z_0 = 3$ and the single point $(0, 0, -3)$ when $z_0 = -3$. If $|z_0| < 3$, $1 - z_0^2/9 > 0$. Let $c = \sqrt{1 - z_0^2/9} > 0$, so the equation for the slice is $x^2 + y^2/4 = c^2$, which can be rewritten in the form

$$\frac{x^2}{c^2} + \frac{y^2}{4c^2} = 1.$$

This defines an ellipse with axes of length $2c$ and $4c$. (See Figure 1.6(a).) A similar analysis holds if we slice $\mathcal{S}$ by planes parallel to the other coordinate planes. (See Figure 1.6(b).)

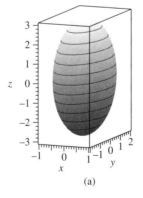
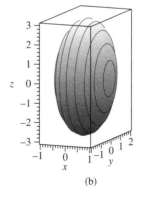

(a)   (b)

Figure 1.6 (a) A plot of the ellipsoid $\mathcal{S} = \{(x, y, z) : \ x^2 + y^2/4 + z^2/9 = 1\}$ showing slices by horizontal planes. (b) The same ellipsoid showing slices by planes parallel to the yz-plane. (See Example 1.4A.)

The set S is an ***ellipsoid*** with major axis of length 6 lying along the z-axis and minor axis of length 2 lying along the x-axis.

B. A Hyperboloid of Two Sheets. Consider the set defined by

$$S = \{(x, y, z) : x^2 - y^2 - z^2 = 1\}.$$

A horizontal slice is given by the intersection of S with a plane $\mathcal{P}$ parallel to the xy-plane, $\mathcal{P} = \{(x, y, z) : z = z_0\}$,

$$S \cap \mathcal{P} = \{(x, y, z) : x^2 - y^2 = 1 + z^2, \ z = z_0\}.$$

Since $1 + z_0^2 \neq 0$, the equation $x^2 - y^2 = 1 + z_0^2$ can be rewritten as

$$\frac{x^2}{1 + z_0^2} - \frac{y^2}{1 + z_0^2} = 1,$$

which is the equation for a hyperbola whose branches open in the x-direction and intersect the x-axis at the points $\pm\sqrt{1 + z_0^2}$. (See Figure 1.7(a).)

Vertical slices of S parallel to the yz-plane have the form

$$S \cap \mathcal{P} = \{(x, y, z) : y^2 + z^2 = x^2 - 1, \ x = x_0\}.$$

When $|x_0| < 1$, the intersection is empty. When $x_0 = 1$, the intersection consists of the point $(1, 0, 0)$, and when $x_0 = -1$, the intersection consists of the point $(-1, 0, 0)$. When $|x_0| > 1$, the intersection consists of a circle of radius $x_0^2 - 1$ centered at the origin in the plane $\mathcal{P}$. (See Figure 1.7(b).) This set is called a ***hyperboloid of two sheets***.

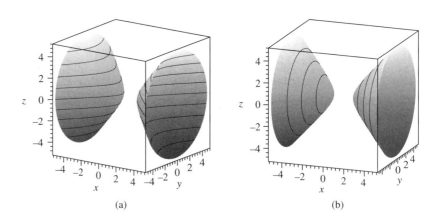

(a)   (b)

Figure 1.7   (a) A plot of the hyperboloid $S = \{(x, y, z) : x^2 - y^2 - z^2 = 1\}$ showing slices by horizontal planes. (b) The same hyperboloid showing slices by planes parallel to the yz-plane. (See Example 1.4B.)

C. A Cylinder. Consider the set $\mathcal{S}$ defined by

$$\mathcal{S} = \{(x, y, z) : \ x^2 + y^2 = 9\}.$$

A horizontal slice of $\mathcal{S}$ is obtained by holding z constant, that is, by intersecting $\mathcal{S}$ with a plane $\mathcal{P} = \{(x, y, z) : \ z = z_0\}$. In symbols,

$$\mathcal{S} \cap \mathcal{P} = \{(x, y, z) : \ x^2 + y^2 = 9, \ z = z_0\}.$$

The equation $x^2 + y^2 = 9$ defines a circle of radius 3 contained in $\mathcal{P}$ with center at $(0, 0, z_0)$ on the z-axis. (See Figure 1.8(a).)

If instead we slice $\mathcal{S}$ by a plane $\mathcal{P}$ parallel to the xz-plane, $\mathcal{P} = \{(x, y, z) : \ y = y_0\}$, the intersection is given by

$$\mathcal{S} \cap \mathcal{P} = \{(x, y, z) : \ x^2 = 9 - y^2, \ y = y_0\}.$$

The equation $x^2 = 9 - y_0^2$ has no solutions if $|y_0| > 3$, in which case the intersection is empty. If $|y_0| = 3$, then $x = 0$, and the intersection consists of the line parallel to the z-axis passing through $(0, 3, 0)$ when $y_0 = 3$ and the line parallel to the z-axis passing through $(0, -3, 0)$ when $y_0 = -3$. If $|y_0| < 3$, then the equation $x^2 = 9 - y_0^2$ has two solutions, $x = \pm\sqrt{9 - y_0^2}$. The intersection consists of the line parallel to the z-axis passing through $(\sqrt{9 - y_0^2}, y_0, 0)$ and the line parallel to the z-axis passing through $(-\sqrt{9 - y_0}, y_0, 0)$. A similar analysis holds if we intersect $\mathcal{S}$ with planes parallel to the yz-plane. (See Figure 1.8(b).) The set is a cylinder of radius 3 centered on the z-axis.

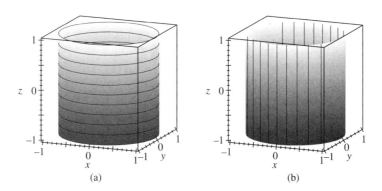

(a)   (b)

Figure 1.8 (a) A plot of the cylinder of radius 3 centered on the z-axis showing slices by horizontal planes. (b) The same cylinder showing slices by planes parallel to the yz-plane. (See Example 1.4C.)

Summary

In this section, we introduced the coordinate systems in the plane and space. The *coordinate plane* $\mathbb{R}^2$ consists of *the set of all ordered pairs of real numbers.* In *set notation* we write

$$\mathbb{R}^2 = \{(x, y) :\ x,\ y \in \mathbb{R}\,\}.$$

The *coordinate axes* divide the plane into *quadrants*. We paid particular attention to *lines parallel to the coordinate axes*. The distance between two points $P = (x_0, y_0)$ and $Q = (x_1, y_1)$ is given by the *distance formula*, $d(P, Q) = \sqrt{(x_1 - x_0)^2 + (y_1 - y_0)^2}$. *Coordinate space* consists of *the set of all ordered triples of real numbers,*

$$\mathbb{R}^3 = \{(x, y, z) :\ x,\ y,\ z \in \mathbb{R}\,\}.$$

The *coordinate axes* are related by the *right-hand rule*. The *coordinate planes* divide space into *octants*. We paid particular attention to *planes parallel to co-ordinate planes*, which we used in the *method of slicing* to analyze subsets of space. In particular, we studied the slices of *quadric surfaces*. The distance between two points $P = (x_0, y_0, z_0)$ and $Q = (x_1, y_1, z_1)$ is given by the *distance formula*, $d(P, Q) = \sqrt{(x_1 - x_0)^2 + (y_1 - y_0)^2 + (z_1 - z_0)^2}$.

Section 1.1 Exercises

1. **Subsets of the Plane.** Use set notation to give a symbolic description of each of the subsets of $\mathbb{R}^2$ shown in Figure 1.9.

2. **Coordinate Axes in Space.** If you look from the positive z-axis toward the origin in space, does the direction from the positive x-axis to the positive y-axis appear to be clockwise or counterclockwise? Suppose instead you look toward the origin from the negative z-axis. What is the direction from the positive x-axis to the positive y-axis? If we interchange the roles of the axes, looking toward the origin along the positive and negative x-axis and then the positive and negative y-axis, what would be the relationship between the remaining axes?

3. **Lines and Planes.** Use set notation to give a symbolic description of each of the following sets.

   (a) The line parallel to the x-axis through (x_0, y_0, z_0).
   (b) The line parallel to the y-axis through (x_0, y_0, z_0).
   (c) The plane parallel to the xz-plane through (x_0, y_0, z_0).
   (d) The plane parallel to the xy-plane through (x_0, y_0, z_0).

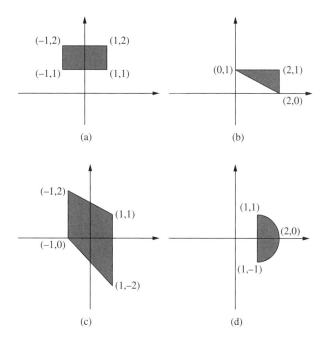

Figure 1.9   Regions for Exercise 1.

4. **Subsets of** $\mathbb{R}^3$ **I.** Describe in words the following subsets of $\mathbb{R}^3$. If the set is defined by more than one condition, say whether it is a union or intersection of sets.

(a) $\{(x, y, z) : x = \pi\}$.
(b) $\{(x, y, z) : x = 5, \ y = 6\}$.
(c) $\{(x, y, z) : (y - 3)(z - 3) = 0\}$.

(d) $\{(x, y, z) : x \geq 3\}$.
(e) $\{(x, y, z) : 1 \leq y \leq 4\}$.
(f) $\{(x, y, z) : x \geq 0, \ z = 5\}$.

5. **Subsets of** $\mathbb{R}^3$ **II.** Use set notation to give a symbolic description of each of the following sets $S \subset \mathbb{R}^3$:

(a) S is the half-space of points with negative y-coordinate.
(b) S is the half-line parallel to the y-axis with positive y-coordinate and passing through $(1, 2, -1)$.
(c) S is the set of points whose z coordinate is greater than 5.
(d) S is the set of points whose z coordinate is greater than 5 or less than -5.
(e) S is the quarter-space with x and y coordinates greater than 2.
(f) S is the set consisting of points of distance less than or equal to 1 from the origin and lying in the first octant.

6. **A Cone.** The quadric surface S in the upper half-space defined by the equation $x^2 + y^2 - z^2 = 0$ is a **cone.** In set notation, $S = \{(x, y, z) : \ x^2 + y^2 - z^2 = 0, \ z \geq 0\}$. Find the equations for and describe the slices of S parallel to the coordinate planes. The horizontal slices of S and one set of vertical slices of S are shown in Figure 1.10.

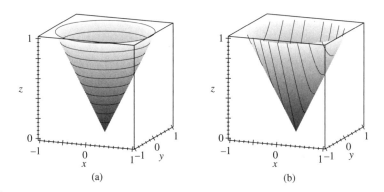

(a)   (b)

Figure 1.10   (a) A plot of a cone with vertex at the origin and centered on the z-axis showing slices by horizontal planes. (b) The same cone showing slices by planes parallel to the yz-plane. (See Exercise 6.)

7. **Positions of a Satellite.** When modeling the motion of a satellite around the earth, it is useful to have a mathematical description of the region of the earth's atmosphere where the satellite is orbiting. Suppose a satellite has a maximum altitude of 200 mi. and that it might pass over any point on the earth's surface. Use set notation to describe the region of the earth's atmosphere where the satellite is orbiting as a subset of coordinate space. (The average diameter of the earth is approximately 7918 mi.)

8. **A Hyperboloid of One Sheet.** The quadric surface $\mathcal{S}$ defined by the equation $x^2 - y^2 - z^2 = -1$ is called a ***hyperboloid of one sheet***. Find the equations for and describe the slices of $\mathcal{S}$ parallel to the coordinate planes. The horizontal slices of $\mathcal{S}$ are shown in Figure 1.11. (*Hint:* Pay particular attention to the vertical slices for $y = \pm 1$ and the horizontal slices for $z = \pm 1$.)

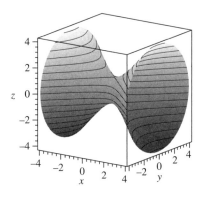

Figure 1.11   The hyperboloid of one sheet defined by $x^2 - y^2 - z^2 = -1$ with its horizontal slices. (See Exercise 8.)

9. **Slices of Quadric Surfaces.** For each of the following sets S and planes P, express the intersection of S and P in set notation and describe the intersection.

   (a) $S = \{(x, y, z) : y^2 + z^2 = 2\}$ and P is the plane parallel to the xz-plane and passing through the point $(4, 1, 3)$.
   (b) $S = \{(x, y, z) : y^2 - z^2 = 4\}$ and P is the plane parallel to the yz-plane and passing through the point $(1, 2, -2)$.
   (c) $S = \{(x, y, z) : x^2 + y^2 + z^2/4 = 1\}$ and P is the plane parallel to the xy-plane and passing through the point $(1, 1, 1)$.
   (d) $S = \{(x, y, z) : x^2 - y^2 + z^2 = 9\}$ and P is the plane parallel to the yz-plane and passing through the point $(1, 0, -1)$.

10. **Quadric Surfaces.** For each of the following sets S, give a complete description of the slices of S by planes parallel to the coordinate planes.

    (a) $S = \{(x, y, z) : x^2 + y^2 + z^2/4 = 1\}$.   (c) $S = \{(x, y, z) : 2x^2 - 4y^2 = 1\}$.
    (b) $S = \{(x, y, z) : x^2 + y^2 - z^2/4 = 1\}$.   (d) $S = \{(x, y, z) : x^2 - 2y^2 + z^2 = 0\}$.

11. **Paraboloids.** Quadric surfaces that have parabolas for slices are called ***paraboloids***. For each of the following sets S describe the slices of S by planes parallel to the coordinate planes. Which sets of slices are parabolas?

    (a) $S = \{(x, y, z) : x^2 + y^2 - z = 0\}$.   (c) $S = \{(x, y, z) : x + y^2 + z^2/4 = 0\}$.
    (b) $S = \{(x, y, z) : x^2 - 4y + z^2 = 0\}$.

12. **The Heart.** In modeling the mechanics of the heart, that is, in modeling the pumping action of the heart, it is necessary to have a mathematical description of the shape of the heart muscle. Because of its irregular shape, it is difficult to work with an exact description of the shape of the heart, and there is an advantage in modeling the heart by simpler geometric shapes.[1] Figure 1.12 shows one such model based on cylinders. The

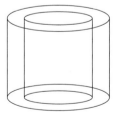

Figure 1.12   Schematic diagram of the shape of the heart for Exercise 12.

---

[1]See J. M. Guccione and A. D. McCulloch, Finite Element Modeling of Ventricular Mechanics, in L. Glass, P. J. Hunter, and A. D. McCulloch, Eds., *Theory of Heart: Biomechanics, Biophysics, and Nonlinear Dynamics of Cardiac Function.* Sprinter-Verlag, New York (1991).

diagram shows the inner surface of the heart and the outer surface of the heart. The heart wall is the region between these surfaces. The heart cavity is the region inside the inner surface.

(a) Use set notation to describe a region in space corresponding to the heart wall for the diagram in Figure 1.12. The region has height h, inner radius r_1, and outer radius r_2. Explain your choice of equations to describe the inner and outer surfaces of the heart.

(b) To make these models more precise, it is important to choose the constants in the defining equations so that the heart cavity and the heart wall have the correct volume. For a canine heart, it is reasonable to assume that the volume of the heart cavity is 40 cc and the volume of the heart wall is 100 cc. Find values of the constants r_1, r_2, and h so that heart cavity and heart wall have these volumes. Explain how you arrived at your answer. Is your answer unique?

13. **Forward-Position Kinematics Problem.** In designing a robotic arm it is important to identify the working envelope, that is, the region of the plane or space that can be reached by the effector given the location of the base and the range of motion at each joint. Figure 1.13 shows three schematic diagrams for a robotic arm that moves in the plane of the page. The base of the arm is located at the point O and the upper link corresponds to the segment OP, the lower link corresponds to the segment PQ, and the effector is located at Q. Assume that l_1, the length of OP, is greater than l_2, the length of PQ. For each of the following combinations of joints, describe the working envelope and give a symbolic description of the region using set notation. (See the collaborative exercise for definitions.)

(a) The joints at O and P are hinge joints free to move through angles of 2π. (See Figure 1.13(a).)

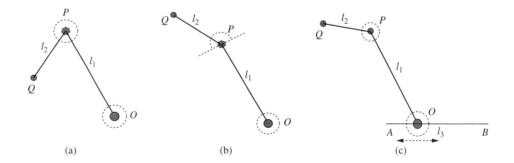

(a)   (b)   (c)

Figure 1.13   Schematic diagrams of a robotic arm. In (a), the segment OP pivots about the point P, and the segment PQ pivots about the point P. In (b), the segment PQ pivots only through an angle of π. In (c), both segments pivot through 2π and the point O is allowed to slide along the track AB. (See Exercises 13 and 14.)

(b) The joint at O is a hinge joint free to move through an angle of 2π, and the joint at P is a hinge joint free to move through an angle of π. (See Figure 1.13(b).)

(c) The joint at O is a combination hinge joint free to move through an angle of 2π and slider free to move a length l_3, and the joint at P is a hinge joint free to move through an angle of 2π. Assume $l_3 > 2(l_1 - l_2)$. (Why?) (See Figure 1.13(c).)

14. Inverse-Position Kinematics Problem. For the each of the robotic arms in Exercise 13:

(a) For each position of the effector, determine the number of different positions of the links that result in this effector location.

(b) Describe subsets of the working envelope for each number of positions in set notation. For example, describe the collection of locations in the working envelope that can be reached by only one position of the links as a subset of the working envelope, then describe the collection of locations in the working envelope that can be reached by two positions of the links as a subset of the working envelope, and so on. (*Hint:* For each combination of hinges, your subsets should be nonintersecting sets that fill the working envelope.)

■ 1.2 Vectors

The physical quantities encountered in one-variable calculus can be described by a single real number. For example, the position of a ball dropped from the top of a building is given by its height above the ground, and its velocity is given by a real number whose magnitude is the speed of the ball and whose sign tells us whether the ball is moving up or down. However, if an object is free to move in a plane, neither its position nor its velocity can be represented by a single real number. To describe its velocity, for example, we would have to know its speed and the direction in which it is moving, which cannot be represented as a sign, plus or minus, attached to its speed. In order to represent quantities like velocity, including acceleration, momentum, and force, we will introduce the notion of a vector.

Graphically, a **vector** can be represented by an ***arrow***, that is, a ***line segment with a direction***. The endpoint of the segment without the arrowhead is called the ***initial point*** of the vector, and the endpoint of the segment with the arrowhead is called the ***endpoint*** of the vector. The ***direction of the vector*** is the direction in which the arrow points. The ***magnitude of the vector*** is the length of the arrow.

We will use the Cartesian coordinate systems in $\mathbb{R}^2$ and $\mathbb{R}^3$ to represent vectors symbolically. We will represent a vector in the plane by the ordered pair $(\Delta x, \Delta y)$, where Δx is the change in x from the initial point to the endpoint of the vector and Δy is the change in y from the initial point to the endpoint of the vector. (See Figure 1.14(a).)

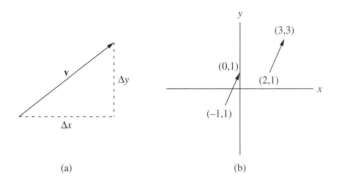

Figure 1.14   (a) The coordinates of vector **v** are the quantities Δx and Δy. (b) Two geometric representations of the vector $(1, 2)$.

We will denote vectors by boldface type. For example, the vector **v** with initial point at $(-1, -1)$ and endpoint at $(0, 1)$ will be represented by the ordered pair $(1, 2)$. Two vectors are the same if their ordered pairs are equal. For example, the vector **w** with initial point $(2, 1)$ and endpoint $(3, 3)$ has $\Delta x = 1$ and $\Delta y = 2$, so it too is represented by the ordered pair $(1, 2)$. We conclude that $\mathbf{v} = \mathbf{w}$. (See Figure 1.14(b).)

On the other hand, if we are given the coordinates $(\Delta x, \Delta y)$ of a vector **v** and an initial point $P = (x_0, y_0)$, we can sketch an arrow representing **v** that starts at P. Its endpoint must have x-coordinate equal to $x_0 + \Delta x$ and y-coordinate equal to $y_0 + \Delta y$. Then **v** can be represented by the arrow from $P = (x_0, y_0)$ to $Q = (x_0 + \Delta x, y_0 + \Delta y)$.

In $\mathbb{R}^3$ we will represent a vector by an ordered triple $(\Delta x, \Delta y, \Delta z)$, where Δx is the change in x from the initial point to the endpoint of the vector, Δy is the change in y from the initial point to the endpoint of the vector, and Δz is the change in z from the initial point to the endpoint of the vector.

Since we will be following the standard convention of allowing ordered pairs and triples to denote either a point or a vector, it will be important to pay attention to the context when we use this notation. The following summarizes the discussion to this point.

Vectors	Every vector **v** in the plane can be represented symbolically by an ordered pair (v_1, v_2), where v_1 is Δx, the change in x from the initial point to the endpoint of **v**, and v_2 is Δy, the change in y from the initial point to the endpoint of **v**. Conversely, given an initial point $P = (x_0, y_0)$, the ordered pair (v_1, v_2) can be represented graphically by the arrow with initial point $P = (x_0, y_0)$ and endpoint $Q = (x_0 + v_1, y_0 + v_2)$. Similarly, we have a correspondence between vectors in $\mathbb{R}^3$ and ordered triples (v_1, v_2, v_3).

| Example 1.5 | **Displacement and Velocity** |

A. Let $P = (x_1, y_1)$ and $Q = (x_2, y_2)$ be two points in the plane. The ***displacement vector*** of an object moved from the point P to the point Q is the vector with initial point P and endpoint Q, which is represented graphically by an arrow with initial point at P and endpoint at Q. (See Figure 1.15(a).) The vector from P to Q, denoted $\overrightarrow{PQ}$, is represented symbolically by the pair $(x_2 - x_1, y_2 - y_1)$.

B. The ***velocity vector*** of an object in motion is the vector that points in the direction of motion and has magnitude equal to the speed of the object. If a billiard ball is struck at an angle α from the positive x-direction and with speed v, we can use trigonometry to determine the coordinates of the velocity vector. The velocity vector $\mathbf{v} = (v \cos \alpha, v \sin \alpha)$. (See Figure 1.15(b).)

The magnitude or length of a vector can be determined directly from the coordinates of the vector in the following way.

Proposition 1.1 If $\mathbf{v} = (v_1, v_2)$ is a vector in the plane, then the magnitude of $\mathbf{v}$ is equal to $\sqrt{v_1^2 + v_2^2}$. If $\mathbf{v} = (v_1, v_2, v_3)$ is a vector in space, then the magnitude of $\mathbf{v}$ is equal to $\sqrt{v_1^2 + v_2^2 + v_3^2}$. We will denote the magnitude of $\mathbf{v}$ by $||\mathbf{v}||$. ◆

Proof: We will prove this proposition for vectors in the plane and leave the proof for vectors in space as an exercise.

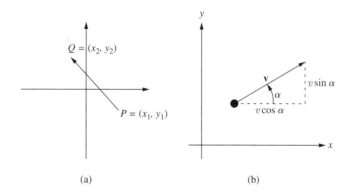

(a)   (b)

Figure 1.15   (a) The displacement vector from P to Q. (b) The velocity vector $\mathbf{v}$ of a billiard ball struck at an angle α from the positive x-direction with speed v, $\mathbf{v} = (v \cos \alpha, v \sin \alpha)$. (See Example 1.5B.)

The pair (v_1, v_2) represents a vector with initial point at (x_0, y_0) and endpoint at $(x_0 + v_1, y_0 + v_2)$. The length of the vector is the distance from the initial point to the endpoint. Using the distance formula, we have

$$\|(v_1, v_2)\| = \sqrt{(x_0 + v_1 - x_0)^2 + (y_0 + v_2 - y_0)^2} = \sqrt{v_1^2 + v_2^2}. \quad \blacksquare$$

Vector Operations

Here we will introduce the algebraic operations of **vector addition** and **scalar multiplication**. We will begin by defining these vector operations algebraically. Later we will relate them to the physical properties of forces.

Definition 1.1 Let $\mathbf{u} = (u_1, u_2)$ and $\mathbf{v} = (v_1, v_2)$ be vectors in the plane. The **vector sum** of $\mathbf{u}$ and $\mathbf{v}$ is the vector $(u_1 + v_1, u_2 + v_2)$, which we denote by $\mathbf{u} + \mathbf{v}$. Thus

$$\mathbf{u} + \mathbf{v} = (u_1, u_2) + (v_1, v_2) = (u_1 + v_1, u_2 + v_2).$$

Similarly, if $\mathbf{u} = (u_1, u_2, u_3)$ and $\mathbf{v} = (v_1, v_2, v_3)$ are vectors in space, the **vector sum**, $\mathbf{u} + \mathbf{v}$, of $\mathbf{u}$ and $\mathbf{v}$ is the vector $(u_1 + v_1, u_2 + v_2, u_3 + v_3)$.

Let $\mathbf{v} = (v_1, v_2)$ be a vector in the plane and c a real number. The **scalar product** of c and $\mathbf{v}$ is the vector (cv_1, cv_2), which we denote by $c\mathbf{v}$. Thus

$$c\mathbf{v} = c(v_1, v_2) = (cv_1, cv_2).$$

Similarly, if $\mathbf{v} = (v_1, v_2, v_3)$ is a vector in space and c a scalar, the **scalar product**, $c\mathbf{v}$, of c and $\mathbf{v}$ is the vector (cv_1, cv_2, cv_3). If $\mathbf{w} = c\mathbf{v}$, we say $\mathbf{w}$ is a **scalar multiple** of $\mathbf{v}$.
♦

The sum of scalar products of two or more vectors is called a **linear combination** of vectors. For example, $(3, 2) = 3(1, 0) + 2(0, 1)$ and $(3, 2) = 2(-1, 1) + \frac{3}{2}(2, 2) - (-2, 3)$ are two ways to write $(3, 2)$ as a linear combination of vectors.

Example 1.6   **Vector Operations**

A. Addition. Let $\mathbf{u} = (3, 4)$ and $\mathbf{v} = (-1, 2)$. The vector sum $\mathbf{u} + \mathbf{v}$ is given by $\mathbf{u} + \mathbf{v} = (3, 4) + (-1, 2) = (3 + (-1), 4 + 2) = (2, 6)$.

B. Scalar Multiplication. Let $\mathbf{v} = (4, -2.2)$ and let $c = 3$. The scalar multiple $c\mathbf{v}$ is given by $c\mathbf{v} = 3(4, -2.2) = (3 \cdot 4, 3 \cdot (-2.2)) = (12, -6.6)$.

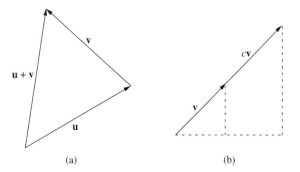

Figure 1.16 (a) Vector addition. (b) Scalar multiplication for $c > 0$.

Each operation has a graphical interpretation as well. To represent the vector sum, we position the arrows representing the vectors $\mathbf{u}$ and $\mathbf{v}$ with the initial point of $\mathbf{v}$ placed at the endpoint of $\mathbf{u}$. Then the vector $\mathbf{u} + \mathbf{v}$ is represented by the arrow joining the initial point of $\mathbf{u}$ to the endpoint of $\mathbf{v}$. Notice that $\mathbf{u}$, $\mathbf{v}$, and $\mathbf{u} + \mathbf{v}$ form the sides of a triangle. This is equivalent to the algebraic definition. (See Figure 1.16(a) and Exercise 2.)

If $\mathbf{v}$ is a vector and c is a scalar, the length of the scalar product $c\mathbf{v}$ is $|c|$ times the length of $\mathbf{v}$, since $|c\mathbf{v}| = \sqrt{(cv_1)^2 + (cv_2)^2} = |c|\sqrt{v_1^2 + v_2^2} = |c|\,\|\mathbf{v}\|$. Also, since the coordinates of $c\mathbf{v}$ and $\mathbf{v}$ are in the same ratio, $(cv_2)/(cv_1) = v_2/v_1$ (for $cv_1 \neq 0$), the arrows for $c\mathbf{v}$ and $\mathbf{v}$ have the same slope. If $c > 0$, then $c\mathbf{v}$ points in the same direction as $\mathbf{v}$, and if $c < 0$, then $c\mathbf{v}$ points in the direction opposite that of $\mathbf{v}$. See Figure 1.16(b) for a diagram when $c > 0$.

Physically, these operations correspond to the behavior of forces such as those that electrically charged particles exert on one another. From experiments we know that particles of like charge repel each other and particles of opposite charge attract each other. Coulomb's law gives the magnitude and direction of these forces. It says that the force $\mathbf{F}$ exerted by one particle on the other acts along the line connecting the two particles and has magnitude that is directly proportional to the product of the charges and inversely proportional to the square of the distance between them. We make this precise in the following example.

<table>
<tr><td>Example 1.7</td><td>**Electric Forces**</td></tr>
</table>

A. Coulomb's Law. Let $\mathcal{P}_1$ be a particle of charge q_1 located at (x_1, y_1, z_1), let $\mathcal{P}_2$ be a particle of charge q_2 located at (x_2, y_2, z_2), and let $\mathbf{v} = (x_2 - x_1, y_2 - y_1, z_2 - z_1)$ be the displacement vector from $\mathcal{P}_1$ to $\mathcal{P}_2$. If q_1 and q_2 have the same sign, the particles will repel each other and the force exerted by $\mathcal{P}_1$ on $\mathcal{P}_2$ will be directed from $\mathcal{P}_1$ to $\mathcal{P}_2$. If q_1 and q_2 have opposite signs, the particles will attract each other and the force exerted by $\mathcal{P}_1$ on $\mathcal{P}_2$ will be directed from $\mathcal{P}_2$ to $\mathcal{P}_1$. The magnitude of the force vector is proportional to $q_1 q_2 / r^2$ where $r = \|\mathbf{v}\|$ is the distance between the particles.

Symbolically, Coulomb's law takes the form

$$\mathbf{F} = \frac{q_1 q_2}{4\pi \epsilon_0 r^3} \mathbf{v},$$

where ϵ_0 is a physical constant called the **permittivity of the vacuum**. Notice that

$$\|\mathbf{F}\| = \left\| \frac{q_1 q_2}{4\pi \epsilon_0 r^3} \mathbf{v} \right\| = \left| \frac{q_1 q_2}{4\pi \epsilon_0 r^3} \right| \|\mathbf{v}\| = \left| \frac{q_1 q_2}{4\pi \epsilon_0 r^3} \right| r = \left| \frac{q_1 q_2}{4\pi \epsilon_0 r^2} \right|,$$

so that the magnitude of the force is proportional to the product of the charges and inversely proportional to the square of the distance between the particles as we claimed. To see the connection to vector operations, observe that if one of the charges is changed by a factor of c, say q_1 is replace by $c q_1$, then the resulting force is $c\mathbf{F}$, a scalar multiple of the original force.

B. The Force Exerted by an Electric Dipole. It can be shown by experiment that forces are additive. Here we write out the force exerted by an electric dipole, which consists of two charged particles a fixed distance apart, on a third charged particle. If the dipole is represented by a particle $\mathcal{P}_1$ of charge q_1 located at (x_1, y_1, z_1) and a particle $\mathcal{P}_2$ of charge q_2 located at (x_2, y_2, z_2), and a third particle $\mathcal{P}$ of charge q is located at (x, y, z), then the **force exerted by the dipole** on $\mathcal{P}$ is

$$\mathbf{F} = \frac{q_1 q}{4\pi \epsilon_0 r_1^3} \mathbf{v}_1 + \frac{q_2 q}{4\pi \epsilon_0 r_2^3} \mathbf{v}_2,$$

where $\mathbf{v}_1 = (x - x_1, y - y_1, z - z_1)$ is the vector from $\mathcal{P}_1$ to $\mathcal{P}$, $\mathbf{v}_2 = (x - x_2, y - y_2, z - z_2)$ is the vector from $\mathcal{P}_2$ to $\mathcal{P}$, $r_1 = \|\mathbf{v}_1\|$, and $r_2 = \|\mathbf{v}_2\|$.

The operations of vector addition and scalar multiplication follow the usual laws for addition and multiplication of numbers, that is, they are commutative, associative, and distributive. These properties can be verified algebraically or graphically. The verification is left as an exercise.

Proposition 1.2   Suppose $\mathbf{u}$, $\mathbf{v}$, and $\mathbf{w}$ are vectors and c and d are scalars. Then

1. $\mathbf{u} + \mathbf{v} = \mathbf{v} + \mathbf{u}$. (*Commutativity*)
2. $(\mathbf{u} + \mathbf{v}) + \mathbf{w} = \mathbf{u} + (\mathbf{v} + \mathbf{w})$. (*Associativity*)
3. $(cd)\mathbf{u} = c(d\mathbf{u})$. (*Associativity*)
4. $(c + d)\mathbf{u} = c\mathbf{u} + d\mathbf{u}$. (*Distributivity*)
5. $c(\mathbf{u} + \mathbf{v}) = c\mathbf{u} + c\mathbf{v}$. (*Distributivity*)   ♦

The Correspondence Between Points and Vectors

While vectors can be represented by arrows placed anywhere in the plane or in space, we will often position them with initial point at the origin. If the initial point of the vector $\mathbf{w} = (a, b)$ is placed at the origin, then the endpoint of $\mathbf{w}$ is at the point $(a, b) \in \mathbb{R}^2$. Thus the vector $\mathbf{w} = (a, b)$ corresponds to the point $P = (a, b)$. Similarly, if the initial point of the vector $\mathbf{w} = (a, b, c)$ is placed at the origin, then the endpoint of $\mathbf{w}$ is at the point $(a, b, c) \in \mathbb{R}^3$. In this way, we establish a correspondence between vectors in the plane or in space and points in $\mathbb{R}^2$ or $\mathbb{R}^3$. *Each point is the endpoint of a unique vector with initial point at the origin.*

This correspondence will enable us to describe subsets of $\mathbb{R}^2$ and $\mathbb{R}^3$ using the language of vectors and to describe sets of vectors using the language of subsets of Euclidean space. In the next example, we will use this correspondence to describe a line and a plane in space.

Example 1.8　　**Lines and Planes**

A. Lines. Suppose $\mathbf{u} = (u_1, u_2)$ and $\mathbf{v} = (v_1, v_2)$ are vectors in $\mathbb{R}^2$ and that $\mathbf{v} \neq \mathbf{0}$, so that v_1 and v_2 are not both zero. Consider the set of vectors

$$\{\mathbf{u} + t\mathbf{v} \ : \ t \in \mathbb{R}\},$$

and let $\mathcal{S}$ be the set of points in $\mathbb{R}^2$ that correspond to these vectors. That is, $\mathcal{S}$ is the set of points that are the endpoints of these vectors when the initial points are placed at the origin. The point (u_1, u_2), which corresponds to the vector $\mathbf{u} = \mathbf{u} + 0\mathbf{v}$, is a point in $\mathcal{S}$. If $Q = (x, y)$ is any point in $\mathcal{S}$, then the vector $\mathbf{q}$ corresponding to the point Q must be of the form $\mathbf{u} + t\mathbf{v}$ for some $t \in \mathbb{R}$. In coordinates,

$$(x, y) = (u_1, u_2) + t(v_1, v_2) = (u_1 + tv_1, u_2 + tv_2).$$

If we sketch some of these vectors and consider the corresponding points, it appears that the points form a straight line in $\mathbb{R}^2$. (See Figure 1.17(a).) Algebraically, if $v_1 \neq 0$, we can verify that the coordinates of the points of $\mathcal{S}$ satisfy the equation

$$y - u_2 = \frac{v_2}{v_1}(x - u_1),$$

which is the equation for a line through the point (u_1, u_2) with slope v_2/v_1. If $v_1 = 0$, then $x = u_1$, which is the equation for a vertical line through the point (u_1, u_2). We say that $\mathbf{v}$ is the *direction vector* of the line, so that this is the *line through the endpoint of* $\mathbf{u}$ *with direction vector* $\mathbf{v}$.

A similar analysis shows that if $\mathbf{u} = (u_1, u_2, u_3)$ and $\mathbf{v} = (v_1, v_2, v_3)$ are vectors in $\mathbb{R}^3$, then the set of points in $\mathbb{R}^3$ corresponding to the set of vectors $\{\mathbf{u} + t\mathbf{v} \ : \ t \in \mathbb{R}\}$ also forms a straight line through (u_1, u_2, u_3) and in the direction of the vector $\mathbf{v}$.

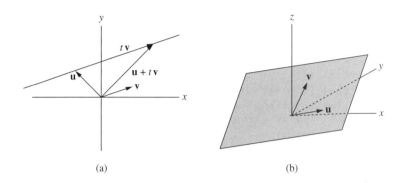

(a)   (b)

Figure 1.17   (a) The line of points in $\mathbb{R}^2$ corresponding to the set of vectors $\{\mathbf{u} + t\mathbf{v} : t \in \mathbb{R}\}$. (b) The plane in $\mathbb{R}^3$ spanned by the vectors $\mathbf{u}$ and $\mathbf{v}$. (See Example 1.8.)

B. Planes Through the Origin. Suppose $\mathbf{u} = (u_1, u_2, u_3)$ and $\mathbf{v} = (v_1, v_2, v_3)$ are two vectors in $\mathbb{R}^3$ that are not scalar multiples of one another. Let us consider the set of vectors,

$$\{s\mathbf{u} + t\mathbf{v} \ : \ s, t \in \mathbb{R}\}.$$

Let $\mathcal{S}$ be the set of points in $\mathbb{R}^3$ that correspond to these vectors. In coordinates, the point (x, y, z) is in $\mathcal{S}$ if

$$(x, y, z) = s(u_1, u_2, u_3) + t(v_1, v_2, v_3) = (su_1 + tv_1, su_2 + tv_2, su_3 + tv_3)$$

for some s and $t \in \mathbb{R}$. In particular, $\mathcal{S}$ contains the points $(0, 0, 0)$, (u_1, u_2, u_3), and (v_1, v_2, v_3). The set $\mathcal{S}$ is the plane in $\mathbb{R}^3$ that contains these three points. We call $\mathcal{S}$ the *plane spanned by* $\mathbf{u}$ *and* $\mathbf{v}$. (See Figure 1.17(b).) We will return to this example in Section 1.4.

Summary   In this section, we introduced **vectors** in the plane and in space to describe quantities that involve a **direction** and a **magnitude**. A vector can be **represented geometrically by an arrow** going from an **initial point** to an **endpoint**. A vector $\mathbf{v}$ in the plane can be **represented algebraically** by an **ordered pair of real numbers** (v_1, v_2) whose coordinates are the changes in x and y from the initial point to the endpoint of **v**. Similarly, a vector **v** in space can be *represented algebraically* by an *ordered triple of real numbers* (v_1, v_2, v_3) whose coordinates are the changes in x, y, and z from the initial point to the endpoint of **v**.

We defined the operations of *vector addition* and *scalar multiplication*, which can be represented geometrically or algebraically. These operations correspond to the properties of physical forces.

We defined the *correspondence between points and vectors*. A point corresponds to the vector whose initial point is the origin and whose endpoint is the given point. We used this correspondence to give a vector description of the *line through the endpoint of* **u** *with direction vector* **v** in the plane or space and of the *plane spanned by* **u** *and* **v** *in space*.

Section 1.2 Exercises

1. **Distance.** Use the distance formula in space to verify that the length of the vector $\mathbf{v} = (v_1, v_2, v_3)$ in $\mathbb{R}^3$ is $\sqrt{v_1^2 + v_2^2 + v_3^2}$.

2. **Vector Sum.** Verify that the graphical interpretation of the sum of two vectors $\mathbf{u} = (u_1, u_2)$ and $\mathbf{v} = (v_1, v_2)$ yields the correct algebraic formula for the sum.

3. **Sketching Vectors.** For each of the following vectors **u** and **v**, sketch the vectors **u**, **v**, $-\mathbf{u}$, $\mathbf{u} + \mathbf{v}$, $\mathbf{u} - \mathbf{v}$, and $\mathbf{v} - 2\mathbf{u}$.

   (a) $\mathbf{u} = (1, 2)$ and $\mathbf{v} = (-1, 1)$.   (c) $\mathbf{u} = (1, 0, 2)$ and $\mathbf{v} = (3, 0, 1)$.
   (b) $\mathbf{u} = (1, \sqrt{3})$ and $\mathbf{v} = (\sqrt{3}, 1)$.   (d) $\mathbf{u} = (\sqrt{2}, \sqrt{2}, 1)$ and $\mathbf{v} = (1, 0, \sqrt{2})$.

4. **Vectors and Angles**

   (a) For each of the following vectors **v**, sketch **v** and compute the length of **v**.

   (i) $\mathbf{v} = (\cos(\pi/6), \sin(\pi/6))$.   (iii) $\mathbf{v} = (\cos(3\pi/4), \sin(3\pi/4))$.
   (ii) $\mathbf{v} = (2\sin(\pi/6), 2\cos(\pi/6))$.   (iv) $\mathbf{v} = (\frac{5}{2}\sin(-3\pi/4), \frac{5}{2}\cos(-3\pi/4))$.

   (b) How would you sketch the vector $(R\cos\alpha, R\sin\alpha)$? What is its length?

5. **Representing Vectors.** Give a graphical and symbolic representation of each of the following vectors:

   (a) The vector representing the acceleration due to gravity of an object falling freely near the surface of the earth.
   (b) The velocity vector of a golf ball that is hit at an angle of $\pi/12$ off the ground with a speed of 70 m/s.
   (c) The vector representing the force exerted by a particle of charge $+1$ located at $(1,0,0)$ on a particle of charge $+2$ located at $(0,1,1)$.
   (d) The vector representing the force exerted by a particle of charge $+1$ located at $(1,0,0)$ on a particle of charge -1 located at $(0,1,1)$.

6. **Vector Sums.** The vectors $\mathbf{u}$, $\mathbf{v}$, and $\mathbf{w}$ are vectors in $\mathbb{R}^3$ with length 2. The vector $\mathbf{u}$ points along the negative x-axis, the vector $\mathbf{v}$ makes an angle of $\pi/3$ with the positive x-axis, and the vector $\mathbf{w}$ makes an angle of $-\pi/3$ with the positive x-axis.

   (a) Sketch the vectors $\mathbf{u}$, $\mathbf{v}$, and $\mathbf{w}$.
   (b) For each of the following vectors, find the coordinates of the vector and, if possible, express it in a different way as a linear combination of $\mathbf{u}$, $\mathbf{v}$, and $\mathbf{w}$.

   (i) $\mathbf{v} + \mathbf{w}$.   (iii) $\mathbf{u} + \mathbf{v} + \mathbf{w}$.
   (ii) $\mathbf{u} + \mathbf{v}$.   (iv) $\frac{1}{2}\mathbf{u} + \mathbf{v} + \mathbf{w}$.

7. **Properties of Vector Sums.** Verify that vector addition is commutative and associative, that is, if $\mathbf{u}$, $\mathbf{v}$, and $\mathbf{w}$ are vectors in $\mathbb{R}^2$ or $\mathbb{R}^3$, then

   (a) $\mathbf{u} + \mathbf{v} = \mathbf{v} + \mathbf{u}$.   (b) $(\mathbf{u} + \mathbf{v}) + \mathbf{w} = \mathbf{u} + (\mathbf{v} + \mathbf{w})$.

   (*Hint*: To verify abstract equalities like these, assign coordinates to each vector, $\mathbf{u} = (u_1, u_2)$, and so on, expand both sides of the equality using the coordinate expressions, then simplify to show that they are equal.)

8. **Properties of Scalar Multiplication.** Verify that if c and d are scalars and $\mathbf{u}$ and $\mathbf{v}$ are vectors in $\mathbb{R}^2$ or $\mathbb{R}^3$, the operations of scalar multiplication and vector addition satisfy:

   (a) $(c + d)\mathbf{u} = c\mathbf{u} + d\mathbf{u}$.   (b) $c(\mathbf{u} + \mathbf{v}) = c\mathbf{u} + c\mathbf{v}$.

   (*Hint*: See the hint to Exercise 7.)

9. **Unit Vectors.** A *unit vector* is a vector of length 1. Often we will want to specify a direction without specifying a magnitude. One way to do this is to specify a unit vector in that direction.

   (a) If $\mathbf{v}$ is any nonzero vector, show that the vector $\frac{1}{\|\mathbf{v}\|}\mathbf{v}$ is a unit vector that points in the same direction as $\mathbf{v}$.

(b) For each of the following vectors $\mathbf{v}$, find a unit vector $\mathbf{u}$ that points in the same direction as $\mathbf{v}$.

(i) $\mathbf{v} = (-1, 2)$.
(ii) $\mathbf{v} = (1, 2, 2)$.
(iii) $\mathbf{v} = (R \sin \theta, R \cos \theta)$, where R is a positive constant and θ is any angle.

(c) Find the unit vector in $\mathbb{R}^2$ that makes an angle θ with the positive x-axis. (Positive angles are measured counterclockwise from the positive x-axis.)

10. **Forces on Charged Particles.** Suppose a particle of charge 5×10^{-6} C is located at the point $(0, 0)$, a particle of charge -3×10^{-6} C is located at the point $(0, 1)$, and a particle of charge 4×10^{-6} C is located at the point $(1, 1)$.

(a) Compute the force exerted by the first particle on the third particle and the force exerted by the second particle on the third particle.
(b) What is the total force on the third particle?

11. **Newton's Law.** Newton's law of gravitation states that every object in the universe attracts every other object with a force that is directly proportional to the product of their masses and inversely proportional to the square of the distance between them.

Thus if two objects have masses m_1 and m_2 and are separated by a distance r, the gravitational force between them has magnitude $G m_1 m_2 / r^2$, where G is a universal constant. The value of G is approximately 6.673×10^{-11} N-m^2/kg^2. The direction of this force is from one object toward the other.

Suppose a particle of mass m_1 is located at the origin and a particle of mass m_2 is located at (x, y, z). Let $r = \sqrt{x^2 + y^2 + z^2}$ denote the distance between the particles.

(a) Find the coordinates of the unit vector $\mathbf{u}$ that points in the direction of the gravitational force exerted by the first particle on the second.
(b) Using Newton's law of gravitation, we know that the gravitational force vector $\mathbf{F}$ is a scalar multiple of $\mathbf{u}$ and that the magnitude of the force is $G m_1 m_2 / r^2$. Give a symbolic representation of the gravitational force vector $\mathbf{F}$.

12. **Gravity of the Earth.** The mass of the earth is approximately 5.98×10^{24} kg, and the radius of the earth is approximately 6.37×10^6 m.

(a) Use Newton's law of gravitation to find the magnitude and direction of the gravitational force exerted by the earth on an object of mass m located h meters above the surface of the earth. (You can assume, as Newton did, that the earth can be treated as an object with its mass all concentrated at the center.)
(b) The mass of the moon is approximately 7.36×10^{22} kg. The distance between the earth and the moon is approximately 3.84×10^8 m. What is the gravitational force exerted by the moon on the earth?
(c) If a rocket of mass m is launched from the earth, directly toward the moon, what is the total gravitational force exerted by the earth and the moon on the rocket when the rocket is halfway to the moon?

(d) At what point will the total gravitational force exerted by the earth and the moon on the rocket be zero?

13. **Force in a Hydrogen Atom.** In a hydrogen atom, the electron and proton are separated by a distance of approximately 5.3×10^{-11} m.

(a) An electron has a charge of approximately -1.6×10^{-19} C, and a proton has a charge of approximately 1.6×10^{-19} C. Find the magnitude of the electrical force between the two particles. Note that $\epsilon_0 \approx 8.9 \times 10^{-12}$ C^2/N-m^2.

(b) An electron has a mass of approximately 9×10^{-31} kg, and a proton has a mass of approximately 1.7×10^{-27} kg. Find the magnitude of the gravitational force between the two particles.

(c) How do these forces compare?

14. **Subsets of Lines.** Let $\mathbf{u}$ and $\mathbf{v}$ be nonzero vectors in $\mathbb{R}^3$ that are not scalar multiples of one another. Use the correspondence between vectors in $\mathbb{R}^3$ and points in $\mathbb{R}^3$ to describe the set of points corresponding to each of the following sets of vectors. (See Example 1.8A.)

(a) $\{t\mathbf{v} : t \in \mathbb{R}\}$.

(b) $\{t\mathbf{v} : t > 0\}$.

(c) $\{t\mathbf{v} : 0 \le t \le 1\}$.

(d) $\{\mathbf{u} + t\mathbf{v} : 0 \le t \le 1\}$.

15. **Subsets of Planes.** Let $\mathbf{u}$ and $\mathbf{v}$ be nonzero vectors in $\mathbb{R}^3$ that are not scalar multiples of one another. Use the correspondence between vectors in $\mathbb{R}^3$ and points in $\mathbb{R}^3$ to describe the set of points corresponding to each of the following sets of vectors. (See Example 1.8B.)

(a) $\{s\mathbf{u} + t\mathbf{v} : 0 \le s \le 1, t \in \mathbb{R}\}$.

(b) $\{s\mathbf{u} + t\mathbf{v} : 0 \le s, t \le 1\}$.

(c) $\{\mathbf{w} + s\mathbf{u} + t\mathbf{v} : 0 \le s \le 1, t \in \mathbb{R}\}$, where $\mathbf{w}$ is another nonzero vector in $\mathbb{R}^3$. Does the structure of the set depend upon the particular $\mathbf{u}$, $\mathbf{v}$, and $\mathbf{w}$ chosen? Explain your answer.

■ 1.3 Vector Products

In this section, we will introduce two vector product operations, the dot product and the cross product. We will begin with the algebraic definition of each product and then interpret the product geometrically. The dot product, which we will consider first, is defined for vectors in $\mathbb{R}^2$ and $\mathbb{R}^3$—in fact, for vectors in $\mathbb{R}^n$ for all n. It is a scalar quantity that can be interpreted as a measurement of the length of a single vector or the angle between two vectors. The cross product is only defined for vectors in $\mathbb{R}^3$. It is a vector with direction perpendicular to the plane spanned by the two vectors and with magnitude equal to the area of the parallelogram formed by the two vectors.

The Dot Product

Definition 1.2 If $\mathbf{u} = (u_1, u_2)$ and $\mathbf{v} = (v_1, v_2)$ are vectors in $\mathbb{R}^2$, the **dot product**, or **inner product**, of $\mathbf{u}$ and $\mathbf{v}$, which we denote by $\mathbf{u} \cdot \mathbf{v}$, is $u_1 v_1 + u_2 v_2$. Thus

$$\mathbf{u} \cdot \mathbf{v} = (u_1, u_2) \cdot (v_1, v_2) = u_1 v_1 + u_2 v_2.$$

If $\mathbf{u} = (u_1, u_2, u_3)$ and $\mathbf{v} = (v_1, v_2, v_3)$ are vectors in $\mathbb{R}^3$, the **dot product**, or **inner product**, of $\mathbf{u}$ and $\mathbf{v}$, which we denote by $\mathbf{u} \cdot \mathbf{v}$, is $u_1 v_1 + u_2 v_2 + u_3 v_3$. Thus,

$$\mathbf{u} \cdot \mathbf{v} = (u_1, u_2, u_3) \cdot (v_1, v_2, v_3) = u_1 v_1 + u_2 v_2 + u_3 v_3. \;\blacklozenge$$

It is worth emphasizing that the **dot product of two vectors is a real number, or scalar, not a vector**. The properties of the dot product are stated in the following proposition. They can be verified by applying the definition. We will verify one of the properties here. The proofs of the remaining properties are left as an exercise. (See Exercise 3.)

Proposition 1.3 If $\mathbf{u}$, $\mathbf{v}$, and $\mathbf{w}$ are vectors in $\mathbb{R}^2$ or $\mathbb{R}^3$ and c is a scalar, then

1. $\mathbf{u} \cdot \mathbf{u} \geq 0$ and $\mathbf{u} \cdot \mathbf{u} = 0$ if and only if $\mathbf{u} = \mathbf{0}$.
2. $\mathbf{u} \cdot \mathbf{v} = \mathbf{v} \cdot \mathbf{u}$.
3. $\mathbf{u} \cdot (\mathbf{v} + \mathbf{w}) = (\mathbf{u} \cdot \mathbf{v}) + (\mathbf{u} \cdot \mathbf{w})$.
4. $\mathbf{u} \cdot c\mathbf{v} = c(\mathbf{u} \cdot \mathbf{v}) = c\mathbf{u} \cdot \mathbf{v}$. $\;\blacklozenge$

Proof of Property 1: Let $\mathbf{u} = (u_1, u_2, u_3)$ be a vector in $\mathbb{R}^3$. The dot product $\mathbf{u} \cdot \mathbf{u} = u_1^2 + u_2^2 + u_3^2$ is a sum of squares, which are nonnegative, so the sum must be nonnegative.

As we prove the second part of the property, it is worth commenting on the meaning of "if and only if." It is the common mathematical way of saying that two statements are equivalent. In this case, it says that each of two statements, first that $\mathbf{u} \cdot \mathbf{u} = 0$ and second that $\mathbf{u} = \mathbf{0}$, implies the other. So to prove an "if and only if" statement, we must make two separate arguments. In this case, "$\mathbf{u} \cdot \mathbf{u} = 0$ if $\mathbf{u} = \mathbf{0}$" means that the hypothesis $\mathbf{u} = \mathbf{0}$ implies the conclusion that $\mathbf{u} \cdot \mathbf{u} = 0$. Of course, if all the entries of $\mathbf{u}$ are zero, the summands of $\mathbf{u} \cdot \mathbf{u}$ are zero, which tells us that $\mathbf{u} \cdot \mathbf{u} = 0$. The converse of the "if" statement is "$\mathbf{u} \cdot \mathbf{u} = 0$ only if $\mathbf{u} = \mathbf{0}$" and means that the hypothesis $\mathbf{u} \cdot \mathbf{u} = 0$ implies the conclusion that $\mathbf{u} = \mathbf{0}$. Since the hypothesis implies that $u_1^2 + u_2^2 + u_3^2 = 0$, and since none of the summands are negative, it must be the case that each summand is equal to zero. Consequently, $u_1 = u_2 = u_3 = 0$ and $\mathbf{u} = \mathbf{0}$. $\;\blacksquare$

<table>
<tr><td>Example 1.9</td><td>**Dot Products**</td></tr>
</table>

A. Length. Let $\mathbf{u} = (u_1, u_2)$ be a vector in $\mathbb{R}^2$. Then

$$\mathbf{u} \cdot \mathbf{u} = u_1^2 + u_2^2 = ||\mathbf{u}||^2.$$

Similarly, if $\mathbf{u} = (u_1, u_2, u_3)$ is a vector in $\mathbb{R}^3$, then

$$\mathbf{u} \cdot \mathbf{u} = u_1^2 + u_2^2 + u_3^2 = ||\mathbf{u}||^2.$$

B. Unit Vectors. We say that a vector $\mathbf{u}$ is a **unit vector** if $||\mathbf{u}|| = 1$. Suppose $\mathbf{u}$ is a unit vector in $\mathbb{R}^2$ that makes an angle of ψ with the positive x-axis and $\mathbf{v}$ is a unit vector in $\mathbb{R}^2$ that makes an angle of ϕ with the positive x-axis. (See Figure 1.18(a)). By convention, angles are measured in radians, with positive angles measured counterclockwise and negative angles measured clockwise from the positive x-axis. Using trigonometry, we can see that $\mathbf{u} = (\cos \psi, \sin \psi)$ and $\mathbf{v} = (\cos \phi, \sin \phi)$. The dot product of $\mathbf{u}$ and $\mathbf{v}$,

$$\mathbf{u} \cdot \mathbf{v} = \cos \psi \cos \phi + \sin \psi \sin \phi.$$

Again using trigonometry,

$$\cos \psi \cos \phi + \sin \psi \sin \phi = \cos(\psi - \phi).$$

With ψ and ϕ defined as above, the angle $\psi - \phi$ is the angle from the vector $\mathbf{v}$ to the vector $\mathbf{u}$. Thus, *the dot product of the unit vectors* $\mathbf{u}$ *and* $\mathbf{v}$ *is the cosine of the angle between* $\mathbf{u}$ *and* $\mathbf{v}$.

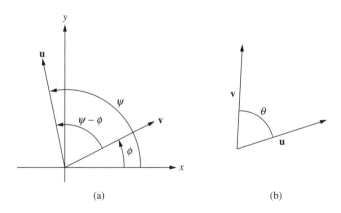

(a)   (b)

Figure 1.18   (a) The dot product, $\mathbf{u} \cdot \mathbf{v}$, of two unit vectors in $\mathbb{R}^2$ is the cosine of the angle between them. (See Example 1.9.) (b) The angle between two vectors is the smaller angle between the vectors when positioned with a common initial point.

The geometric interpretations of the dot product in Example 1.9A as a length and in Example 1.9B as the cosine of an angle extends to the dot product of any two nonzero vectors in the following proposition.

Proposition 1.4   Suppose $\mathbf{u}$ and $\mathbf{v}$ are nonzero vectors in $\mathbb{R}^2$ or $\mathbb{R}^3$. Then

$$\mathbf{u} \cdot \mathbf{v} = ||\mathbf{u}||\,||\mathbf{v}|| \cos \theta,$$

where θ, $0 \leq \theta \leq \pi$, is the angle between the vectors $\mathbf{u}$ and $\mathbf{v}$.   ◆

For this result to make sense geometrically, we must assume that the vectors $\mathbf{u}$ and $\mathbf{v}$ are represented by arrows that have a common initial point. The angle between the vectors is understood to be the angle less than or equal to π between these arrows. (See Figure 1.18(b).) The results of Example 1.9 are special cases of the proposition. If $\mathbf{u} = \mathbf{v}$, then by Example 1.9A, we have $\mathbf{u} \cdot \mathbf{u} = ||\mathbf{u}||^2$. Also, if $\mathbf{u} = \mathbf{v}$, then $\theta = 0$, and we have

$$||\mathbf{u}||\,||\mathbf{u}|| \cos \theta = ||\mathbf{u}||\,||\mathbf{u}|| \cos(0) = ||\mathbf{u}||^2.$$

Thus the proposition is true if $\mathbf{u} = \mathbf{v}$. If $\mathbf{u}$ and $\mathbf{v}$ are unit vectors, then by Example 1.9B $\mathbf{u} \cdot \mathbf{v} = \cos \theta$, where θ is the angle between $\mathbf{u}$ and $\mathbf{v}$. Since $||\mathbf{u}|| = ||\mathbf{v}|| = 1$, we have

$$||\mathbf{u}||\,||\mathbf{v}|| \cos \theta = \cos \theta,$$

which verifies the result for unit vectors. The proof of the proposition for any vectors $\mathbf{u}$ and $\mathbf{v}$ follows.

Proof: First, we recall the law of cosines from trigonometry. In any triangle with sides of lengths a, b, and c, we have

$$c^2 = a^2 + b^2 - 2ab \cos \theta,$$

where θ is the angle opposite the side of length c.

If we position $\mathbf{u}$ and $\mathbf{v}$ at the same initial point, then we can form a triangle with sides $\mathbf{u}$, $\mathbf{v}$, and $\mathbf{u} - \mathbf{v}$, where $\mathbf{u} - \mathbf{v}$ is positioned with its initial point at the endpoint of $\mathbf{v}$. Applying the law of cosines, we have

$$||\mathbf{u} - \mathbf{v}||^2 = ||\mathbf{u}||^2 + ||\mathbf{v}||^2 - 2||\mathbf{u}||\,||\mathbf{v}|| \cos \theta,$$

where θ is the angle between $\mathbf{u}$ and $\mathbf{v}$. We have already shown that the dot product of a vector with itself is the length of the vector squared. Thus we have

$$(\mathbf{u} - \mathbf{v}) \cdot (\mathbf{u} - \mathbf{v}) = \mathbf{u} \cdot \mathbf{u} + \mathbf{v} \cdot \mathbf{v} - 2||\mathbf{u}||\,||\mathbf{v}|| \cos \theta.$$

We can expand and simplify the left-hand side using the distributive and commutative properties of the dot product to obtain

$$\mathbf{u} \cdot \mathbf{u} - 2\mathbf{u} \cdot \mathbf{v} + \mathbf{v} \cdot \mathbf{v} = \mathbf{u} \cdot \mathbf{u} + \mathbf{v} \cdot \mathbf{v} - 2||\mathbf{u}||\,||\mathbf{v}|| \cos \theta.$$

Simplifying this equation, we have

$$\mathbf{u} \cdot \mathbf{v} = ||\mathbf{u}||\,||\mathbf{v}|| \cos \theta,$$

which proves the proposition. ∎

The proposition allows us to use the dot product to compute the angle between two vectors when they are positioned with a common initial point. In particular we can use the dot product to determine when two vectors are perpendicular, or **orthogonal.**

Definition 1.3 Nonzero vectors $\mathbf{u}$ and $\mathbf{v}$ are **orthogonal** if the angle between $\mathbf{u}$ and $\mathbf{v}$ is $\pi/2$. ◆

Since the angle θ between two vectors satisfies $0 \le \theta \le \pi$, $\theta = \pi/2$ if and only if $\cos \theta = 0$. The following corollary follows from the definition.

Corollary 1.1 If $\mathbf{u}$ and $\mathbf{v}$ are nonzero vectors in $\mathbb{R}^2$ or $\mathbb{R}^3$, then $\mathbf{u} \cdot \mathbf{v} = 0$ if and only if $\mathbf{u}$ and $\mathbf{v}$ are orthogonal. ◆

In the example, we will exploit the geometric interpretation of the dot product.

Example 1.10   **Angles and Orthogonality**

A. The Angle Between Vectors. Let $\mathbf{u} = (1, 1, 2)$, $\mathbf{v} = (2, 0, -1)$, and $\mathbf{w} = (-1, 1, 1)$. The dot product of $\mathbf{u}$ and $\mathbf{v}$, $\mathbf{u} \cdot \mathbf{v} = 2 + 0 + (-2) = 0$. Thus $\mathbf{u}$ and $\mathbf{v}$ are orthogonal. The dot product of $\mathbf{u}$ and $\mathbf{w}$, $\mathbf{u} \cdot \mathbf{w} = -1 + 1 + 2 = 2$. Thus $\mathbf{u}$ and $\mathbf{w}$ are not orthogonal. The angle θ between $\mathbf{u}$ and $\mathbf{w}$ satisfies

$$\cos \theta = \frac{\mathbf{u} \cdot \mathbf{w}}{||\mathbf{u}||\,||\mathbf{w}||} = \frac{2}{(\sqrt{6})(\sqrt{3})} = \frac{2}{3\sqrt{2}} = \frac{\sqrt{2}}{3}.$$

The angle between $\mathbf{u}$ and $\mathbf{w}$ is $\theta = \arccos(\sqrt{2}/3)$, which is approximately 1.08.

B. The Line Orthogonal to a Vector. Let $\mathbf{v} = (v_1, v_2)$ be a nonzero vector in $\mathbb{R}^2$. The vector $\mathbf{x} = (x_1, x_2)$ is orthogonal to $\mathbf{v}$ if and only if $\mathbf{v} \cdot \mathbf{x} = v_1 x_1 + v_2 x_2 = 0$, so that x_1 and x_2 satisfy the equation $v_1 x_1 = -v_2 x_2$. The vector $\mathbf{x} = (-v_2, v_1)$ is one solution to this equation. Any scalar multiple $c\mathbf{x} = (-cv_2, cv_1)$ of $\mathbf{x}$ is also a solution, since $\mathbf{v} \cdot c\mathbf{x} = c(\mathbf{v} \cdot \mathbf{x}) = 0$. In fact, every solution (y_1, y_2) to the equation $v_1 y_1 = -v_2 y_2$ is of the form $\mathbf{y} = c\mathbf{x}$. (Why?) Thus the set of all vectors orthogonal to $\mathbf{v}$ is $\{c\mathbf{x} \;:\; c \in \mathbb{R}$ and $\mathbf{x} = (-v_2, v_1)\}$, which is the line through the origin with direction vector $\mathbf{x}$. We will say that this is the *line orthogonal to the nonzero vector* $\mathbf{v}$. (See Example 1.8A of Section 1.2.)

C. The Plane Orthogonal to a Vector. Let $\mathbf{v} = (1, 2, 1)$. The vector $\mathbf{x} = (x_1, x_2, x_3)$ is orthogonal to $\mathbf{v}$ if and only if $\mathbf{v} \cdot \mathbf{x} = x_1 + 2x_2 + x_3 = 0$. One solution to this equation is $\mathbf{x} = (1, 0, -1)$. A second solution to this equation is $\mathbf{y} = (0, 1, -2)$. In this case, $\mathbf{y}$ is not simply a scalar multiple of $\mathbf{x}$, it is a "new" solution. If $\mathbf{u} = c\mathbf{x} + d\mathbf{y}$ for any scalars c and d, then

$$\mathbf{v} \cdot \mathbf{u} = \mathbf{v} \cdot (c\mathbf{x} + d\mathbf{y}) = c(\mathbf{v} \cdot \mathbf{x}) + d(\mathbf{v} \cdot \mathbf{y}) = 0.$$

Thus $\mathbf{u}$ is also orthogonal to $\mathbf{v}$. Conversely, suppose $\mathbf{w}$ satisfies $w_1 + 2w_2 + w_3 = 0$, so that $w_3 = -w_1 - 2w_2$. Then $\mathbf{w} = (w_1, w_2, -w_1 - 2w_2) = w_1(1, 0, -1) + w_2(0, 1, -2) = w_1\mathbf{x} + w_2\mathbf{y}$, and we see that all solutions are of the form $c\mathbf{x} + d\mathbf{y}$. Thus the set of all vectors orthogonal to $\mathbf{v}$ is $\{c\mathbf{x} + d\mathbf{y} \;:\; c, d \in \mathbb{R}\}$. This is the plane spanned by the vectors $\mathbf{x}$ and $\mathbf{y}$. We will say that this is the *plane through the origin orthogonal to the nonzero vector* $\mathbf{v}$, and $\mathbf{v}$ *is orthogonal or normal to the plane spanned by* $\mathbf{x}$ *and* $\mathbf{y}$. (See Example 1.8B of Section 1.2).

Projections

There is an additional geometric interpretation of the dot product if $\mathbf{v}$ is any vector and $\mathbf{u}$ is a unit vector. Then

$$\mathbf{v} \cdot \mathbf{u} = ||\mathbf{v}|| \, ||\mathbf{u}|| \cos\theta = ||\mathbf{v}|| \cos\theta,$$

where θ is the angle between $\mathbf{v}$ and $\mathbf{u}$. If $\mathbf{v}$ and $\mathbf{u}$ are represented by arrows with a common initial point, we can form a right triangle by extending a line L through the vector $\mathbf{u}$ and dropping a perpendicular line from the endpoint of $\mathbf{v}$ to the line L. (See Figure 1.19.) The absolute value of the quantity $||\mathbf{v}|| \cos\theta$ is the length of the side of the right triangle adjacent to θ. We call $\mathbf{v} \cdot \mathbf{u} = ||\mathbf{v}|| \cos\theta$ the *component of* $\mathbf{v}$ *in the direction of* $\mathbf{u}$.

The vector $(\mathbf{v} \cdot \mathbf{u})\mathbf{u}$ is a scalar multiple of $\mathbf{u}$ with length equal to $|\mathbf{v} \cdot \mathbf{u}|$. This vector is a positive multiple of $\mathbf{u}$ if $\cos\theta > 0$, that is, if $0 \leq \theta < \pi/2$, and is a negative multiple of $\mathbf{u}$ if $\cos\theta < 0$, that is, if $\pi/2 < \theta \leq \pi$. If this vector is represented by an arrow with its initial point at the common initial point of $\mathbf{u}$ and $\mathbf{v}$, its endpoint is at the point of intersection of the line L and the perpendicular from the endpoint of $\mathbf{v}$. This vector is called the *projection of* $\mathbf{v}$ *onto* $\mathbf{u}$ and denoted $P_{\mathbf{u}}(\mathbf{v})$. (See Figure 1.20.)

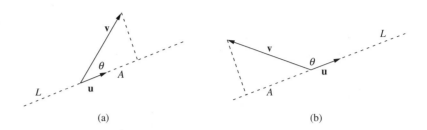

(a)   (b)

Figure 1.19   (a) The component of **v** in the direction of **u** is the length of side A. (b) The component of **v** in the direction of **u** is the negative of the length of side A .

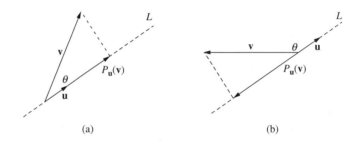

(a)   (b)

Figure 1.20   The projection of **v** onto a unit vector **u**.

In our analysis above, we assumed that the vector **v** was projected onto a unit vector **u**. In the example below, we see that we can in fact compute the projection of **v** onto any nonzero vector.

Example 1.11   **Components and Projections.** Let **v** and **w** be any two nonzero vectors. In order to compute the component of **v** in the direction of **w** and the projection of **v** onto **w**, we first note that the vector

$$\mathbf{u} = \frac{1}{\|\mathbf{w}\|}\mathbf{w}$$

is a unit vector that points in the same direction as **w**. (See Exercise 9 of Section 1.2.) The component of **v** in the direction of **w** is the same as the component of **v** in the direction of **u**. (See Figure 1.21.) Thus the component of **v** in the direction of **w** is

$$\mathbf{v} \cdot \mathbf{u} = \mathbf{v} \cdot \frac{\mathbf{w}}{\|\mathbf{w}\|} = \frac{\mathbf{v} \cdot \mathbf{w}}{\|\mathbf{w}\|}.$$

The projection of **v** onto **w** is the vector from the common initial point of **v** and **w** to the point of intersection of the line L extended through **w** and the perpendicular line

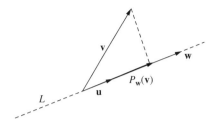

Figure 1.21   The projection of **v** onto a vector **w** of arbitrary length.

from the endpoint of **v** to L. This vector is the same as the projection of **v** onto **u**. Thus we have

$$P_{\mathbf{w}}(\mathbf{v}) = P_{\mathbf{u}}(\mathbf{v}) = (\mathbf{v} \cdot \mathbf{u})\mathbf{u} = \left(\frac{\mathbf{v} \cdot \mathbf{w}}{||\mathbf{w}||} \right) \frac{\mathbf{w}}{||\mathbf{w}||} = \left(\frac{\mathbf{v} \cdot \mathbf{w}}{||\mathbf{w}||^2} \right) \mathbf{w}.$$

To summarize, we have the following definition.

Definition 1.4   Let **v** and **w** be any two nonzero vectors. The ***component of* v *in the direction of* w** is $\frac{\mathbf{v} \cdot \mathbf{w}}{||\mathbf{w}||}$. The ***projection of* v *onto* w** is the vector

$$P_{\mathbf{w}}(\mathbf{v}) = \left(\frac{\mathbf{v} \cdot \mathbf{w}}{||\mathbf{w}||^2} \right) \mathbf{w}. \quad \blacklozenge$$

Projections have a number of applications in the sciences and mathematics. In Section 1.4, we will see how to use projections to compute the distance between a point and a plane. Here we will use the idea of the component of a vector in a given direction to define the physical concept of work. In one-variable calculus, we saw that if a constant force F acting in the direction of a line moves an object along the line through a distance d, then the work done by the force in moving the object is defined to be Fd. More generally, if an object is moved in a straight line in the plane or in space by a constant force vector **F**, we define work as follows.

Definition 1.5   The ***work*** done by a constant force vector **F** in moving a particle a distance d in the direction of the vector **v** is defined to be the product of the component of **F** in the direction of **v** with the distance d. That is,

$$W = \left(\frac{\mathbf{F} \cdot \mathbf{v}}{||\mathbf{v}||} \right) d. \quad \blacklozenge$$

In the following example, we consider an object set into motion by the force of gravity.

Example 1.12   **Work Done by Gravity.** If an object is dropped from a point near the surface of the earth, gravity accelerates the object in a downward direction at 9.8 m/s². This can be represented by the vector $\mathbf{a}_g = (0, 0, -9.8)$. If the object does not fall straight down, the acceleration of the object is the projection of $\mathbf{a}_g$ onto the direction of motion of the object.

A. Acceleration. Suppose you are the object and you are sitting at the top of a straight slide which is 3 meters high, and suppose the angle between the slide and the ground is $\pi/6$. (See Figure 1.22.) To simplify matters, let us assume that all of your acceleration as you move down the slide is due to gravity, so, for example, we are ignoring friction. The acceleration vector $\mathbf{a}$ of your motion is the projection of $\mathbf{a}_g$ onto the direction of the slide. The direction of motion is given by the unit vector $\mathbf{u} = (-\cos(\pi/6), 0, -\sin(\pi/6)) = (-\sqrt{3}/2, 0, -1/2)$. We can calculate that $\mathbf{a} = P_{\mathbf{u}}(\mathbf{a}_g) = 4.9(-\sqrt{3}/2, 0, -1/2)$.

B. Work. The force of gravity on an object of mass m is the vector $\mathbf{F}_g = (0, 0, -9.8m)$. If m is your mass, we can use this to compute the work done by gravity in moving you from the top of the slide to the bottom. The component of $\mathbf{F}_g$ in the direction of motion is $(0, 0, -9.8m) \cdot (-\sqrt{3}/2, 0, -1/2) = 4.9m$. The length of the slide is 6 m. Thus the work done by gravity in moving you from the top of the slide to the bottom of the slide is $29.4m$.

C. More Work. Suppose that instead of sliding down the slide, you jump to the ground and then walk to the bottom of the slide. The direction of motion as you jump is given by the vector $(0, 0, -1)$, so that the component of $\mathbf{F}_g$ in the direction of motion is $9.8m$. The distance traveled from the top of the slide to the ground is 3 m so that the work done by gravity in moving you from the top of the slide to the ground is $29.4m$. As you walk to the bottom of the slide, your direction of motion is given by the vector $(-1, 0, 0)$. The component of $\mathbf{F}_g$ in the direction of motion is 0 since $\mathbf{F}_g$ is orthogonal to the direction of motion. It follows that the work done by gravity as you walk to the bottom of the slide is 0. We conclude that the work done by gravity in this scenario is also $29.4m$.

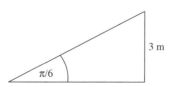

Figure 1.22 A schematic diagram of the slide of Example 1.12.

The fact that the work done by gravity is $29.4m$ in Example 1.12B and C is not a coincidence. In Chapter 6, we will see that gravitational force has the property that the work done in moving an object from one point to another depends only on the two points and not on the path traveled between the two points.

The Cross Product

As we stated in the introduction to this section, the cross product is a vector quantity, not a scalar quantity like the dot product. In addition, the cross product is only defined for vectors in $\mathbb{R}^3$. We will define the cross product symbolically, but as was the case for the dot product, we will see that the cross product vector has important geometric interpretations.

Definition 1.6 Let $\mathbf{u} = (u_1, u_2, u_3)$ and $\mathbf{v} = (v_1, v_2, v_3)$ be vectors in $\mathbb{R}^3$. The **cross product** of $\mathbf{u}$ and $\mathbf{v}$ is

$$\mathbf{u} \times \mathbf{v} = (u_1, u_2, u_3) \times (v_1, v_2, v_3) = (u_2 v_3 - u_3 v_2, \, u_3 v_1 - u_1 v_3, \, u_1 v_2 - u_2 v_1). \; \blacklozenge$$

Before establishing the algebraic properties of the cross product, we will compute several numerical examples.

Example 1.13 | **Cross Product Calculations**

A. Let $\mathbf{u} = (1, 1, 0)$ and $\mathbf{v} = (-1, 2, 0)$. The cross product of $\mathbf{u}$ and $\mathbf{v}$ is

$$\mathbf{u} \times \mathbf{v} = ((1)(0) - (0)(2), (0)(-1) - (1)(0), (1)(2) - (1)(-1)) = (0, 0, 3).$$

The cross product of $\mathbf{v}$ and $\mathbf{u}$ is

$$\mathbf{v} \times \mathbf{u} = ((2)(0) - (0)(1), (0)(1) - (-1)(0), (-1)(1) - (2)(1)) = (0, 0, -3).$$

B. Let $\mathbf{u} = (1, -1, 3)$ and $\mathbf{v} = (2, 1, 0)$. The cross product of $\mathbf{u}$ and $\mathbf{v}$ is

$$\mathbf{u} \times \mathbf{v} = ((-1)(0) - (3)(1), (3)(2) - (1)(0), (1)(1) - (-1)(2)) = (-3, 6, 3).$$

The cross product of $\mathbf{v}$ and $\mathbf{u}$ is

$$\mathbf{v} \times \mathbf{u} = ((1)(3) - (0)(-1), (0)(1) - (2)(3), (2)(-1) - (1)(1)) = (3, -6, -3).$$

C. Let $\mathbf{u} = (1, -1, 3)$ and $\mathbf{v} = (2, -2, 6)$. The cross product of $\mathbf{u}$ and $\mathbf{v}$ is

$$\mathbf{u} \times \mathbf{v} = ((-1)(6) - (3)(-2), (3)(2) - (1)(6), (1)(-2) - (-1)(2)) = (0, 0, 0).$$

The cross product of $\mathbf{v}$ and $\mathbf{u}$ is

$$\mathbf{v} \times \mathbf{u} = ((-2)(3) - (6)(-1), (6)(1) - (2)(3), (2)(-1) - (-2)(1)) = (0, 0, 0).$$

These examples lead to two observations about the cross product. The first is that the cross product is not commutative, $\mathbf{u} \times \mathbf{v} \neq (\mathbf{v} \times \mathbf{u})$. Rather, the examples indicate that it is **anticommutative**, that is, $\mathbf{u} \times \mathbf{v} = -(\mathbf{v} \times \mathbf{u})$. The second is that when $\mathbf{v}$ is a scalar multiple of $\mathbf{u}$, as in Example 1.13C, the cross product is the zero vector. These are examples of more general algebraic properties of the cross product, which are given in the following proposition. The proof is left as an exercise.

Proposition 1.5   Suppose $\mathbf{u}$, $\mathbf{v}$, and $\mathbf{w}$ are vectors in $\mathbb{R}^3$ and a and b are scalars.

1. $\mathbf{u} \times \mathbf{v} = -(\mathbf{v} \times \mathbf{u})$.
2. $\mathbf{u} \times \mathbf{u} = \mathbf{0}$.
3. $a\mathbf{u} \times b\mathbf{v} = ab(\mathbf{u} \times \mathbf{v})$.
4. $\mathbf{u} \times (\mathbf{v} + \mathbf{w}) = (\mathbf{u} \times \mathbf{v}) + (\mathbf{u} \times \mathbf{w})$.   ◆

We can also describe the cross product geometrically. To do this, we must determine the direction and the length of $\mathbf{u} \times \mathbf{v}$. The following proposition is a first step toward determining the direction of the cross product vector.

Proposition 1.6   If $\mathbf{u}$ and $\mathbf{v}$ are vectors in $\mathbb{R}^3$, then the cross product of $\mathbf{u}$ and $\mathbf{v}$, $\mathbf{u} \times \mathbf{v}$, is orthogonal to $\mathbf{u}$ and $\mathbf{v}$.   ◆

Proof: We must show that $\mathbf{u} \cdot (\mathbf{u} \times \mathbf{v}) = 0$ and $\mathbf{v} \cdot (\mathbf{u} \times \mathbf{v}) = 0$.

$$
\begin{aligned}
\mathbf{u} \cdot (\mathbf{u} \times \mathbf{v}) &= (u_1, u_2, u_3) \cdot (u_2v_3 - u_3v_2, u_3v_1 - u_1v_3, u_1v_2 - u_2v_1) \\
&= u_1u_2v_3 - u_1u_3v_2 + u_2u_3v_1 - u_2u_1v_3 + u_3u_1v_2 - u_3u_2v_1 \\
&= 0.
\end{aligned}
$$

Similarly, we have

$$
\begin{aligned}
\mathbf{v} \cdot (\mathbf{u} \times \mathbf{v}) &= (v_1, v_2, v_3) \cdot (u_2v_3 - u_3v_2, u_3v_1 - u_1v_3, u_1v_2 - u_2v_1) \\
&= v_1u_2v_3 - v_1u_3v_2 + v_2u_3v_1 - v_2u_1v_3 + v_3u_1v_2 - v_3u_2v_1 \\
&= 0. ∎
\end{aligned}
$$

Since $\mathbf{u} \times \mathbf{v}$ is orthogonal to $\mathbf{u}$ and $\mathbf{v}$, it is orthogonal to any vector of the form $s\mathbf{u} + t\mathbf{v}$, where $s, t \in \mathbb{R}$. (See Exercise 11.) The set of vectors of the form $s\mathbf{u} + t\mathbf{v}$ is the plane

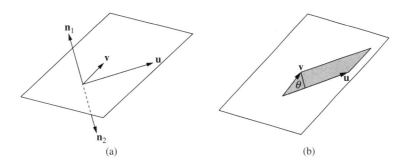

(a)   (b)

Figure 1.23   (a) The vectors $\mathbf{n}_1$ and $\mathbf{n}_2$ are both perpendicular to the plane spanned by $\mathbf{u}$ and $\mathbf{v}$. The ordered triple of vectors $\mathbf{u}$, $\mathbf{v}$, and $\mathbf{n}_1$ satisfies the right-hand rule, while the ordered triple $\mathbf{u}$, $\mathbf{v}$, and $\mathbf{n}_2$ does not. (b) The parallelogram spanned by $\mathbf{u}$ and $\mathbf{v}$ has height equal to $\sin(\theta)\|\mathbf{v}\|$, where θ is the angle between $\mathbf{u}$ and $\mathbf{v}$. It is the length of the perpendicular from the endpoint of $\mathbf{v}$ to the line whose direction vector is $\mathbf{u}$.

spanned by $\mathbf{u}$ and $\mathbf{v}$. (See Example 1.8B of Section 1.2.) This means that $\mathbf{u}\times\mathbf{v}$ is orthogonal to this plane and can point in one of the two possible directions that are orthogonal to the plane. These directions can be represented by unit vectors $\mathbf{n}_1$ and $\mathbf{n}_2$ with $\mathbf{n}_2 = -\mathbf{n}_1$. (See Figure 1.23(a).) The cross product $\mathbf{u}\times\mathbf{v}$ "chooses" one of these directions according to the ***right-hand rule***. Three vectors in space, $\mathbf{u}$, $\mathbf{v}$, and $\mathbf{w}$, satisfy the right-hand rule if, ***when you place your right hand on the plane spanned by $\mathbf{u}$ and $\mathbf{v}$ so that your fingers curl from the vector $\mathbf{u}$ toward the vector $\mathbf{v}$ through an angle less than π, the vector $\mathbf{w}$ points to the same side of the plane spanned by $\mathbf{u}$ and $\mathbf{v}$ as your thumb***. Vectors oriented according to the right-hand rule are oriented in the same way as the positive coordinate axes. (See Section 1.1.) For example, the vectors $(1,0,0)$, $(0,1,0)$, and $(0,0,1)$, which point along the positive coordinate axes, satisfy the right-hand rule for vectors, while the vectors $(1,0,0)$, $(0,1,0)$, and $(0,0,-1)$ do not. The vectors $\mathbf{u}$, $\mathbf{v}$, and $\mathbf{u}\times\mathbf{v}$ satisfy the right-hand rule. If we consider the vectors $\mathbf{u}$ and $\mathbf{v}$ in Figure 1.23(a), the cross product $\mathbf{u}\times\mathbf{v}$ must point in the direction of the vector $\mathbf{n}_1$.

Next we must consider the length of the cross product vector,

$$\|\mathbf{u}\times\mathbf{v}\|^2 = (u_2v_3 - u_3v_2)^2 + (u_3v_1 - u_1v_3)^2 + (u_1v_2 - u_2v_1)^2.$$

If we expand this expression, regroup terms, and simplify, it is equal to the expression

$$(u_1^2 + u_2^2 + u_3^2)(v_1^2 + v_2^2 + v_3^2) - (u_1v_1 + u_2v_2 + u_3v_3)^2$$

$$= \|\mathbf{u}\|^2\|\mathbf{v}\|^2 - (\mathbf{u}\cdot\mathbf{v})^2.$$

As we saw previously, $\mathbf{u} \cdot \mathbf{v} = ||\mathbf{u}|| \, ||\mathbf{v}|| \cos\theta$, where θ is the angle between $\mathbf{u}$ and $\mathbf{v}$. If we substitute this into the expression above, we have

$$
\begin{aligned}
||\mathbf{u} \times \mathbf{v}||^2 &= ||\mathbf{u}||^2 ||\mathbf{v}||^2 - ||\mathbf{u}||^2 ||\mathbf{v}||^2 \cos^2\theta \\
&= ||\mathbf{u}||^2 ||\mathbf{v}||^2 (1 - \cos^2\theta) \\
&= ||\mathbf{u}||^2 ||\mathbf{v}||^2 \sin^2\theta.
\end{aligned}
$$

This leads to the following proposition.

Proposition 1.7   The length of the cross product of $\mathbf{u}$ and $\mathbf{v}$ is given by

$$
||\mathbf{u} \times \mathbf{v}|| = ||\mathbf{u}|| \, ||\mathbf{v}|| \sin\theta,
$$

where θ is the angle between $\mathbf{u}$ and $\mathbf{v}$.   ♦

Notice we are assuming that $0 \le \theta \le \pi$, so $\sin\theta \ge 0$ and the expression $||\mathbf{u}|| \, ||\mathbf{v}|| \sin\theta$ is always nonnegative.

This formula for the length of the cross product vector also has a geometric interpretation. If $\mathcal{P}$ is the parallelogram with sides $\mathbf{u}$ and $\mathbf{v}$, the height of the parallelogram is $||\mathbf{v}|| \sin\theta$. (See Figure 1.23(b).) The area of $\mathcal{P}$ is *(length of base)* $\times$ *height* $= ||\mathbf{u}|| \, ||\mathbf{v}|| \sin\theta$. Thus the length of the cross product of $\mathbf{u}$ and $\mathbf{v}$ is the area of the parallelogram formed by $\mathbf{u}$ and $\mathbf{v}$.

In sum, we could *define the cross product of* $\mathbf{u}$ *and* $\mathbf{v}$ *geometrically as the vector that is orthogonal to the plane spanned by* $\mathbf{u}$ *and* $\mathbf{v}$*, pointing in the direction given by the right-hand rule, and with length equal to the area of the parallelogram spanned by* $\mathbf{u}$ *and* $\mathbf{v}$.

The cross product occurs in a number of different settings in physics. In the following example, we represent the force on a charged particle moving through a magnetic field in terms of the cross product.

Example 1.14   **Magnetic Force.** If a charged particle is in motion in a magnetic field, then the field will exert a force on the particle that acts in a direction perpendicular to the direction of motion of the particle. This can be expressed in terms of the cross product of the velocity vector of the particle and the magnetic field vector. If the particle has charge q and velocity $\mathbf{v}$ at a point in space and the magnetic field is represented by the vector $\mathbf{B}$ at the point, then the force $\mathbf{F}_m$ exerted by the field on the particle is equal to $q\mathbf{v} \times \mathbf{B}$. Let us apply this to the particular example of the magnetic field that results from an electric current moving through a wire.

Let us suppose that a current of magnitude I is in an infinite wire that coincides with the z-axis and that the current is moving in the direction of positive z. At a point

$P = (x, y, z)$ with $x^2 + y^2 > 0$, so that P does not lie on the z-axis, the magnetic field $\mathbf{B}$ is given by

$$\mathbf{B} = \frac{\mu_0 I}{2\pi} \left(\frac{-y}{x^2 + y^2}, \frac{x}{x^2 + y^2}, 0 \right),$$

where μ_0 is a constant, called the ***permeability of a vacuum*** or ***magnetic constant***. This vector has magnitude

$$\frac{\mu_0 I}{2\pi \sqrt{x^2 + y^2}}.$$

Notice that $\mathbf{B}$ is orthogonal to the z-axis, the direction of the current. If a particle of charge q has velocity $\mathbf{v} = (v_1, v_2, v_3)$, then the force exerted by the magnetic field on the particle is

$$\mathbf{F}_m = q\mathbf{v} \times \mathbf{B} = \frac{q\mu_0 I}{2\pi(x^2 + y^2)} (-v_3 x, -v_3 y, v_1 x + v_2 y).$$

If the particle is moving in the direction of the magnetic field, so that $\mathbf{v}$ is a multiple of $\mathbf{B}$ at P, the force exerted by the field is $\mathbf{0}$.

Summary

In this section, we introduced the ***dot product*** of two vectors in the plane or space and the ***cross product*** of two vectors in space.

The dot product is given by $\mathbf{u} \cdot \mathbf{v} = u_1 v_1 + u_2 v_2$ in the plane and $\mathbf{u} \cdot \mathbf{v} = u_1 v_1 + u_2 v_2 + u_3 v_3$ in space. We investigated the properties of the dot product and showed that ***the angle θ between two vectors*** is related to their dot product by the formula

$$\cos \theta = \frac{\mathbf{u} \cdot \mathbf{v}}{\|\mathbf{u}\| \, \|\mathbf{v}\|}.$$

Two nonzero vectors are ***orthogonal*** if their dot product is zero. We used the dot product to construct the ***component*** of $\mathbf{v}$ in the direction of $\mathbf{u}$ and the ***projection*** of $\mathbf{v}$ in the direction of $\mathbf{u}$. We employed this when calculating the ***work*** done by a force in moving an object along a straight path.

The cross product is given by $\mathbf{u} \times \mathbf{v} = (u_2 v_3 - u_3 v_2, u_3 v_1 - u_1 v_3, u_1 v_2 - u_2 v_1)$. We investigated the properties of the cross product and showed, in particular, that if $\mathbf{u}$ and $\mathbf{v}$ are not multiples of one another, then $\mathbf{u} \times \mathbf{v}$ ***is orthogonal to both $\mathbf{u}$ and $\mathbf{v}$ and the plane spanned by $\mathbf{u}$ and $\mathbf{v}$***. Further, $\mathbf{u}$, $\mathbf{v}$, and $\mathbf{u} \times \mathbf{v}$ are related by the ***right-hand rule***. The length of $\mathbf{u} \times \mathbf{v}$ is the ***area of the parallelogram spanned by $\mathbf{u}$ and $\mathbf{v}$***. Finally, we showed that ***force due to a magnetic field*** is given by a cross product.

Section 1.3 Exercises

1. **Dot Product Calculations.** For each pair of vectors, compute $\mathbf{u}\cdot\mathbf{v}$ and the angle between $\mathbf{u}$ and $\mathbf{v}$.

   (a) $\mathbf{u} = (1,2)$ and $\mathbf{v} = (-1,1)$.
   (b) $\mathbf{u} = (1,\sqrt{3})$ and $\mathbf{v} = (\sqrt{3},1)$.
   (c) $\mathbf{u} = (1,0,2)$ and $\mathbf{v} = (3,0,1)$.
   (d) $\mathbf{u} = (\sqrt{2},\sqrt{2},1)$ and $\mathbf{v} = (1,0,\sqrt{2})$.
   (e) $\mathbf{u} = (\cos\alpha,\sin\alpha)$ and $\mathbf{v} = (\sin\alpha,\cos\alpha)$, $0 < \alpha < \pi/2$.
   (f) $\mathbf{u} = (\cos\alpha,\sin\alpha,1)$ and $\mathbf{v} = (\sin\alpha,\cos\alpha,1)$, $0 < \alpha < \pi/2$.

2. **Components and Projections.** Let $\mathbf{v}$ be a nonzero vector and $\mathbf{w} = c\mathbf{v}$, $c \neq 0$.

   (a) Let $\mathbf{x}$ be any vector. How does the component of $\mathbf{x}$ in the direction of $\mathbf{v}$ compare to the component of $\mathbf{x}$ in the direction of $\mathbf{w}$? (Be sure to consider $c > 0$ and $c < 0$.)
   (b) How does the projection of $\mathbf{x}$ onto $\mathbf{v}$ compare to the projection of $\mathbf{x}$ onto $\mathbf{w}$? (Be sure to consider $c > 0$ and $c < 0$.)

3. **Properties of the Dot Product.** Prove each of the following properties of the dot product. (*Hint:* For each equality, assign coordinates to each vector, for example, $\mathbf{u} = (u_1, u_2)$, expand the left-hand side using the coordinate expressions, and rearrange the coordinate expressions to show that they equal the coordinate expressions for the right-hand side.)

   (a) Prove that the dot product is *commutative*, that is, prove that $\mathbf{u}\cdot\mathbf{v} = \mathbf{v}\cdot\mathbf{u}$ for any vectors $\mathbf{u}$ and $\mathbf{v}$.
   (b) Prove that the dot product is *bilinear*, that is, prove that

   (i) $\mathbf{u}\cdot(\mathbf{v}+\mathbf{w}) = (\mathbf{u}\cdot\mathbf{v}) + (\mathbf{u}\cdot\mathbf{w})$ for any vectors $\mathbf{u}$, $\mathbf{v}$, and $\mathbf{w}$.
   (ii) $(c\mathbf{u})\cdot\mathbf{v} = c(\mathbf{u}\cdot\mathbf{v}) = \mathbf{u}\cdot(c\mathbf{v})$ for any vectors $\mathbf{u}$ and $\mathbf{v}$ and scalars c.

4. **Cauchy–Schwarz.** Prove the Cauchy–Schwarz inequality: For any vectors $\mathbf{u}$ and $\mathbf{v}$,

$$|\mathbf{u}\cdot\mathbf{v}| \leq ||\mathbf{u}||\,||\mathbf{v}||.$$

5. **Triangle Inequality.** Prove the triangle inequality for vectors:

$$||\mathbf{u}+\mathbf{v}|| \leq ||\mathbf{u}|| + ||\mathbf{v}||$$

   for any vectors $\mathbf{u}$ and $\mathbf{v}$. (*Hint:* Square both sides of the inequality and use Exercise 4.)

6. **Work.** Suppose you are sitting at the top of a straight frictionless slide that is 2 m high, and suppose the angle between the slide and the ground is $\pi/4$.

   (a) What is your acceleration vector when you slide down the slide?
   (b) Compute the work done by gravity in moving you down the slide.

7. **Cross Product Calculations.** For each of the following pairs of vectors $\mathbf{u}$ and $\mathbf{v}$, find the vector $\mathbf{u} \times \mathbf{v}$ and compute the area of the parallelogram spanned by $\mathbf{u}$ and $\mathbf{v}$.

   (a) $\mathbf{u} = (1, 2, 0)$ and $\mathbf{v} = (-1, 2, 1)$.
   (b) $\mathbf{u} = (3, 1, 2)$ and $\mathbf{v} = (1, -1, -1)$.
   (c) $\mathbf{u} = (1, 0, 0.5)$ and $\mathbf{v} = (-2, 1, 2)$.
   (d) $\mathbf{u} = (2, -4, 1)$ and $\mathbf{v} = (-\sqrt{2}, 2\sqrt{2}, -0.5\sqrt{2})$.

8. **Properties of the Cross Product.** Let $\mathbf{u}$ and $\mathbf{v}$ be two vectors in $\mathbb{R}^3$.

   (a) Use the symbolic definition or the geometric definition of the cross product to show that $\mathbf{u} \times \mathbf{v} = -(\mathbf{v} \times \mathbf{u})$.
   (b) Show that $\mathbf{u} \times \mathbf{u} = \mathbf{0}$.

9. **Methane.** A tetrahedron is regular if all of its edges are the same length.

   (a) Show that the tetrahedron with vertices at $(0, 0, 0)$, $(1, 1, 0)$, $(1, 0, 1)$, and $(0, 1, 1)$ is a regular tetrahedron.
   (b) Find the angle between any two adjacent edges of the tetrahedron.
   (c) The methane molecule CH_4 is arranged with its four hydrogen molecules at the vertices of a regular tetrahedron and the carbon atom at its center. Suppose the hydrogen molecules are arranged at the vertices of the tetrahedron from part (a) and the carbon atom is located at the center, which is at $(0.5, 0.5, 0.5)$. What is the bonding angle between the bonds from the carbon atom to two of the hydrogen atoms?

10. **Properties of the Cross Product.** Let $\mathbf{u}$, $\mathbf{v}$, and $\mathbf{w}$ be vectors in $\mathbb{R}^3$ and let a and b be scalars.

    (a) Use the definition to show that $a\mathbf{u} \times b\mathbf{v} = (ab)(\mathbf{u} \times \mathbf{v})$.
    (b) Use the definition to show that $\mathbf{u} \times (\mathbf{v} + \mathbf{w}) = (\mathbf{u} \times \mathbf{v}) + (\mathbf{u} \times \mathbf{w})$.

11. **Orthogonality of $\mathbf{u} \times \mathbf{v}$.** Show that the vector $\mathbf{u} \times \mathbf{v}$ is orthogonal to every vector in the plane spanned by $\mathbf{u}$ and $\mathbf{v}$. That is, show that the vector $\mathbf{u} \times \mathbf{v}$ is orthogonal to $s\mathbf{u} + t\mathbf{v}$ for any scalars s and t.

12. **Nonassociativity of the Cross Product.** Let $\mathbf{u}$, $\mathbf{v}$, and $\mathbf{w}$ be vectors in space. Show by example that the cross product is not associative, that is,

$$\mathbf{u} \times (\mathbf{v} \times \mathbf{w}) \neq (\mathbf{u} \times \mathbf{v}) \times \mathbf{w}.$$

13. **Magnetic Force.** In Example 1.14, we said that the force on a charged particle moving through a magnetic field is given in terms of the cross product of the velocity of the particle and the magnetic field. That is, $\mathbf{F}_m = q\mathbf{v} \times \mathbf{B}$. Suppose that the speed $\|\mathbf{v}\|$ is held constant. For which directions of motion, that is, for which directions of $\mathbf{v}$, will

the magnitude of $\mathbf{F}_m$ be greatest? For which directions will the magnitude of $\mathbf{F}_m$ be smallest? Explain your answer.

■ 1.4 Lines and Planes

Although lines and planes are among the simplest subsets of Euclidean space, it is important for us to take the time to describe them carefully. When we introduce a new concept, it will often be the case that we will first apply the concept to lines and planes. More than this, however, a thorough understanding of lines and planes is necessary before we can develop more sophisticated ideas of multivariable calculus later in the text.

In this section, we will expand upon the definition of a line and the definition of a plane given in Example 1.8 of Section 1.2, where we used the correspondence between vectors with initial point at the origin and points in $\mathbb{R}^2$ or $\mathbb{R}^3$ to describe lines and planes as sets of points. In particular, we will use the dot product and the cross product to explore the geometry of lines and planes.

Lines

The slope of a line in the plane is the quotient of the change in y divided by the change in x, $\Delta y / \Delta x$, for any pair of distinct points in the line. Consequently, the displacement vectors between any two pairs of points in the line are multiples of one another. (See Exercise 3.) We will call any vector that is a displacement vector between two distinct points on a line a **direction vector** for the line. Any two direction vectors for a line are scalar multiples of one another. If the multiple is negative, the vectors point in opposite directions. The slope, however, is well defined. If $\mathbf{v} = (v_1, v_2)$ is a direction vector for L and $v_1 \neq 0$, the slope of L is v_2 / v_1.

Suppose that Q is a point on the line L through the point P with direction vector $\mathbf{v}$. From above, the displacement vector $\vec{PQ}$ is a scalar multiple t of $\mathbf{v}$. That is, $\mathbf{q} - \mathbf{p} = t\mathbf{v}$, where $\mathbf{q}$ is the vector that corresponds to Q and $\mathbf{p}$ is the vector that corresponds to P. Rewriting this equation, we have that $\mathbf{q} = \mathbf{p} + t\mathbf{v}$. We will call this the **vector** or **parametric description** of L. The scalar t is the **parameter**. Conversely, if Q is a point that satisfies $\mathbf{q} = \mathbf{p} + t\mathbf{v}$ for some scalar t, then the displacement vector from $\mathbf{p}$ to $\mathbf{q}$ is $t\mathbf{v}$, and it follows that Q is a point on L. The same can be said for points and lines in $\mathbb{R}^3$. We summarize this in the following proposition.

Proposition 1.8   Let P be a point and let $\mathbf{v}$ be a nonzero vector, $\mathbf{v} \neq \mathbf{0}$. The **line passing through the point P with direction vector** $\mathbf{v}$ is the set L of points corresponding to vectors of the form $\mathbf{p} + t\mathbf{v}$, where $\mathbf{p}$ is the vector corresponding to P and t is a real number. Using set notation,

$$L = \{\, \mathbf{x} : \ \mathbf{x} = \mathbf{p} + t\mathbf{v}, \ t \in \mathbb{R} \,\}. \ \blacklozenge$$

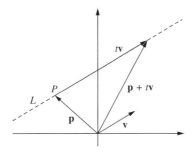

Figure 1.24   The line through P with direction vector $\mathbf{v}$. (See Proposition 1.8.)

A plot of the line through a point P in the direction of a vector $\mathbf{v}$ is shown in Figure 1.24. Notice that for $t = 0$, $\mathbf{x} = \mathbf{p} + 0\mathbf{v} = \mathbf{p}$, so that P is contained in this set. At times it will be useful to write out the coordinates of points on the line through P with direction $\mathbf{v}$. In $\mathbb{R}^2$, if $P = (x_0, y_0)$ and $\mathbf{v} = (v_1, v_2)$, then we have

$$\mathbf{p} + t\mathbf{v} = (x_0, y_0) + t(v_1, v_2) = (x_0 + tv_1, y_0 + tv_2).$$

In $\mathbb{R}^3$, if $P = (x_0, y_0, z_0)$ and $\mathbf{v} = (v_1, v_2, v_3)$, then we have

$$\mathbf{p} + t\mathbf{v} = (x_0, y_0, z_0) + t(v_1, v_2, v_3) = (x_0 + tv_1, y_0 + tv_2, z_0 + tv_3).$$

Let us consider a particular example of a vector equation for a line.

Example 1.15   **A Line in $\mathbb{R}^3$**

A.   Let L be the line through the point $(1, 2, -1)$ with direction vector $(1, -1, 2)$. The points on this line are the points corresponding to vectors $\mathbf{x}$ of the form

$$\mathbf{x} = (1, 2, -1) + t(1, -1, 2) = (1 + t, 2 - t, -1 + 2t),$$

for $t \in \mathbb{R}$.

B.   In order to determine if a given point (x_1, y_1, z_1) lies in the line L, we must determine if there is a real number t such that

$$x_1 = 1 + t$$
$$y_1 = 2 - t$$
$$z_1 = -1 + 2t.$$

For example, the point $(1/2, 5/2, -2)$ does lie on L, since $t = -1/2$ solves the system of equations. However, the point $(2, 1, 3)$ does not lie on the line, because solving the first equation requires $t = 1$, but solving the third equation requires $t = 2$, so that no single value of t yields the required point.

To write down the equation for a line L, we can start with any point in L and any direction vector $\mathbf{v}$ for L. As the following example shows, we can also start with two points that lie in the line.

Example 1.16 **The Line Between Two Points.** From geometry, we know that if P and Q are distinct points, there is a unique line L that passes through both P and Q. We would like to write out the vector description for this line. Let $\mathbf{p}$ be the vector corresponding to P, let $\mathbf{q}$ be the vector corresponding to Q, and let $\mathbf{v} = \overrightarrow{PQ} = \mathbf{q} - \mathbf{p}$ be the displacement vector from P to Q. Let L be the line through P with direction vector $\mathbf{v}$, $L = \{\mathbf{x} : \mathbf{x} = \mathbf{p} + t\mathbf{v}, \ t \in \mathbb{R}\}$. As above, when $t = 0$, $\mathbf{x} = \mathbf{p}$, so that $P \in L$. When $t = 1$, $\mathbf{x} = \mathbf{p} + \mathbf{v} = \mathbf{p} + (\mathbf{q} - \mathbf{p}) = \mathbf{q}$, so that $Q \in L$. Since L contains P and Q, it must be the unique line passing through P and Q.

We will say that two lines are **parallel** if their direction vectors are scalar multiples of each other. That is, if L is the line through P with direction vector $\mathbf{v} \neq \mathbf{0}$ and $\tilde{L}$ is the line through $\tilde{P}$ with direction vector $\tilde{\mathbf{v}} \neq \mathbf{0}$, then L and $\tilde{L}$ are parallel if and only if there is a scalar c so that $\tilde{\mathbf{v}} = c\mathbf{v}$. (See Exercise 5.)

We will say that two lines are **orthogonal** if they intersect and if their direction vectors are orthogonal. That is, with the same notation $\tilde{L}$ and L are orthogonal if $\tilde{\mathbf{v}} \cdot \mathbf{v} = 0$. (See Exercise 6.) Note that in the plane, two lines either intersect or are parallel. In space, lines that do not intersect need not be parallel. In this case, we say the lines are **skew**.

Planes

The algebraic definition of a plane in space is similar to our description of a line in the plane or space. Because a plane is "two-dimensional," we will need two vectors that lie in the plane rather than the single direction vector in the case of a line. We will require that these two vectors be linearly independent. We will say that two vectors $\mathbf{u}$ and $\mathbf{v}$ are **linearly independent** if they are nonzero and if they are not scalar multiples of one another. (See Exercise 10 for an alternative formulation of the definition.) It follows that if two vectors are linearly independent, they are direction vectors for two distinct lines. We are now ready to define a plane.

Definition 1.7 Let P be a point in space and let $\mathbf{p}$ be the corresponding vector. Let $\mathbf{u}$ and $\mathbf{v}$ be linearly independent vectors. The **plane through P spanned by $\mathbf{u}$ and $\mathbf{v}$** is the set of points $\mathcal{P}$

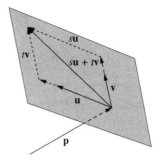

Figure 1.25 The plane corresponding to the set of vectors $\{\mathbf{p} + s\mathbf{u} + t\mathbf{v} : s, t \in \mathbb{R}\}$. (See Definition 1.7.)

in $\mathbb{R}^3$ corresponding to vectors of the form $\mathbf{p} + s\mathbf{u} + t\mathbf{v}$, where s and t are real numbers. We will call this the *vector* or *parametric description* of $\mathcal{P}$. The scalars s and t are the parameters. Using set notation,

$$\mathcal{P} = \{\mathbf{x} : \mathbf{x} = \mathbf{p} + s\mathbf{u} + t\mathbf{v}, \ s, t \in \mathbb{R}\}. \ \blacklozenge$$

A plot of the plane $\mathcal{P}$ through P spanned by $\mathbf{u}$ and $\mathbf{v}$ is shown in Figure 1.25. There are several observations to make about $\mathcal{P}$: When $s = 0$ and $t = 0$, $\mathbf{x} = \mathbf{p} + 0\mathbf{u} + 0\mathbf{v} = \mathbf{p}$, so that $\mathcal{P}$ contains P. The vectors $\mathbf{u}$ and $\mathbf{v}$ lie in the plane in the sense that the line segments representing $\mathbf{u}$ and $\mathbf{v}$ are subsets of the plane. It follows that the lines through P with direction vectors $\mathbf{u}$ and $\mathbf{v}$ are subsets of the plane. The line with direction vector $\mathbf{u}$ corresponds to the points with $t = 0$, and the line with direction vector $\mathbf{v}$ corresponds to the points with $s = 0$. Notice that since every point in the plane corresponds to a vector of the form $\mathbf{p} + s\mathbf{u} + t\mathbf{v}$, the endpoint of $\mathbf{p} + s\mathbf{u} + t\mathbf{v}$ is in the plane. The vector $\mathbf{p} + s\mathbf{u} + t\mathbf{v}$ does not lie in the plane. In fact, one can show that all vectors in the plane are of the form $s\mathbf{u} + t\mathbf{v}$ for $s, t \in \mathbb{R}$. (See Exercise 12.)

Example 1.17	**A Plane**

A. The plane through the point $(1, 2, -1)$ spanned by the vectors $\mathbf{u} = (2, 1, 3)$ and $\mathbf{v} = (-1, 1, 0)$ is the set of points corresponding to the set of vectors

$$\{\mathbf{x} : \mathbf{x} = (1, 2, -1) + s(2, 1, 3) + t(-1, 1, 0), \ s, t \in \mathbb{R}\}.$$

B. To determine if a point (x, y, z) lies in a plane, we must determine if there are real numbers s and t such that the vector $\mathbf{x} = (x, y, z)$ can be written $\mathbf{x} = \mathbf{p} + s\mathbf{u} + t\mathbf{v}$. For example, to determine if the point $(0, 3, -2)$ lies in the plane from part A, we must find real numbers s and t so that

$$(0, 3, -2) = (1, 2, -1) + s(2, 1, 3) + t(-1, 1, 0).$$

This vector equation is equivalent to the three coordinate equations

$$0 = 1 + 2s - t$$
$$3 = 2 + s + t$$
$$-2 = -1 + 3s.$$

Using the third equation, we have $s = -1/3$, and substituting this value into the second equation, we have $t = 4/3$. These values do not solve the first equation. Thus $(0, 3, -2)$ does not lie in the plane.

As was the case for lines, the vector description of a plane is not unique. The point P and the vectors $\mathbf{u}$ and $\mathbf{v}$ can be replaced by any other point and pair of vectors that lie in the plane.

From geometry, we know that given three noncollinear points in space, P, Q, and R, there is exactly one plane $\mathcal{P}$ containing P, Q, and R. Let us give a vector description of $\mathcal{P}$ as a plane passing through the point P. First, we need to find a pair of linearly independent vectors that will span $\mathcal{P}$. Let $\mathbf{p}$, $\mathbf{q}$, and $\mathbf{r}$ be the vectors corresponding to P, Q, and R. Since the three points lie in $\mathcal{P}$, the displacement vectors $\mathbf{u} = \overrightarrow{PQ} = \mathbf{q} - \mathbf{p}$ from P to Q and $\mathbf{v} = \overrightarrow{PR} = \mathbf{r} - \mathbf{p}$ from P to R lie in $\mathcal{P}$. Since the three points are noncollinear, the vectors $\mathbf{u}$ and $\mathbf{v}$ do not lie in the same line, that is, they are linearly independent. Let $\mathcal{P}$ be the plane through P spanned by $\mathbf{u}$ and $\mathbf{v}$,

$$\mathcal{P} = \{\, \mathbf{x} : \mathbf{x} = \mathbf{p} + s\mathbf{u} + t\mathbf{v}, \ s, t \in \mathbb{R} \,\}.$$

Because $\mathbf{p} + 1\mathbf{u} + 0\mathbf{v} = \mathbf{p} + (\mathbf{q} - \mathbf{p}) = \mathbf{q}$, $\mathcal{P}$ contains Q, and because $\mathbf{p} + 0\mathbf{u} + 1\mathbf{v} = \mathbf{p} + (\mathbf{r} - \mathbf{p}) = \mathbf{r}$, $\mathcal{P}$ contains R. Since $\mathcal{P}$ contains P, Q, and R, it must be the unique plane containing these points.

Another important description of a plane utilizes a vector orthogonal to the plane. Suppose that $\mathcal{P}$ is the plane through P spanned by the linearly independent vectors $\mathbf{u}$ and $\mathbf{v}$. We make the following definition.

Definition 1.8 The vector $\mathbf{n}$ is *orthogonal* or *normal* to the plane $\mathcal{P}$ if $\mathbf{n}$ is orthogonal to every vector in $\mathcal{P}$. (See Figure 1.26.) ◆

Every vector in a plane $\mathcal{P}$ is of the form $s\mathbf{u} + t\mathbf{v}$. (See Exercise 12.) Thus $\mathbf{n}$ is orthogonal to $\mathcal{P}$ if and only if it is orthogonal to every vector of the form $s\mathbf{u} + t\mathbf{v}$. We have

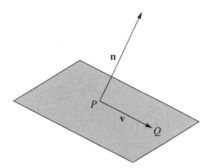

Figure 1.26 The plane normal to $\mathbf{n}$ and passing through P contains the vector $\mathbf{v} = \overrightarrow{PQ}$. (See Definition 1.8.)

already seen that $\mathbf{u} \times \mathbf{v}$ is orthogonal to $s\mathbf{u} + t\mathbf{v}$. Therefore if $\mathcal{P}$ is the plane through P spanned by $\mathbf{u}$ and $\mathbf{v}$, $\mathbf{u} \times \mathbf{v}$ is normal to $\mathcal{P}$.

Conversely, if we are given a point P and a nonzero vector $\mathbf{n}$, it is possible to construct a plane through P that is orthogonal to $\mathbf{n}$. In fact, from geometry we know that there can be only one such plane. In Example 1.10C of Section 1.3, we carried out an explicit calculation of a plane through $P = (0,0,0)$ normal to a given vector. More generally, a point Q is in the plane if and only if the displacement vector $\overrightarrow{PQ}$ is orthogonal to $\mathbf{n}$. Thus a point $\mathbf{x} = (x, y, z)$ is an element of $\mathcal{P}$ if and only if it satisfies the equation $\mathbf{n} \cdot (\mathbf{x} - \mathbf{p}) = 0$. In coordinates, this equation becomes

$$n_1(x - x_0) + n_2(y - y_0) + n_3(z - z_0) = 0.$$

We will call this the **coordinate equation** of the plane. The coordinate equation can be rewritten in the form

$$n_1 x + n_2 y + n_3 z = d,$$

where $d = n_1 x_0 + n_2 y_0 + n_3 z_0$. Conversely, suppose we are given a coordinate equation of the form $n_1 x + n_2 y + n_3 z = d$ and a point $P = (x_0, y_0, z_0)$ that solves the equation. Working backwards, the set of solutions to this equation is the plane through P with normal vector $\mathbf{n} = (n_1, n_2, n_3)$. Using the above construction, it is possible to find vectors $\mathbf{u}$ and $\mathbf{v}$ that span $\mathcal{P}$.

Now let us turn to a numerical example.

Example 1.18   **A Plane Normal to a Vector**

A. The plane through the point $(1, -2, 1.5)$ with normal vector $(0.5, 1, -3)$ is given by the equation

$$(0.5, 1, -3) \cdot (x - 1, y + 2, z - 1.5) = 0.$$

This can be rewritten as

$$0.5(x-1) + (y+2) - 3(z-1.5) = 0,$$

or

$$0.5x + y - 3z = -6.$$

B. To determine if a point lies in the plane given in part A, we must determine if the coordinates of the point satisfy the equation of the plane. For example, the point $(0, 3, 3)$ is in the plane since $0.5(0) + 3 - 3(3) = -6$. On the other hand, the point $(2, -5, 1)$ is not in the plane since $0.5(2) + (-5) - 3(1) = -7 \neq -6$.

C. We have seen that the point $(2, -5, 1)$ is not in the plane given by the equation $0.5x + y - 3z = -6$. However, we can find a plane parallel to this plane that contains $(2, -5, 1)$. *Two planes are parallel if their normal vectors have the same or opposite directions.* Thus two planes are parallel if the normal vectors of the planes are scalar multiples of one another. The plane $0.5x + y - 3z = -6$ has normal vector $(0.5, 1, -3)$. The plane parallel to this plane and containing the point $(2, -5, 1)$ is given by the equation $0.5(x-2) + (y+5) - 3(z-1) = 0$, or $0.5x + y - 3z = -7$.

As with the vector description of a plane, the coordinate equation of a plane $\mathcal{P}$ is not unique. The normal vector $\mathbf{n}$ can be replaced by any scalar multiple of $\mathbf{n}$, and the point P can be replaced by any point in $\mathcal{P}$. It is the case, however, that the equation of the plane is unique up to multiplication by a constant. (See Exercise 12 of Section 1.5.)

It is possible to move from the vector description of a plane to the coordinate equation of a plane and vice versa. Here we demonstrate how to go from the vector description to the coordinate equation of a plane. Suppose that we are given a vector description of a plane $\mathcal{P}$. That is, we are given a point $P = (x_0, y_0, z_0)$ and linearly independent vectors $\mathbf{u} = (u_1, u_2, u_3)$ and $\mathbf{v} = (v_1, v_2, v_3)$, so that $\mathcal{P}$ is the set of points corresponding to vectors of the form $\mathbf{p} + s\mathbf{u} + t\mathbf{v}$, $s, t \in \mathbb{R}$. We know that $\mathbf{u} \times \mathbf{v}$ is a normal vector for $\mathcal{P}$, so that the coordinate equation for $\mathcal{P}$ is

$$\mathbf{u} \times \mathbf{v} \cdot (x - x_0, y - y_0, z - z_0) = 0.$$

The following example illustrates these calculations.

Example 1.19   **The Coordinate Equation of a Plane**

A. Let $\mathcal{P}$ be the plane containing the vectors $\mathbf{u} = (1, -2, 1)$ and $\mathbf{v} = (2, 1, -2)$ and the point $P = (1, 0, 1)$. A normal vector to this plane is the vector $\mathbf{u} \times \mathbf{v} = (3, 4, 5)$. The plane consists of all points $(x, y, z) \in \mathbb{R}^3$ that satisfy the equation

$$3(x - 1) + 4(y - 0) + 5(z - 1) = 0,$$

or, after simplifying, $3x + 4y + 5z = 8$.

B. Let $\mathcal{P}$ be the plane containing the points $P = (1, 0, 1)$, $Q = (1, 2, -1)$, and $R = (2, -1, 1)$. We know that $\mathbf{u} = \vec{PQ} = (0, 2, -2)$ and $\mathbf{v} = \vec{PR} = (1, -1, 0)$ lie in $\mathcal{P}$. It follows that $\mathbf{n} = \mathbf{u} \times \mathbf{v} = (-2, -2, -2)$ is normal to $\mathcal{P}$. Thus a coordinate equation for $\mathcal{P}$ is

$$-2(x - 1) - 2(y - 0) - 2(z - 1) = 0,$$

or, after simplifying, $x + y + z = 2$.

Finally, we demonstrate how to calculate the distance between a point and a plane.

| Example 1.20 | **The Distance from a Point to a Plane** |

A. Suppose that $\mathcal{P}$ is the plane through P that is normal to the vector $\mathbf{n}$. The distance from a point Q to $\mathcal{P}$ is the length of the line segment from Q to $\mathcal{P}$ that is perpendicular to $\mathcal{P}$. Suppose that R is the endpoint of the segment that lies in $\mathcal{P}$. (See Figure 1.27.) Then the distance from Q to $\mathcal{P}$ is $\|\vec{QR}\|$, which is the absolute value of the component of $\vec{PQ}$ in the direction of $\mathbf{n}$. Thus

$$\|\vec{QR}\| = \left| \vec{PQ} \cdot \frac{\mathbf{n}}{\|\mathbf{n}\|} \right|.$$

(See Example 1.11 of Section 1.3.)

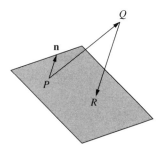

Figure 1.27 The distance from Q to the plane through P normal to the vector $\mathbf{n}$ is the length of QR, where R is the point in the plane that makes QR orthogonal to the plane. (See Example 1.20.)

B. Let us use the formula in part A to compute the distance from $Q = (4, 0, 0)$ to the plane through $P = (1, 2, 1)$ that is normal to the vector $\mathbf{n} = (1, -1, 3)$. Since $\overrightarrow{QP} = (1, 2, 1) - (4, 0, 0) = (-3, 2, 1)$, the distance from Q to $\mathcal{P}$ is

$$\left| (-3, 2, 1) \cdot \frac{(1, -1, 3)}{\|(1, -1, 3)\|} \right| = \left| (-3, 2, 1) \cdot \frac{1}{\sqrt{11}} (1, -1, 3) \right| = \frac{2}{\sqrt{11}}.$$

Summary

In this section, we studied the algebra and geometry of *lines* in the plane and space and *planes* in space.

We defined the *line through a point P in the direction* $\mathbf{v}$ and gave a *vector* or *parametric description* and a *coordinate description* of the line. We showed how to determine if a point lies on a line, and we defined what it means for lines to be *parallel, orthogonal,* and *skew*.

When considering planes, we defined what it means for a pair of vectors $\mathbf{u}$ and $\mathbf{v}$ to be *linearly independent*. We then defined the *plane through a point P spanned by the linearly independent vectors* $\mathbf{u}$ *and* $\mathbf{v}$ and gave a *vector* or *parametric description* of the plane. We showed how to determine if a point lies in a plane and how to construct the plane passing through three noncollinear points.

We also showed how to construct a *plane through P orthogonal to a nonzero vector* $\mathbf{n}$ and that *the plane through P spanned by* $\mathbf{u}$ *and* $\mathbf{v}$ *is the plane through P orthogonal to* $\mathbf{u} \times \mathbf{v}$. We used this construction to write out the *coordinate equation* for a plane. Finally, we showed how to compute the *distance from a point to a plane*.

Section 1.4 Exercises

1. **Descriptions of Lines.** Give the vector description of each of the following lines and the coordinate form for a point on the line.

    (a) The line through $P = (1, 2)$ with direction vector $(-1, 1)$.
    (b) The line through $P = (-1, 3, 2)$ with direction vector $(0, -1, 1)$.
    (c) The line through $P = (3, 2, 0)$ and parallel to the line $\{\mathbf{x} : \mathbf{x} = (-1, 3, 2) + t(3, 1, 2),\ t \in \mathbb{R}\}$.

2. **Line Calculations.** Determine if each of the given points Q lies on the given line. If not, find a line through Q parallel to the given line.

    (a) $Q = (1, -3)$ and L is the line $\{\mathbf{x} : \mathbf{x} = (0, -1) + t(3, 2),\ t \in \mathbb{R}\}$.
    (b) $Q = (-2, 4)$ and L is the line $\{(x, y) : x = 1 + 3t, y = 3 - t,\ t \in \mathbb{R}\}$.

(c) $Q = (-1, -3, 4)$ and L is the line $\{\mathbf{x} : \mathbf{x} = (0, -1, 3) + t(2, 4, -2), \ t \in \mathbb{R}\}$.

(d) $Q = (1, 2, 1)$ and L is the line $\{(x, y, z) : x = 1 + t, y = 3 - 2t, z = -1 - t, \ t \in \mathbb{R}\}$.

3. **Direction Vectors of a Line.** Suppose that the pairs of points $\mathbf{x} = (x_1, x_2)$, $\mathbf{y} = (y_1, y_2)$ and $\tilde{\mathbf{x}} = (\tilde{x}_1, \tilde{x}_2)$, $\tilde{\mathbf{y}} = (\tilde{y}_1, \tilde{y}_2)$ lie in the line L and that the line has slope m. Use the fact that m can be computed using either pair of points to show that the direction vector from $\tilde{\mathbf{x}}$ to $\tilde{\mathbf{y}}$ is a multiple of the direction vector from $\mathbf{x}$ to $\mathbf{y}$.

4. **The Point-Slope Equation.** Let L be the line passing through $P = (x_0, y_0)$ with direction vector $\mathbf{v} = (v_1, v_2) \neq \mathbf{0}$. If $v_1 \neq 0$, show that points in L satisfy the equation

$$y - y_0 = m(x - x_0),$$

where $m = v_2/v_1$. This is the **point-slope** form of the equation of the line passing through P with slope m. If $v_1 = 0$, show that points on L satisfy the equation $x = x_0$. In this case, the line is vertical.

5. **Parallel Lines.** Let L be the line in the plane through $P = (x_0, y_0)$ with direction vector $\mathbf{v} = (v_1, v_2) \neq \mathbf{0}$ and let $\tilde{L}$ be the line in the plane through $\tilde{P} = (\tilde{x}_0, \tilde{y}_0)$ with direction vector $\tilde{\mathbf{v}} = (\tilde{v}_1, \tilde{v}_2) \neq \mathbf{0}$.

(a) Show that if L and $\tilde{L}$ are parallel, that is, there is a scalar $c \neq 0$ so that $\tilde{\mathbf{v}} = c\mathbf{v}$, then $\tilde{L}$ and L have the same slope or are both vertical. (See Exercise 4.)

(b) Conversely, suppose that $\tilde{L}$ and L have the same slope or are both vertical. Show that there is a scalar $c \neq 0$ so that $\tilde{\mathbf{v}} = c\mathbf{v}$.

6. **Orthogonal Lines I.** Let L be the line in the plane through $P = (x_0, y_0)$ with direction vector $\mathbf{v} = (v_1, v_2) \neq \mathbf{0}$ and let $\tilde{L}$ be the line in the plane through $\tilde{P} = (\tilde{x}_0, \tilde{y}_0)$ with direction vector $\tilde{\mathbf{v}} = (\tilde{v}_1, \tilde{v}_2) \neq \mathbf{0}$.

(a) Show that if L and $\tilde{L}$ are orthogonal, then either their slopes are negative reciprocals of each other, or one line is vertical and one line is horizontal. (See Exercise 4.)

(b) Conversely, show that if the slopes of L and $\tilde{L}$ are negative reciprocals of each other, then L and $\tilde{L}$ are orthogonal.

7. **Intersecting Lines**

(a) Given two lines L_1 and L_2 in $\mathbb{R}^3$, carefully describe how you would determine if the two lines intersect.

(b) Do the following pairs of lines intersect?

    (i) The line through $P = (-1, 3, 2)$ with direction vector $(0, -1, 1)$ and the line through $Q = (3, 4, -1)$ with direction vector $(2, 1, -2)$.

    (ii) The line through $P = (1, 0, -2)$ with direction vector $(-1, 2, 1)$ and the line through $Q = (0, 2, 1)$ with direction vector $(2, 1, -1)$.

8. **Angles Between Lines.** Two distinct lines in $\mathbb{R}^3$ are parallel, skew, or intersect in a single point. If they intersect, the angle between the lines is defined to be the angle θ, $0 \leq \theta \leq \pi/2$, made by the lines at the point of intersection.

   (a) If L_1 and L_2 intersect, how would you determine the angle between the two lines?
   (b) Determine if the following pairs of lines in $\mathbb{R}^3$ are parallel, skew, or intersecting. If the lines intersect, find the angle between them.

      (i) The line through $(-1, 2, 1)$ with direction vector $(3, -1, 2)$ and the line through $(2, 1, 3)$ with direction vector $(0, 2, 1)$.
      (ii) $\{(x, y, z) : x = 1 + s, y = 2 - s, z = 3 + 2s, \ s \in \mathbb{R}\}$ and $\{(x, y, z) : x = 3t, y = 1 - 3t, z = 3 + 6t, \ t \in \mathbb{R}\}$.
      (iii) $\{\mathbf{x} : \mathbf{x} = (1, 2, 3) + s(1, -1, 2), \ s \in \mathbb{R}\}$ and $\{\mathbf{x} : \mathbf{x} = (0, 3, 1) + t(1, 0, 2), \ t \in \mathbb{R}\}$.
      (iv) The line through $(-1, 2, 1)$ with direction vector $(3, -1, 2)$ and $\{(x, y, z) : x = 3t, y = 1 - 3t, z = 3 + 6t, \ t \in \mathbb{R}\}$.

9. **Orthogonal Lines II.** Determine if the following pairs of lines are orthogonal.

   (a) The line through $P = (1, 0, -2)$ with direction vector $(-1, 2, 1)$ and the line through $Q = (0, 2, 1)$ with direction vector $(2, 1, -1)$.
   (b) The line through $P = (-1, 3, 2)$ with direction vector $(0, -2, -2)$ and the line through $Q = (0, 2, 1)$ with direction vector $(2, 1, -1)$.
   (c) The line through $P = (2, 1, 3)$ with direction vector $(-1, 2, 1)$ and the line through $Q = (0, 5, 5)$ with direction vector $(1, 1, -1)$.

10. **Linear Independence.** Let $\mathbf{u}$ and $\mathbf{v}$ be nonzero vectors in the plane or space. Let c and d be scalars.

   (a) Suppose that $c \neq 0$, $d \neq 0$, and $c\mathbf{u} + d\mathbf{v} = \mathbf{0}$. Show that $\mathbf{u}$ and $\mathbf{v}$ are scalar multiples of one another.
   (b) Suppose now that $\mathbf{u}$ and $\mathbf{v}$ are linearly independent. Show that if $c\mathbf{u} + d\mathbf{v} = \mathbf{0}$, then both $c = 0$ and $d = 0$. (*Hint:* Assume first that $c \neq 0$ or $d \neq 0$, and then show that the vectors are not linearly independent. This is called a *proof by contradiction.*)

   This exercise shows $\mathbf{u}$ and $\mathbf{v}$ being linearly independent is equivalent to saying that whenever $c\mathbf{u} + d\mathbf{v} = \mathbf{0}$, both c and d must be zero.

11. **Equations of Planes.** Find a coordinate equation for each of the planes described below.

   (a) The plane containing the point $P = (1, -1, 2)$ with normal vector $\mathbf{n} = (1, 0, 1)$.
   (b) The plane containing the point $P = (3, 1, -1)$ with normal vector $\mathbf{n} = (-1, 2, 1)$.
   (c) The plane containing the point $P = (1, 1, 2)$ and the vectors $\mathbf{u} = (1, -1, 1)$ and $\mathbf{v} = (0, 1, -1)$.
   (d) The plane containing the points $P = (1, 0, 2)$, $Q = (-2, 3, 2)$, and $R = (1, -1, 0)$.

12. **Vectors in a Plane.** Let $\mathcal{P}$ be the plane through P spanned by the linearly independent vectors $\mathbf{u}$ and $\mathbf{v}$. A vector $\mathbf{w}$ is contained in $\mathcal{P}$ if its initial point and endpoint lie in $\mathcal{P}$. Show that $\mathbf{w}$ is equal to $s\mathbf{u} + t\mathbf{v}$ for s and $t \in \mathbb{R}$. (*Hint:* Use the vector description of $\mathcal{P}$ to express the initial point and endpoint of $\mathbf{w}$ in terms of $\mathbf{u}$ and $\mathbf{v}$.)

13. **Distance of a Point from a Plane.** For each of the following, determine if the point Q is on the plane. If not, find the distance from the point Q to the plane.

    (a) The plane is given by the equation $2x - y + 3z = 4$ and the point $Q = (1, 2, -1)$.
    (b) The plane is given by the equation $2x - y + 3z = 4$ and the point $Q = (1, 1, 1)$.
    (c) The plane is given by the equation $x + y - 2z = 2$ and the point $Q = (1, 2, -1)$.

14. **Parallel Planes.** Let $\mathcal{P}_1$ be a plane with normal vector $\mathbf{n}_1 \neq \mathbf{0}$ and $\mathcal{P}_2$ be a plane with normal vector $\mathbf{n}_2 \neq \mathbf{0}$.

    (a) How would you determine if the two planes are parallel?
    (b) If the two planes are parallel, how would you find the distance between them?

■ 1.5 End of Chapter Exercises

1. **Subsets of $\mathbb{R}^3$.** Describe in words the following subsets of $\mathbb{R}^3$. If the set is defined by more than one condition, say whether it is a union or an intersection of sets.

    (a) $\{(x, y, z) : z^2 = 4\}$.
    (b) $\{(x, y, z) : y = 3, xz = 0\}$.
    (c) $\{(x, y, z) : (x - 1)(y - 2)(z - 3) = 0\}$.
    (d) $\{(x, y, z) : z^2 \leq 4\}$.
    (e) $\{(x, y, z) : |z| \leq 4, |y| = 3\}$.
    (f) $\{(x, y, z) : |x - 1| \leq 1, |y - 1| \leq 1, |z - 1| \leq 1\}$.

2. **An Ellipsoid.** In Example 1.4A of Section 1.1, we studied the slices of the ellipsoid $\mathcal{S}$ defined by $x^2 + y^2/4 + z^2/9 = 1$. Consider instead the set $\tilde{\mathcal{S}}$ defined by the equation $(x - 3)^2 + y^2/4 + z^2/9 = 1$.

    (a) Describe the horizontal slices of $\tilde{\mathcal{S}}$. How do they differ from the horizontal slices of $\mathcal{S}$?
    (b) Describe the vertical slices of $\tilde{\mathcal{S}}$. How do they differ from the vertical slices of $\mathcal{S}$? (Consider both slices parallel to the xz-plane and parallel to the yz-plane.)
    (c) Describe $\tilde{\mathcal{S}}$. How does it differ from $\mathcal{S}$?
    (d) Suppose instead $\tilde{\mathcal{S}}$ had been defined by the equation $x^2 + (y - 3)^2/4 + z^2/9 = 1$. Summarize how this would change your answers to (a), (b), and (c).

(e) Suppose that x_0, y_0, and z_0 are constants, and let $\tilde{S}$ be defined by $(x - x_0)^2 + (y - y_0)^2/4 + (z - z_0)^2/9 = 1$. Describe $\tilde{S}$.

3. **Slices of Surfaces.** For each of the following sets S, give a complete description of the slices of S by planes parallel to the coordinate planes.

(a) $S = \{(x, y, z) : (x - 2)^2 + y^2 + z^2 = 1\}$.
(b) $S = \{(x, y, z) : x^2 + (y - 4)^2 + (z + 1)^2 = 1\}$.
(c) $S = \{(x, y, z) : z = (x + 1)^2 + y^2\}$.
(d) $S = \{(x, y, z) : (x - 3)^2 + 2y^2 = 1\}$.
(e) $S = \{(x, y, z) : (x - 4)^2 - (y + 1)^2/9 - z^2 = 0\}$.
(f) $S = \{(x, y, z) : y = -x^2 - 4(z - 1)^2\}$.

4. **Shapes in the Plane.** Given a collection of points in the plane or in space, it is often useful to be able to locate the points in a subset that has a simple geometric description, for example, a rectangle or a circle. For each of the following, sketch the set of the specified type, use set notation to describe it symbolically, and explain how you arrived at your answer.

(a) The smallest rectangle that contains the points and whose sides are vertical and horizontal. The points are $(0, 0)$, $(1.2, 0.2)$, $(-0.3, -0.4)$, $(0.5, 1.7)$, and $(0.25, 0.75)$.
(b) The smallest rectangle that contains the points and whose sides are on diagonals 45 degrees from the vertical. Use the point set from (a).
(c) The smallest disk that contains the points. The points are $(1, 1)$, $(1, 3)$, and $(1.5, 1.25)$.
(d) The smallest disk that contains the points. The points are $(1, 1)$, $(1, 3)$, $(1.5, 1.25)$, and $(2, -1)$.

5. **Epidemic Models.** In Example 1.1C of Section 1.1, in a model for the spread of an epidemic, we identified the coordinates of a point in space with the number of susceptible, infected, and recovered individuals. Of course, the number of individuals in each of the three subpopulations must be a nonnegative number. If further assumptions are made about the model, it is possible to say more about sizes of the subpopulations and the set of points in space whose coordinates represent these sizes. For each of the following assumptions about the behavior of the epidemic and the sizes of the populations, describe the corresponding set of points in space in words and using set notation.

(a) If the model is concerned with a short period of time, it is common to assume that the total number of individuals is constant. Suppose that the total population is always equal to 1000.
(b) If the disease is fatal, or if the period of time is long enough so that the natural death rate must be taken into account, it is common to assume that the total population never exceeds a maximum value. Suppose that the total population is always less than or equal to 1000.

6. **Positions of a Satellite.** Suppose a satellite has a maximum altitude of 250 mi. and that it might pass over any point on the earth's surface between the Tropic of Cancer

and the Tropic of Capricorn. (The Tropics are located at 23.5° north and 23.5° south latitude, respectively.) Use set notation to describe the region of the earth's atmosphere where the satellite is orbiting as a subset of coordinate space. (The average diameter of the earth is approximately 7918 mi.)

7. **A Spherical Model of the Heart.** (See Exercise 12 of Section 1.1.) A more realistic model of the heart wall and cavity is based on bases of spheres. A *base* of a sphere centered at the origin is the portion of a sphere lying below a horizontal plane. Here we are interested in bases that lie inside a sphere of radius R and below a plane at height $z = h$, $0 < h < R$. The volume of such a base is $\frac{1}{6}\pi(R + h)(3(R^2 - h^2) + (R + h)^2)$.

   (a) Let $\mathcal{S}$ denote the sphere of radius R centered at the origin and let $\mathcal{P}$ be the plane $z = h$, $0 < h < R$. Find the intersection of $\mathcal{S}$ and $\mathcal{P}$ and sketch the region $\mathcal{R}$ that lies inside $\mathcal{S}$ and below $\mathcal{P}$. Label your sketch.
   (b) Let $\mathcal{R}_1$ and $\mathcal{R}_2$ denote the bases of the spheres of radii $R = R_1$ and $R = R_2$, respectively, lying below the horizontal plane of height $z = h$, $h < R_1 < R_2$. Describe and sketch the region $\tilde{\mathcal{R}}$ that lies inside $\mathcal{R}_2$ and outside $\mathcal{R}_1$. What is the volume of $\tilde{\mathcal{R}}$?
   (c) We want to use $\tilde{\mathcal{R}}$ to model the heart wall and the region enclosed by $\tilde{\mathcal{R}}$ to model the heart cavity. Let $\bar{\mathcal{R}}$ denote the region enclosed by $\tilde{\mathcal{R}}$. It lies inside $\mathcal{R}_1$. Find values of the constants R_1, R_2, and h so that the heart cavity and heart wall have volumes 100 cc and 40 cc, respectively. Explain how you arrived at your answer. Is your answer unique?

8. **Volume and the Cross Product.** Let $\mathbf{u}$, $\mathbf{v}$, and $\mathbf{w}$ be fixed vectors in space. Let us consider the set of vectors of the form

$$\{a\mathbf{u} + b\mathbf{v} + c\mathbf{w} \ : \ a, b, c \in [0, 1]\,\}.$$

Using our identification of vectors and points in $\mathbb{R}^3$, this set of vectors corresponds to a parallelepiped in space.

   (a) Sketch the parallelepiped determined by $\mathbf{u}$, $\mathbf{v}$, and $\mathbf{w}$.
   (b) What is the volume of a parallelepiped?
   (c) Use properties of the dot product and the cross product to show that the volume of the parallelepiped determined by $\mathbf{u}$, $\mathbf{v}$, and $\mathbf{w}$ is $|\mathbf{u} \cdot (\mathbf{v} \times \mathbf{w})|$.
   (d) Compute the volume of the parallelepiped spanned by $\mathbf{u} = (1, 2, 1)$, $\mathbf{v} = (-1, 2, 2)$, and $\mathbf{w} = (0, 1, 1)$.

9. **Current Forces.** If two current-carrying wires are brought near each other, the magnetic field caused by the second current will exert a force on the first wire and vice versa. In the idealized version, where the wires are short straight segments, it is possible to express this force in terms of cross products. Suppose that the segments of wire are represented by the vectors $\mathbf{v}_1$ and $\mathbf{v}_2$ located at $\mathbf{x}_1$ and $\mathbf{x}_2$, respectively, and that the wires carry currents of magnitude I_1 and I_2. (See Figure 1.28.) The direction of the vectors $\mathbf{v}_1$ and

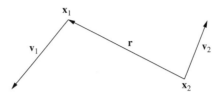

Figure 1.28   The vectors $\mathbf{v}_1$ and $\mathbf{v}_2$ represent short straight segments of current-carrying wires located, respectively, at $\mathbf{x}_1$ and $\mathbf{x}_2$. Note that $\mathbf{r} = \mathbf{x}_1 - \mathbf{x}_2$. (See Exercise 9.)

$\mathbf{v}_2$ denote the directions of the currents I_1 and I_2, respectively. The force $\mathbf{F}_1$ exerted on the first wire is given by

$$\mathbf{F}_1 = \frac{\mu_0 I_1 I_2}{4\pi r^3} \left(\mathbf{v}_1 \times (\mathbf{v}_2 \times \mathbf{r})\right),$$

where $\mathbf{r} = \mathbf{x}_1 - \mathbf{x}_2$ and $r = ||\mathbf{r}||$. The expression for the force $\mathbf{F}_2$ exerted by the first wire by the current in the second is obtained by reversing the roles of the indices in the above formula. (*Hint:* In each of the following, think about the geometry of the vectors before you calculate anything.)

(a) If the segments of wire lie in the same line, what is the force exerted by one current on the other wire? Explain your answer.
(b) Suppose that $\mathbf{v}_1$ and $\mathbf{v}_2$ are parallel and point in the same direction. Find the directions of $\mathbf{F}_1$ and $\mathbf{F}_2$.
(c) Suppose that $\mathbf{v}_1$ and $\mathbf{v}_2$ are parallel and point in opposite directions. Find the directions of $\mathbf{F}_1$ and $\mathbf{F}_2$.
(d) Suppose that $\mathbf{v}_1$ and $\mathbf{v}_2$ are orthogonal. Can we make a general statement about the directions of $\mathbf{F}_1$ and $\mathbf{F}_2$?

10. **Descriptions of Lines.** Give the vector description of each of the following lines and the coordinate form for a point on the line.

(a) The line through $P = (1,3)$ and $Q = (-1,3)$.
(b) The line through $P = (1,1,0)$ and $Q = (2,-1,3)$.
(c) The line through $P = (1,3,4)$ and $Q = (-1,3,3)$.

11. **Equations of Planes.** Find a coordinate equation for each of the planes described below.

(a) The plane containing the points $P = (1,-1-1)$ and $Q = (2,0,3)$ and the vector $\mathbf{u} = (1,1,2)$.
(b) The plane containing the point $P = (0,-1,1)$ and the line that passes through $(1,0,-2)$ with direction vector $\mathbf{u} = (1,-1,2)$.
(c) The plane containing the point $P = (1,2,0)$ and the vector $\mathbf{u} = (3/2,2,1)$, which is orthogonal to the plane given by the equation $x + 2y - z = 3$.

12. **The Coordinate Equation of a Plane.** Let $\mathcal{P}$ be the plane containing the point P with normal vector $\mathbf{n}$.

 (a) Suppose Q is another point in $\mathcal{P}$. Does the coordinate equation for the plane change if we use Q instead of P to determine it? Explain your answer.
 (b) Does the coordinate equation for the plane change if we use $k\mathbf{n}$, $k \neq 0$, to determine it instead of $\mathbf{n}$? Explain your answer.

13. **A Line and a Plane**

 (a) Suppose L is a line in $\mathbb{R}^3$ and $\mathcal{P}$ is a plane in $\mathbb{R}^3$. What are the possible geometric relationships between L and $\mathcal{P}$?
 (b) Explain how you would determine what the geometric relationship is between a given line and plane.

14. **Two Lines and a Plane.** If L_1 and L_2 are two intersecting lines in $\mathbb{R}^3$, there is a unique plane that contains both lines. How would you determine the coordinate equation of this plane?

15. **Collinear Points**

 (a) Given three points in $\mathbb{R}^3$, how would you determine if the points are collinear?
 (b) For each of the following sets of points, determine if they are collinear. If so, find the line that contains them. If not, find the unique plane that contains them.

 (i) $P = (2, 1, 0)$, $Q = (1, 3, -1)$, $R = (2, 4, -3)$.
 (ii) $P = (1, 2, -3)$, $Q = (2, 0, -1)$, $R = (-1, 6, -7)$.
 (iii) $P = (1, 2, -3)$, $Q = (2, 1, 1)$, $R = (-3, 4, 1)$.

16. **Coplanar Points**

 (a) Given four points in $\mathbb{R}^3$, how would you determine if the points are coplanar, that is, if there is a plane containing all four points?
 (b) For each of the following sets of points, determine if they are coplanar. If so, find the unique plane that contains them. If not, find the distance from the point S to the plane containing P, Q, and R.

 (i) $P = (2, 1, 0)$, $Q = (1, 3, -1)$, $R = (2, 4, -3)$, $S = (1, -1, 3)$.
 (ii) $P = (1, 2, -3)$, $Q = (2, 0, -1)$, $R = (-1, 6, -7)$, $S = (0, 2, 2)$.
 (iii) $P = (1, 1, 0)$, $Q = (2, 0, -1)$, $R = (2, 3, 2)$, $S = (4, 1, 3)$.

17. **The Distance Between Two Lines.** Let L_1 be the line through the point P with direction vector $\mathbf{u}_1$, and let L_2 be the line through the point Q with direction vector $\mathbf{u}_2$.

 (a) Find a vector $\mathbf{n}$ that is orthogonal to L_1 and L_2.
 (b) Use vector methods to find a formula for the distance between L_1 and L_2.
 (c) Suppose L_1 is the line through the point $(1, 2, 1)$ with direction vector $(1, -1, 2)$ and L_2 is the line through the point $(2, -1, 2)$ with direction vector $(1, 1, 0)$. Use your method from part (b) to determine the distance between L_1 and L_2.

18. **Dot Products and Planes**

    (a) Suppose that a plane passes through the point P and is perpendicular to the vector $\mathbf{n}$. The plane divides space into two half-spaces, one in the direction of $\mathbf{n}$ and one in the opposite direction. If Q is a point in space that does not lie in this plane, how can we use the dot product to determine which of these half-spaces contains Q?

    (b) If $\mathbf{u}$, $\mathbf{v}$, and $\mathbf{w}$ are three vectors in space, how can we determine if $\mathbf{u}$ is in the plane spanned by $\mathbf{v}$ and $\mathbf{w}$?

19. **The Right-Hand Rule**

    (a) If $\mathbf{u}$, $\mathbf{v}$, and $\mathbf{w}$ are three vectors in space, then taken in the order $\mathbf{u}$-$\mathbf{v}$-$\mathbf{w}$, the vectors either satisfy the right-hand rule or they do not. How can we determine if they satisfy the right-hand rule?

    (b) Use your answer to (a) to determine whether $(1,1,1)$, $(-1,-1,1)$, and $(-1,1,1)$ satisfy the right-hand rule in the given order.

20. **The Triple Product.** Let $\mathbf{u}$, $\mathbf{v}$, and $\mathbf{w}$ be vectors in space. The quantity $(\mathbf{u} \times \mathbf{v}) \cdot \mathbf{w}$ is called the ***triple product*** of $\mathbf{u}$, $\mathbf{v}$, and $\mathbf{w}$. Under what circumstances is the triple product positive? Under what circumstances is the triple product negative? Under what circumstances is the triple product zero?

Parametric Curves and Vector Fields

Having set the stage in the first chapter, we are now ready to begin the study of calculus. In Section 2.1, we introduce functions, called parametrizations, that are defined on subsets of $\mathbb{R}$ and take values in $\mathbb{R}^2$ or $\mathbb{R}^3$. The images of these functions are curves in the plane or in space, and their coordinates are themselves real-valued functions. If the coordinate functions are quantities associated with physical or biological systems, the function gives rise to a mathematical model of the system. In Section 2.2, we use the intuitive idea of velocity to motivate the definition of the derivative of a parametrization. We also show how we can use techniques from one-variable calculus to calculate derivatives. In Section 2.4, we use an intuitive understanding of the motion of a fluid flow to introduce vector fields. Vector fields are functions whose domain is a subset of the plane or space and whose image can be thought of as a set of vectors in the plane or vectors in space. We also introduce the concepts of a flow line of a vector field and of a critical point of a vector field, which connect the study of vector fields to parametrizations and their derivatives.

In Sections 2.3 and 2.5, we apply the concept of a parametrization to explore models from epidemiology, physics, physiology, and ecology. These are distinguished by the fact that they model phenomena that can be described by two or three quantities that vary in time. Our goal here is to use mathematics to understand the qualitative and quantitative behavior of these systems as they change in time. In each case, we provide sufficient background information to construct the model from the physical characteristics of the system. As is often the case when developing mathematical models, the descriptions of the systems lead to conditions on the rates of change of the functions, that is, on their derivatives. We will rely on a computer algebra system to generate symbolic, numerical, or graphical solutions to the equations we construct.

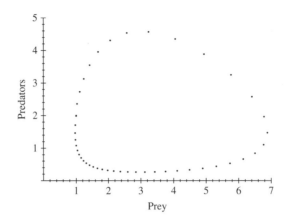

Figure 2.1   Changing populations of a predator–prey system. Consecutive points are separated by the same fixed time interval. (See Question 1.)

A Collaborative Exercise—Time-Dependent Biological Systems

Time-dependent plots of the parameters of interest are fundamental tools in the mathematical study of biological systems. Unfortunately, the data required to produce such plots are hard to come by. For example, in population biology, there is no way to know the size of two interacting wild populations at every instant in time. In human biology, specifically immunology, cell and pathogen counts in blood or tissue can only be obtained by invasive procedures, dramatically restricting one's ability to obtain data. Consequently, abstract or idealized models of biological systems can be important surrogates for working with real data. The following examples illustrate this.

A Predator–Prey System

The predator–prey model of Lotka and Volterra was one of the first theoretical models in population biology.[1] Here we focus on the output of their model. We suppose we have two interacting species that interact as predator and prey. Although much simplified—for example, the predator population only preys on this prey species, the prey is only preyed upon by this predator species, and the prey population's food supply is not taken into account—the model does shed some light on the population dynamics.

Figure 2.1 contains a simulation of a predator–prey system, where each dot represents the populations at an instant in time. The first coordinate represents the number of prey (in thousands), and the second represents the number of predators in (in tens). The dots should be read counterclockwise, and the time interval between consecutive dots

---

[1] Alfred J. Lotka, an American mathematician and physicist, and Vito Volterra, an Italian mathematician and physicist, developed this model independently in the 1920s.

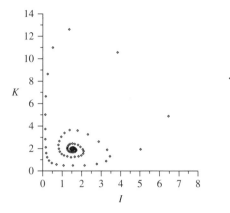

Figure 2.2   Changing cell counts for killer T-cell-infected cell system in units of 100 cells per ml. Consecutive points are separated by the same fixed time interval. The initial value is $(5, 2)$. (See Question 2.)

is assumed to be the same, but will not be specified. Here are several questions about this plot.

1. a. Where in the plot do the prey and predator populations reach their maxima and minima?
   b. Where in the plot are the prey and predator populations changing most rapidly? Least rapidly? Explain your answer.
   c. Describe the long-term simultaneous behavior of the two populations. (*Hint:* Use the language of increasing, decreasing, maximum, and minimum.)

A Killer T-Cell-Virus System

The human immune system is enormously complicated, but it is possible to construct reasonable models of particular features. An important example is that of the adaptive immune response of cytotoxic or killer T-cells to an antigen, for example, a virus. Briefly, killer T-cells are formed in bone marrow and migrate to the thymus (the T is for thymus), where they differentiate in order to react to a specific antigen. When a person is infected by a virus, through an intricate chain of cellular events, antigen-specific killer T-cells are signaled to replicate and to recognize and kill infected cells. If the immune response is successful, the virus will be eliminated. If the immune response is unsuccessful, the infected cell population will grow. A third possibility is that of chronic infection, where the virus is not eliminated, but the growth of the infected cell population is limited by the immune response.

Figure 2.2 contains a simulation of a killer T-cell-infected cell system.[2] The first coordinate of a point represents the number of infected cells per milliliter in hundreds, and

---

[2]See Section 2.2 of Dominik Wodarz, *Killer Cell Dynamics*, Springer, 2007.

the second represents the number of killer T-cells per milliliter in hundreds. As above, the dots should be read counterclockwise, and the time interval between consecutive dots is assumed to be the same and is not specified. The initial dot in the sequence is located at $(5, 2)$.

Here are several questions about this plot.

2. a. Where in the plot do the infected cell and killer T-cell counts reach their maxima and minima?
    b. Where in the plot are the infected cell and killer T-cell counts changing most rapidly? Least rapidly? Explain your answer.
    c. Describe the long-term simultaneous behavior of the two cell counts. (*Hint:* Use the language of increasing, decreasing, maximum, and minimum.)
    d. Would you characterize the immune response as successful, unsuccessful, or resulting in chronic infection? Explain your answer.

■ 2.1 Parametric Representations of Curves

In this section, we will introduce a way to use the language of functions to represent curves in the plane or in space. At first, we will think of curves that are traced by an object in motion. In this example, we will consider the motion of a satellite across the surface of the earth.

| Example 2.1 | **Satellite Tracks.** Figure 2.3 shows the path of the Hubble Space Telescope and MIR Space Station on a flat map of the earth. As time elapses, we see, for example, that the space telescope traces a curve that oscillates roughly between the latitude of the southern edge of Brazil and the latitude of the northern coast of Africa. At each time, we can specify the position of the space telescope by giving its coordinates in latitude and longitude. As time changes, the coordinates change, so that we can think of each coordinate as being a function of time. We will represent the position of the space telescope by an ordered pair of functions of time: the first, or x-coordinate, giving the longitude and the second, or y-coordinate, giving the latitude. |

Now let us express the ideas of the example in symbolic terms. If an object moves through the plane with xy-coordinates over an interval of time $a \leq t \leq b$, we will describe the position of the object at time t by an ***ordered pair*** $(x(t), y(t))$, whose coordinates are functions of time. This defines a function α whose domain is the interval $[a, b]$ and which takes values in $\mathbb{R}^2$. We write $\alpha : [a, b] \rightarrow \mathbb{R}^2$, where $\alpha(t) = (x(t), y(t))$.

If an object moves through space with xyz-coordinates over a time interval $[a, b]$, we will describe the position of the object at time t by an ***ordered triple***, $(x(t), y(t), z(t))$, whose coordinates are functions of time. We will then write $\alpha : [a, b] \rightarrow \mathbb{R}^3$, where $\alpha(t) = (x(t), y(t), z(t))$.

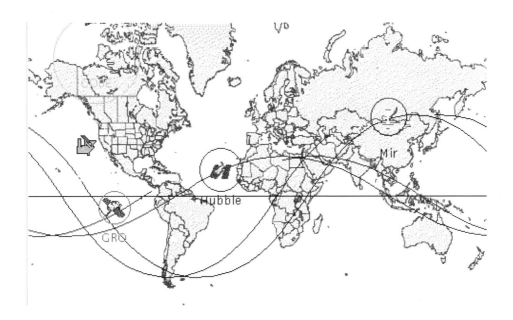

Figure 2.3 The track of the orbits of the Hubble Space Telescope and the MIR Space Station on a map of the earth. Image courtesy of National Aeronautics and Space Administration. (See Example 2.1.)

Example 2.2

Examples from Biology. As we saw in the collaborative exercise at the beginning of this chapter, it is also possible use time-dependent functions to model physical systems that are not related to motion. In population biology or ecology, the number of organisms or animals in a population is considered a function of time. Identifying a population with the coordinate of time-dependent function, we are able to study the interactions of two or more populations. Over time, the sizes of the population vary, tracing a curve in Euclidean space. Since populations are nonnegative, the curve will lie in the first quadrant (or octant). Similarly, in cell biology and immunology, cell or antigen concentrations per unit volume can be considered functions of time and can be identified with the coordinates of a time-dependent function that takes values in the first quadrant (or octant).

The following definition introduces the terminology we will use when referring to functions of time with values in $\mathbb{R}^2$ or $\mathbb{R}^3$.

Definition 2.1 A time-dependent function $\alpha : [a, b] \to \mathbb{R}^2$ given by $\alpha(t) = (x(t), y(t))$, or $\alpha : [a, b] \to \mathbb{R}^3$ given by $\alpha(t) = (x(t), y(t), z(t))$, that traces a portion of a curve in the plane or in space is called a *parametrization* or *parametric representation* of the curve.

The independent variable t is called the **parameter** of α. The **image** of α, which we denote by $\mathrm{Im}(\alpha)$, is the portion of the curve that is traced by α. In set notation,

$$\mathrm{Im}(\alpha) = \{(x, y) : (x, y) = (x(t), y(t)),\ t \in [a, b]\}$$

for curves in the plane. For curves in space,

$$\mathrm{Im}(\alpha) = \{(x, y, z) : (x, y, z) = (x(t), y(t), z(t)),\ t \in [a, b]\}.$$

The point $\alpha(a)$ is called the **initial point** of α and the point $\alpha(b)$ is called the **final point** or **endpoint** of α. ◆

We will assume throughout the text that our parametrizations are continuous in the sense that their coordinate functions are continuous. We will represent parametrizations of curves by the lowercase Greek letters alpha, beta, and gamma, which are written α, β, and γ, respectively.

It will be useful for us to have available a list of parametrizations of familiar and frequently used curves in the plane and in space. In the following examples, we will consider parametrizations of lines and line segments, circles and ellipses, and graphs of functions. Our first example, parametrizing lines and line segments, makes use of the language of vectors. Here we will work in the plane, so that a parametrization α has two coordinates. However, the construction also can be used in space, where the parametrization will have three coordinates.

Example 2.3   **Linear Parametrizations**

A. Lines. The parametric description of a line introduced in Section 1.4 with parameter t is a parametrization in the sense of Definition 2.1. Let $P = (1, -1)$ and $Q = (2, 3)$. Following Section 1.4, the line L through P and Q is the set of points corresponding to vectors of the form $\mathbf{p} + t\mathbf{v}$, where $\mathbf{p}$ is the vector corresponding to P, $\mathbf{q}$ is the vector corresponding to Q, $\mathbf{v} = \mathbf{q} - \mathbf{p}$, and $t \in \mathbb{R}$. Expanding $\mathbf{p} + t\mathbf{v}$ in coordinates, we obtain a parametrization for L,

$$\alpha(t) = \mathbf{p} + t\mathbf{v} = (1, -1) + t\,((2, 3) - (1, -1)) = (1 + t, -1 + 4t).$$

More generally, if $P = (x_0, y_0)$ and $Q = (x_1, y_1)$, the line through P and Q can be parametrized by

$$\alpha(t) = \mathbf{p} + t\mathbf{v} = (x_0, y_0) + t((x_1, y_1) - (x_0, y_0)) = (x_0 + t(x_1 - x_0),\ y_0 + t(y_1 - y_0)).$$

B. Line Segments. If we restrict the domain of the parametrization α of part A to the time interval $[0, 1]$, we obtain a parametrization of the line segment extending from P to Q. That is, $\mathrm{Im}(\alpha)$ is the line segment from P to Q. The initial point is $\alpha(0) = P$ and the endpoint is $\alpha(1) = Q$.

More generally, restricting α to the interval $[a, b]$ parametrizes the line segment in L extending from $\mathbf{p} + a\mathbf{v}$ to $\mathbf{p} + b\mathbf{v}$.

In our next example, we will make use of the trigonometric identity, $\cos^2 t + \sin^2 t = 1$ to parametrize circles and ellipses.

Example 2.4 | **Trigonometric Parametrizations**

A. Circles. The circle of radius r_0 centered at the origin satisfies the polynomial equation $x^2 + y^2 = r_0^2$. We can parametrize this circle by

$$\alpha(t) = (r_0 \cos t, r_0 \sin t).$$

Notice that if we substitute the coordinates of α into the equation for the circle, the equation is satisfied,

$$(r_0 \cos t)^2 + (r_0 \sin t)^2 = r_0^2(\cos^2 t + \sin^2 t) = r_0^2.$$

The parametrization α traces the circle in a counterclockwise manner. If we restrict the domain to the interval $[0, 2\pi]$, then α traces the circle once and the initial point and final point are both equal to the point $(r_0, 0)$. (See Figure 2.4(a).) By choosing different intervals for the domain of α, we can parametrize different portions of the circle or we can trace it more than once.

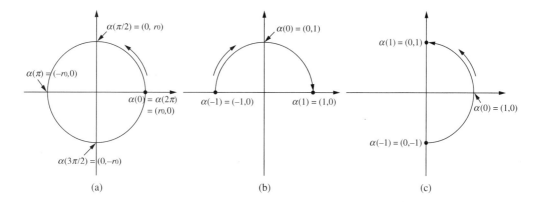

(a) (b) (c)

Figure 2.4 (a) The image of the function of Example 2.4A. Notice that the initial point and endpoint of α are $(r_0, 0)$ and the circle is traced counterclockwise in 2π units of time. (b) The image of the function of Example 2.5A. Notice that the initial point α is $(-1, 0)$ and the endpoint is $(1, 0)$. The semicircle is traced clockwise in 2 units of time. (c) The image of the function of Example 2.5B. Notice that the initial point α is $(0, -1)$ and the endpoint is $(0, 1)$. The semicircle is traced counterclockwise in 2 units of time.

B. Ellipses. The polynomial equation for an ellipse centered at the origin with axes of length $2a$ lying along the x-axis and length $2b$ lying along the y-axis is

$$\frac{x^2}{a^2} + \frac{y^2}{b^2} = 1.$$

By modifying the parametrization α of the circle of part A, we can produce a parametrization of this ellipse. Define α by

$$\alpha(t) = (a \cos t, b \sin t).$$

A calculation, which we leave to the reader, shows that the coordinates of α satisfy the equation of the ellipse. (See Exercise 4.) As above, if we restrict the domain to the interval $[0, 2\pi]$, the ellipse is traced once in a counterclockwise manner with initial point and final point equal to $(a, 0)$.

Unlike the previous examples, which used particular functions for the coordinates of the parametrization, the following construction applies to the graph of any function.

Example 2.5 **Graphs of Functions of One Variable**

A. Graph of a Function $y = f(x)$. The graph of a function $y = f(x)$ of one variable is a curve in the plane. It can be parametrized by

$$\alpha(t) = (t, f(t)),$$

where t is restricted to the domain of f. Notice that this uses the x-coordinate as the parameter for α. For example, the top half of the unit circle centered at the origin is the graph of the function $f(x) = \sqrt{1 - x^2}$, which is defined on the interval $[-1, 1]$. Thus we can parametrize the top half of the unit circle by

$$\alpha(t) = (t, \sqrt{1 - t^2}) \text{ for } t \in [-1, 1].$$

This parametrizes the semicircle in a clockwise direction with initial point $(-1, 0)$ and final point $(1, 0)$. (See Figure 2.4(b).)

B. Graph of a Function $x = g(y)$. In a similar manner, we can parametrize the graph of a function $x = g(y)$ by

$$\alpha(t) = (g(t), t),$$

where t is restricted to the domain of g. This uses the y-coordinate as the parameter for α. For example, the portion of the unit circle centered at the origin lying in the first and fourth quadrants is the graph of $x = g(y) = \sqrt{1 - y^2}$. It can be parametrized by

$$\alpha(t) = (\sqrt{1 - t^2}, t) \text{ for } t \in [-1, 1].$$

This parametrizes this semicircle in a counterclockwise direction with initial point $(0, -1)$ and final point $(0, 1)$. (See Figure 2.4(c).)

As the last two examples show, there are different ways to parametrize the same curve. In fact, if we restrict the parametrization of Example 2.4A (with $r_0 = 1$) to the interval $[0, \pi/2]$, the parametrization of Example 2.5A to the interval $[0, 1]$, and the parametrization of Example 2.5B also to the interval $[0, 1]$, we have three different parametrizations of the quarter circle centered at the origin in the first quadrant.

In the following example, we will employ a parametrization of an ellipse to model the motion of a simple pendulum. In this case, the coordinates in the plane will represent the position and velocity of the pendulum. Keep in mind that the coordinate plane we use for the model is not the physical plane in which the pendulum is swinging.

Example 2.6 **The Simple Pendulum.** A simple pendulum is an idealized version of a real pendulum and is one of the first mechanical systems studied in physics. It consists of a mass suspended from a rod that is free to swing in a vertical plane under the influence of gravity. Physically, it is assumed that the mass is concentrated at a point, that the rod has no mass, and that the rod is free to swing without friction. This is why we say it is an "idealized" version of a pendulum rather than a real pendulum. Swinging back and forth, the pendulum traces an arc of a circle, and its position can be specified by giving the displacement of the pendulum mass from the vertical as a distance x along its arc. See Figure 2.5(a). The distance x is related to the angle θ by $x = l\theta$, where θ is measured in radians and l is the length of the pendulum rod. If the pendulum is displaced a distance x_0 from the vertical along its arc and released with no initial velocity, its position is given by $x(t) = x_0 \sin(\sqrt{\frac{9.8}{l}} t + \pi/2)$. (The coefficient of t ensures that Newton's second law is satisfied.) The velocity v is the derivative of x, so we have $v(t) = x_0 \sqrt{\frac{9.8}{l}} \cos(\sqrt{\frac{9.8}{l}} t + \pi/2)$. A calculation shows that

$$\frac{x(t)^2}{x_0^2} + \frac{v(t)^2}{x_0^2(9.8/l)} = 1,$$

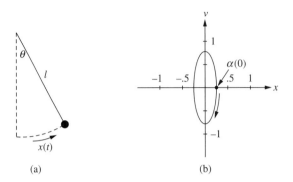

(a)   (b)

Figure 2.5   (a) A simple pendulum. The displacement of the pendulum is measured along the arc of its motion from the rest (vertical) position. Positive displacement is measured counter-clockwise from the vertical, and negative displacement is measured clockwise from the vertical. (b) The ellipse in the xv-plane for a pendulum with length $l = 1$ and initial displacement $x_0 = 0.25$. It is the image of the parametric representation α for the motion of the pendulum in Example 2.29.

for all values of t. This is the equation of an ellipse in the xv-plane with axes of length $2x_0$ and $2x_0\sqrt{9.8/l}$. Figure 2.5(b) is a plot of the ellipse for $l = 1$ and $x_0 = 0.25$. The parametrization α of this ellipse is

$$\alpha(t) = (0.25\sin(\sqrt{9.8}\,t + \pi/2),\ 0.25\sqrt{9.8}\cos(\sqrt{9.8}\,t + \pi/2)).$$

The first coordinate of α gives the displacement of the pendulum from the vertical, and the second coordinate is the velocity of the pendulum mass along its arc. As t increases from 0, we see that $\sin(\sqrt{9.8}\,t + \pi/2)$ oscillates from 1 to -1: thus the position of the pendulum oscillates from 0.25 to -0.25. At the same time, the velocity of the pendulum changes from 0 to $-0.25\sqrt{9.8}$ to 0 to $0.25\sqrt{9.8}$ to 0, reflecting the motion of the pendulum to the left and then the right with changing speeds. (This example is explored further in the exercises.)

The formula that we used in Example 2.6 for the displacement of the pendulum is a mathematical model for the motion of a pendulum. This model is more accurate for small oscillations of the pendulum, and it becomes less accurate as the size of the oscillations increase. Nevertheless, it provides a good qualitative tool for understanding the motion of a pendulum.

The Algebra of Parametrizations

We will devote the remainder of this section to the algebra of parametrizations. For the most part, we will be concerned with manipulating a single parametrization, either by changing the parameter or by moving the image of the parametrization. First, we will look at two ways to alter the parameter.

Example 2.7 **Transformations of the Parameter**

A. Shifting. Consider the parametrization $\alpha(t) = (\cos(t), \sin(t))$, $0 \leq t \leq \pi$, of the top half of the unit circle centered at the origin. The initial point is $(1, 0)$ and the final point is $(-1, 0)$. If we define a new parametrization by

$$\beta(t) = \alpha(t + \pi/2) = (\cos(t + \pi/2), \sin(t + \pi/2)) \text{ for } 0 \leq t \leq \pi,$$

the initial point of β is $(0, 1)$ and the final point of β is $(0, -1)$. The image of β is the portion of the unit circle lying in the second and third quadrants. That is, adding $\pi/2$ to the parameter of α has the effect of shifting the image of α by a distance that is traversed in $\pi/2$ units of time.

More generally, if α is a parametrization of a curve, **shifting** the parameter of a parametrization by c to produce a parametrization $\beta(t) = \alpha(t + c)$ has the effect of shifting the image of α by a distance that is traversed in c units of time. If $c > 0$, this shift is forward along the curve: if $c < 0$, this shift is backward along the curve. Alternatively, we can think of β as a parametrization of the same curve over a different time interval. That is, if α traces a portion of a curve over the time interval $[a, b]$, then β traces the same portion over the time interval $[a - c, b - c]$.

B. Scaling. Starting with the same parametrization α, consider a parametrization β defined by

$$\beta(t) = \alpha(t/2) = (\cos(t/2), \sin(t/2)) \text{ for } 0 \leq t \leq \pi.$$

The initial point of β is $(1, 0)$ and the endpoint of β is $(0, 1)$. The image of β is the quarter-circle lying in the first quadrant. Multiplying the parameter by $1/2$ has the effect of slowing down the parametrization by a factor of 2 so that over the same time period β traces out half the image of α. Similarly, if we had multiplied the parameter by 2, the parametrization would speed up by a factor of 2. Over the same time interval, we would trace the entire circle.

More generally, if α is a parametrization of a curve defined on an interval $[0, b]$, **scaling** the parameter by a factor of c to produce a parametrization $\beta(t) = \alpha(ct)$ has the effect of stretching the image of the parametrization along the curve by a time factor of c.

It is worth noting that the parametrization of the ellipse that arose in the model for a simple pendulum in Example 2.6 is obtained from a parametrization of the form of Example 2.4B by shifting and scaling the parameter. (See Exercise 13.)

Both of the transformations we introduced in Example 2.7 are examples of the composition of functions. In each case, we composed a parametrization α with a function

$g = g(t)$ of one variable to produce a function $\beta(t) = \alpha(g(t))$, which is the **composition** of α with g. In coordinates, if $\alpha(t) = (x(t), y(t))$, then in coordinates, $\beta(t) = (x(g(t)), y(g(t)))$. We will denote the composition of α with g by $\alpha \circ g$. In Example 2.7A, $g(t) = t + c$, and in Example 2.7B, $g(t) = ct$.

It is also possible to move or translate the image of a parametrization by adding a constant vector. Here we will work with a parametrization of the unit circle centered at the origin.

Example 2.8

Transforming the Image: Translation. Let $\alpha(t) = (\cos t, \sin t)$, $0 \le t \le 2\pi$. Let $\mathbf{p} = (2, 1)$ and define a new parametrization β by

$$\beta(t) = \alpha(t) + \mathbf{p} = (\cos t, \sin t) + (2, 1) = (\cos t + 2, \sin t + 1)$$

for $0 \le t \le 2\pi$. A calculation, which we leave to the reader, shows that the coordinates of β satisfy the equation

$$(x - 2)^2 + (y - 1)^2 = 1,$$

which is the equation for a circle of radius 1 centered at the point $(2, 1)$. That is, adding $(2, 1)$ to α has the effect of moving or translating the image of α by $(2, 1)$.

In general, if β is related to α by $\beta(t) = \alpha(t) + \mathbf{p}$ for a vector $\mathbf{p}$, then the image of β is the **translation** of the image α by the vector $\mathbf{p}$.

Let us put the ideas of shifting, scaling, and translation together to construct a parametrization of a person walking around a running track at a constant speed. While the verbal description is simple, translating it into mathematics takes some effort.

Example 2.9

A Compound Motion. We would like to parametrize the motion of a person who walks with constant speed counterclockwise around a 400 m track in four minutes. The straightaways of the track are 100 m in length, and the ends of the track are semicircles 100 m in length. Thus the ends are semicircles of radius $100/\pi$ m. Choose the origin to be in the middle of the track and orient the track so that the sides of the track are parallel to the x-axis as in Figure 2.6. The track consists of four curves of equal length: the segments C_1 and C_3 and the semicircles C_2 and C_4. If the path is traveled at constant speed, it takes one minute to travel each portion of the track. Suppose that the motion begins at the right-hand endpoint of C_1, $(50, 100/\pi)$ in our chosen coordinate system. We will give the details of the construction of the parametrization of C_1 and C_2 and leave the constructions for C_3 and C_4 for the exercises. (See Exercise 17.)

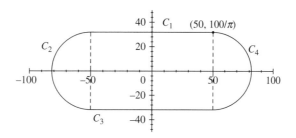

Figure 2.6 The image of the parametrization of Example 2.9.

$\mathcal{C}_1$: Since $\mathcal{C}_1$ is the line segment with endpoints $(50, 100/\pi)$ and $(-50, 100/\pi)$, we can apply the formula of Example 2.3B to produce a parametrization. Denoting this parametrization by α_1, we have that

$$\alpha_1(t) = (50, 100/\pi) + t(-100, 0) = (50 - 100t, 100/\pi)$$

for $t \in [0, 1]$. Notice that $\alpha_1(0) = (50, 100/\pi)$ and $\alpha_1(1) = (-50, 100/\pi)$; thus the first segment is traveled in the first minute.

$\mathcal{C}_2$: The curve $\mathcal{C}_2$ is the left-hand semicircle of a circle of radius $100/\pi$ centered at $(-50, 0)$. We will build the parametrization of $\mathcal{C}_2$ in several steps. Eventually, the domain of the parametrization must be $[1, 2]$ to reflect that this part of the track is walked during the second minute. To parametrize $\mathcal{C}_2$, we will first parametrize a semicircle centered at the origin, scale and shift the parameter to get the correct time interval, and then translate the parametrization to the left end of the track.

Using the formula from Example 2.4A, we start with the parametrization of the semicircle of radius $100/\pi$ in the third and fourth quadrants centered at the origin

$$((100/\pi) \cos t, (100/\pi) \sin t) \text{ for } \pi/2 \leq t \leq 3\pi/2.$$

Next, scale the parameter by π so that the time interval has length 1 rather than π,

$$((100/\pi) \cos(\pi t), (100/\pi) \sin(\pi t)) \text{ for } 1/2 \leq t \leq 3/2.$$

Then shift the time parameter by $-1/2$ so that the interval is $[1, 2]$,

$$((100/\pi) \cos(\pi(t - 1/2)), (100/\pi) \sin(\pi(t - 1/2))) \text{ for } 1 \leq t \leq 2.$$

Finally, to obtain the expression for α_2, translate the image so that it parametrizes the left half of a circle centered at $(-50, 0)$,

$$\alpha_2(t) = ((100/\pi) \cos(\pi(t - 1/2)) - 50, (100/\pi) \sin(\pi(t - 1/2))) \text{ for } 1 \leq t \leq 2.$$

This is the desired parametrization of $\mathcal{C}_2$.

$\mathcal{C}_3$: The line segment $\mathcal{C}_3$ can be parametrized the way we parametrized $\mathcal{C}_1$. However, we will have to shift the parameter so that the time interval is $[2,3]$. Carrying out these calculations, we get

$$\alpha_3(t) = (-50, -100/\pi) + (t-2)(100, 0) = (-50 + (t-2)100, -100/\pi) \text{ for } 2 \le t \le 3.$$

$\mathcal{C}_4$: As we did for $\mathcal{C}_2$, we can parametrize a circle of radius $100/\pi$ centered at the origin, scale the parameter, shift the parameter, and translate the image. These steps yield the parametrization

$$\alpha_4(t) = ((100/\pi)\cos(\pi(t-7/2)) + 50, (100/\pi)\sin(\pi(t-7/2))), \ t \in [3,4].$$

Putting these four pieces together, we can parametrize the walk around the track by $\alpha : [0,4] \to \mathbb{R}^2$, where

$$\alpha(t) = \begin{cases} (50 - 100t, 100/\pi) & t \in [0,1], \\ ((100/\pi)\cos(\pi(t-1/2)) - 50, (100/\pi)\sin(\pi(t-1/2))) & t \in [1,2], \\ (-50 + (t-2)100, -100/\pi) & t \in [2,3], \\ ((100/\pi)\cos(\pi(t-7/2)) + 50, (100/\pi)\sin(\pi(t-7/2))) & t \in [3,4]. \end{cases}$$

Summary   In this section, we introduced the language of **parametrizations** to represent the motion of an object along a curve in the plane or in space and to model the evolution of a physical system over time. In examples, we developed explicit parametrizations of **lines**, **line segments**, **circles**, and **ellipses**. We studied the effects of **shifting** and **scaling** the **parameter** of the parametrization and of **translating** the **image** of the parametrization. We also introduced a model for the motion of a **simple pendulum**.

Section 2.1 Exercises

1. **Line Segments.** For each of the following pairs of points and intervals, give a parametrization of the line segment between P and Q whose domain is the given interval $[a,b]$. (*Hint:* Begin with the standard parametrization of a line segment given in Example 2.3, and then shift or scale the parameter as necessary.)

   (a) $P = (1,0)$, $Q = (0,1)$, and $[a,b] = [0,1]$.
   (b) $P = (1,0,1)$, $Q = (0,1,-1)$, and $[a,b] = [0,2]$.
   (c) $P = (-2,-2)$, $Q = (1,6)$, and $[a,b] = [3,4]$.
   (d) $P = (-2,2,1)$, $Q = (3,-1,2)$, and $[a,b] = [-1,2]$.

2. **Lines.** Suppose that L is a line in the xy-plane that passes through the point (x_0, y_0) and has slope m.

   (a) Find a direction vector for the L.

(b) Find a parametrization α of L that satisfies $\alpha(0) = (x_0, y_0)$.
(c) Find a parametrization of the line with slope $m = 5/2$ that passes through $(x_0, y_0) = (-1, 0.5)$ at time $t = 0$.

3. **Arcs of Circles.** Sketch the images of the following parametrizations α. Label the end-points of α and indicate the direction of α on the sketch.

(a) $\alpha(t) = (\cos t, \sin t)$ for $\pi/4 \leq t \leq 5\pi/4$.
(b) $\alpha(t) = (\cos 2\pi t, \sin 2\pi t)$ for $0 \leq t \leq 1/2$.
(c) $\alpha(t) = (\cos 3t, \sin 3t)$ for $-\pi/6 \leq t \leq \pi/6$.
(d) $\alpha(t) = (\cos \pi t, \sin \pi t)$ for $1/2 \leq t \leq 1$.

4. **Ellipses.** Consider the parametrization $\alpha(t) = (a\cos t, b\sin t)$, $0 \leq t \leq 2\pi$, where $a > 0$ and $b > 0$. (See Example 2.4B.)

(a) Show that the coordinates of α satisfy the equation

$$\frac{x^2}{a^2} + \frac{y^2}{b^2} = 1.$$

(b) Use α to construct a parametrization β of the same ellipse with the property that $\beta(0) = (0, b)$.
(c) Use α to construct a parametrization β of the same ellipse so that the ellipse is traced once in 2 units of time.

5. **Arcs of Ellipses.** Sketch the images of the following parametrizations α. Label the endpoints of α and indicate the direction of the α on the sketch.

(a) $\alpha(t) = (2\cos t, \sin t)$ for $0 \leq t \leq 3\pi/4$.
(b) $\alpha(t) = (\cos \pi t, \frac{1}{2}\sin \pi t)$ for $1 \leq t \leq 2$.
(c) $\alpha(t) = (\frac{1}{3}\cos(2\pi t), \sin(2\pi t))$ for $1/4 \leq t \leq 3/4$.
(d) $\alpha(t) = (\frac{1}{2}\cos(3t), \frac{2}{3}\sin(3t))$ for $-\pi/2 \leq t \leq 0$.

6. **More Ellipses.** In this exercise, we want to introduce some terms concerning ellipses and consider another useful representation of an ellipse. Let us begin with an ellipse defined by

$$\frac{x^2}{a^2} + \frac{y^2}{b^2} = 1,$$

where a and b are positive constants. If $a > b$, we call the interval from $-a$ to a on the x-axis the **major axis** and the interval from $-b$ to b on the y-axis the **minor axis** of the ellipse. The **semimajor axis** is the interval from the origin to $(a, 0)$ and the **semiminor axis** is the interval from the origin to $(0, b)$. The **eccentricity** e is defined to be $e = \sqrt{1 - (b^2/a^2)}$. The **foci** of the ellipse are located a distance $\sqrt{a^2 - b^2}$ from its center on the major axis. (See Figure 2.7.)

(a) Suppose that the major axis of an ellipse has length 6 and lies along the x-axis and the minor axis has length 4 and lies along the y-axis. Using the above form for the ellipse, locate the foci.

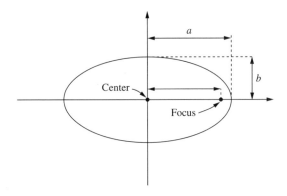

Figure 2.7   An ellipse with semimajor axis of length a and semiminor axis of length b. The center of the ellipse is at the origin, and the foci of the ellipse are located at a distance $\sqrt{a^2 - b^2}$ from the focus along the major axis. The ellipse satisfies the equation $x^2/a^2 + y^2/b^2 = 1$. (See Exercise 6.)

    (b) By shifting the x-coordinate, rewrite the above equation to find the equation of an ellipse with a focus located at the origin. (This will make the minor axis parallel to the y-axis instead of on the y-axis.)

    (c) Construct a parametrization of the shifted ellipse from part (b).

7. **Circles and Ellipses.** It is possible to parametrize circles and ellipses using a sine function for the first coordinate and a cosine function for the second coordinate. Sketch the image of each of the following parametrizations α. Label the endpoints of α and indicate the direction of the α on the sketch.

    (a) $\alpha(t) = (\sin t, \cos t)$ for $0 \le t \le \pi/2$.
    (b) $\alpha(t) = (\sin 2t, \cos 2t)$ for $-\pi/4 \le t \le \pi/2$.
    (c) $\alpha(t) = (\sin t, 2\cos t)$ for $0 \le t \le \pi/2$.
    (d) $\alpha(t) = (2\sin(t/2), 3\cos(t/2))$ for $-\pi \le t \le \pi$.

8. **Parametrizing Arcs of Circles.** Find a parametrization α for each of the following arcs of a circle that satisfies the given conditions.

    (a) α parametrizes the unit circle centered at the origin counterclockwise with $\alpha(0) = (0,1)$ and $\alpha(\pi) = (0,-1)$.
    (b) α parametrizes the unit circle centered at the origin clockwise with $\alpha(0) = (-1,0)$ and $\alpha(1) = (1,0)$.
    (c) α parametrizes the circle of radius 2 centered at the origin counterclockwise with $\alpha(0) = (-2,0)$ and $\alpha(1) = (-2,0)$.
    (d) α parametrizes the circle of radius 3 centered at $(2,0)$ counterclockwise with $\alpha(0) = (2,3)$ and $\alpha(2) = (2,-3)$.

$C.$

9. **Helices.** The curve in space parametrized by $\alpha(t) = (\cos(t), \sin(t), t)$ is called a **helix**. Notice that the first two coordinates of α are the coordinates of a parametrization of a circle of radius 1 centered at the origin.

   (a) Describe in words the image of α for the time interval $[0, 2\pi]$.

   (b) For each of the following parametrizations β, describe the image of β.

      (i) $\beta(t) = (2\cos t, 2\sin t, t)$ for $0 \leq t \leq 2\pi$.
      (ii) $\beta(t) = \alpha(2t)$ for $0 \leq t \leq 2\pi$.
     (iii) $\beta(t) = (\cos t, t, \sin t)$ for $0 \leq t \leq 2\pi$.

10. **Reversing the Direction of a Parametrization**

   (a) The parametrization $\alpha(t) = (t, t^3)$, $1 \leq t \leq 3$, parametrizes the portion of the graph of $y = x^3$ from $(1, 1)$ to $(3, 27)$. By using shifting and/or scaling of the parameter of α, produce a parametrization β of the same curve with the same domain, but with $\beta(1) = (3, 27)$ and $\beta(3) = (1, 1)$.

   (b) Suppose that $\alpha(t)$, $a \leq t \leq b$, parametrizes motion along a curve from $P = \alpha(a)$ to $Q = \alpha(b)$. Use α to construct a parametrization β of motion along the same curve in the opposite direction, that is, with $\beta(a) = Q$ and $\beta(b) = P$. Explain your answer.

11. **Graphs of Functions.** For each of the following parametrizations of curves in the plane, find an expression for the curve as a portion of the graph of a function $y = f(x)$ or a portion of the graph of a function $x = g(y)$. Be sure to say which form you found.

   (a) $\alpha(t) = (e^t, e^t)$, $0 \leq t < \infty$.
   (b) $\alpha(t) = (e^t, e^{2t})$, $-\infty < t \leq 0$.
   (c) $\alpha(t) = (e^{-t}, e^t)$, $-\infty < t < \infty$.
   (d) $\alpha(t) = (\sin t, \sin^3 t)$, $0 \leq t \leq \pi/2$.

12. **Interacting Populations.** In the collaborative exercise at the beginning of the chapter, we introduced a plot modeling the populations of two species that interact as predator and prey. Figure 2.8 shows a similar curve. Assume that it takes 3 years for the populations to trace this curve once and let $\alpha(t) = (x(t), y(t))$ denote a parametrization of this curve with initial point $\alpha(0) = (540, 15)$.

   (a) Locate the points where the predator population reaches its maximum and minimum values and the points where the prey population reaches its maximum and minimum values. Approximately, what are the coordinates of these points?

   (b) Describe what happens to the individual populations $x = x(t)$ and $y = y(t)$ as α traces the curve in Figure 2.8.

   (c) Use your descriptions from part (a) to sketch the graphs of x and y as functions of t on the same set of coordinate axes. (Plot t on the horizontal axis and x and y on the vertical axis.)

   (d) Why do you think the maximum values for x and y occur at different times?

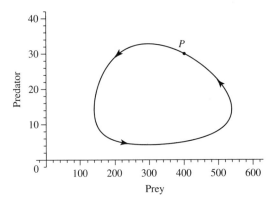

Figure 2.8 A curve whose coordinates are the number of prey and the number of predators in two interacting populations. For example, the point P represents the time when there are 400 prey and 30 predators. The arrows indicate the direction in which the curve is traced over time. (See Exercise 12.)

13. **The Simple Pendulum.** In Example 2.6, we parametrized the motion of a simple pendulum of length l with initial position and velocity $(x_0, 0)$ by

$$\alpha(t) = (x_0 \sin(\sqrt{\tfrac{9.8}{l}}t + \pi/2), \; x_0\sqrt{\tfrac{9.8}{l}} \cos(\sqrt{\tfrac{9.8}{l}}t + \pi/2)).$$

Use shifting and scaling to obtain α from the parametrization

$$\beta(t) = (x_0 \sin(t), \; x_0\sqrt{\tfrac{9.8}{l}} \cos(t)).$$

Explain.

14. **The Period of a Simple Pendulum.** When set in motion, a simple pendulum oscillates back and forth. The time it takes to complete one full oscillation and return to initial position is called the **period** of its motion.

    (a) What is the period of a pendulum of length l? (Use the formula of Exercise 13.)
    (b) What is the effect of changing the length of the pendulum on the motion of the pendulum? Explain.
    (c) How does the image of α depend on l?

15. **The Velocity of a Simple Pendulum.** In Example 2.6, we parametrized the motion of a simple pendulum of length l with initial position and velocity $(x_0, 0)$.

    (a) Describe in words the velocity of the pendulum through one period of its motion. In particular, where does it reach its minimum and maximum speeds?

(b) How does the motion of the pendulum depend on x_0? How is this reflected in the parametrization α? In particular, how does the image of α depend on x_0?

16. **Motion under the Influence of Gravity.** A simplified model of the motion of an object near the surface of the earth takes into account only the force of gravity and neglects other forces, for example, drag due to air resistance. Gravity acts in a downward direction with magnitude $g = 9.8$ m/s^2. Thus the vertical component of the force is $-g = -9.8$ m/s^2. Based on Newton's second law, the formula for the vertical component of the motion of an object is given by

$$y(t) = -\frac{g}{2}t^2 + v_0 t + y_0,$$

where v_0 is the initial vertical velocity of the object and y_0 is the initial vertical position of the object. If an object is launched so that its motion also has a horizontal component, $x(t)$, the motion can be modeled by the function $\alpha(t) = (x(t), y(t))$.

(a) Suppose an object is launched from the ground so that its horizontal velocity is 10 m/s, so that $x(t) = 10t$. How far will the object travel over the ground if its initial vertical velocity is also 10 m/s?

(b) Suppose an object is launched from the ground so that its horizontal velocity is 10 m/s and it travels a total of 100 m before hitting ground. What was its initial vertical velocity?

17. **Compound Motion.** Use shifting, scaling, and translation to construct the parametrizations of $\mathcal{C}_3$ and $\mathcal{C}_4$ of Example 2.9. Explain your construction.

18. **Compound Motion—A Slide.** A slide is in the shape of a helix of radius 5 ft and height 50 ft that makes five complete turns as it descends to the ground. To get to the top of the slide, you must climb a ladder that goes from the ground up through the center of the slide to a platform at the top. Suppose that it takes you 2 minutes to climb to the platform at the top of the slide, 15 sec to walk to the edge of the platform, and 30 sec to slide down the slide. Assume that each portion of the trip is made at a (different) constant rate. Construct a parametrization of this motion. (See Exercise 9 for a parametrization of a helix.)

■ 2.2 The Derivative of a Parametrization

If an object is in motion, then at each instant in time, the object will have a speed and a direction of motion. For example, we might imagine a car being driven along a winding mountain road. At each time, we can read the speed of the car off the speedometer and tell the direction of motion by looking straight ahead through the windshield. We would, of course, read the speed as a nonnegative number. On the other hand, it would be convenient to represent the direction by a vector having three coordinates, with the x-coordinate representing the east-west component of the direction, the y-coordinate representing the north-south component of the direction, and the z-coordinate representing the vertical component of the direction, that is, how much the car is ascending or descending. If we scale this vector to have magnitude equal to the speed of the car, then we will have constructed the velocity vector of the car at the instant in time. This leads us to make the following informal definition of velocity.

Definition 2.2 The **velocity** of the motion of an object at a particular time is the vector whose direction is the direction of motion of the object at that time and whose magnitude is the speed of the object at that time. ◆

If an object in motion traces a curve in the plane or in space, we might imagine sketching vectors that represent its velocity at points on the curve. At a particular point on the curve, the velocity can be represented by a vector that begins at the point, points tangent to the curve in the direction of motion, and has magnitude equal to the speed of the object. For example, Figure 2.9(a) shows velocity vectors for the motion of an object moving in a counterclockwise direction with speed 1 around a circle of radius 1 centered at the origin.

Our immediate goal is to develop a rigorous definition of velocity, which, among other features, will allow us to compute the velocity vector of an object from a parametrization of its motion. We begin with the notion of the average velocity of an object.

Suppose that the motion of an object is parametrized by α. In the plane, $\alpha(t) = (x(t), y(t))$, and in space, $\alpha(t) = (x(t), y(t), z(t))$. Let t_0 be the time at which we want

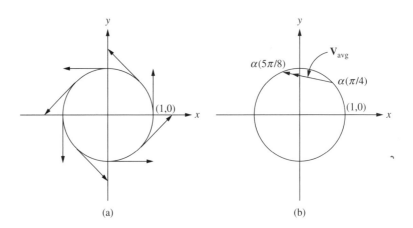

(a)   (b)

Figure 2.9   (a) Velocity vectors that represent motion around the unit circle in a counterclockwise direction at speed equal to 1 at all times. (b) The average velocity of $\alpha(t) = (\cos t, \sin t)$ at $t = \pi/4$.

to compute the velocity. The average velocity of the object over a time interval $[t_0, t]$ is the displacement of the object multiplied by $1/\Delta t$. Let us state this more formally.

Definition 2.3 If the motion of an object is parametrized by the function α, then *average velocity*, $\mathbf{v}_{\text{avg}}$, of the object over the time interval $[t_0, t]$ of length $\Delta t = t - t_0$ is the vector

$$\mathbf{v}_{\text{avg}} = \frac{1}{\Delta t}(\alpha(t) - \alpha(t_0)) = \frac{\alpha(t) - \alpha(t_0)}{\Delta t}. \quad \blacklozenge$$

Note that $(\alpha(t) - \alpha(t_0))/\Delta t$ is called a *difference quotient*. Let us consider an example of this calculation.

Example 2.10 **Average Velocity.** Suppose that the motion of an object is parametrized by $\alpha(t) = (\cos t, \sin t)$, so that the object traces the unit circle centered at the origin. Let us compute the average velocity of the object over the time interval $[\pi/4, 5\pi/8]$. This is

$$\mathbf{v}_{\text{avg}} = \frac{\alpha(5\pi/8) - \alpha(\pi/4)}{5\pi/8 - \pi/4}$$

$$= \frac{(\cos(5\pi/8), \sin(5\pi/8)) - (\cos(\pi/4), \sin(\pi/4))}{3\pi/8}$$

$$\approx (-0.925, 0.184)$$

The displacement vector $\alpha(5\pi/8) - \alpha(\pi/4)$ and the average velocity vector are shown on a plot of the image of α in Figure 2.9(b). Notice that since the average velocity is a positive multiple of the displacement vector, it points in the same direction as the displacement vector.

The average velocity at a point can be thought of as an *approximation* to the velocity at the point. By keeping t_0 fixed and choosing smaller time intervals, we will obtain better approximations to the velocity. For example, based upon Figure 2.9(b), we would expect that by keeping $\pi/4$ fixed and choosing times closer to $\pi/4$, we would obtain a vector that more accurately reflects the true velocity vector at $t = \pi/4$. To obtain an exact value for the velocity vector, we will take the limit of the average velocity as the length of the time interval $[t_0, t]$ approaches 0, that is, as Δt approaches 0. The length of this vector will be the speed of the object at time t_0. In order to express this limit in terms of t_0 and Δt alone, we will use the fact that $t = t_0 + \Delta t$. Writing this out carefully, we have the following definition.

Definition 2.4   If the motion of an object is parametrized by the function α, then the **velocity** of the object at time t_0 is the vector $\mathbf{v}(t_0)$ defined by

$$\mathbf{v}(t_0) = \lim_{\Delta t \to 0} \frac{\alpha(t_0 + \Delta t) - \alpha(t_0)}{\Delta t}.$$

The **speed** $s(t_0)$ of the object at time t_0 is $s(t_0) = \|\mathbf{v}(t_0)\|$.   ◆

Notice that since speed is the length of a vector, it will always be a nonnegative real number. Before we can carry out an explicit calculation of the velocity of an object from a parametrization of its motion, we will need to make the connection between velocity and differentiation. This is the subject of the next subsection.

Differentiable Functions and the Derivative

While our intuition tells us that at each instant in time a moving object must have a velocity, this does not follow automatically from the definition. The difficulty is that before the fact, we do not know if the limit of the average velocities will exist. For example, because the denominator of the expression for the average velocity is approaching zero, the entire expression might grow in magnitude without bound. That is, the limit might not converge to a vector with finite entries. Fortunately, there is a large collection of functions for which this limit exists. We say that these functions are differentiable. Let us put this in the form of a definition.

Definition 2.5   Let $\alpha : [a, b] \to \mathbb{R}^2$ (or $\mathbb{R}^3$) and let $t_0 \in (a, b)$. The function α is said to be **differentiable** at t_0 if the following limit exists:

$$\lim_{\Delta t \to 0} \frac{\alpha(t_0 + \Delta t) - \alpha(t_0)}{\Delta t}.$$

If this limit exists, we call it the **derivative** of α at t_0, or the **tangent vector** at t_0, and we denote it by $\alpha'(t_0)$.

If this limit exists for every t, $a < t < b$, we say that α is **differentiable** on (a, b), and we define the **derivative** of α on (a, b) to be the function whose value at each point $t \in (a, b)$ is this limit.   ◆

If α parametrizes the motion of an object, the **velocity vector** of the motion at t_0 is the tangent vector at t_0.

It turns out that the derivative of α can be expressed in terms of the derivative of the coordinate functions of α. This reduces the problem of computing derivatives of parametrizations to a problem in one-variable calculus. In order to see why this is possible, let us apply the limit definition of the derivative to the coordinate form of α. Here we will work in $\mathbb{R}^2$, but the argument is also valid in $\mathbb{R}^3$.

Let us begin with a parametrization α that is differentiable at t_0, so that

$$\alpha'(t_0) = \lim_{\Delta t \to 0} \frac{\alpha(t_0 + \Delta t) - \alpha(t_0)}{\Delta t}$$

exists. This limit should be interpreted as the vector whose entries are the limits of the difference quotients of the coordinates:

$$\left(\lim_{\Delta t \to 0} \frac{x(t_0 + \Delta t) - x(t_0)}{\Delta t}, \lim_{\Delta t \to 0} \frac{y(t_0 + \Delta t) - y(t_0)}{\Delta t} \right).$$

The first entry of this vector is the derivative of the coordinate function x with respect to t at t_0, $x'(t_0)$, and the second coordinate is the derivative of the coordinate function y at t_0, $y'(t_0)$. So, we have shown that $\alpha'(t_0) = (x'(t_0), y'(t_0))$.

It is possible to reverse this argument. If $x = x(t)$ and $y = y(t)$ are differentiable functions, then $\alpha(t) = (x(t), y(t))$ is differentiable. These calculations are summarized in the following proposition.

Proposition 2.1 Let $\alpha : [a, b] \to \mathbb{R}^2$ be a function with coordinates $\alpha(t) = (x(t), y(t))$. Then

1. α is a differentiable function if and only if the coordinate functions of α, $x = x(t)$ and $y = y(t)$, are differentiable functions $[a, b] \to \mathbb{R}$.
2. If α is differentiable, then the derivative of α is given by

$$\alpha'(t) = (x'(t), y'(t)). \quad \blacklozenge$$

Returning to the discussion of derivatives, the proposition also holds for functions $\alpha : [a, b] \to \mathbb{R}^3$. If $\alpha(t) = (x(t), y(t), z(t))$, then

$$\alpha'(t) = (x'(t), y'(t), z'(t)).$$

If α parametrizes the motion of an object and is a differentiable function, we can now compute the velocity and the speed of the object in terms of the coordinates of α.

Definition 2.6 If the motion of an object is parametrized by α, the **velocity** **v** of the object is α', $\mathbf{v}(t) = \alpha'(t)$, and the **speed** s of the object is the magnitude of the velocity, $s(t) = \|\mathbf{v}(t)\|$. In coordinates in the plane,

$$\mathbf{v}(t) = (x'(t), y'(t)) \text{ and } s(t) = \sqrt{x'(t)^2 + y'(t)^2}.$$

In space,

$$\mathbf{v}(t) = (x'(t), y'(t), z'(t)) \text{ and } s(t) = \sqrt{x'(t)^2 + y'(t)^2 + z'(t)^2}. \quad \blacklozenge$$

Now let us carry out several explicit calculations of velocity and speed for motions that we have considered previously. In each case, the coordinates of the parametrization will be differentiable, so that the parametrization will also be differentiable. We can then calculate the derivative of the parametrization α by differentiating its coordinates.

Example 2.11   **Velocity and Speed**

A. Linear Motion. Suppose that an object moves along the line from $P = (x_0, y_0)$ to $Q = (x_1, y_1)$ according to the parametrization $\alpha(t) = \mathbf{p} + t\mathbf{v}_0$, where the vector $\mathbf{p}$ corresponds to the point P and $\mathbf{v}_0 = \overrightarrow{PQ}$ is the displacement vector from P to Q. (See Example 2.3.) In coordinates, $\alpha(t) = (x_0 + t(x_1 - x_0),\ y_0 + t(y_1 - y_0))$. Then

$$\alpha'(t) = ((x_0 + t(x_1 - x_0))',\ (y_0 + t(y_1 - y_0))')$$
$$= (x_1 - x_0,\ y_1 - y_0)$$
$$= \mathbf{v}_0.$$

This calculation shows that derivative of α is the direction vector $\mathbf{v}_0$ of the line. So, we also have that the velocity $\mathbf{v}(t) = \mathbf{v}_0$. This says that the velocity vector of this motion is constant. It follows that the speed is also constant, $s(t) = \|\mathbf{v}_0\|$. We will say that motion along a straight line that has **constant speed** is **linear motion**.

B. Uniform Circular Motion. Suppose that an object moves around a circle of radius 1 centered at the origin and the motion is given by

$$\alpha(t) = (\cos(s_0 t), \sin(s_0 t)),$$

where s_0 is a positive constant. Then

$$\alpha'(t) = (\cos(s_0 t)', \sin(s_0 t)')$$
$$= (-s_0 \sin(s_0 t), s_0 \cos(s_0 t)).$$

The velocity of the object is

$$\mathbf{v}(t) = (-s_0 \sin(s_0 t), s_0 \cos(s_0 t)),$$

and the speed of the object is

$$s(t) = \sqrt{(-s_0 \sin(s_0 t))^2 + (s_0 \cos(s_0 t))^2} = s_0.$$

This shows that the speed of the object is constant. We will say that motion around a circle that has **constant speed** is ***uniform circular*** motion. Velocity vectors for uniform circular motion around a circle of radius 1 with speed 1 are shown in Figure 2.9(a).

Now let's consider an example where the speed is not constant.

Example 2.12   **Motion Around an Ellipse.** Suppose that an object in motion in the plane moves according to the $\alpha(t) = (3\cos t, \sin t)$, so that it traces an ellipse with major axis of length 6 lying along the x-axis and minor axis of length 2 lying along the y-axis. Then $\mathbf{v}(t) = \alpha'(t) = (-3\sin t, \cos t)$ and $s(t) = \sqrt{9(\sin t)^2 + (\cos t)^2}$. (See Figure 2.10.) Since the speed is not constant, we would like to find out where the object is moving fastest and where it is moving slowest. That is, we want to find the maximum and minimum values of speed. This is a problem from one-variable calculus. First, we find the critical points of s by computing $s'(t)$ and finding solutions to $s'(t) = 0$. Differentiating s, we have that

$$s'(t) = \frac{8\sin(t)\cos(t)}{\sqrt{9(\sin t)^2 + (\cos t)^2}},$$

which is 0 if the numerator is zero. Thus $s'(t) = 0$ if $\sin(t)\cos(t) = 0$, which is the case for $t = 0, \pi/2, \pi, 3\pi/2, \ldots$. Computing the corresponding speeds, we have

$$s(0) = 1,\ s(\pi/2) = 3,\ s(\pi) = 1,\ s(3\pi/2) = 3, \ldots.$$

The object is moving fastest when $t = \pi/2$ and $t = 3\pi/2$, which occurs at the points $(0, 1)$ and $(0, -1)$, when it is closest to the origin. It is moving slowest when $t = 0$ and $t = \pi$, which occurs at the points $(3, 0)$ and $(-3, 0)$, when it is furthest from the origin.

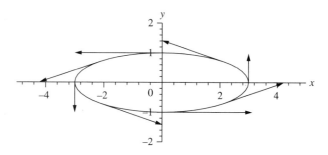

Figure 2.10   Velocity vectors for the parametrization $\alpha(t) = (3\cos t, \sin t)$ of the ellipse $x^2/9 + y^2 = 1$. (See Example 2.12.)

Before we present the next example, we must make the following definition.

Definition 2.7   If the motion of an object is parametrized by α, the **_acceleration_** **a** of the object is the derivative of the velocity of the object. That is, the acceleration of an object is the second derivative of the parametrization of its motion,

$$\mathbf{a}(t) = \mathbf{v}'(t) = \alpha''(t). \; \blacklozenge$$

Combined with Proposition 2.1, this definition tells us that the coordinates of the acceleration of an object are obtained by twice differentiating the coordinates of the parametrization of the motion. If we know the acceleration of an object, we can integrate the coordinates of the acceleration to obtain the coordinates of the velocity. We can recover the coordinates of the parametrization by integrating the coordinates of the velocity. The following example illustrates an important application of this construction.

Example 2.13   **Motion under the Influence of Gravity.** Consider the motion of an object of unit mass in a vertical plane near the surface of the earth, so that there is one horizontal direction and one vertical direction. The force of gravity is given by the vector $\mathbf{a} = (0, -9.8)$ in units of meters per second squared. If we neglect other forces on the object in motion, like air resistance, then this is also the acceleration of the object. Since the velocity of the object and the acceleration are related by $\mathbf{v}'(t) = \mathbf{a}(t)$, we know that $\mathbf{v}'(t) = (0, -9.8)$. Integrating each coordinate, we have $\mathbf{v}(t) = (v_1, -9.8t + v_2)$, where v_1 and v_2 are constants of integration. Since $\mathbf{v}(0) = (v_1, v_2)$, these constants are the coordinates of the velocity of the object at time 0. We call $\mathbf{v}(0)$ the **_initial velocity_** of the object.

Since the parametrization α of the motion and the velocity $\mathbf{v}$ are related by $\alpha'(t) = \mathbf{v}(t)$, we know that $\alpha'(t) = (v_1, -9.8t + v_2)$. Integrating each coordinate, we can find α:

$$\alpha(t) = (v_1 t + x_1, -9.8t^2/2 + v_2 t + x_2),$$

where x_1 and x_2 are constants of integration. Since $\alpha(0) = (x_1, x_2)$, the constants are the position of the object at time 0. We call $\alpha(0)$ the **_initial position_** of the object.

We will explore this example further in Exercise 8.

Tangent Lines

The tangent vector of a parametrization at a point can be used to construct a line tangent to the curve at the point. In one-variable calculus, the tangent line to the graph of a function at a point is the line passing through the point with slope equal to the slope of the graph, or derivative of the function at the point. The analogous statement is true for parametrizations: The tangent line to the image of α at $\alpha(t_0)$ is the line passing through $\alpha(t_0)$ whose direction vector is equal to the tangent vector, or derivative of the parametrization at the point, $\alpha'(t_0)$. It is convenient to represent the line as the image of a linear parametrization β. We do this in the following definition.

Definition 2.8 Suppose that α is a differentiable parametrization, t_0 is in the domain of α, and $\alpha'(t_0) \neq \mathbf{0}$. We define the **tangent line** to the image of α at $\alpha(t_0)$ to be the line parametrized by

$$\beta(t) = \alpha(t_0) + t\,\alpha'(t_0), \quad t \in \mathbb{R}. \; \blacklozenge$$

Example 2.14

The Tangent Line to a Helix. The image of the parametrization $\alpha(t) = (\cos t, \sin t, t)$ is a helix that winds counterclockwise about the z-axis. (See Exercise 9 of Section 2.1.) The tangent vector is $\alpha'(t) = (-\sin t, \cos t, 1)$. It follows that the tangent line to the helix at the point $\alpha(t_0)$ is given by

$$\beta(t) = (\cos t_0, \sin t_0, t_0) + t(-\sin t_0, \cos t_0, 1).$$

For example, if we select the time $t_0 = 3\pi/2$, we have

$$\beta(t) = (\cos(3\pi/2), \sin(3\pi/2), 3\pi/2) + t(-\sin(3\pi/2), \cos(3\pi/2), 1)$$
$$= (0, -1, 3\pi/2) + t(-1, 0, 1).$$

Figure 2.11(a) contains a plot of the image of α and this tangent line. Notice that the tangent line and the image of α come together at $\alpha(3\pi/2)$ and that they have the same direction there.

In order to apply the tangent line formula, we required the derivative to be nonzero. If $\alpha'(t_0) = \mathbf{0}$, the formula $\beta(t) = \alpha(t_0) + t\alpha'(t_0)$ yields $\beta(t) = \alpha(t_0)$. The image of β is

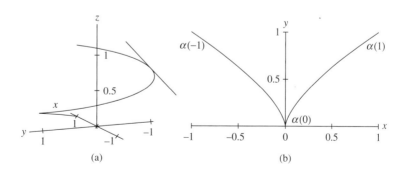

(a)   (b)

Figure 2.11 (a) The tangent line to the helix $\alpha(t) = (\cos t, \sin t, t)$ at $t = 3\pi/2$. (See Example 2.14.) (b) The image of $\alpha(t) = (t^3, t^2)$. (See Example 2.15.)

the point $\alpha(t_0)$, not a line. Even though α is differentiable at t_0, it does not guarantee that there is a tangent line. This should be contrasted with the one-variable case. Since the derivative of a function $y = f(x)$ at a point x_0 is the slope of the tangent line, being differentiable is equivalent to the graph of f having a tangent line. The following example shows what can "go wrong" when $\alpha'(t_0) = \mathbf{0}$.

Example 2.15 **The Tangent Behavior at a Cusp.** The parametrization $\alpha(t) = (t^3, t^2)$ is differentiable for all t. Computing the derivative, $\alpha'(t) = (3t^2, 2t)$, we see that $\alpha'(0) = (0,0)$. The coordinates of α satisfy the equation $y = x^{2/3}$. Since $x = t^3$ takes all possible values, the image of α is the graph of the function $y = f(x) = x^{2/3}$, which, as a function of one variable, fails to be differentiable at $x = 0$. The graph of f, hence the image of α, comes to a sharp point, or **cusp**, at the origin. (See Figure 2.11(b).) Since f is not differentiable at the origin, its graph does not have a tangent line at the origin. Consequently, the image of α does not have a tangent line at the origin, even though α is differentiable.

Chain Rule for Parametrizations

If a parametrization β is the composition of a function g and a parametrization α, $\beta = \alpha \circ g$, so that $\beta(t) = \alpha(g(t))$, the derivative of β can be expressed in terms of the derivative of α and the derivative of g. This result is known as the **chain rule**, and its form is analogous to the chain rule for functions of one variable. In one variable, if $h(x) = f(g(x))$, then $h'(x) = f'(g(x))g'(x)$. We will state the chain rule as a proposition.

Proposition 2.2 **The Chain Rule.** Let $\alpha : R \to \mathbb{R}^2$ or $\alpha : R \to \mathbb{R}^3$ be a differentiable parametrization, and let $g : \mathbb{R} \to \mathbb{R}$ be a differentiable function. Then the composition $\alpha \circ g$ is a differentiable parametrization and

$$(\alpha \circ g)'(t) = \alpha'(g(t))g'(t). \ \blacklozenge$$

The proof of the chain rule has a form common to many of the equalities involving vectors. Working from the side of the equality that we want to expand or simplify, $(\alpha \circ g)'$, we express it in coordinates, apply facts about functions of one variable in each of the coordinates, and then reorganize the coordinate form to obtain the vector form we want.

Proof: We will write out the steps for a parametrization of a curve in the plane. The steps for a parametrization of a curve in space are analogous. Let us begin with a differentiable parametrization $\alpha = \alpha(s)$, where $\alpha(s) = (x(s), y(s))$, and a differentiable function $g = g(t)$. The composition $\alpha \circ g$ is given by

$$(\alpha \circ g)(t) = (x(g(t)), y(g(t))).$$

Since α is differentiable, the coordinate functions x and y are differentiable. Then, since g is differentiable, the compositions $x(g(t))$ and $y(g(t))$ are differentiable. This means that when we compute the derivative of the composition, we may apply the one-variable chain rule to its entries.

$$(\alpha \circ g)'(t) = (x(g(t))', y(g(t))')$$
$$= (x'(g(t))g'(t), \ y'(g(t))g'(t)) .$$

Factoring out the term $g'(t)$, we have

$$(x'(g(t)), \ y'(g(t))) \, g'(t) = \alpha'(g(t))g'(t).$$

This is the desired result. ∎

Two instances of the chain rule are of particular importance for us: scaling and shifting the independent variable. (See Example 2.7.)

Example 2.16   **Transforming the Parameter**

A. Shifting Revisited. Suppose $\alpha : \mathbb{R} \to \mathbb{R}^3$ is differentiable and $g(t) = t + k$, where k is a constant. If $\beta = \alpha \circ g$, then

$$\beta'(t) = (\alpha \circ g)'(t) = \alpha'(g(t))g'(t) = \alpha'(t+k)1 = \alpha'(t+k).$$

Shifting the independent variable of a parametrization by k has the effect of shifting the tangent vectors so the tangent vector to β at t_0 corresponds to the tangent vector to α at $t_0 + k$. If the parameter represents time, this corresponds to our sense that shifting the parameter corresponds to shifting the behavior of the system k units forward or backward in time.

B. Scaling Revisited. Now let $g(t) = kt$, where k is a constant. If $\beta = \alpha \circ g$, then

$$\beta'(t) = (\alpha \circ g)'(t) = \alpha'(g(t))g'(t) = \alpha'(kt)k = k\alpha'(kt).$$

Scaling the independent variable of the parametrization by k also scales the derivative by a factor of k.

In terms of the motion, if α is a parametrization of the motion of an object, we know that $\alpha(kt)$ represents motion along the same curve that in 1 unit of time covers as much of the curve as the original motion does in k units of time. The chain rule says that the velocity of the new motion is $k\mathbf{v}(kt)$, so that the velocity is also k times as large as it is for the original motion.

Summary   We defined the **velocity** of an object to be the **limit** of its **average velocity**. The **speed** of an object is the length of its velocity vector. The discussion of velocity motivated the definition of the **derivative of a parametrization**. At any point the derivative of a parametrization is a vector that is tangent to the image of the parametrization. It is called the **tangent vector** of the parametrization. If the parametrization represents motion, the derivative is the velocity of the motion and its length is the speed of the motion. We showed how to compute the derivative of a parametrization in terms of the derivatives of its coordinate functions. If $\alpha(t) = (x(t), y(t))$, then $\alpha'(t) = (x'(t), y'(t))$. We defined the **acceleration** of an object in motion to be the derivative of its velocity, and we defined the **tangent line** to the image of α at t_0 to be the image of $\beta(t) = \alpha(t_0) + t\alpha'(t_0)$. Finally, we stated and proved the **chain rule** for parametrizations, $(\alpha \circ g)'(t) = \alpha'(g(t))g'(t)$.

Section 2.2 Exercises

1. **Velocity and Speed.** Each of the following functions α is a parametrization for the motion of an object in the plane or in space. Find the velocity vector and the speed of the motion at the given time t_0.

    (a) $\alpha(t) = (e^t, e^{3t})$ at $t_0 = 0$.
    (b) $\alpha(t) = (2\cos t, 3\sin t)$ at $t_0 = \pi/2$.
    (c) $\alpha(t) = (\cos 2t, t, \sin 2t)$ at $t_0 = \pi/4$.
    (d) $\alpha(t) = (1, t, \sqrt{1 + t^2})$ at $t_0 = 0$.

2. **Tangent Lines.** For each of the following parametrizations α, find the tangent vector and the tangent line to the curve parametrized by α at the point t_0.

    (a) $\alpha(t) = (1 + \cos 2t, 2 + \sin 2t)$ at $t_0 = \pi/4$.
    (b) $\alpha(t) = (te^{2t}, e^{t^2})$ at $t_0 = 1$.
    (c) $\alpha(t) = (1 + 2t, 1 - t, 2 + t)$ at $t_0 = 1$.
    (d) $\alpha(t) = ((2 + \sin 2t)\cos t, (2 + \sin 2t)\sin t, \cos 3t)$ at $t_0 = 0$.

3. **The Algebra of Differentiation.** Each of the following algebraic facts can be verified by the type of coordinate calculation that we used to prove the chain rule. Carry out the calculation for each of the following identities involving parametrizations. In each case, say which property of derivatives of functions of one variable you use to verify the identify. Assume that all the parametrizations are differentiable and take values in $\mathbb{R}^3$:

    (a) $(c\alpha(t))' = c\alpha'(t)$.
    (b) $(\alpha(t) + \beta(t))' = \alpha'(t) + \beta'(t)$.
    (c) $(\alpha(t) \cdot \beta(t))' = \alpha'(t) \cdot \beta(t) + \alpha(t) \cdot \beta'(t)$.
    (d) $(\alpha(t) \times \beta(t))' = \alpha'(t) \times \beta(t) + \alpha(t) \times \beta'(t)$.

4. **The Tangent Line to a Graph.** Suppose C is a curve in $\mathbb{R}^2$ that is the graph of a differentiable function $f : R \to \mathbb{R}$.

   (a) What is the point-slope equation for the tangent line to this curve at the point $(x_0, f(x_0))$?
   (b) Parametrize this curve and then use the techniques developed in this section to find a parametrization of the tangent line to this curve at the point $(x_0, f(x_0))$.
   (c) Show that the parametrization of part (b) is a parametrization of the line from part (a).

5. **The Cycloid.** Consider a wheel of radius 1 that rests on the x-axis and rolls without slipping in the positive direction along the x-axis with unit speed. If we mark a point on the wheel, it will trace a curve in the plane as the wheel rolls. This curve is called a *cycloid*. If we assume that the motion starts with the wheel resting on the origin and that the marked point is at the top of the wheel, one unit from the center at the beginning of the motion, the motion of the point is parametrized by

$$\alpha(t) = (t + \sin t, 1 + \cos t).$$

   Figure 2.12 is a plot of the image of α for $0 \le t \le 4\pi$.

   (a) Compute the velocity vector $\mathbf{v}(t)$ and the speed $s(t)$ for the motion of the point.
   (b) At which time t is the velocity vector horizontal? At which time t is the velocity vector vertical?
   (c) Where is the point moving fastest? Where is it moving slowest?
   (d) Describe what is happening at the points $t = \pi$ and $t = 3\pi$.

6. **Acceleration.** Suppose $\alpha(t)$ is a differentiable function that parametrizes the motion of an object and that $\mathbf{v}(t)$ and $\mathbf{a}(t)$ are the corresponding velocity and acceleration vectors.

   (a) Suppose $\alpha(t)$ parametrizes linear motion along the line segment from $P = (x_0, y_0, z_0)$ to $Q = (x_1, y_1, z_1)$, with

$$\alpha(t) = (x_0, y_0, z_0) + ct(x_1 - x_0, y_1 - y_0, z_1 - z_0),$$

   where c is a positive constant. What is the acceleration vector $\mathbf{a}(t)$ for this motion?

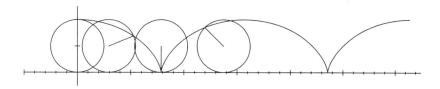

Figure 2.12 The cycloid of Exercise 5.

(b) Suppose $\alpha(t) = (R\cos(ct), R\sin(ct))$, so that α parametrizes uniform circular motion around a circle of radius R centered at the origin. What is the acceleration vector $\mathbf{a}(t)$ for this motion?

(c) Suppose $\alpha(t) = (R\cos t, R\sin t, ct)$ parametrizes motion along a helix. What is the acceleration vector $\mathbf{a}(t)$ for this motion?

7. **Constant Speed Motion.** Suppose α is a differentiable function that parametrizes the motion of an object that is traveling at a constant speed. That is, $s(t) = k$. What is the angle between the velocity vector $\mathbf{v}(t)$ and the acceleration vector $\mathbf{a}(t)$ for this motion? (*Hint:* Use the facts that (i) $s(t) = \sqrt{\mathbf{v}(t) \cdot \mathbf{v}(t)}$ and (ii) $s'(t) = 0$.)

8. **Projectile Motion.** Let us consider the motion of an object near the surface of the earth that is subject only to the force of gravity. As we saw in Example 2.13, the motion of the object can be parametrized by

$$\alpha(t) = (v_1 t + x_1, -9.8t^2/2 + v_2 t + x_2),$$

where $\alpha(0) = (x_1, x_2)$ is the initial position of the object and (v_1, v_2) is its initial velocity.

(a) Suppose the initial position of the object is $\alpha(0) = (0,0)$ and the initial velocity of the object is $\mathbf{v}(0) = (5, 10)$. How long will the object remain in motion, how high will it travel, and how far will it travel?

(b) Suppose an object is launched with initial speed of 50 m/s and it reached a maximum height of 20 m. How far did it travel?

9. **More Projectile Motion.** Suppose the motion of an object is parametrized by the function α from Exercise 8.

(a) If the initial position of the object is $(0,0)$ and the initial velocity is (v_1, v_2), when will the object hit the ground? What horizontal distance does it travel before hitting the ground?

(b) If the initial position of the object is $(0,0)$, when will it reach its maximum height? What is the maximum height?

(c) At which angle θ should the object be launched in order to maximize the horizontal distance traveled? (*Hint:* Express $\mathbf{v}(0)$ as $v_0(\cos\theta, \sin\theta)$, where v_0 is a constant and $(\cos\theta, \sin\theta)$ is a unit vector.)

10. **Halley's Comet.** In this exercise, we will investigate the velocity of Halley's comet. Although its orbit is an ellipse, it does not have a simple parametrization. Thus we will work directly with position data. We will use astronomical units (AU) in our calculations. An astronomical unit is approximately 1.496×10^8 km, the average distance from the earth to the sun.

The orbit of the comet is an ellipse with eccentricity 0.9674 with the sun located at one of its foci. (See Exercise 6 in Section 2.1.) The distance at perihelion, the closest point to the sun, is approximately 0.59 astronomical units. The comet takes 76 years to orbit the sun. The table in Figure 2.13 shows the location of Halley's comet at four-year

t	x	y	t	x	y	t	x	y	t	x	y
0	−0.59	0	20	30.16	3.28	40	35.54	−0.38	60	27.24	−3.86
4	11.77	4.37	24	32.38	2.61	44	35.03	−1.15	64	23.46	−4.33
8	18.53	4.58	28	33.98	1.90	48	33.98	−1.90	68	18.53	−4.58
12	23.46	4.33	32	35.03	1.15	52	32.38	−2.61	72	11.77	−4.35
16	27.24	3.86	36	35.54	.38	56	30.16	−3.28	76	−0.59	0

Figure 2.13   The location of Halley's comet at four-year intervals in xy-coordinates. The sun is located at the origin of the coordinate system, and the major axis coincides with the x-axis. Distances are in astronomical units. (See Exercise 10.)

intervals using a Cartesian coordinate system with the sun at the origin and the x-axis aligned with the major axis of the orbit. Figure 2.14 contains a plot of the data points.

(a) How would you use the data to approximate the velocity of Halley's comet?
(b) Carry out your approximation of the velocity at four-year intervals.
(c) Where is Halley's comet moving fastest? Where is it moving slowest?
(d) Compare the motion of Halley's comet to the motion corresponding to a parametrization of the type given in Example 2.4B. Explain how they differ.

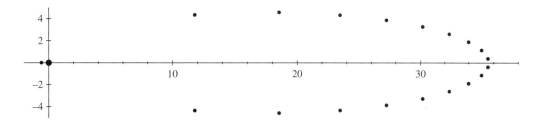

Figure 2.14   The plot of the data points for the orbit of Halley's comet given in Figure 2.13. The sun is located at the origin. (See Exercise 10.)

■ 2.3 Modeling with Parametric Curves

In this section, we would like to introduce two different uses of parametric curves. The first concerns the velocity of a robotic arm. It builds on the collaborative exercise at the beginning of Chapter 1. The other is a model for the spread of influenza. This develops ideas introduced in the collaborative exercise at the beginning of Chapter 2.

Velocity and Inverse-Velocity Kinematics of a Robotic Arm

In the collaborative exercise at the beginning of Chapter 1, we considered the forward- and inverse-position kinematic problems for a two-link (two-segment) robotic arm. As in Chapter 1, the robot will consist of links, one of length 1 extending from a point O on the x-axis to a point P, and the second, also of length 1, extending from P to Q. (See Figure 2.15.) The *effector* of the robot is at the point Q. Knowing the lengths of the links and the properties of the joints, it was possible to determine the working envelope of the robot. That is, it was possible to determine all possible positions of the effector of the robot. This is the forward-position kinematics problem. We were also able to provide a qualitative answer to the question of how many positions of the links give the same position of the effector. This is the inverse-position kinematics problem.

Here we want to use parametric curves to answer similar questions for the velocity and speed of this robot. In particular, the very practical problem we will solve is how to make the effector move at constant speed. Regardless of the configuration of the joints, we must first construct a parametrization for the motion of the effector. We can then determine the velocity of the motion of the effector by computing the derivative of this parametrization. Of course, the speed is the length of the velocity. The parametrization of the motion of the effector will be a sum of a parametrization of the motion of P and a parametrization of the motion of Q relative to P. We can further modify either of these summands by composition with a function of t. This will allow us to adjust the speed of P or Q and thus to determine how to make Q (the effector) move at a constant speed.

In each case, it will be a sum of a parametrization of the motion of P and a parametrization of the motion of Q relative to P. We can further modify either of these summands by composing it with a function of t. This will allow us to adjust the speed of P or Q. To find the speed of the effector, we must first find its velocity. This can be accomplished by combining the rule for the derivative of the sum and the chain rule. We begin with a single hinge joint.

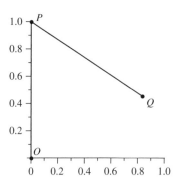

Figure 2.15 A robot with two arms of length 1. (See Example 2.17.)

Example 2.17	**A Single Hinge Joint.** If $\overline{OP}$ is fixed as shown in Figure 2.15 and the joint at P is a hinge joint free to move through an angle of 2π, then the working envelope of the robot is the circle of radius 1 centered at $(0, 1)$. Let $g = g(t)$ be the angle made by the segment $\overline{PQ}$ with the segment $\overline{OP}$ measured counterclockwise from the position where Q coincides with O. Then the position of the effector is given by

$$\alpha(t) = (0, 1) + (\sin(g(t)), -\cos(g(t))) = (\sin(g(t)), 1 - \cos(g(t))).$$

Using the chain rule, the speed of the effector is

$$s(t) = \|\alpha'(t)\| = \|(\cos(g(t)), \sin(g(t)))g'(t)\| = 1 \cdot |g'(t)| = |g'(t)|.$$

If g is an increasing function of t, that is, $g'(t) \geq 0$, then the speed of the effector is simply $g'(t)$. Consequently, for the effector to move at constant speed, the rate of change of the angle must be constant, $g'(t) = k$, so that g must be a linear function of t, $g(t) = kt + \theta_0$, where θ_0 is the initial angle at the hinge joint.

Now let us consider a more complicated example. Suppose there is a slider joint located at O allowing O to slide along the x-axis while $\overline{OP}$ remains perpendicular to the axis, in addition to a hinge joint located at P that is free to rotate through an angle of 2π. At time t, the hinge joint at P will be located at $(f(t), 1)$, where we assume that f is a differentiable function of t. If $g = g(t)$ again represents the angle made by $\overline{PQ}$ with $\overline{OP}$, then position of the effector is given by

$$\alpha(t) = (f(t), 1) + (\sin(g(t)), -\cos(g(t))) = (f(t) + \sin(g(t)), 1 - \cos(g(t))).$$

In Example 2.17, f was held constant. In the general case, we can calculate $\alpha'(t)$ and $s(t)$. First,

$$\alpha'(t) = (f'(t) + \cos(g(t))g'(t), \sin(g(t))g'(t)).$$

Then, the speed of the effector is

$$\begin{aligned} s(t) &= \|\alpha'(t)\| = \|(f'(t) + \cos(g(t))g'(t), \sin(g(t))g'(t))\| \\ &= \sqrt{f'(t)^2 + 2f'(t)\cos(g(t))g'(t) + \cos^2(g(t))g'(t)^2 + \sin^2(g(t))g'(t)^2} \\ &= \sqrt{f'(t)^2 + 2f'(t)\cos(g(t))g'(t) + g'(t)^2}. \end{aligned}$$

Let us consider the special case when f and g are linear functions of t.

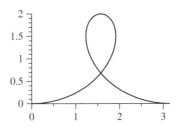

Figure 2.16   The path of the effector for a slide joint moving linearly and a hinge joint rotating with constant angular speed. (See Example 2.18.)

Example 2.18   **A Hinge Plus Slider Forward-Velocity Kinematics Problem.** Assume that $f(t) = \frac{1}{2}t$ and $g(t) = t$. It follows from our calculation that the speed of the effector is $s(t) = \sqrt{\frac{5}{4} + \cos(t)}$. Figure 2.16 shows the path traced by the effector.

Notice the speed is a maximum when $t = 0$ and a minimum when $t = \pi$. This makes intuitive sense because when $t = 0$, the velocity of the base and effector are in the same direction, and the magnitude of the sum is the sum of the magnitudes. On the other hand, when $t = \pi$, the base is moving to the right while the effector has reached its highest position and is moving to the left, and the magnitude of the sum is the difference of magnitudes.

An interesting and natural question is how to move the effector at constant speed. Or, we might say, what conditions does the requirement that $s(t) = s_0$ place on f and g? This problem does not have a unique answer. However, if f is required to be linear, it is possible to derive an ordinary differential equation for g, which can be solved using numerical techniques from calculus. The following example demonstrates this.

Example 2.19   **A Hinge Plus Slider Inverse-Velocity Kinematics Problem.** To begin, assume that f is linear, $f(t) = v_0 t$, so that the slider joint moves with constant velocity $v_0 > 0$. Then $s(t) = \sqrt{v_0^2 + 2v_0 \cos(g(t))g'(t) + g'(t)^2}$. Squaring both sides, it suffices to work with the equation:

$$s_0^2 = v_0^2 + 2v_0 \cos(g(t))g'(t) + g'(t)^2.$$

To simplify this further, we write s_0 as a product, $s_0 = s_1 v_0$. Using this substitution, moving the $s_0^2 = (s_1 v_0)^2$ term to the right and rearranging, we obtain the following quadratic equation in $g'(t)$:

$$0 = g'(t)^2 + 2v_0 \cos(g(t))g'(t) + v_0^2(1 - s_1^2).$$

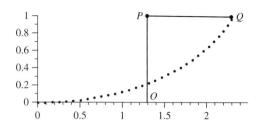

Figure 2.17 A constant speed solution to moving the effector from the origin to height 1 as the slider moves at constant speed 1 to the right. The robot is shown in its final position. (See Example 2.19.)

Applying the quadratic formula gives

$$g'(t) = \frac{1}{2}\left(-2v_0 \cos(g(t)) \pm \sqrt{4v_0^2 \cos^2(g(t)) + 4v_0^2(s_1^2 - 1)}\right)$$

$$= -v_0 \cos(g(t)) \pm v_0 \sqrt{\cos^2(g(t)) + s_1^2 - 1}.$$

Choosing the positive square root yields a formula for $g'(t)$ that is always positive. This is a first order differential equation for g:

$$\frac{dg}{dt} = -v_0 \cos(g(t)) + v_0 \sqrt{\cos^2(g(t)) + s_1^2 - 1}.$$

This equation cannot be solved explicitly, but if we choose values for v_0 and s_1, we can use Euler's method or another numerical method from one-variable calculus to solve for g. We can then substitute the numerically computed values for $g(t)$ into the formula for α to plot the position of the effector. Figure 2.17 shows a point plot of the position of the effector for $v_0 = 1$, $s_0 = 2$, and $g(0) = 0$ (the initial position of the effector at the origin). The displayed points are 0.042 units of time apart, and the final time is approximately 1.29 units of time. This represents a constant speed solution to the problem of lifting the effector from the origin to height 1 as the slider joint moves to the right at constant speed. The final location of O is approximately 1.3 and of Q is approximately $(2.3, 1)$.

Alternately, it is also possible to require that g be linear and solve for f. This is explored in Exercise 5.

SIR Epidemic Model

An **SIR** model for the spread of an infectious disease can be used to model diseases, like measles or a particular strain of influenza, for which the memory feature of the immune system prevents reinfection. Thus the population can be divided into three categories:

the **susceptible** people, the **infected** people, and the **recovered** people.[3] If a susceptible person contracts the disease, he or she will move to the infected group, and then, after a period of time, to the recovered group. Since recovery from the disease confers immunity, the recovered population is distinct from the susceptible population. For diseases that do not confer immunity, an *SI* model with only susceptible and infected populations would be appropriate. Here we will focus on a disease that confers immunity.

We will think of the sizes of each of the three populations as being functions of time alone. Let $S = S(t)$ be the number of susceptible individuals at time t, $I = I(t)$ be the number of infected individuals at time t, and $R = R(t)$ be the number of recovered individuals at time t. If the time period for the model is relatively short, which it is for the outbreak we will model, we can assume that the overall population is constant. Thus for any t, the three populations satisfy the equation

$$S(t) + I(t) + R(t) = N,$$

where the total number of people N in the population is constant. If we know S, I, and N, we can recover R from the fact that $R(t) = N - S(t) - I(t)$. This allows us to concentrate on the relationship between S and I and compute R if we need to.

Since susceptible individuals contract the disease from infected individuals and then move out of the susceptible category into the infected category, there is a relationship between these two categories. It makes sense, then, to plot the size of the susceptible population, $S(t)$, on one coordinate axis in the plane, and the size of the infected population, $I(t)$, on the other. For a particular outbreak of a disease, as time evolves, the values of S and I will trace a curve in the SI-plane. Denote this curve by α, where $\alpha : [0, b] \to \mathbb{R}^2$ and $\alpha(t) = (S(t), I(t))$.

If we knew the values of S and I for each t, we could immediately plot the image of α. Of course, to do so, we would have to wait until the end of an epidemic to produce a mathematical model for it! Instead we are going to assume that **the spread of the disease behaves according to a simple set of rules**. Our job will be to come up with mathematical versions of these rules. Since the rules are for the spread of the disease, these will involve S' and I', the rates of change of S and I.

We will concentrate on formulating rules for influenza. Influenza is transmitted through contact between a susceptible person and an infected person. The change in the susceptible population is due entirely to individuals becoming infected, that is, to individuals moving from the susceptible population to the infected population. It follows that S', the rate of change of S, is always negative. We will assume that this rate is jointly proportional to the number of susceptible individuals and the number of infected individuals.

---

[3]For a general reference on SIR models, see Chapter 19 of J.D. Murray, *Mathematical Biology*, Springer-Verlag, 1993.

Symbolically,

$$S'(t) = -rS(t)I(t),$$

where $r > 0$ is a constant, which we call the **rate of transmission**. This equation makes sense because if there are more infected people, we expect that an individual susceptible person should come into contact with more infected people, and so should have a greater chance of becoming infected. On the other hand, if there are more susceptible people, there are more people who could contract the disease from contact. In both cases, the rate should increase.

Now consider I', the rate of change of the infected population. Since every decrease in the susceptible population results in an equal increase in the infected population, I' must contain the term $rI(t)S(t)$. As people recover, they move from the infected population to the recovered population. We will assume that this rate is proportional to the size of the infected population. This will be reflected by a term of the form $-aI(t)$, where $a > 0$ is a constant, which we call the **rate of recovery**. (See Exercise 11.) Assuming there are no other factors to consider, I' is given by

$$I'(t) = rS(t)I(t) - aI(t).$$

Below is a summary of this discussion.

SIR Epidemic Model

Denote the susceptible population by $S = S(t)$, the infected population by $I = I(t)$, and the recovered population by $R = R(t)$, and assume the total population is constant, then

$$S(t) + I(t) + R(t) = N.$$

The **SIR**, or **susceptible-infected-recovered**, model of an epidemic is given by the pair of differential equations for the rate of change of S and the rate of change of I,

$$S'(t) = -rS(t)I(t)$$
$$I'(t) = \ \ rS(t)I(t) - aI(t),$$

where the constant $r > 0$ is the **rate of transmission** and the constant $a > 0$ is the **rate of recovery**.

We would like to solve this pair of equations for I and for S. Here is one way to do this. First, we can eliminate t from these equation by applying the chain rule from one-variable calculus. It says that

$$\frac{dI}{dS} = \frac{dI/dt}{dS/dt}.$$

Substituting the equations of the model into this quotient, we have

$$\frac{dI}{dS} = \frac{rSI - aI}{-rSI} = -\frac{rS - a}{rS} = -1 + \frac{a}{rS}.$$

Separating variables, we obtain

$$dI = (-1 + \frac{a}{rS})dS.$$

Integrating both sides of this equation, we obtain an equation for I in terms of S,

$$I = -S + \frac{a}{r}\ln S + k,$$

where k is a constant of integration. If we start the model at time $t = 0$, we can express k in terms of the initial sizes of the susceptible and infected populations.

$$k = I(0) + S(0) - \frac{a}{r}\ln S(0).$$

Let us apply this model for I to the data from a real epidemic.

Example 2.20   **Influenza at a British Boarding School**

A.   **The Parametric Curve.** In 1978, the *British Medical Journal*[4] reported on an outbreak of influenza at a British boys' boarding school. There were 763 students at the school, and the outbreak began with one infected student, so that $S(0) = 762$ and $I(0) = 1$. In the report it was determined that $r \approx 2.18 \times 10^{-3}$/day and $a \approx 4.40 \times 10^{-1}$/day, so that $a/r \approx 202$ and $k \approx 763 - 202\ln(762) \approx -577$. The equation for I becomes

$$I = -S + 202\ln S - 577.$$

---

[4]News and Notes, *British Medical Journal*, 1:6112.586, March 4, 1978.

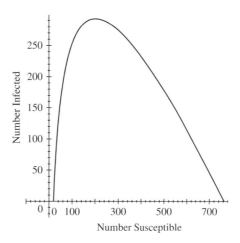

Figure 2.18   The plot of α , $\alpha(t) = (S(t), I(t))$, for the SIR model of Example 2.20.

This defines a curve in the first quadrant of the SI-plane that contains the image of α. (See Figure 2.18.)

B. **An Analysis of the Model.** The initial point of α is $\alpha(0) = (762, 1)$, the right endpoint of the curve. From the model, $S' < 0$ as long as there are susceptible *and* infected people remaining. Thus S must be decreasing as long as there are susceptible and infected people remaining, so that α traces the curve in Figure 2.18 from right to left.

From the model, $I'(t) = I(t)\,(rS(t) - a)$. If $I(t) > 0$, then $I'(t) > 0$ when $rS(t) - a > 0$, or when $S(t) > a/r \approx 202$. So as the susceptible population decreases from 762 to 202, the infected population increases. When the susceptible population falls below 202, $I'(t) < 0$ and the infected population decreases. That is, the epidemic begins to subside. This is because the number of susceptible people has decreased to the point that the rate at which the susceptible population becomes infected, $rS(t)I(t)$, is exceeded by the rate at which the infected population recovers, $aI(t)$. This epidemic ends when $I = 0$. This occurs with 19 people still uninfected.

It is important to note that we learned a good deal about the spread of the epidemic in Example 2.20 without having an explicit formula for I or S in terms of t. The equations of the model by themselves were sufficient for us to determine when the populations were increasing or decreasing and when they had a maximum or a minimum.

Summary   Our primary goals in developing these models is to illustrate different ways in which parametric curves may be used to analyze the behavior of a real-world system.

In modeling the motion of a simple robot, we used **sums** and **compositions** to construct a formula for the position of the effector at time t. We computed the velocity and speed of the effector. We also demonstrated how to find a constant speed motion of the effector by reducing the problem to solving an ordinary differential equation. Since the equation could not be solved symbolically, it was necessary to use Euler's method or a similar method to produce a numerical solution. It was then possible to plot the position of the effector at the numerically computed times.

The second application concerned an **SIR**, or **susceptible-infected-recovered**, model for the spread of an influenza-like epidemic. We developed a model for the **relationship between the susceptible and infected populations**. By making assumptions about **the spread of the disease**, we constructed a pair of differential equations for the **rates of change of these populations**. Using techniques from one-variable calculus, we were able to solve these equations to find the image of the parametrization and were able to analyze the data from a particular outbreak of influenza.

Section 2.3 Exercises

1. **Forward-Velocity Kinematics.** In Example 2.18, we considered the motion of the effector of a simple robot when the slider joint moved with constant positive velocity and the hinge joint rotated with constant positive angular velocity. Using the notation of the example, the effector moves according to

$$\alpha(t) = (f(t) + \sin(g(t)), 1 - \cos(g(t))),$$

where $f(t) = k_1 t$ and $g(t) = k_2 t$ for constants k_1 and k_2. In Example 2.18, $k_1 = \frac{1}{2}$ and $k_2 = 1$. Here we consider other possibilities for these coefficients. In each case, describe the motion of the effector on the interval $0 \leq t \leq 2\pi$. (*Hint:* Examine the signs $x'(t)$ and $y'(t)$ on $[0, 2\pi]$.)

   (a) Suppose $k_1 > k_2$, for example, $k_1 = 2$ and $k_2 = 1$.
   (b) Suppose $k_1 = k_2$, for example, $k_1 = k_2 = 1$.
   (c) Suppose $k_1 > |k_2|$, for example, $k_1 = 2$ and $k_2 = -1$.
   (d) Suppose $k_1 = -k_2$, for example, $k_1 = 1$ and $k_2 = -1$.
   (e) Suppose $k_1 < |k_2|$, for example, $k_1 = 1$ and $k_2 = -2$.

2. **Inverse-Velocity Kinematics I.** In Example 2.19, we considered a solution to the problem of moving the effector of a simple robot at constant speed $s_0 = 2$ to height 1 for horizontal velocity $v_0 = 1$.

(a) Suppose that $v_0 = 1.5$ while s_0 remains equal to 2. Without doing any calculations, how would this change the solution? In particular:

(i) Would the final position of O be to the right or left of 1.3 (the final position in Example 2.19)? Explain.

(ii) Would the values of g be greater or less than in Example 2.19? Explain.

(b) Suppose that $v_0 = 0.5$ while s_0 remains equal to 2. Without doing any calculations, how would this change the solution? In particular:

(i) Would the final position of O be to the right or left of 1.3 (the final position in Example 2.19)? Explain.

(ii) Would the values of g be greater or less than in Example 2.19? Explain.

3. **Inverse-Velocity Kinematics II.** Consider the robot of Example 2.19. The example exhibited a constant speed motion of the effector in which the final position of the effector was approximately $(2.3, 1)$. How would the robot have to move from the initial position of both O and Q at the origin (so the angle is 0) in order to lift the effector to the following points? That is, should f' and g' be positive or negative? Explain your answer.

(a) $(\frac{1}{2}, \frac{1}{2})$.   (b) $(0, \frac{3}{2})$.   (c) $(-\frac{1}{2}, 1)$.

4. **A Numerical Solution I.** Consider the robot of Example 2.19. Here we consider what happens if the other root is chosen.

(a) Use a numerical differential equation solver to solve the differential equation using the negative square root:

$$\frac{dg}{dt} = v_0 \cos(g(t)) - v_0 \sqrt{\cos^2(g(t)) + s_1^2 - 1}.$$

Use the same speed for $\overline{OP}$, $v_0 = 1$, the same speed for the effector, $s_0 = 2$, and the same initial condition, $g(0) = 0$. The result should be a list of times and values for g at those times.

(b) The motion of the effector can be obtained from the numerical solution in (a) by the following procedure. If the times are $(t_0, t_1, \ldots, t_n)$ and the corresponding g values are $(g_0, g_1, \ldots, g_n)$, then plot the list of points

$$(v_0 t_i + \sin(g_i), 1 - \cos(g_i)), \quad i = 1, \ldots, n.$$

Describe the resulting plot and motion.

5. **A Numerical Solution II.** In Example 2.19, the horizontal velocity of the slider was held constant. This allowed us to find a differential equation for g', which we solved numerically. Suppose instead that g' is held constant. That is, set $g(t) = v_0 t$.

(a) Following the steps in Example 2.19, derive a differential equation for $f'(t)$ whose solution will solve the constant speed problem $s(t) = s_0$.

(b) How does this equation differ from the one for $g'(t)$ in Example 2.19?

(c) Choosing the positive square root [your solution to (a) should involve a square root], integrate the differential equation for $f(t)$ using a computer algebra system to find $f(t)$ for the values $s_0 = 2$ and $v_0 = 1$.

(d) How long does it take the effector to reach height 1?

6. **A Cartesian Robot.** A Cartesian or gantry robot consists of an effector suspended from a beam. (See Figure 2.19.) The effector is able to move vertically, and its distance from the beam is denoted by z with $z = 0$ corresponding to the effector in the raised position and $z = h_0$ corresponding to the effector on the floor (or ground). The point of suspension of the effector is free to slide horizontally along the beam. This position is denoted by x with $0 \leq x \leq w_0$. The beam is in turn free to slide along two rails. The position of the beam along the rails is denoted by y with $0 \leq y \leq l_0$. We will say the robot has height h_0, width w_0, and length l_0.

(a) The possible positions of the effector lie in a box in the first octant (all coordinates are nonnegative) of a Euclidean space with coordinates $\{(z, x, y) : 0 \leq x \leq w_0,$ $0 \leq y \leq l_0,$ and $0 \leq z \leq h_0\}$. Sketch and label this box and the coordinate axes so that positions with the effector retracted $(z = 0)$ are at the top of the box as shown in Figure 2.19.

(b) Are the coordinates in part (a) right-handed or left-handed? Explain.

(c) One method of moving the effector at constant speed between two locations at the same height, say from (z_1, x_1, y_1) to (z_1, x_2, y_2), is to use a linear parametrization of the motion. Write a formula for such a parametrization so that it has constant speed v_0 ft/s.

(d) Suppose the dimensions of a Cartesian robot in feet are $h_0 = 20$, $w_0 = 50$, and $l_0 = 100$. (Cartesian robots used in manufacturing can be quite large.) Construct a parametrization in three parts to lift an object on the floor at location $(20, 20, 0)$ to a height of 10 ft with unit speed, move it horizontally and linearly to $(10, 30, 100)$

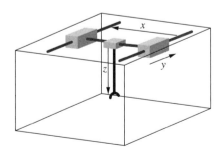

Figure 2.19   A schematic diagram of a Cartesian or gantry robot. (See Exercise 6.)

with constant speed 2 ft/s, and then lower the object to height of 5 ft above the floor, again with unit speed.

(e) Assuming no time is spent accelerating to constant speed or changing velocity at the "corners" of the motion, how long will the entire motion last?

7. **A SCARA Robot I.** Figure 2.20(a) shows a simplified rendering of a Selective Compliant Articulated Robot Arm, or SCARA robot. The joints at P and Q are hinge joints (ideally) free to move through 2π, and the effector at R moves vertically on a slider joint. SCARA robots are capable of the same tasks as a Cartesian robot, but have a smaller footprint. However, to move the effector of a SCARA robot in a straight line at a fixed height, it is necessary to solve the inverse kinematics problem.

In this problem, we will assume the horizontal links have length 1 and will work in a horizontal plane containing P and Q with coordinates as shown in Figure 2.20(b). The angle between the link $\overline{PQ}$ and the positive x-axis is θ measured counterclockwise, and the angle between the link $\overline{PQ}$ and the link $\overline{QR}$ is ϕ measured clockwise. Thus, for example, when $\theta = 0$ and $\phi = 0$, the projection of R to this plane is located at the origin. With this notation, the projection of R to this plane has coordinates

$$R(\theta, \phi) = (\cos\theta - \cos(\theta - \phi), \sin\theta - \sin(\theta - \phi)).$$

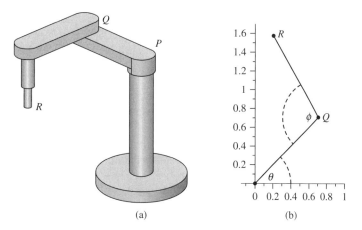

(a)   (b)

Figure 2.20   (a) A SCARA robot with hinge joint at P and Q and a slider joint moving the effector R. (b) xy-coordinates for the effector of a SCARA robot projected to the horizontal plane containing P and Q. (See Exercise 7.)

(a) Using either Figure 2.20(b) or the above formula, what are the xy-coordinates for

    (i) $R(0, \pi/2)$.   (ii) $R(\pi/4, \pi/2)$.   (iii) $R(\pi/2, \pi/2)$.

(b) For a given angle $\theta < \pi/2$, use the formula to show that $\phi = 2\theta$ locates the projection of the effector R on the y-axis.

(c) If the effector is to be positioned on a line that makes an angle τ with the positive x-axis, what is the relationship between θ and ϕ?

8. **A SCARA Robot II.** Here we solve the inverse kinematics problem for the SCARA robot of Exercise 7 when the projection of the effector moves along the y-axis.

    (a) Suppose that the angle θ is a function of time, $\theta = \theta(t)$. Using Exercise 7(b) and the formula given in Exercise 7, write a parametric formula $r(t)$ for the projection of the projector in terms of $\theta(t)$ when the projection of the effector is located on the y-axis.

    (b) Find the velocity and speed of $r(t)$ in terms of θ.

    (c) Suppose that the speed of the projection of the effector along the y-axis is a constant s_0. Use your formula for speed in part (b) to derive a differential equation for $\theta(t)$.

    (d) Solve your differential equation in part (c) for θ. Note: Since the solution may be rotated, this solves the inverse-velocity kinematics problem for a SCARA robot moving in a radial direction at constant speed.

9. **When Does an Epidemic End?** In Example 2.20, we noted that when the epidemic ended there were 19 individuals remaining who had not contracted influenza.

    (a) If there were still susceptible individuals, why did the epidemic end? Refer to the SIR model in your explanation.

    (b) This behavior is a common behavior of epidemics. In general, why is it that not all susceptible individuals become infected in an epidemic?

    (c) What does this imply about childhood diseases like the measles and chicken pox? Why is this a health concern?

10. **Initial Conditions.** The SIR model in Example 2.20 described a flu epidemic at a boarding school of 763 students that began with initial populations of 762 susceptible students and 1 infected student. We analyzed the progress of the epidemic by considering a parametric curve whose coordinates gave the susceptible and infected populations at any time t. Figure 2.21(a) is a plot of the parametric curves for this epidemic that can be derived from the SIR model if we assume different initial populations of susceptibles and infected and assume that the total population is 763. The initial populations used are $(762, 1)$ (this gives the curve that was analyzed in Example 2.20), $(700, 63)$, $(600, 163)$, $(500, 263)$, $(400, 363)$, $(300, 463)$, and $(200, 563)$. For each curve, the infection rate is $r = 2.18 \times 10^{-3}$ and the recovery rate is $a = 4.40 \times 10^{-1}$ as in Example 2.20.

    (a) Generally describe the spread of the disease as modeled by these curves.

    (b) In general, what is the effect of changing the number of students initially infected by the flu?

(c) For each of these curves, at what point does the epidemic begin to subside? How does that point depend on the initial point? Use the equation for $I'(t)$ and the implicit equation for the curve to explain your answer.

11. **Rate of Recovery.** The rate of recovery, a, in the SIR model is the coefficient of $I(t)$ in the differential equation for $I'(t)$. The quantity $1/a$ can be interpreted as the average length of time it takes to recover from the disease. Does this make sense for the particular model of Example 2.20? Explain why this makes sense generally.

12. **Rate of Transmission.** In the construction of the SIR model, we assumed that the rate of change of the susceptible population depended on the size of the susceptible population and the size of the infected population, $S'(t) = -rS(t)I(t)$, where the constant r is the rate of transmission. Another way to read this equation is that S' depends on nothing else about the populations other than their size. Is this a realistic assumption for all outbreaks of a disease? What other factors might affect the rate of change of the susceptible population? Explain.

13. **SIR with Immigration.** The SIR model in Example 2.20 assumed that the total population of students was constant. We would like to change the basic model to include the regular infusion of new members to the susceptible population. For example, if the disease were a long-lasting one, the model should reflect new people moving into the community.

    (a) The plot in Figure 2.21(b) is the parametric curve that represents the susceptible and infected populations for this modified scenario. Describe the spread of the disease as modeled by this curve.

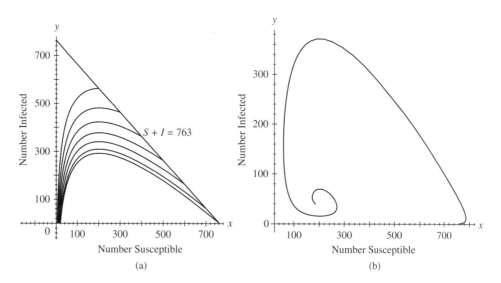

Figure 2.21 (a) The SIR model for different initial conditions. (See Exercise 10.) (b) An SIR model with immigration. (See Exercise 13.)

(b) How might the equations of the SIR model be modified to take into account immigration? Explain your answer.

14. **Vaccination.** Suppose that a vaccination for the disease is available, which is the case for some strains of influenza. Further, suppose that a community is able to vaccinate a fixed number of people each day. How would you take this into account in your SIR model? Explain your answer.

15. **Transmission Without Symptoms.** Suppose that individuals who are infected with a disease are contagious before they show any symptoms, but once they display symptoms they are quarantined for two weeks. How might you incorporate this into the basic SIR model? Explain your answer.

■ 2.4 Vector Fields

In this section, we develop the concept of a **vector field**. Our motivation is to model **the motion of a moving fluid or gas**, which we call a **flow**. In the next section, we use vector fields to model other physical phenomena.

We will assume that fluid or gas is moving through a region of the plane or space so that at each location the velocity of the flow does not vary in time. For example, we might think of the steady flow of water in a stream, the prevailing winds in a geographic region, or the regular flow of air across the surface of an airplane wing. In each of these examples, we can imagine focusing on one location and observing that the speed and direction of the water or air do not vary in time. In contrast, there are flows where the direction and speed of the water at a location will vary in time and, in some cases, change unpredictably from one instant to another. These more complicated flows are beyond the scope of our discussion.

For the moment, consider the movement of water on the surface of a stream. First, choose a coordinate system on the surface of the water so that a point on the surface is represented by an ordered pair (x, y). The velocity of the stream at a point (x, y) can be represented by a vector (u, v). Since the coordinates of (u, v) depend on the location and *not* the time, we will think of u and v as being functions of x and y, $u = u(x, y)$, and $v = v(x, y)$. Further, we will think of the collection of velocity vectors as the values of a function $\mathbf{F}$, $\mathbf{F}(x, y) = (u(x, y), v(x, y))$, so that $\mathbf{F}(x, y)$ is the velocity vector of the flow at (x, y). If the fluid is moving through space, the velocity vector is in $\mathbb{R}^3$, and we will write $\mathbf{F}(x, y, z) = (u(x, y, z), v(x, y, z), w(x, y, z))$. However, our primary interest in this chapter is in flows in the plane. Based on this discussion, we have the following definition.

Definition 2.9 A *vector field* on a subset $\mathcal{D}$ of the plane, $\mathcal{D} \subset \mathbb{R}^2$, is a function $\mathbf{F} : \mathcal{D} \to \mathbb{R}^2$. In coordinates, $\mathbf{F}(x, y) = (u(x, y), v(x, y))$, where u and v are real-valued functions, $u : \mathcal{D} \to \mathbb{R}$ and $v : \mathcal{D} \to \mathbb{R}$.

If the vector field consists of velocity vectors of a flow, we call it a *velocity field*. ◆

To "plot" a vector field, we will plot sufficiently many of the vectors $\mathbf{F}(x, y)$ to indicate the qualitative behavior of the corresponding flow. Naturally, when we plot a particular vector $\mathbf{F}(x, y)$, we will place the arrow that represents the vector so that its initial point is located at (x, y). The plots of vector fields that you see in the figures are computer generated. In order to keep the plots from becoming hopelessly cluttered, the plotting software plots a regular pattern or grid of vectors and scales the lengths of the vectors so that the vectors do not overlap. The following example illustrates how we might interpret the plot of a vector field.

Example 2.21

The Jet Stream. The jet stream is a fast-moving current of air high in the atmosphere that determines large-scale weather patterns. Figure 2.22 contains a plot of the velocity of

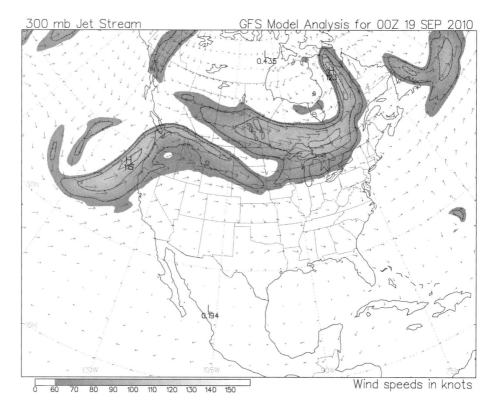

Figure 2.22 A plot of the velocity vector field of the jet stream over North America. Courtesy of the Meteorology program at San Francisco State University. (See Example 2.21.)

the flow of air in the upper atmosphere over North America. The jet stream corresponds to the swatch of longer vectors that meanders from west to east across Canada and the northeastern United States. From the plot, we can see that it moves north over the Pacific Ocean, dips south over British Columbia, moves north to the Arctic, south over Hudson Bay into the United States, and then sweeps north along the Eastern seaboard. The maximum speed of the jet stream, which is represented by the longest vectors, is 145 knots per hour.

We will also find it useful to work with a symbolic expression for the coordinates of a vector field. For example, the following gives an idealized model for the flow of fluid in a pipe.

Example 2.22 **Flow Through a Pipe**. When fluid moves through a pipe at a constant rate, that is, at a constant volume per unit time, the velocity of the fluid is not constant. Because fluid adheres to the walls of the pipe, fluid near the center of the pipe moves more rapidly than fluid near the walls. A flow of this type is called a ***Poiseuille flow***. If we choose a coordinate system with the z-axis aligned with the center of the pipe, there is a simple symbolic expression for the velocity vector field of the flow. If the pipe has radius r_0, the speed of the fluid at the center of the pipe is v_0, and the fluid is moving in the direction of positive z direction, the velocity field of the flow is

$$\mathbf{F}(x, y, z) = \left(0, 0, v_0(1 - \frac{x^2 + y^2}{r_0^2})\right)$$

for $x^2 + y^2 \leq r_0^2$. The speed of the fluid at a point (x, y, z) is $\|\mathbf{F}(x, y, z)\|$. In this case, $\|\mathbf{F}(x, y, z)\| = v_0(1 - (x^2 + y^2)/r_0^2)$.

Flow Lines of Vector Fields

A useful way to understand the motion of a fluid is to think about the motion of an individual particle in the flow. This time, imagine placing a cork in a moving stream at point $P = (x_0, y_0)$ at time $t = 0$. The cork will move downstream with the flow, and the velocity of the cork at a point will be the velocity of the fluid at the point. If the motion of the cork is parametrized by a function α, its velocity at time t is $\alpha'(t)$. Since the cork is located at $\alpha(t) = (x(t), y(t))$, the vector $\alpha'(t)$ must equal $\mathbf{F}(x(t), y(t))$. Focusing on the mathematical content of this discussion, we are led to the following definition.

Definition 2.10 A ***flow line*** of a vector field $\mathbf{F}(x, y) = (u(x, y), v(x, y))$ with ***initial point*** (x_0, y_0) is a parametrization $\alpha : [0, b] \to \mathbb{R}^2$ such that $\alpha(0) = (x_0, y_0)$ and

$$\alpha'(t) = \mathbf{F}(\alpha(t)), \quad t \in [0, b].$$

In coordinates, if $\alpha(t) = (x(t), y(t))$, then $\alpha(0) = (x(0), y(0)) = (x_0, y_0)$ and

$$(x'(t), y'(t)) = (u(x(t), y(t)), \ v(x(t), y(t))). \ \blacklozenge$$

Intuitively, if α is a flow line of $\mathbf{F}$, the tangent vector α' must be a vector of the vector field. In order to verify that a parametrization α is a flow line of $\mathbf{F}$, it is necessary to show that $\alpha'(t) = \mathbf{F}(\alpha(t))$ for all $t \in [0, b]$. That is, we must show that the coordinates of α satisfy the equations

$$x'(t) = u(x(t), y(t)) \quad \text{and} \quad y'(t) = v(x(t), y(t)).$$

Let us apply these ideas to a flow in the plane.

Example 2.23   **A Circular Flow.** Here we investigate the flow lines of the vector field $\mathbf{F}(x, y) = (-y, x)$. A plot of $\mathbf{F}$ near the origin is contained in Figure 2.23(a).

A. **An Intuitive Argument.** Imagine that $\mathbf{F}$ is the velocity field of a fluid flow. Based on the plot, we might conjecture that the fluid is circulating around the origin in a counterclockwise manner. This is supported by a simple calculation. The dot product of $\mathbf{F}(x, y) = (-y, x)$ and the direction vector (x, y) from the origin is always zero, $(-y, x) \cdot (x, y) = 0$. This says that $\mathbf{F}$ is always perpendicular to the vector (x, y), so that at each location in the plane the fluid is moving in a direction perpendicular to the direction to the origin, which is the case when a particle moves in a circle about the origin. (See Figure 2.23(b).)

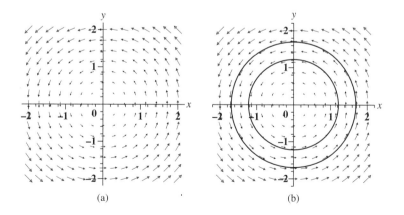

(a)   (b)

Figure 2.23   (a) The vector field $\mathbf{F}(x, y) = (-y, x)$. (b) Flow lines of the vector field $\mathbf{F}(x, y) = (-y, x)$. (See Example 2.23.)

B.   A Calculation. Our conjecture is, in fact, correct and can be verified symbolically. Define, for example, $\alpha(t) = (R\cos t, R\sin t)$, for $t \in \mathbb{R}$, where $R > 0$. We know from our work with curves that α is a constant speed parametrization of a circle of radius R centered at the origin. It remains to show that α satisfies the equations of a flow line of $\mathbf{F}$. First, $\alpha'(t) = (-R\sin t, R\cos t)$. Substituting $\alpha(t)$ into the formula for $\mathbf{F}$, we have

$$\mathbf{F}(\alpha(t)) = (-y(t), x(t))$$
$$= (-R\sin t, R\cos t)$$
$$= \alpha'(t).$$

Further, $\alpha(0) = (R, 0)$. These calculations show that α is a flow line of $\mathbf{F}$ with initial point $\alpha(0) = (R, 0)$.

As the example demonstrates, given formulas for α and $\mathbf{F}$, it is possible to check if α is a flow line of $\mathbf{F}$. We need only substitute the formulas for $x = x(t)$ and $y = y(t)$ into the formulas for u and v and check that the resulting expressions are equal to x' and y'. Unfortunately, it is more often the case that we need to analyze flow lines of $\mathbf{F}$ knowing only the formulas for u and v. In order to find x and y, we have to solve the differential equations

$$x'(t) = u(x(t), y(t))$$
$$y'(t) = v(x(t), y(t))$$

for x and for y. We will call these equations the *flow line equations* of the vector field $\mathbf{F}$. If the expressions for u and v are sufficiently simple, it is possible to solve these equations explicitly using ideas from calculus. The next example demonstrates this for the vector field $\mathbf{F}(x, y) = (2x, y)$.

Example 2.24

A Symbolic Solution to the Flow Line Equations. Let $\mathbf{F}(x, y) = (2x, y)$. A plot of $\mathbf{F}$ is given in Figure 2.24. Let us find the flow line α of $\mathbf{F}$ whose initial point is $(\frac{1}{4}, \frac{1}{2})$. The coordinate functions $x = x(t)$ and $y = y(t)$ of α must satisfy $\alpha(0) = (x(0), y(0)) = (\frac{1}{4}, \frac{1}{2})$ and the flow line equations for $\mathbf{F}$:

$$x'(t) = 2x(t)$$
$$y'(t) = y(t).$$

Each one is a differential equation for exponential growth. To solve these equations, we use the fact from one variable calculus that if a function f satisfies the differential equation $f'(t) = kf(t)$ for $k \neq 0$, then $f(t) = f(0)e^{kt}$. The solution to the first equation

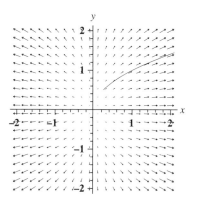

Figure 2.24 The vector field $\mathbf{F}(x, y) = (2x, y)$ and the flow line with initial point $(\frac{1}{4}, \frac{1}{2})$. (See Example 2.24.)

is $x(t) = x(0)e^{2t}$, so that $x(t) = \frac{1}{4}e^{2t}$, and the solution to the second equation is $y(t) = y(0)e^t$ or $y(t) = \frac{1}{2}e^t$. The flow line is given by

$$\alpha(t) = (\tfrac{1}{4}e^{2t}, \tfrac{1}{2}e^t).$$

Notice that as t approaches ∞, each of the coordinates of α approaches ∞. In this case, the x-coordinate of α is the square of the y-coordinate, $x(t) = y(t)^2$. This means that the image of the flow line α traces a portion of the graph of the parabola $x = y^2$. The image of α is shown in Figure 2.24.

For most vector fields, it is not possible to find symbolic solutions to the flow line equations. Nevertheless, we can use intuitive graphical techniques to sketch the curves traced by flow lines. The idea is to sketch a curve starting at the initial point whose tangent vectors are vectors of the vector field. Since vector field plots contain a relatively small number of vectors, this is a rough process when done by hand. It will not give us quantitative information, that is, it will not tell us the location of $\alpha(t)$ for particular values of t, but it is potentially useful when trying to sort out the qualitative behavior of a flow. We can also use computer software packages to generate pictures of flow lines that give qualitative information about the flow.

Example 2.25

Computer-Generated Flow Lines. Consider the vector field $\mathbf{F}(x, y) = (-y + 0.1(x^2 - 1), x^2 - 1 + 0.1y)$, which is plotted in Figure 2.25(a). The vectors of the field near $(1, 0)$ indicate a counterclockwise flow around this point. Figure 2.25(b) is a plot of this vector field along with the computer-generated flow lines for the initial points $(1.5, 0)$, $(0, 1.2)$, $(0, 1.044)$, and $(-1.75, -1.25)$. Initially, the flow line beginning at $(1.5, 0)$ circles the point $(1, 0)$ in a counterclockwise manner. However, as t increases, the flow line moves into the

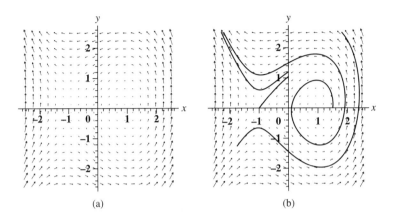

(a)   (b)

Figure 2.25   (a) The vector field $\mathbf{F}(x, y) = (-y + 0.1(x^2 - 1), x^2 - 1 + 0.1y)$. (b) Flow lines of $\mathbf{F}$ with initial points $(1.5, 0)$, $(0, 1.2)$, $(0, 1.044)$, and $(-1.75, -1.25)$. (See Examples 2.25 and 2.26.)

second quadrant and eventually moves toward ∞. The flow line beginning at $(0, 1.2)$ begins by moving toward the point $(-1, 0)$ but then moves away from this point and off toward ∞ in the second quadrant. The flow line beginning at $(0, 1.044)$ appears to approach the point $(-1, 0)$. The flow line beginning at $(-1.75, -1.25)$ moves toward the point $(-1, 0)$ and then moves off into the fourth quadrant, following the first flow line. Since the length of $\mathbf{F}$ increases further from the origin, the flow lines are traversed more rapidly further from the origin. Notice that each of the flow lines is tangent to very few of the plotted vectors $\mathbf{F}(x, y)$. Nevertheless, the flow lines appear to closely follow the directions of the plotted vectors.

As useful as a computer plot of flow line can be, the plot itself does not give the quantitative information of the location of $\alpha(t)$ for particular t values. Depending on the software, however, it is possible to generate values for the coordinates of $\alpha(t)$ for particular t.

Critical Points of Vector Fields

In trying to understand the behavior of a flow, points where the velocity of the flow are zero play a special role. First, let us make a definition.

Definition 2.11   A *critical point* or *equilibrium point* of a vector field $\mathbf{F}(x, y)$ is a point (x_0, y_0) such that $\mathbf{F}(x_0, y_0) = (0, 0)$.   ◆

Intuitively, a critical point of fluid flow corresponds to a point where the fluid is not moving. If we were to place a cork in a fluid flow at a critical point, it would remain at that point. For this reason, critical points are also called equilibrium points. More formally, if (x_0, y_0) is a critical point of $\mathbf{F}$, we can define a parametrization α by $\alpha(t) = (x_0, y_0)$ for all t. Then $\alpha'(t) = (0,0)$ for all t. Substituting into the flow line equations, we have $\mathbf{F}(\alpha(t)) = \mathbf{F}(x_0, y_0) = (0,0) = \alpha'(t)$. This says that α is a flow line of $\mathbf{F}$ for the initial point (x_0, y_0). Of course, this α describes the motion of an object at rest at the point (x_0, y_0). In contrast, if we drop a cork *near* a critical point, its subsequent behavior can be quite interesting. Let's return to Example 2.25 for a moment.

Example 2.26

The Flow near a Critical Point. Consider the flow that gives rise to the vector field $\mathbf{F}(x, y) = (-y + 0.1(x^2 - 1), x^2 - 1 + 0.1y)$ of Example 2.25. To find where $\mathbf{F}$ is $(0,0)$, we must solve the equations

$$-y + 0.1(x^2 - 1) = 0$$
$$x^2 - 1 + 0.1y = 0.$$

Solving the first equation for y, we get $y = 0.1(x^2 - 1)$. Substituting this into the second equation, we get $x^2 - 1 + 0.1(0.1(x^2 - 1)) = 0$, which simplifies to $1.01(x^2 - 1) = 0$. The solutions to this equation are $x = 1$ and $x = -1$. Substituting these into the first equation and solving for y, we find that this vector field is zero at $(1, 0)$ and $(-1, 0)$. If we were to place a cork in the flow at either of these points, it would remain there forever. From Example 2.25, if we were to place a cork in the flow at $(0, 1.044)$ it would approach the critical point at $(-1, 0)$. If we were to place the cork at nearby points, for example, $(0, 1.2)$ the cork would approach $(-1, 0)$ and then veer off. In contrast, if we put a cork into the flow at any point near $(1, 0)$, it would trace a spiral-like curve around $(1, 0)$ and then move off to infinity. (See Figure 2.25(b).)

Intuitively, if we choose an initial point near a critical point, the long-term behavior of the flow line falls into two categories. It might be the case that the flow line remains close to the critical point for all time, or it might be the case that the flow line eventually moves away from the critical point. These two types of behavior can also be used to assign critical points to two categories, that is, to *classify* critical points. We will say that a critical point is *stable* if every flow line that starts near the critical point remains near the critical point for all time. If this is not the case, we say that a critical point is *unstable*. In other words, a critical point is unstable if for at least some initial points near the critical point, the flow lines eventually move away from the critical point.

Returning to our cork–fluid analogy, if a cork is placed in a fluid flow near a stable critical point, it will remain near the critical point for all time. On the other hand, it can be placed near an unstable critical point so that it will eventually move away from the critical point.

In order to determine if a critical point is stable or unstable, we will plot flow lines for initial points near the critical point. In most of the examples we will consider, it will be sufficient to examine only a few flow lines in order to make a determination about the critical point.

| Example 2.27 | **Stable and Unstable Critical Points** |

A. The vector field $\mathbf{F}(x, y) = (-y, x)$ of Example 2.23 has only one critical point, which is located at the origin, $(0, 0)$. The flow lines of $\mathbf{F}$ are circles centered at $(0, 0)$ that are traversed counterclockwise. Since flow lines that start near $(0, 0)$ remain near $(0, 0)$ for all time, it is a stable critical point.

B. The vector field $\mathbf{F}(x, y) = (2x, y)$ of Example 2.24 also has a critical point at the origin. In this case, since the flow lines of $\mathbf{F}$ that begin near the origin flow away from the origin, it is an unstable critical point. (See Exercise 3.)

C. In Examples 2.25 and 2.26, we considered the vector field $\mathbf{F}(x, y) = (-y + 0.1(x^2 - 1), x^2 - 1 + 0.1y)$. It has two critical points, $(-1, 0)$ and $(1, 0)$. We saw that some flow lines that start near $(-1, 0)$ flow away from the point, while one flow line appears to flow toward it. On the other hand, all flow lines that start near $(1, 0)$ flow away from the point. (See Figure 2.25(b).) It follows that each critical point is unstable.

D. Consider the vector field, $\mathbf{F}(x, y) = (-2x - y, 4x - 7y)$, which we have not considered before. If we solve the system of equations

$$-2x - y = 0$$

$$4x - 7y = 0,$$

we see that there is a single critical point at $(0, 0)$. Figure 2.26 shows the flow lines of $\mathbf{F}$ for the initial points $(0.5, 2)$, $(0.5, -2)$, $(-0.5, -2)$, $(2, 0)$, $(-2, 0)$, and $(-1, 2)$. Each flow line moves toward the critical point at the origin. Since all the vectors near the origin point toward the origin, it is reasonable to conclude that every flow line of $\mathbf{F}$ that starts near the origin will move toward the origin. Thus we conclude that $(0, 0)$ is a stable critical point.

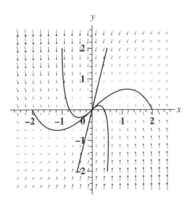

Figure 2.26   Flow lines of the vector field $\mathbf{F}(x, y) = (-2x - y, 4x - 7y)$. (See Example 2.27D.)

Based on Example 2.27, we can refine our classification of critical points by dividing each category, stable and unstable, into two subcategories. In Example 2.27A, the flow lines of $\mathbf{F}(x,y) = (-y,x)$ near the stable critical point at the origin circle the origin, but do not flow in toward the origin. In Example 2.27D, the flow lines near the stable critical point at the origin all flow toward the origin. For unstable critical points, we also see two options. In Example 2.27C, all the flow lines that start near $(1,0)$ flow away from $(1,0)$, while only some of the flow lines that start near $(-1,0)$ flow away from $(-1,0)$. These distinctions are summarized in the following definition.

Definition 2.12   A stable critical point is called a ***sink*** if all flow lines that begin near the critical point flow toward the critical point. A stable critical point is called a ***center*** if all flow lines that begin near the critical point trace closed curves around the critical point.

An unstable critical point is called a ***source*** if all flow lines that begin near the critical point flow away from the critical point. An unstable critical point is called a ***saddle*** if some flow lines that begin near the critical point flow away from the critical point and others flow toward the critical point.   ◆

Using this language, we see that the stable critical point of $\mathbf{F}(x,y) = (-y,x)$ at $(0,0)$ is a center and the stable critical point of $\mathbf{F}(x,y) = (-2x-y, 4x-7y)$ at $(0,0)$ is a sink. The unstable critical point of $\mathbf{F}(x,y) = (-y + 0.1(x^2-1), x^2-1+0.1y)$ at $(1,0)$ is a source, and the unstable critical point at $(-1,0)$ is a saddle.

Summary

The discussion in this section was motivated by an intuitive understanding of the behavior of a moving fluid, that is, of a ***fluid flow***. Focusing on flows in the plane, we defined a ***vector field*** to be a function $\mathbf{F}$ that assigns a vector, $\mathbf{F}(x,y) = (u(x,y), v(x,y))$ to points (x,y) in the plane. A ***flow line*** α of a vector field $\mathbf{F}$ is a parametrization whose tangent vector α' is a vector of the field $\mathbf{F}$ at all times. If the vector field is the velocity field of a fluid flow, a flow line describes the motion of a particle placed in the flow at the initial point of the flow line.

To understand the behavior of a vector field on an intuitive or geometric level, we studied the behavior of flow lines that start near ***critical*** or ***equilibrium points*** of the vector field. These are points (x,y) where the field is zero, that is, where $\mathbf{F}(x,y) = (0,0)$. Based on the behavior of nearby flow lines, we ***classified*** critical points into two categories, ***stable*** and ***unstable***. Then we further classified stable critical points into ***sinks*** and ***centers*** and unstable critical points into ***sources*** and ***saddles***.

Section 2.4 Exercises

1. **Flow Lines of Vector Fields.** For each of the following vector fields, proceed as follows.
   (i) Sketch the flow lines starting at the initial points $(\pm 1, 0)$, $(0, \pm 1)$, $(1, \pm 1)$, and $(-1, \pm 1)$. (These may also be plotted using a computer algebra system, if one is available

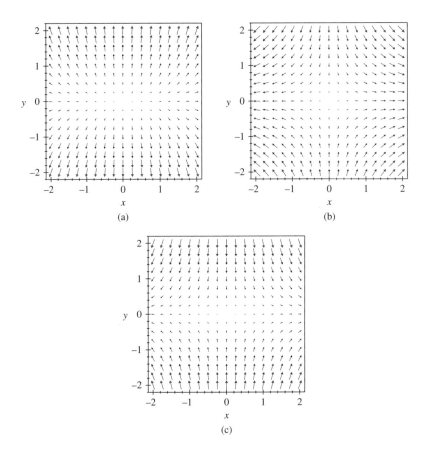

Figure 2.27 Vector fields for Exercise 1 (a)–(c).

to you.) (ii) Use your sketch to classify the critical point of **F** at the origin as stable or unstable and as a center, sink, source, or saddle.

(a) $\mathbf{F}(x, y) = (x, 3y)$. (See Figure 2.27(a).)
(b) $\mathbf{F}(x, y) = (x, -y)$. (See Figure 2.27(b).)
(c) $\mathbf{F}(x, y) = (x, -3y)$. (See Figure 2.27(c).)

2. **Verifying Parametrizations of Flow Lines.** For each of the following vector fields and parametrizations, verify by a direct calculation that the function α parametrizes a flow line of the vector field and find the initial point of the flow line.

(a) $\mathbf{F}(x, y) = (3x, 2y)$.

  (i) $\alpha(t) = (e^{3t}, e^{2t})$.
  (ii) $\alpha(t) = (-2e^{3t}, 4e^{2t})$.

(b) $\mathbf{F}(x, y) = (x - 1, y)$.

   (i) $\alpha(t) = (2e^t + 1, 3e^t)$.
   (ii) $\alpha(t) = (1 - e^t, -e^t)$.

3. **Symbolic Solutions to the Flow Line Equations.** Let $\mathbf{F}(x, y) = (2x, y)$ be the vector field of Example 2.24.

   (a) Following the method of Example 2.24, use techniques from one-variable calculus to construct parametrizations of the flow lines of $\mathbf{F}$ with initial points at $(-2, 1)$, $(-1, 3)$, and $(0, 1)$.
   (b) Each of the flow lines from part (a) traces a portion of the graph of a function of y. For each flow line in (a), write out the expression for this function in the form $x = g(y)$.
   (c) Does every flow line of $\mathbf{F}$ trace part of the graph of a function $x = g(y)$? Explain your answer.

4. **Flow Lines and Initial Points.** In each of the following parts, the general form for a flow line α of the vector field $\mathbf{F}$ is given in terms of constants A and B. Verify by a direct calculation that the function α parametrizes a flow line of the vector field. For each of the given initial points (x_0, y_0), find the values of A and B that give the flow line with (x_0, y_0) as the initial point. (*Hint:* Set $\alpha(0) = (x_0, y_0)$ and solve for A and B.)

   (a) $\mathbf{F}(x, y) = (-x, 2y)$. $\alpha(t) = (Ae^{-t}, Be^{2t})$. Initial points: (i) $(x_0, y_0) = (2, 1)$.
   (ii) $(x_0, y_0) = (-3, 3)$.
   (b) $\mathbf{F}(x, y) = (x + 2, y + 1)$. $\alpha(t) = (Ae^t - 2, Be^t - 1)$. Initial points: (i) $(x_0, y_0) = (1, 1)$.
   (ii) $(x_0, y_0) = (0, 1)$.
   (c) $\mathbf{F}(x, y) = (-2x + y, -x)$. $\alpha(t) = (Ae^{-t} + Bte^{-t}, (A + B)e^{-t} + Bte^{-t})$. Initial points:
   (i) $(x_0, y_0) = (1, 1)$. (ii) $(x_0, y_0) = (1, -1)$.
   (d) $\mathbf{F}(x, y) = (2x + 2y, -x)$. $\alpha(t) = (Ae^t(\cos t + \sin t) + Be^t(\sin t - \cos t),$
   $-Ae^t \sin t + Be^t \cos t)$. Initial points: (i) $(x_0, y_0) = (1, 0)$. (ii) $(x_0, y_0) = (-1, 1)$.

5. **Critical Points of Vector Fields.** Determine the critical points of each of the following vector fields on the indicated domain, and based on the accompanying plot, determine the type of the critical points. (If a computer algebra system is available to you, you may find it helpful to plot several flow lines of the given vector fields.)

   (a) $\mathbf{F}(x, y) = (2x + y - 1, -y + x)$, $[-1.5, 1.5] \times [-1.5, 1.5]$. (See Figure 2.28(a).)
   (b) $\mathbf{F}(x, y) = (xy - x, xy - y)$, $[-1.5, 1.5] \times [-1.5, 1.5]$. (See Figure 2.28(b).)
   (c) $\mathbf{F}(x, y) = (\cos(x - y), xy)$, $[-2, 2] \times [-2, 2]$. (See Figure 2.28(c).)

6. **Exponential Functions and Flow Lines.** From one-variable calculus, we know that the function $f(t) = Ae^{kt}$ is the unique solution to the differential equation $f'(t) = kf(t)$ subject to the initial condition $f(0) = A$.

   (a) Use this fact to construct the flow line of the vector field $\mathbf{F}(x, y) = (3x, 2y)$ with initial point $(x_0, y_0) = (1, -2)$.

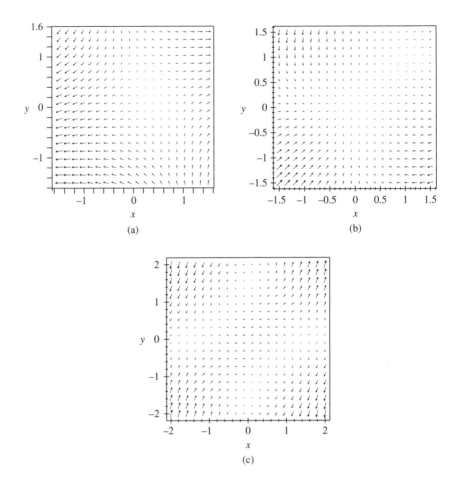

Figure 2.28   Vector fields for Exercise 5.

(b) Use this fact to construct the general form for a flow line of the vector field $\mathbf{F}(x, y) = (kx, ly)$, where k and l are constants, with initial point (x_0, y_0).

(c) Give a qualitative description of the behavior of the flow lines of $\mathbf{F}(x, y) = (kx, ly)$ as $t \to \infty$, if (i) $k > 0$, $l > 0$, (ii) $k > 0$, $l < 0$, (iii) $k < 0$, $l > 0$, and (iv) $k < 0$, $l < 0$.

7. **Shifting and Scaling.** In Example 2.23, we saw that $\alpha(t) = (R \cos t, R \sin t)$ is a flow line of the vector field $\mathbf{F}(x, y) = (-y, x)$.

(a) Verify that $\beta(t) = \alpha(t + k)$, where k is a constant is also a flow line of $\mathbf{F}$.

(b) Suppose that α is replaced by $\beta(t) = \alpha(ct)$, where c is a constant, $c \neq 0$, 1. Is β a flow line of $\mathbf{F}$? Why or why not?

8. **Shifting and Scaling: The General Case.** Suppose that $\alpha(t) = (x(t), y(t))$ is a flow line of the vector field $\mathbf{F}(x, y) = (u(x, y), v(x, y))$ with initial point (x_0, y_0).

   (a) Suppose that α is replaced by $\beta(t) = \alpha(t + k)$, where k is a constant. Is β a flow line of $\mathbf{F}$? Why or why not? If so, what is the initial point of β? If not, is β the flow line of a vector field related to $\mathbf{F}$?

   (b) Suppose that α is replaced by $\beta(t) = \alpha(ct)$, where c is a constant, $c \neq 0$, 1. Is β a flow line of $\mathbf{F}$? Why or why not? If so, what is the initial point of β? If not, is β the flow line of a vector field related to $\mathbf{F}$?

9. **The Vector Field $-\mathbf{F}$.** Consider the vector field $\mathbf{F}(x, y) = (2x, y)$ of Example 2.24 and the vector field $\mathbf{G}(x, y) = -\mathbf{F}(x, y) = (-2x, -y)$.

   (a) Sketch the vector field $\mathbf{G}$ and several of its flow lines. How do the vector fields $\mathbf{F}$ and $\mathbf{G}$ differ?

   (b) What, if any, is the relationship between the flow lines of $\mathbf{F}$ and $\mathbf{G}$? Explain. (These may also be plotted using a computer algebra system, if one is available to you.)

   (c) What, if any, is the relationship between the critical points of $\mathbf{F}$ and the critical points of $\mathbf{G}$?

10. **$-\mathbf{F}$, the General Case.** Consider an arbitrary vector field $\mathbf{F}(x, y) = (u(x, y), v(x, y))$ and the vector field $\mathbf{G}(x, y) = (-u(x, y), -v(x, y))$.

    (a) How do the vector fields $\mathbf{F}$ and $\mathbf{G}$ differ?

    (b) What, if any, is the relationship between the flow lines of $\mathbf{F}$ and $\mathbf{G}$? Explain.

11. **Critical Points of $-\mathbf{F}$.** Suppose that $\mathbf{F}(x, y)$ has a critical point at (x_0, y_0).

    (a) Show that $\mathbf{G}(x, y) = -\mathbf{F}(x, y)$ also has a critical point at (x_0, y_0).

    (b) For each of the types for the critical point of $\mathbf{F}$ at (x_0, y_0) (sink, center, source, saddle), what is the type of the critical point of $\mathbf{G}$ at (x_0, y_0)? Explain.

12. **Intersecting Flow Lines?** Let α and β be parametrizations of two different flow lines of a vector field $\mathbf{F}$. Is it possible for the image of α and the image of β to intersect? Explain why or why not.

13. **Limit Cycles.** Define a vector field $\mathbf{F}$ by

$$\mathbf{F}(x, y) = (x \sin(\pi\sqrt{x^2 + y^2}) + y \cos(\pi\sqrt{x^2 + y^2}),$$
$$y \sin(\pi\sqrt{x^2 + y^2}) - x \cos(\pi\sqrt{x^2 + y^2})).$$

   (a) Show that unit circle parametrized $\alpha(t) = \cos(\pi t), \sin(\pi t))$ is a flow line of $\mathbf{F}$.

   (b) $\mathbf{F}$ has a critical point at the origin. Use a computer algebra system to plot $\mathbf{F}$. What is the type of critical point at the origin?

   (c) Use a computer algebra system to plot flow lines of $\mathbf{F}$ with initial point near the origin. The unit circle is called a *limit cycle* for this vector field. How does your polot justify this terminology? (*Hint:* Be sure to use a sufficiently large time interval.)

■ 2.5 Modeling with Vector Fields

In Section 2.4, we used fluid flows to motivate our discussion of vector fields, flow lines, and critical points. Here we would like to turn things around and use the mathematical objects to help us to think about real-world phenomena. For example, we already know that the velocity field of a fluid flow can be used to understand the motion of the fluid. In this section, we will use vector fields to construct mathematical models of two "real-world" systems that evolve in time: a predator–prey model and a pendulum model.

In each case, we will use two quantities to describe the behavior of the system. We will think of these quantities as coordinates of a coordinate system in the plane. For now, call them x and y. Since x and y depend on time, we will write them as functions of time, $x = x(t)$ and $y = y(t)$. If we know values of x and y at time t, we will say that we know the **state of the system at time** t. In each case, we will start with enough information to write out formulas for x' and y', the rates of change of x and y, but not with enough information to write out formulas for x and y.

In order to proceed, we will have to make an important assumption about x' and y'. We will assume that at time t the rates of change x' and y' can be expressed in terms of the values of x and y at time t. We express this by writing x' and y' as functions of x and y, which in turn depend on t:

$$x'(t) = u(x(t), y(t))$$
$$y'(t) = v(x(t), y(t)).$$

These are the flow line equations for the vector field $\mathbf{F}$ whose coordinate functions are u and v, $\mathbf{F}(x, y) = (u(x, y), v(x, y))$. We will call the differential equations for x' and y' a **mathematical model** for the physical system described by x and y.

As in Section 2.4, a flow line α for the initial point (x_0, y_0) will tell us what happens to the quantities x and y if they start at the values $x = x_0$ and $y = y_0$ and time is allowed to evolve. Since x and y describe the original physical system, knowledge of the flow line starting at (x_0, y_0) will tell us what happens to the system when it starts in the state (x_0, y_0). For example, if (x_0, y_0) is a critical point of $\mathbf{F}$, we know that a flow line that starts at (x_0, y_0) remains there for all time since its velocity is always zero. In physical terms, (x_0, y_0) describes an **equilibrium state of the system**. If the system starts in an equilibrium state, it remains there for all time. As we explore the models in this section, we will see that the type of a critical point has important implications for the behavior of the physical system.

As we did in Section 2.3, we will describe the physical system in some detail before we construct our model. This is necessary in order for us to be able to draw meaningful conclusions about the physical system.

Predator–Prey Model

Here we want to construct a mathematical model for the population sizes of two species that interact as predators and their prey. (See the collaborative exercise at the beginning of Chapter 2.) For example, we might think of sharks and fish or coyote and prairie dogs. Let $x = x(t)$ represent the number of prey and $y = y(t)$ represent the number of predators. We will begin by making several assumptions about these populations and their interactions. For each rate of change, we will end up with a sum of positive and negative terms. The positive terms represent factors that cause the population to increase, and the negative terms represent factors that cause the population to decrease.

Our primary assumption is that the only factors affecting the rates of growth of predator and prey can be expressed in terms of the number of predators and the number of prey. This means that we will neglect interactions with other species or with the surrounding environment. Now let us focus on x', the rate of growth of the prey population.

We will assume that in the absence of predators, the number of prey grows at a rate proportional to the number of prey,

$$x'(t) = ax(t).$$

This is the differential equation that describes exponential growth. If $x(0) = x_0$, the solution to this equation is given by $x(t) = x_0 e^{at}$. Thus in the absence of predators, we are assuming that the prey population will grow exponentially. In the presence of predators, the rate of change of the prey must reflect the number of incidents of predation in a unit of time. Intuitively, we would expect that there will be more contacts between predator and prey, hence more predation, if either the number of prey increases or the number of predators increases. The simplest way to express this is to say that the number of contacts is proportional to the product of the number of the predators and the number of prey. Since this should cause a decrease in the rate of growth of the prey, we conclude that the presence of predators contributes a term of the form $-bxy$ to the total rate of change of the prey. Assuming that there are no other factors affecting the rate of change of the prey population, we have

$$x'(t) = ax(t) - bx(t)y(t),$$

where a and b are positive constants.

The situation for predators is somewhat different. The predator population should increase as the number of contacts between predator and prey increases. Thus the expression for y' should include a term of the form dxy, where d is a positive constant. On the other hand, since predators compete for the same food source—the prey—the more predators there are, the less food that is available for each predator, thus driving down the population of predators. Thus the rate of change of predators should also contain a negative term that increases as the predator population increases. We will assume that this term is proportional to the predator population and is of the form $-cy$, where c is a positive constant. Assuming there are no other factors affecting the rate of change of the predator population, we have

$$y'(t) = -cy(t) + dx(t)y(t),$$

where c and d are positive constants.

Below is a summary of this discussion.

The Predator– *Prey Model*	Denote the prey population by $x = x(t)$ and the predator population by $y = y(t)$. The **predator–prey** model for the interaction of the two species is given by the pair of differential equations for the rate of change of x and the rate of change of y, $$x'(t) = \ \ ax(t) - bx(t)y(t)$$ $$y'(t) = -cy(t) + dx(t)y(t),$$ where a, b, c, and d are positive constants.

It is important to note that these equations are **autonomous**. That is, the time variable does not appear other than as the independent variable for x and y. For example, in one variable, the differential equation $x'(t) = (t + 1)x(t)$, is **nonautonomous**. Autonomous systems of equations are represented by vector fields that do not vary in time. The vector field **F** corresponding to the predator–prey equations is

$$\mathbf{F}(x, y) = (ax - bxy, -cy + dxy).$$

Since x and y must be nonnegative, we will focus on the first quadrant. A flow line α of **F** with initial point $\alpha(0) = (x_0, y_0)$ will represent the behavior of the two populations for all time if the initial value of the prey population is x_0 and the initial value of the predator population is y_0.

In the next example, we assign numerical values to the coefficients so that we can analyze the plot and flow lines of **F**.

Example 2.28   **A Predator-Prey Vector Field.** Figure 2.29 is a plot of the predator–prey vector field $\mathbf{F}(x, y) = (1.5x - xy, -3y + xy)$, where the prey or x-axis is marked in units of 1000 and the predator or y-axis is marked in units of 100. The initial point of the flow line is $(1, 2)$, which corresponds to 1000 prey and 200 predators.

Beginning with populations of 1000 prey and 200 predators, we see that the predator population decreases initially while the prey population remains fairly stable. This indicates that there were insufficient prey to support a population of 200 predators. The prey population begins to increase in response to the decrease in the predator population, when the predator population dips below 100. The prey population continues to increase until approximately 6800. During this period, the predator population decreases until the

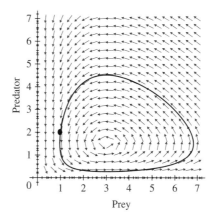

Figure 2.29 The predator–prey model of Example 2.28. The horizontal axis is marked in units of 1000 and the vertical axis is marked in units of 100. The flow line shown is for the initial conditions $(x_0, y_0) = (1, 2)$ or 1000 prey and 200 predators. (Note the vectors are not to scale.)

prey population reaches approximately 3000, when it begins to respond to the increase in prey. The increase in predators is at first gradual, but then becomes more rapid. At the point of maximum prey population, the predator population is approximately 150. After this point, the predator population continues to increase, which causes the prey population to decrease. As the prey population decreases, the predator population increases more slowly and reaches a maximum of approximately 440, when it begins to decrease. At this time the prey population is down to 3000 and continues to decrease even as the number of predators decrease. Both populations continue to decrease and return to their initial values and the cycle begins again.

Notice that the predator and prey populations change in response to each other but that there is a time delay in the response of one population to changes in the other.

If we examine the plot in Figure 2.29, we can see that the vector field $\mathbf{F}$ appears to have a critical point when there are approximately 3000 prey and 150 predators. Let us return to the general equations to analyze the critical points of $\mathbf{F}$. The critical points are solutions to the pair of equations given by $\mathbf{F}(x, y) = (0, 0)$:

$$ax - bxy = 0$$
$$-cy + dxy = 0,$$

where a, b, c, and d are positive constants. Since $ax - bxy = x(a - by)$, either $x = 0$ or $y = a/b$. If $x = 0$, then from the second equation we must also have $y = 0$. If $y = a/b$, then from the second equation, $x = c/d$. We have found two critical points, $(0, 0)$ and $(c/d, a/b)$. In the above example, we speculated on the presence of the second critical point, $(3/1, 1.5/1) = (3, 1.5)$, which corresponds to 3000 prey and 150 predators.

Let us analyze the critical point at the origin. If there are no predators initially, we would expect the prey population to increase exponentially. In fact, if $y_0 = 0$, then the flow line α of $\mathbf{F}$ with initial point $(x_0, 0)$ is given by $\alpha(t) = (x_0 e^{at}, 0)$, which parametrizes the x-axis in the direction of increasing x. No matter how small the initial value of x_0, the flow line will eventually move away from the origin. This means that the origin is an unstable critical point of $\mathbf{F}$. On the other hand, if there are no prey initially, it can be shown that the number of predators will decrease to zero. (See Exercise 3.) This would model a scenario when the entire prey population is eradicated, say by disease.

A symbolic analysis of the type of the second critical point is more involved, so we will instead rely on a graphical analysis. It appears in Figure 2.29 and is indeed the case that every flow line that has positive initial conditions, $x_0 > 0$ and $y_0 > 0$, is a closed curve that circles the critical point $(c/d, a/b)$. Further, if a flow line starts close to the critical point, it will remain close to the critical point for all t. We conclude that $(c/d, a/b)$ is a stable critical point and is a center. (See Exercise 1 for an analysis of these closed curves.)

In the exercises, you will have the opportunity to explore this and other models of interacting populations. Vito Volterra used the predator–prey model to explain the populations of selachians (sharks) and fish in the Mediterranean Sea during World War I.[5] It has been used to model a variety of predator–prey systems since that time. It is worth noting that the cycles that are seen in the predator–prey model occur in epidemiology where microbes are the predators and humans are the prey.[6]

The Simple Pendulum

Here we want to use the language of vector fields to reconsider the simple pendulum, which we introduced in Example 2.6 of Section 2.1. The pendulum is an example of a mechanical system, that is, it consists of a finite number of objects or parts that are subject to forces that behave according to Newton's second law. The second law says that the vector sum of all the forces acting on an object is equal to the scalar product of the mass of the object and its acceleration vector. In symbols,

$$\mathbf{F} = m\mathbf{a}.$$

(Note that we will use $\mathbf{F}$ to represent the force vector and $\mathbf{G}$ to represent the vector field of the model.) A consequence of Newton's second law is that if we are given the initial position and velocity of the objects in a mechanical system, we can, in principle, determine the behavior of the system from Newton's equation. To construct the vector field that models the system, we must first write out Newton's equation and then turn

---

[5]Vito Volterra, "Fluctuations in the Abundance of a Species Considered Mathematically," *Nature*, Vol. 118, pp. 558-560.

[6]In their comprehensive text *Infectious Diseases of Humans*, Oxford University Press, 1992, Roy M. Anderson and Robert M. May comment (p. 128) that "sustained host-microparasite cycles ... are the clearest examples of predator-prey cycles that ecologists are likely to find."

Newton's equation into a pair of equations for position and velocity. As above, we will have to make simplifying assumptions about the nature of the physical system, in this case the action of the forces, in order to write out the equations. We will illustrate this and the construction of the vector field with the simple pendulum.

We will assume that the pendulum rod has no mass, so that it can be neglected when analyzing the forces acting on the pendulum. We will also assume that the pendulum swings without friction or air resistance, so that the only force acting on the pendulum is gravity. Gravity acts vertically in a downward direction with magnitude equal to mg, where g is the scalar acceleration due to gravity. If we place the pendulum in a coordinate plane with the first coordinate representing the horizontal direction and the second coordinate representing the vertical direction, the force acting on the pendulum is given by $\mathbf{F} = (0, -mg)$. We will express the position of the pendulum in terms of its displacement x along its arc. If the pendulum has length l, $x = l\theta$, where θ is the angle the pendulum makes with the vertical. (See Figure 2.30(a).)

We are interested in the component of $\mathbf{F}$ in the direction of motion of the pendulum, which is equal to $\mathbf{F} \cdot \mathbf{t}$, where $\mathbf{t}$ is the unit tangent vector to the motion. Since $\mathbf{F} = m\mathbf{a}$, it follows that $\mathbf{F} \cdot \mathbf{t} = m\mathbf{a} \cdot \mathbf{t}$, where $\mathbf{a}$ is the acceleration of the pendulum. A calculation shows that this equality implies that

$$-g\sin(x(t)/l) = x''(t).$$

(See Exercise 5.) Since $v(t) = x'(t)$, we can rewrite this equation as a pair of first order equations in position and velocity:

$$x'(t) = v(t)$$
$$v'(t) = -g\sin(x(t)/l).$$

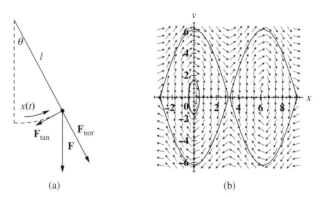

(a)   (b)

Figure 2.30 (a) The forces acting on a pendulum. (b) The vector field $\mathbf{G}$ for Example 2.29 showing flow lines with initial points $(0.5, 0)$, $(\pi - 0.1, 0)$, and $(\pi + 0.1, 0)$.

These are the equations that we have been seeking. Notice that, like the system of equations for the predator–prey system, this system is autonomous.

Let us summarize what we have done so far.

The Simple Pendulum Model	The ***simple pendulum model*** for a pendulum of length l and mass m moving under the influence of gravity and no other forces is $$x'(t) = v(t)$$ $$v'(t) = -g\sin(x(t)/l),$$ where $x = x(t)$ is the displacement of the pendulum bob along its arc and $v = v(t)$ is the velocity of the pendulum bob. The constant g is the acceleration due to gravity.

The simple pendulum equations give rise to a vector field **G** in the plane with position-velocity coordinates. It is given by

$$\mathbf{G}(x, v) = (v, -g\sin(x/l)).$$

A flow line α of **G** with initial point (x_0, v_0) models the motion of the pendulum when it is set in motion with initial displacement x_0 and initial velocity v_0. For a mechanical system, the position-velocity plane is called the ***phase plane*** of the mechanical system and a flow line is called a ***phase plane trajectory*** of the system. In the following example, we will analyze a particular **G**.

Example 2.29   **A Simple Pendulum.** Figure 2.30(b) contains a plot of **G** for a pendulum of length $l = 1$ m. (Note $g = 9.8 \text{m/s}^2$.) Since $l = 1$, $x = \theta$.

The critical points of **G** satisfy the equations

$$v = 0$$

$$-g\sin(x) = 0.$$

The second equation is satisfied when $x = k\pi$, where k is an integer. It follows that the critical points of **G** are of the form $(k\pi, 0)$ for k, an integer. These correspond to two distinct states of the pendulum: when the pendulum is positioned vertically downward and at rest and when the pendulum is positioned vertically upward and at rest.

We will use the plot in Figure 2.30(b) to determine the type of each critical point. Let us start at the origin. Initial points near the origin correspond to giving the pendulum a small initial displacement and a small initial velocity, which will cause the pendulum to undergo small periodic oscillations close to the vertical downward position. The corresponding flow line α is a closed curve that encircles the origin. (See Figure 2.30(b).) Since this is the case for all initial points near the origin, the origin is a stable critical point and is a center. (See Exercise 9.) Now let us analyze the critical point at $(\pi, 0)$. An initial point $(x_0, 0)$ with x_0 close to π corresponds to releasing the pendulum from near the upward vertical position with zero initial velocity. The corresponding flow line α is a closed curve around the origin, or $(2\pi, 0)$. No matter how close the initial point is to the critical point, the flow line will leave a small neighborhood of $(\pi, 0)$. Thus $(\pi, 0)$ is an unstable critical point of **G**. In Exercise 9, you will be asked determine whether this is a source or saddle.

It is worth noting that we were able to model the simple pendulum by a vector field in the plane because the pendulum had "one degree of freedom" in which to move. Thus the position and velocity could each be described by a single real number, and the position and velocity together could be described by a pair of real numbers. This, of course, is not always the case. For example, a Foucault pendulum, which can be used to demonstrate the rotation of the earth, is free to swing in any direction. So its motion has two degrees of freedom, and its position is described by giving two angles rather than one. Similarly, the velocity is described by two real numbers, the velocity in each of the angular directions. Together, then, it takes four real numbers to describe the position and velocity of a Foucault pendulum.

Summary   In this section, we developed the idea of a ***mathematical model*** for a time-dependent physical system that can be described by two time-dependent quantities. The values of the two quantities at a particular time give the ***state of the system*** at that time. A model consists of a ***pair of first order ordinary differential equations*** for the two quantities. The pair of equations can be ***represented*** by a ***vector field*** in the plane. If the system of equations is ***autonomous***, the vector field will not depend on time. A ***flow line*** of the vector field with initial point (x_0, y_0) describes the behavior of the system when it is placed in the state (x_0, y_0) at time $t = 0$.

We introduced mathematical models for a ***predator–prey*** system of interacting species and for a ***simple pendulum.*** We then used the ideas that we developed in Section 2.4 to give a qualitative analysis of the behavior of each system. In particular, we used ***graphical methods to analyze flow lines*** and ***classify the critical points*** of the corresponding vector field.

Section 2.5 Exercises

1. **Extreme Values of the Populations.** The nonconstant flow lines in the first quadrant of the predator–prey model are closed curves that enclose the critical point. On each flow line, there are points where the populations reach maximum and minimum values.

   (a) How would you characterize the points where prey population is at a maximum or a minimum? Explain your answer. (*Hint:* Think about the flow line equations.)

(b) How would you characterize the points where the predator population is at a maximum or a minimum? Explain your answer. (*Hint:* Think about the flow line equations.)

(c) Is it possible to find the coordinates of these points exactly? Explain your answer.

(d) What, if any, is the relationship between these points and the equilibrium point? Explain your answer.

2. **Average Population Values.** Figure 2.29 contains the plot of the flow line for the particular predator–prey model of Example 2.28 with initial populations of 1000 prey and 200 predators.

(a) Sketch the flow lines for initial populations of 500 prey and 200 predators and of 2000 prey and 200 predators. (These may also be plotted using a computer algebra system, if one is available to you.) In each case, describe how the evolution of the predator system differs from the evolution of the system with the original initial conditions.

(b) Let us assume that one cycle of the predator–prey model takes T units of time to complete. The *average population of prey* over the cycle is $x_{avg} = \frac{1}{T}\int_0^T x(t)dt$. Solving the predator equation for $x(t)$, we see that

$$x(t) = \frac{1}{d}\left(\frac{y'(t)}{y(t)} + c\right).$$

Substitute this expression for $x(t)$ into the integral for x_{avg} and evaluate the integral.

(c) Similarly, the *average population of predators* is given by the integral $y_{avg} = \frac{1}{T}\int_0^T y(t)dt$. Solve for $y(t)$ in the prey equation, and use this expression to compute y_{avg}.

(d) Do these results make sense in light of flow lines you sketched? Explain why or why not.

3. **Predators Without Prey.** In the general predator–prey model, suppose that the initial number of prey is zero and the initial number of predators is y_0.

(a) Find a symbolic form for the flow line of the predator–prey vector field **F** with initial conditions $\alpha(0) = (0, y_0)$.

(b) We have already concluded that the critical point at the origin is unstable. What is the type of the unstable critical point at the origin?

4. **Competition for Resources.** A more realistic model of interacting populations of predator and prey should include competition within a species for resources, which would depend on the number of interactions between members of the same species. We can incorporate this into our basic model by including a degree 2 term, x^2, with a negative

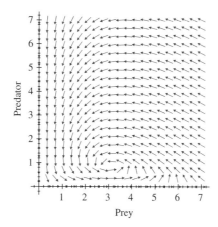

Figure 2.31 The predator–prey model with competition within species. The axes are labeled in units of 1000 for prey and units of 100 for predators. (See Exercise 4.)

coefficient in the prey equation, and a degree 2 term, y^2, with a negative coefficient in the predator equation. This gives us

$$x' = ax - bxy - ex^2,$$
$$y' = -cy + dxy - fy^2,$$

where e, $f > 0$. The vector field for this model is plotted in Figure 2.31, with $a = 1.5$, $b = d = 1, c = 3$, and $e = f = 0.2$.

(a) Sketch the flow lines of this vector field for initial populations of 6000 prey and 0, 200, 400, and 600 predators. (These may also be plotted using a computer algebra system, if one is available to you.)

(b) Describe what happens to the predator–prey system for each of the initial populations of part (a).

5. **The Tangential Component of Force.** In this exercise, we will verify that the equation

$$-g \sin(x(t)/l) = x''(t)$$

for the motion of a simple pendulum is a consequence of Newton's law. (See Figure 2.30(a).) Place the pendulum of length l in the xy-plane so that its pivot is at the origin and the downward vertical direction is aligned with the negative y-axis. Since $\theta = x/l$, the position of the pendulum bob displaced through an angle θ at time t is given by

$$\beta(t) = (l \cos((x(t)/l) - \pi/2), l \sin((x(t)/l) - \pi/2)).$$

Notice that when $x(t) = 0$, $\beta(t) = (0, -l)$ and that $\beta(t) \cdot \beta(t) = l^2$.

(a) Differentiate the equation $\beta(t) \cdot \beta(t) = l^2$ and use the result to show that $\beta'(t) \cdot \beta(t) = 0$.

(b) Show that the acceleration $\mathbf{a} = \beta''$ of the pendulum can be written

$$\mathbf{a}(t) = -\left(\frac{x'(t)}{l}\right)^2 \beta(t) + \frac{x''(t)}{x'(t)}\beta'(t)$$

if $x'(t) \neq 0$.

(c) Let

$$\mathbf{t}(t) = \frac{1}{\|\beta'(t)\|}\beta'(t)$$

be the unit tangent vector to the motion. Show that

(i) $m\mathbf{a}(t) \cdot \mathbf{t}(t) = mx'(t)x''(t)/\|\beta'(t)\|$.

(ii) $\mathbf{F} \cdot \mathbf{t}(t) = -mg\cos((x(t)/l) - \pi/2)x'(t)/\|\beta'(t)\|$.

(d) Use Newton's law, $\mathbf{F} = m\mathbf{a}$, and the two parts of (c) to show that $-g\cos((x(t)/l) - \pi/2) = x''(t)$.

(e) Finally, using part (d) and the fact that $\cos(\theta - \pi/2) = \sin\theta$, show that we arrive at the equation $-g\sin(x(t)/l) = x''(t)$ used in the text.

6. **The Small Displacement Approximation I.** In Example 2.6 of Section 2.1, we modeled the motion of a simple pendulum by the function

$$\alpha(t) = (x_0\sin(\sqrt{\frac{9.8}{l}}t + \pi/2), x_0\sqrt{\frac{9.8}{l}}\cos(\sqrt{\frac{9.8}{l}}t + \pi/2)),$$

and we commented that this function is *an approximation to the motion of the pendulum that is better when the displacement is small.* The model we have developed in this section is not an approximation, but you should keep in mind that a simple pendulum is not a real pendulum. Based on the discussion of the simple pendulum in this section, is the comment about Example 2.6 of Section 2.1 justified? Explain your answer.

7. **The Small Displacement Approximation II.** Using material from one-variable calculus, we can derive the approximation α in Example 2.6 of Section 2.1 from the equation $x''(t) = -g\sin(x(t)/l)$. We will start with the facts that the Taylor series for the sine function is given by

$$\sin(u) = \sum_{j=0}^{\infty} \frac{(-1)^j u^{2j+1}}{(2j+1)!}$$

and that the first Taylor polynomial of sine is just the first term of this series, $P_1(u) = u$. If we replace $\sin(u)$ by u in the differential equation for x, we obtain the simpler equation

$$x''(t) = -(g/l)x(t).$$

(a) Show that the first coordinate of α given in Exercise 6 satisfies this equation.
(b) Use this construction of the first coordinate of α to explain why α is a good approximation when x is small. (*Hint:* Think about the relationship between sine and its Taylor series for small values of u.)

8. **Nonclosed Flow Lines.** Are all the flow lines of the vector field **G** of Example 2.29 closed curves? Explain your answer and describe the behavior of the pendulum that corresponds to any flow lines that are not closed curves.

9. **Stable and Unstable Critical Points**

(a) In Example 2.29, we claimed that the origin is a stable critical point of the vector field **G**. Using your physical understanding about the motion of the pendulum, explain why this satisfies the definition of a stable critical point.
(b) In Example 2.29, we argued that the critical point of **G** at $(\pi, 0)$ is unstable. Is this critical point a saddle or a source? Explain your answer by referring to Figure 2.30(b).

10. **Critical Points of Mechanical Systems.** In the predator–prey model, the vector field **F** that we constructed had a critical point in the interior of the first quadrant, in particular, not on the horizontal axis. In physical terms, what would it mean for a vector field that models a mechanical system to have a critical point that is not on the x-axis? Explain your answer.

11. **Nonautonomous Systems.** All the vector fields that we have studied so far have represented *autonomous* systems. This means that the coordinate functions of the vector field do not depend explicitly on time. For example, we would say that $\sin(x(t))$ does not depend explicitly on time, but that t^2 or $\sin(t)$ does. Suppose that a vector field **F** is of the form

$$\mathbf{F}(x, y) = (u(x, y, t), v(x, y, t)),$$

meaning that the vector field depends explicitly on time. Vector fields of this form are called *nonautonomous*.

(a) Describe a physical system that can only be modeled by a nonautonomous vector field. What necessitates that your example be nonautonomous as opposed to autonomous?
(b) How would you define flow line of a nonautonomous vector field?
(c) Intuitively, what are the differences between flow lines of nonautonomous and autonomous vector fields? Explain your answer.

■ 2.6 End of Chapter Exercises

---

1. **Arcs of Circles and Ellipses.** Sketch the images of the following parametrizations α and indicate the direction of α on the sketch.

   (a) $\alpha(t) = (\cos 2\pi t, \sin 2\pi t)$ for $1 \leq t \leq 3/2$.
   (b) $\alpha(t) = (\sin \pi t, 4 \cos \pi t)$ for $-1/4 \leq t \leq 1/4$.
   (c) $\alpha(t) = (\cos(2t - \pi), \sin(2t - \pi))$ for $0 \leq t \leq \pi/2$.
   (d) $\alpha(t) = (2 \sin(2\pi(t - 1)), \cos(2\pi(t - 1)))$ for $0 \leq t \leq 2$.

2. **Parametrizing Arcs of Circles.** Find a parametrization α of each of the following arcs of a circle that satisfies the given conditions.

   (a) α parametrizes the unit circle centered at the origin clockwise with $\alpha(0) = (0, 1)$ and $\alpha(\pi) = (0, -1)$.
   (b) α parametrizes the unit circle centered at the origin counterclockwise with $\alpha(0) = (\sqrt{2}/2, \sqrt{2}/2)$ and $\alpha(1) = (-\sqrt{2}/2, \sqrt{2}/2)$.
   (c) α parametrizes the circle of radius 1 centered at $(0, 1)$ counterclockwise with $\alpha(0) = (0, 0)$ and $\alpha(1) = (2, 0)$.
   (d) α parametrizes the circle of radius 2 centered at $(-1, -1)$ clockwise with $\alpha(0) = (-3, -1)$ and $\alpha(1) = (1, -1)$.

3. **A Simple Pendulum.** Consider a simple pendulum of length 1 (as in Example 2.6 of Section 2.1) whose motion is parametrized by

$$\alpha(t) = (A \sin(\sqrt{9.8}t + 0.246), \ A\sqrt{9.8} \cos(\sqrt{9.8}t + 0.246)),$$

   where $A = (\pi/4)(1/\sin(0.246))$. The image of the parametrization is shown in Figure 2.32.

   (a) What are the initial position and the initial velocity of the pendulum?
   (b) What is the period of the pendulum?
   (c) Describe the motion of the pendulum from time $t = 0$ until the pendulum completes one period of its motion.

4. **Vector Sum of Parametrizations.** Certain motions of objects can be written as the vector sum of two different motions. For example, if α parametrizes the motion of a person walking in a straight line and β parametrizes the motion of a yo-yo being spun in a vertical plane at the end of its string, $\gamma = \alpha + \beta$ will represent the motion of the yo-yo being spun in a vertical plane as the person spinning the yo-yo walks in a straight line. In coordinates in the plane, $\alpha(t) = (x_1(t), y_1(t))$ and $\beta(t) = (x_2(t), y_2(t))$,

$$\gamma(t) = \alpha(t) + \beta(t) = (x_1(t), y_1(t)) + (x_2(t), y_2(t)) = (x_1(t) + x_2(t), y_1(t) + y_2(t)).$$

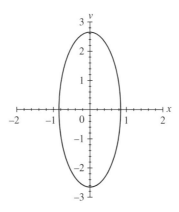

Figure 2.32   The image of the parametrization α of Exercise 3.

(a) Suppose that a yo-yo is being spun in a circle on the end of its string with constant speed so that it traces a circle in a vertical plane (the xy-plane). If one rotation takes 1 s, the string has length 0.75 m, and the string is being held at a point 1 m off the ground, how would you parametrize the motion?

(b) Suppose a person walks at a speed of 1 m/s along the x-axis in the xy-plane. How would you parametrize the motion?

(c) How would you parametrize the motion of a yo-yo being spun as in part (a) by a person walking at a speed of 1 m/s?

5. **Vector Sum of Parametrizations.** A carousel is 40 ft in diameter and makes one revolution every 30 s. The carousel horses move up and down a distance of 1 ft twice during each revolution of the carousel. Parametrize the motion of a person riding the carousel on a horse that is 2 ft from the edge of the carousel and whose seat is 4 ft off the ground at its lowest.

6. **The Orbit of the Moon about the Sun.** This exercise investigates the motion of the moon about the sun. For the purposes of this exercise, we will assume that the orbit of the earth about the sun is a circle and the orbit of the moon around the earth is a circle.

(a) Construct a parametrization of the orbit of the earth about the sun that places the sun at the origin. Assume that the orbit is a circle of radius 149.5×10^8 km. It takes the earth approximately 365.26 days to orbit the sun.

(b) Construct a parametrization of the orbit of the moon about the earth that places the earth at the origin. Assume that the orbit is circle of radius $384,405$ km (the mean distance from the earth to the moon). It takes the moon approximately 27.32 days to orbit the earth.

(c) Construct a parametrization of the motion of the moon about the sun that places the sun at the origin. (Neglect the 5.1 deg tilt in the plane of the orbit of the moon relative to the plane of the orbit of the earth about the sun.) Describe this motion.

7. **Retrograde Motion of Mars.** In this exercise, we will analyze the direction from the earth to Mars as a function of time. If α parametrizes the orbit of the earth and β parametrizes the orbit of Mars, the direction vector at time t is equal to $(\beta - \alpha)(t) = \beta(t) - \alpha(t)$. The motion represented by this direction is the motion of Mars relative to the backdrop of the stars in the sky. For this problem, let us assume that the orbits of the earth and Mars are both circles centered at the origin with the sun located at the origin and that these circles lie in the xy-plane. The radius of the orbit of the earth is approximately 92.9 million miles and that of Mars is approximately 141.5 million miles. Mars takes 686.8 earth days to orbit the sun.

   (a) Construct parametrizations α of the orbit of the earth and β of the orbit of Mars.
   (b) Use a computer to plot the parametrization $\beta - \alpha$ of the direction vector from the earth to Mars for a period of at least 10 years.
   (c) Use the plot in (b) to analyze the direction of Mars from the earth. In particular, describe how Mars would move as it crossed the sky over this time period, were we to look at Mars from earth. Explain your answer.

8. **Velocity and Speed.** Each of the following functions α is a parametrization for the motion of an object in the plane or in space. Find the velocity vector and the speed of the motion at the given time t_0.

   (a) $\alpha(t) = (t^2 - 1, 4t^3)$ at $t_0 = 0$.
   (b) $\alpha(t) = (\cos 2t, \sin t, t - \pi)$ at $t_0 = \pi/2$.
   (c) $\alpha(t) = (e^{2t}, e^{-2t}, 2t)$ at $t_0 = 1$.
   (d) $\alpha(t) = (\frac{1}{1+t^2}, t)$ at $t_0 = 0$.

9. **Tangent Lines.** For each of the following parametrizations α, find the tangent vector and the tangent line to the curve parametrized by α at the point t_0.

   (a) $\alpha(t) = (\cos(2\pi t), 3\sin(2\pi t))$ at $t_0 = 1$.
   (b) $\alpha(t) = (e^{t/2}, e^{t^2/4})$ at $t_0 = 0$.
   (c) $\alpha(t) = (t - 1, 3t + 1, -4t)$ at $t_0 = 2$.
   (d) $\alpha(t) = (2 - \sin(\pi t), 2 + \sin(\pi t), t)$ at $t_0 = -1$.

10. **Cycloids.** As in Exercise 5 of Section 2.2, assume that a wheel of radius 1 rests on the x-axis and rolls without slipping in the positive direction along the x-axis with unit speed. Assume that the marked point lies a distance b units from the center of the wheel along a radial segment (or spoke). If the motion starts with the wheel resting on the origin and the radial segment lies along the y-axis at the beginning of the motion, the curve traced by the point is parametrized by $\alpha(t) = (t + b\sin t, 1 + b\cos t)$. If $b = 1$, the curve is the cycloid of Exercise 5 of Section 2.2.

   (a) Compute the velocity vector $\mathbf{v}(t)$ for the motion of the point.
   (b) Suppose $b < 1$. At which time(s) t is the velocity vector horizontal? At which time(s) t is the velocity vector vertical?

(c) Suppose $b > 1$. At which time(s) t is the velocity vector horizontal? At which time(s) t is the velocity vector vertical?

(d) Sketch the image of $\alpha(t)$ for $b = 2$ and $b = 1/2$.

11. **Verifying Parametrizations of Flow Lines.** For each of the following vector fields and parametrizations, verify by a direct calculation that the function α parametrizes a flow line of the vector field and find the initial point of the flow line.

(a) $\mathbf{F}(x, y) = (y, x)$.

  (i) $\alpha = ((e^t + e^{-t})/2, (e^t - e^{-t})/2)$.
  (ii) $\alpha = ((3e^t + e^{-t})/2, (3e^t - e^{-t})/2)$.

(b) $\mathbf{F}(x, y) = (y, -4x)$.

  (i) $\alpha(t) = (\cos(2t)/2, -\sin(2t))$.
  (ii) $\alpha(t) = (\cos(2t) + \sin(2t)/2, -2\sin(2t) + \cos(2t))$.

12. **Flow Lines of Vector Fields.** For each of the following vector fields, proceed as follows. (i) Sketch the flow lines starting at the initial points $(\pm 1, 0)$, $(0, \pm 1)$, $(1, \pm 1)$, and $(-1, \pm 1)$. (These may also be plotted using a computer algebra system, if one is available to you.) (ii) Use your sketch to classify the critical point of $\mathbf{F}$ at the origin as stable or unstable and as a center, sink, source, or saddle.

(a) $\mathbf{F}(x, y) = (y, 2x)$. (See Figure 2.33(a).)
(b) $\mathbf{F}(x, y) = (3y, x)$. (See Figure 2.33(b).)
(c) $\mathbf{F}(x, y) = (x + y, y)$. (See Figure 2.33(c).)

13. **Critical Points of Vector Fields.** Determine the critical points of each of the following vector fields on the indicated domain, and based on the accompanying plot, determine the type of the critical points. (If a computer algebra system is available to you, you may find it helpful to plot several flow lines of the given vector fields.)

(a) $\mathbf{F}(x, y) = (x + y - 1, y - x - 1)$, $[-1.5, 1.5] \times [-1.5, 1.5]$. (See Figure 2.34(a).)
(b) $\mathbf{F}(x, y) = (y - x + x^2 - 1, y - x)$, $[-1.5, 1.5] \times [-1.5, 1.5]$. (See Figure 2.34(b).)
(c) $\mathbf{F}(x, y) = (e^{xy}((y - x)^2 + x^2 - 0.25), e^{xy}(y - x)^2)$, $[-1, 1] \times [-1, 1]$. (See Figure 2.34(c).)

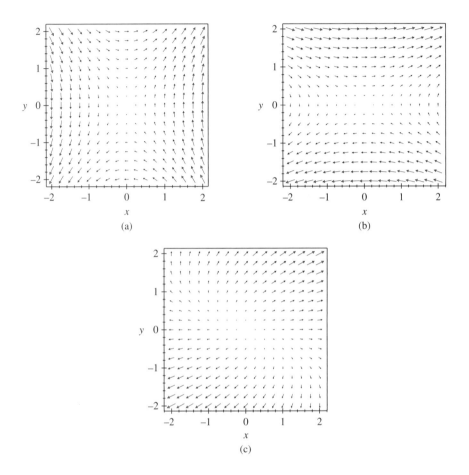

Figure 2.33 Vector fields for Exercise 12.

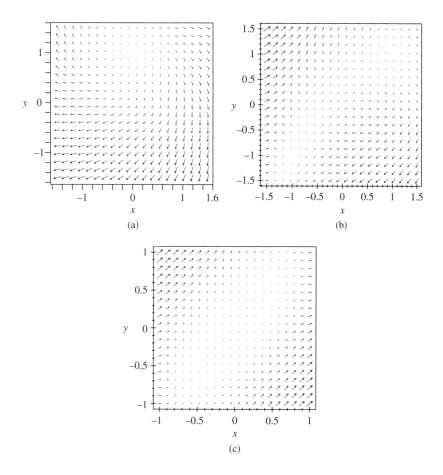

Figure 2.34   Vector fields for Exercise 13.

3

Differentiation of Real-Valued Functions

In this chapter, we will begin to explore real-valued functions of two and three variables, that is, functions whose domain is a subset of the plane $\mathbb{R}^2$ or space $\mathbb{R}^3$ and whose range is a subset of $\mathbb{R}$. It is important to keep in mind that we are developing a collection of tools that will allow us to answer basic questions about the behavior of functions. These questions arise naturally in the context of applications in the sciences, where the language of functions is an efficient way to present quantitative information and to describe relationships among the data.

We will begin in Section 3.1 by developing an intuitive understanding of functions defined on the plane or space. We will focus on different ways to represent functions and provide a number of examples of functions that occur in the sciences. In Section 3.2, we will introduce directional derivatives and partial derivatives. We will apply symbolic techniques from one-variable calculus to evaluate these derivatives. In Section 3.3, we introduce the concepts of limit and continuity for functions of two variables. In Section 3.4, we use limits to develop the concept of differentiability. In Section 3.5, we will introduce the chain rule and the gradient vector field of a differentiable function. The gradient vector field is defined on the domain of a function. We will apply our work in Sections 2.4 and 2.5 on vector fields to the gradient vector field in order to further understand the behavior of functions.

A Collaborative Exercise—Contour Plots

Functions of several variables occur naturally in a number of different contexts in the sciences and mathematics. One of the first functions one encounters is altitude (above sea level) on the surface of the earth, which is commonly represented by a geographic contour plot. In this exercise, we will investigate a single contour plot in order to determine what we can learn about a function from its contour plot.

First, we define what we mean by a contour plot.

Definition 3.1 The *contour curve* or *level set* of a function $f : \mathcal{D} \subset \mathbb{R}^2 \to \mathbb{R}$ for the value c is the subset of $\mathcal{D}$, given by

$$\{(x, y) : (x, y) \in \mathcal{D} \text{ and } f(x, y) = c\}.$$

A *contour plot* for f is a plot of a portion of the domain of f showing several level sets of f for different values c. ◆

Usually the collection of values c that are used to determine the contour curves are evenly spaced values, for example, 5, 10, 15, 20, and so on. Often only a selection of the values are indicated on the plot. For example, if the contours are separated by 5 units, then it might be the case that only every tenth contour is labeled. Thus we would only see the values 0, 50, 100, 150, and so on marked on the plot.

These questions refer to the accompanying contour plot of Doublehead Mountain in New Hampshire. (See Figure 3.1.) In this case, the function f under consideration is altitude above sea level measured in feet. Of course, there is no symbolic form for such a function. Note that the contours on the plot are for values c that are separated by 20 ft.

1. Locate the local maximum values for the altitude function on the plot. How would you characterize the contours near the local maxima? Will your characterization always hold near a maximum? Explain why or why not. If not, explain under what circumstances your characterization applies.
2. In addition to Trails 202 and 204 to the summit of North Doublehead Mountain, we can imagine a straight-line path from the junction of Trails 202 and 204 just below the 1900-ft contour heading directly to the summit of North Doublehead Mountain. Comparing this new path to Trail 202, which would you prefer to take and why? What features of the contour plot indicate that your choice is the preferred route?
3. Let Q be the point between North Doublehead and South Doublehead where Trail 204 ends on Trail 202.

   a. Carefully describe the path along Trail 202 from North Doublehead through the point Q to South Doublehead.
   b. Suppose Trail 204 continued through the point Q to the point R marked on the eastern edge of the plot. Carefully describe the path along this new trail from the point just below the 1900-ft contour where Trails 202 and 204 meet, through the point Q to the point R.
   c. What does the surface of Doublehead Mountain look like around the point Q?

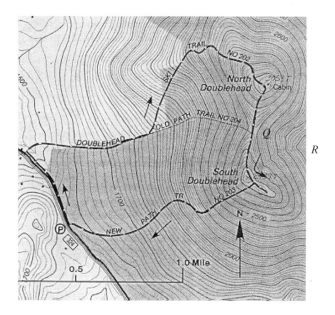

Figure 3.1   A contour plot of the region near Doublehead Mountain in New Hampshire. Notice that the contours correspond to a 20-ft change in altitude, and the 100-ft contours are labeled on the plot. Map courtesy of the U.S. Geological Survey.

■ 3.1 Representation and Graphical Analysis of Functions

In this section, we will begin to study *real-valued functions of two and three variables*, that is, functions defined on a subset $\mathcal{D}$ of the plane or space and that take values in $\mathbb{R}$. Initially, we will focus on the question of how to represent these functions. As is the case in one-variable calculus, we want to represent functions numerically, graphically, and symbolically. Before we illustrate the different representations, let us introduce a number of examples of the type we will consider and some of the questions that we might want to be able to answer.

Example 3.1	**A Catalogue of Applications**

A. "Geographic" Functions. The first type of example we will consider is quantities that can be described as *functions of a geographic location*. For example, as in the collaborative exercise, the altitude above sea level can be thought of as a function of the latitude and longitude of the location. In addition to being interested in the values of this function (what is the altitude at a particular location), we would like to be able to determine the locations of its maximum and minimum values and to determine the rate of change of the altitude at a particular location. Other quantities, such as temperature, precipitation, and population density, can also be thought of as functions of latitude and longitude. Similarly, we might be interested in the extreme values of these functions.

B. Physical Laws. Physical laws can sometimes be expressed as functions. For example, in chemistry, the ***perfect gas law*** describes the relationship between the temperature, pressure, and volume of a fixed amount of a gas. This can be written symbolically in the form

$$T = \frac{1}{nR}pV,$$

where T is the temperature in kelvin, p is the pressure in pascals or newtons per square meter, V is the volume in cubic meters, n is the amount of gas in moles, and R is the ***gas constant***, which does not depend on the gas. This expresses the temperature of a fixed amount of gas as a function of the pressure and volume of the gas, thus p and V are the independent variables and T is the dependent variable. We may be interested in the relationship of any two of temperature, volume, and pressure, if the third quantity is held fixed. Although there are no "perfect" gases, that is, gases that behave exactly according to this equation, the lower the pressure of a real gas, the more closely its behavior is described by this equation.

C. Time Dependent Systems. Another type of example we will consider is quantities for which time is one of the independent variables. For example, the blood pressure in a capillary is a function of the position along the capillary and the time in the heart cycle. The motion of a vibrating string, say a violin string, is a function of time and position on the string. One model of the spread of an infectious disease, such as measles, describes the number of measles cases as a function of the time of year and the age of the population. We will want to analyze the function at fixed times or for a fixed value of the other independent variable. For example, we will consider the position of a vibrating string at a fixed time.

D. Surfaces. Surfaces in $\mathbb{R}^3$ can be described using functions as well. In this case, the coordinates of points in the surface are the independent and dependent variables of a function of two variables. For example, the z-coordinate of a point in $\mathcal{S}$ can be described as a function of the x- and y-coordinates of the point.

The questions that arise naturally in the context of the examples are basic questions about the corresponding functions. For example, what is the ***value*** of a function at a particular point? What are its ***maximum*** and ***minimum values***? Where is the function ***increasing*** or ***decreasing***? What is the ***rate of change*** of the function as the independent quantities change? What is the ***behavior of the function as one variable changes while the other is held fixed***?

In order to formulate and answer these questions, we will use the standard mathematical terms and symbols for functions. In the examples above, a single dependent variable is described as a function of two or three independent variables. Thus the functions are defined on a subset $\mathcal{D}$ of the plane or space and take values in the real line. We will write $f : \mathcal{D} \to \mathbb{R}$ and $f : \mathcal{D} \subset \mathbb{R}^2 \to \mathbb{R}$ or $f : \mathcal{D} \subset \mathbb{R}^3 \to \mathbb{R}$ when we want to emphasize that f is a function of two or three variables, respectively. A point in $\mathcal{D}$ will be represented by an ordered pair or an ordered triple of real numbers, for example, $P = (x_0, y_0)$. By convention, we will use the somewhat abbreviated notation $f(x_0, y_0)$ rather than $f((x_0, y_0))$ in referring to $f(P)$. In the following definition, we will use this terminology to define maxima and minima for functions.

Definition 3.2 Let f be a real-valued function defined on a domain $\mathcal{D}$ in the plane or space, and let P and Q denote points in $\mathcal{D}$.

1. If $f(P)$ is greater than or equal to $f(Q)$ for all points Q in $\mathcal{D}$, we say that P is a ***maximum*** for f and that $f(P)$ is a ***maximum value*** for f on $\mathcal{D}$. In symbols,

$$f(P) \geq f(Q) \text{ for } Q \in \mathcal{D}.$$

   If $f(P) \geq f(Q)$ for all points Q that are near P, we say $f(P)$ is a ***local maximum*** for f.
2. If $f(P)$ is less than or equal to $f(Q)$ for all points Q in $\mathcal{D}$, we say that P is a ***minimum*** for f and that $f(P)$ is a ***minimum value*** for f on $\mathcal{D}$. In symbols,

$$f(P) \leq f(Q) \text{ for } Q \in \mathcal{D}.$$

   If $f(P) \leq f(Q)$ for all points Q near P, P is a ***local minimum*** for f. ◆

Maxima and minima are collectively called ***extrema*** or ***extreme values*** for a function. In Section 3.3, we will give a precise definition of what it means for Q to be "near" P. In the meantime, we will intuitively interpret "near" to mean that the distance from Q to P is small.

Representing Functions of Two Variables

We will represent functions in three ways: numerically, graphically, and symbolically. Here we will focus on functions of two variables. Functions in the sciences often appear in numerical form, that is, as a data set. Usually the information is arranged in a table or spreadsheet. Symbolically, functions are given by a formula that expresses the dependent quantity in terms of the independent variables.

There are three different graphical representations for functions of two variables: ***density plots***, ***contour plots***, and ***graphs***. Density plots and contour plots "live" in the two-dimensional domain of the function. Graphs of functions of two variables "live" in three-dimensional space, that is, in a dimension one higher than that of the domain, which is also the case for graphs of functions of one variable.

A ***density plot*** for $f : \mathcal{D} \subset \mathbb{R}^2 \to \mathbb{R}$ is a shaded plot of the domain. Usually, the domain is covered by a grid consisting of small rectangles, and each rectangle is shaded entirely in one color according to the values that the function takes in the rectangle. We will shade density plots according to ***gray levels***, that is, shades of gray ranging from black to white. Of course, more elaborate color schemes are possible and are easily generated by computer. If the function takes more than one value in a rectangle, it will be necessary to make a choice of values in order to assign the level of shading. This is a particularly useful tool when analyzing data sets, because it produces a visual image from the digital information.

As we described in the collaborative exercise at the beginning of the chapter, a ***contour plot*** for a function $f : \mathcal{D} \subset \mathbb{R}^2 \to \mathbb{R}$ consists of a collection of contour curves in $\mathcal{D}$. A ***contour curve***, or ***level set***, of f for the value c is the set

$$\{(x, y) : f(x, y) = c\}.$$

Since f is defined at every point of $\mathcal{D}$, every point in the domain is contained in a level set. However, it suffices to show contours that correspond to values of the function separated by a fixed interval. (Indeed, if all the level sets were shown, the contour plot would be a single color.)

A ***graph*** of f on the domain $\mathcal{D}$ is a plot of the collection of points of the form $(x, y, f(x, y))$ in $\mathbb{R}^3$. The graph may be thought of as forming a ***surface*** in $\mathbb{R}^3$.

Our ability to answer the previous questions will depend on the representation of the function. For example, if the function is represented numerically as a data set, we know its exact value at some points but not at every point. If the function is represented graphically, it may be easy to approximately locate maximum and minimum points, but we may not be able to determine the exact value of the function at any point. If the function is represented symbolically, we can usually compute its values, but we often cannot locate maxima and minima. Because of these difficulties, it may be useful to consider more than one representation of the same function.

In the following example, we consider a function from meteorology: the depth of snow coverage in the Northeast on the day after the blizzard of February 1978. We will consider both the numerical representation of this function and the density plot.

Example 3.2   **The Blizzard of February 1978**

A. The Data. The data set in Figure 3.2 contains readings for the depth of snow at a number of sites in the northeastern United States on February 8, 1978, the day after the blizzard of February 1978. In the chart, snow depth is represented as a function of latitude and longitude and the values are given in inches. The data is arranged in columns, so that reading across a row we see a location in latitude and longitude and the corresponding snow depth. The maximum recorded depth was 41 in, which occurred at 43.63° north latitude and 72.32° west longitude, and the minimum recorded depth was 5 in at 41.17° north latitude and 71.58° west longitude. Of course, there may have been locations not appearing on this chart where the snow depth was greater than 41 in or less than 5 in. Notice also that it is difficult to see any patterns in the data when it is presented in this form. It is more useful for us to plot the data over a map of the region.

B. The Density Plot. Figure 3.3 is a density plot for the data in Figure 3.2. The plot is superimposed on a map of the northeastern states showing state borders. A key for the shadings appears at the right of the figure. Immediately we can see the patterns in the snow coverage that were hidden in the chart. The areas of greater snow depth occur in bands that run north-south. There were two regions with significant snow coverage: one in western Pennsylvania and one in New England. The greatest snow depths were recorded along the southern half of the border between Vermont and New Hampshire.

lat.	long.	in.	lat.	long.	in.	lat.	long.	in.	lat.	long.	in.
42.22	−75.98	12	41.80	−78.63	32	41.18	−78.90	30	40.77	−73.90	14
42.93	−78.73	23	40.70	−74.17	17	42.75	−73.80	16	40.30	−78.32	12
40.65	−75.43	17	42.37	−71.03	29	41.93	−72.68	21	44.47	−73.15	14
43.20	−71.50	16	43.35	−73.62	14	40.22	−76.85	17	42.22	−71.12	33
44.27	−71.30	21	40.82	−82.52	16	41.63	−73.88	19	40.90	−78.08	34
43.65	−70.32	16	41.73	−71.43	27	43.12	−77.67	30	43.12	−76.12	11
40.22	−74.77	12	41.33	−75.73	12	41.25	−76.92	18	40.03	−74.35	18
42.15	−70.93	31	40.20	−75.15	19	41.42	−81.87	17	40.00	−82.88	7
42.42	−83.02	14	41.02	−83.67	17	42.97	−83.75	14	42.27	−84.47	26
42.78	−84.60	22	43.53	−84.08	20	41.25	−80.67	8	42.08	−80.18	13
44.20	−72.57	23	44.93	−74.85	24	40.78	−73.97	18	42.27	−71.87	30
43.63	−72.32	41	40.65	−73.78	14	44.00	−76.02	19	41.17	−71.58	5
43.15	−75.38	11	44.37	−84.68	20	40.50	−80.22	7	41.60	−83.80	18

Figure 3.2   Data for snow depth in the northeastern United States on February 8, 1978, the day after the blizzard of February 1978. The snow depth is given as a function of latitude and longitude. Note that west longitude is written as a negative number. Data is courtesy of the National Climatic Data Center. (See Example 3.2A.)

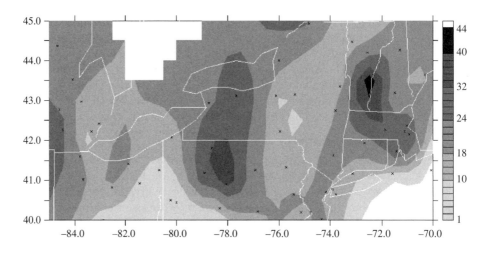

Figure 3.3   A density plot of the snow cover on February 8, 1978, in the northeastern United States the day after the blizzard of February 1978. The depth is measured in inches. This plot was generated by interpolating the data in Figure 3.2. Map courtesy of the National Climatic Data Center. (See Example 3.2B.)

In the following example, we will use a contour plot to analyze an altitude function of the type described in Example 3.1A. In the process, we will introduce the idea of a saddle point of a function.

| Example 3.3 | **The Sandia Mountains** |

A. How High? Figure 3.4 is a geographic contour plot for an area of the Sandia peaks in the Sandia Mountain Wilderness east of Albuquerque, New Mexico. (Figure 3.5 is a photograph of the Sandia peaks from the west.) The contour curves on the plot are for 40-ft intervals in altitude; however, only the curves at 200-ft intervals are labeled. If a location lies on a contour, we can immediately read its altitude off the plot. If a location lies between contours, we can only approximate its altitude depending on its distance from the closest contours.

There are two peaks, or maximum values, for the altitude function on this plot. They occur just to the left or west of Trail 130. One is located to the north of Trail 140 at an altitude of approximately 9600 ft, and the other is located to the south of Trail 140 at an altitude of approximately 9560 ft. This second peak is only a local maximum for the altitude function since it is higher than the nearby points on the plot, but it is not as high as the other maximum. Notice that the contour curves closest to the maximums are loops or closed curves. This is a characteristic of the contours near any local maximum that is strictly greater than nearby function values.

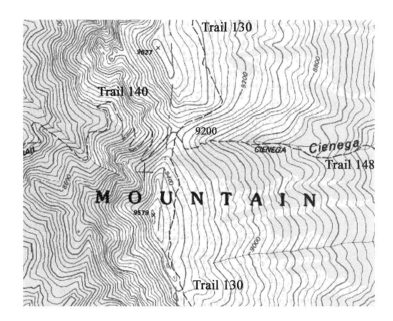

Figure 3.4 A geographic contour plot for an area of the Sandia peaks in the Sandia Mountain Wilderness east of Albuquerque, New Mexico. Courtesy of the USGS. (See Example 3.3A.)

There is a second type of point, called a ***saddle point***, that is of interest on this plot. This is located at the junction of Trail 130 and Trail 148 at an altitude between 9200 and 9240 ft. Notice that the junction is the highest point on Trail 148, but that it is a low point on Trail 130, so that the point is a pass between two regions of higher elevation. Intuitively, the ground is "saddle-shaped" at such a point, thus the name saddle point. Near the saddle point on the plot, the contours resemble a family of hyperbolas, which, in this case, open to the north and south to regions of higher elevation and to the east and west to regions of lower elevation. This behavior is typical of contours near a saddle point.

B. How Steep? We should also notice the relative steepness on different portions of the plot. On the western side of the plot (which is the side of the Sandia peaks shown in Figure 3.5), the contours are closer together than on the eastern side of the plot. This means that the 40-ft height change between adjacent contours occurs over a smaller horizontal distance; that is, the slope is steeper. Generally, if contours correspond to evenly spaced function values, the closer they are together, the more rapidly the function is changing. Conversely, the farther apart the contours, the less rapidly the function is changing.

If we look at the trails on the map, we can determine the difficulty of the trails by considering the number of contours the trails cross over a given horizontal distance. For example, if we look at Trail 140, the trail to the west of the intersection with Trail 130 is steeper than the trail to the east of the intersection with Trail 130. In fact, the slope of

Figure 3.5   A photograph of the Sandia peaks from the west. Photograph courtesy of De-Ping Yang. (See Example 3.3B.)

the mountain is so great to the west of the intersection with Trail 130 that it is necessary for trail to follow several switchbacks rather than ascending the ridge directly as it does to the east. Trail 130, on the other hand, lies primarily between the contours at levels 9200 and 9400 feet for its entire length. Thus the trail is generally level with only gradual inclines and declines.

In Section 1.1, we introduced the method of slicing to better understand subsets of space. Here we want to apply it to the graph of a function $f : \mathcal{D} \subset \mathbb{R}^2 \to \mathbb{R}$. The graph is the subset of $\mathbb{R}^3$ given by $\{(x, y, f(x, y)) : (x, y) \in \mathcal{D}\}$. A ***vertical slice*** of the graph parallel to the xz-plane is the intersection of the graph of f with the plane given by $y = k$, where k is a constant. Using set notation, the slice is given by

$$\{(x, k, z) : (x, k) \in \mathcal{D} \text{ and } z = f(x, k)\},$$

where $\mathcal{D}$ is the domain of f. The slice is the graph of z as a function of x holding y constant, which we will express by writing $x \to f(x, k)$. In a similar manner, a vertical slice of the graph of f parallel to the yz-plane is the intersection of f with a plane of the form $\{(x, y, z) : x = k\}$ for k a fixed real number. In symbolic form, the slice is given by

$$\{(k, y, z) : (k, y) \in \mathcal{D} \text{ and } z = f(k, y)\}.$$

The slice is the graph of z as function of y, $y \to f(k, y)$.

The horizontal slices of the graph of f are obtained by intersecting the graph of f with horizontal planes $z = c$. Thus the horizontal slice of the graph of f for the value $z = c$ is the subset of $\mathbb{R}^3$ given by

$$\{(x, y, c) : f(x, y) = c\}.$$

The horizontal slice for $z = c$ corresponds to the contour curve or level set of f for the value c. Intuitively, the slice $z = c$ is obtained by lifting the level set given by $f(x, y) = c$, which is a subset of the xy-plane, to height $z = c$ in space.

Let us use these ideas to analyze the graph of $f(x, y) = -x^4 + 2x^2 - \frac{1}{2}x - 1 - y^2$.

Example 3.4

The Graph of a Polynomial. Figure 3.6(a) shows a contour plot of $f(x, y) = -x^4 + 2x^2 - \frac{1}{2}x - 1 - y^2$ on the domain $\mathcal{D} = [-1.5, 1.5] \times [-1.5, 1.5]$. The level sets correspond to the values c of f satisfying $-4.4 \leq c \leq 0.4$ at intervals of 0.2 units. The level sets that are shown on the plot are closed curves or parts of closed curves. The contours at the edge of the plot correspond to lower values of f. As we move toward the center of the plot, the values of f on the contours increase. The highest contour values in this plot occur as we move toward the centers of the two families of concentric contours. We also note that the contours are closer together near the edges of the plot, indicating that the function is changing more rapidly there.

We can see that f has local extrema in the center of the families of concentric contours. Based on the contour values, we classify these extrema as local maxima. One is located at approximately $(-1.15, 0)$, where the function value is between 0.4 and 0.6, and the second

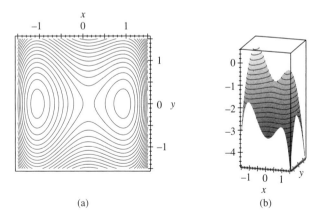

(a) (b)

Figure 3.6   (a) A contour plot of $f(x, y) = -x^4 + 2x^2 - \frac{1}{2}x - 1 - y^2$ on the domain $-1.5 \leq x \leq 1.5$, $-1.5 \leq y \leq 1.5$. (b) The graph of $f(x, y) = -x^4 + 2x^2 - \frac{1}{2}x - 1 - y^2$ on the same domain. (See Example 3.4.)

located at approximately $(0.9, 0)$, where the function value is between -0.6 and -0.4. It is important to realize that the contour plot does not give us sufficient information to be sure of the exact location or value of the maxima. We can also observe that there is a saddle point of f located near $(0.1, 0)$. As was the case for the local maxima of f, we cannot determine the exact location of the saddle point from the contour plot.

Figure 3.6(b) is a plot of the graph of f on $\mathcal{D}$ that shows the horizontal slices of the graph corresponding to the level sets in Figure 3.6(a). We can see the local maximum and the saddle point that we identified on the contour plot. It can be productive to use the contour plot and the graph of a function together because it may be the case that certain features of a function can be hidden on one of these plots but visible on the other. (See Exercise 10.)

Application: A Vibrating String

It is often the case that one of the independent variables of a function represents time. In this way, functions of several variables can be used to represent quantities that vary in time and space. In the next example, we consider the motion of an idealized taut string. While this is a simplified model of the behavior of a real string, it is an important prototype for systems that exhibit wavelike behavior or periodic behavior.

Example 3.5

A Standing Wave. This example is motivated by the behavior of a vibrating string whose endpoints are held fixed; for example, we might imagine a piano or guitar string. The simplest motions of a string can be represented by a function u that is a product of functions of one variable, $u(x, t) = f(x)g(t)$. The string lies along the x-axis from 0 to π. As the string vibrates, it moves vertically in the plane while the endpoints at 0 and π remain fixed. The function u gives the vertical displacement of the string: At time t, location x on the string is displaced from $z = 0$ to $z = u(x, t)$.

The position of the string at a fixed time t_0 is the graph of the function of x defined by $x \to u(x, t_0)$. This is a vertical slice of the graph of u parallel to the xz-plane. For example, suppose $u(x, t) = \sin(2x) \cos(3t)$. As t changes, the vertical slices model the motion of the string. Let us work out the details.

Since $\cos(3t)$ is a periodic function with period $2\pi/3$, the motion of the string is periodic with period $2\pi/3$. Let us describe one period of the motion of the string. (See Figure 3.7.) Since $\cos 0 = 1$, the $t = 0$ slice is the graph of $f(x) = \sin(2x)$. As t increases, $\cos(3t)$ decreases, so the string moves toward the x-axis, that is, downward on the interval $[0, \pi/2]$ and upward on the interval $[\pi/2, \pi]$. When $t = \pi/6$, the string lies along the axis. The string continues to move in this manner until $t = \pi/3$, when $u(x, \pi/3) = -\sin(2x)$. The string then reverses its direction of motion and returns to its original position when $t = 2\pi/3$, the period of the cosine function. Notice that throughout the motion of the string, the point $\pi/2$ remains fixed. We call such a point a *node*.

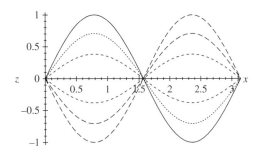

Figure 3.7   Slices of the graph of the function $u(x,t) = \sin(2x)\cos(3t)$ of Example 3.5. Vertical slices of the graph for $t = 0, \pi/12, \pi/8, 5\pi/24, \pi/4, \pi/3$ are shown on the plot. (See Example 3.5.)

Representing Functions of Three Variables

Functions of three variables, that is, functions $f : \mathcal{D} \subset \mathbb{R}^3 \to \mathbb{R}$, may also be given numerically, symbolically, or graphically. Numerically, a function may be given by a data set, that is, as a collection of function values, $f(x, y, z)$, for points (x, y, z) in the domain of f. Symbolically, these functions may be given by a formula that expresses the dependent variable in terms of the three independent variables. Graphically, the contour plot will be useful for functions of three variables.

A density plot for a function of three variables would consist of a three-dimensional grid of shaded boxes colored according to the values of the function in the box. While it is possible to imagine such an object, it would not be easy to put it to effective mathematical use. The graph of a function of three variables would consist of a plot of the collection of points of the form $(x, y, z, f(x, y, z))$ in four-dimensional Euclidean space, $\mathbb{R}^4$. The graph makes perfectly good mathematical sense, but it again would be difficult to use effectively. On the other hand, a level set of a function of three variables is a surface in $\mathbb{R}^3$. More precisely, a ***level surface*** or ***contour surface*** of $f : \mathcal{D} \subset \mathbb{R}^3 \to \mathbb{R}$ for the value c is the set of points $(x, y, z) \in \mathcal{D}$ with the property that $f(x, y, z) = c$. In symbols, the level surface for the value c is

$$\{(x, y, z) \in \mathcal{D} \ : \ f(x, y, z) = c\}.$$

We can plot several level surfaces in a single plot as we did for level curves of functions of two variables. Working by hand, we can use the method of slices to sketch and analyze individual level surfaces. Since the expression $f(x, y, z) = c$ specifies the ***relationship*** between the coordinates, but does not explicitly specify the coordinates themselves, we say the surface is defined ***implicitly*** by the equation $f(x, y, z) = c$.

Application: Acoustics

An interesting example of level surfaces comes from functions that represent *sound pressure levels* in space. In particular, we will analyze the level surfaces of the pressure function in order to understand the sound produced by an acoustic source. These surfaces are called *isobars* of pressure. Sound is a variation in air pressure levels that is sensed by our auditory system. The changes in air pressure are the result of "pressure waves" that are generated by the source of the sound. As a pressure wave passes a point in space, the air is first compressed, increasing pressure, and then decompressed, decreasing pressure. In practice, because the change in air pressure determines the loudness of the sound, when we speak of the acoustic pressure of the source we mean the average (over time) absolute value of the change in pressure. Since the absolute value is computed by taking the square root of the square of the pressure difference, pressure is called the *root-mean square*, or *rms pressure*.

Example 3.6 **An Acoustic Monopole.** In this example, we will consider an idealized version of sound produced by a small sphere pulsating in and out at a constant frequency. This is called a *monopole* source of sound. Collections of acoustic monopoles can be used to model more complicated sources of sound.

If we assume that a monopole source is located at the origin, then for points $\mathbf{x} = (x, y, z)$ away from the origin, the rms pressure at $\mathbf{x}$ can be approximated by a function of the form

$$p_a(\mathbf{x}) = \frac{A}{\sqrt{x^2 + y^2 + z^2}},$$

where A is a positive constant that depends on the source, the density of air, and the speed of sound. The subscript a is to remind us that this represents the average variation. For convenience, we will let $r(\mathbf{x}) = \sqrt{x^2 + y^2 + z^2}$, the distance of the $\mathbf{x}$ from the source, and write $p_a(\mathbf{x}) = A/r(\mathbf{x})$. Notice that p_a is not defined at the origin and is positive for $\mathbf{x} \neq \mathbf{0}$.

Since p_a can be expressed as a function of r, p_a is constant on sets for which r is constant. Of course, since r is the distance from $\mathbf{x}$ to the origin, it is constant on each sphere centered at the origin. Therefore, p_a is also a constant on each sphere centered at the origin, and the level surfaces are concentric spheres centered at the origin. (See Figure 3.8.) If the radius of the sphere is r_0, then $p_a(\mathbf{x}) = A/r_0$ on the sphere. Further, as the radius increases, $p_a(\mathbf{x})$ decreases, and $p_a(\mathbf{x}) = p_a(\tilde{\mathbf{x}})$ only if $\mathbf{x}$ and $\tilde{\mathbf{x}}$ are the same distance from the origin, that is, $r(\mathbf{x}) = r(\tilde{\mathbf{x}})$. This says that p_a is a strictly decreasing function of the distance from the origin. It follows that the isobar for the pressure level c,

$$\left\{ (x, y, z) \ : \ \frac{A}{r} = c, \ r = \sqrt{x^2 + y^2 + z^2} \ \right\},$$

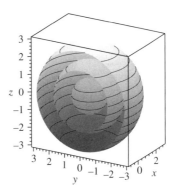

Figure 3.8 Isobars of an acoustic monopole are concentric spheres. Notice the plot is cut off at $x = -1$ to expose the nesting of the isobars. (See Example 3.6.)

is a sphere of radius A/c. In Exercise 15, you will investigate the effect of changing the value of A on the isobars of p_a.

Summary

In this section, we introduced the study of *functions of several variables.* These are functions whose *domain is contained in the plane or space* and that *take values in the real numbers.* We focused on different ways to *represent functions.*

As is the case for functions of one variable, we can represent a function of several variables *numerically*, *symbolically*, or *graphically.* In two variables, our graphical representations include *density plots*, *contour plots*, and *graphs.* In three variables, we are limited to plotting the *level* or *contour surfaces* of a function in space.

We posed fundamental questions that will guide our work in this and the next chapter. These concerned the *values* and *rate of change* of a function. In particular, we defined what it means for a point to be a *maximum, local maximum, minimum,* and *local minimum* value of a function.

Along the way we introduced a variety of different examples of functions that come from outside mathematics, which we will return to in later sections.

Section 3.1 Exercises

1. **Worth Mountain.** Figure 3.9 is a contour plot of Worth Mountain in Central Vermont. As in Example 3.3, the contours on the plot represent the altitude function. On this plot, the darker contours are evenly spaced every 100 ft. A hiking trail is marked on this

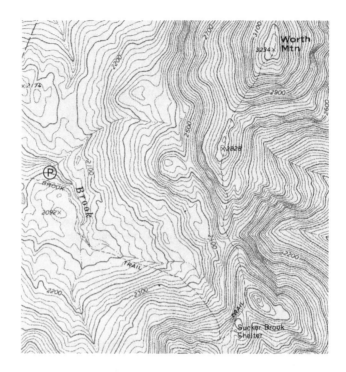

Figure 3.9   A contour plot of Worth Mountain in Vermont showing Sucker Brook Trail and Long Trail. Courtesy of the USGS. (See Exercise 1.)

contour plot. The first portion of the hike follows Sucker Brook Trail from the parking area at 2020 ft to the intersection with Long Trail at 2680 ft. The hike then follows Long Trail to the peak of Worth Mountain at 3234 ft. Suppose you are asked to write a description of this hike for a guide book.

(a) Carefully describe the first portion of the hike along Sucker Brook Trail. Be sure to describe where the hike is steepest and where it is less steep. Approximate the distance from the parking area to the intersection with Long Trail.

(b) Carefully describe the portion of the hike along Long Trail. Be sure to describe any interesting points the trail passes through.

2. **Spruce Mountain.** Figure 3.10 is a contour plot of Spruce Mountain in Northern Vermont. The contours are the level curves of the altitude function of this region. The darker contours are evenly spaced 100 ft apart. Consider the region on this plot that is enclosed by the 2300-ft contour curve that surrounds Spruce Mountain. Carefully describe and sketch the surface that corresponds to this portion of the contour plot.

3. **A Perfect Gas.** In this exercise, we will investigate the graph of the temperature function of a "perfect gas" that was introduced in Example 3.1B. Assume there is a mole of gas,

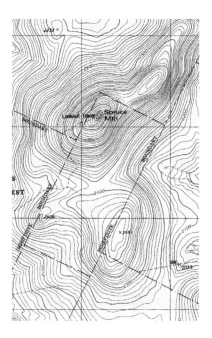

Figure 3.10 A contour plot of Spruce Mountain in Vermont showing the region of the peak inside the 2300-ft contour. Courtesy of the USGS. (See Exercise 2.)

$n = 1$, so that the temperature is given as a function g of its pressure p and volume V by

$$T = g(p, V) = \frac{1}{nR}pV = \frac{1}{8.3145}pV.$$

Note $R = 8.3145$ is the gas constant. Since temperature is a function of the physical quantities pressure and volume, the domain of g should be restricted to the values of p and V that are physically meaningful, $p \geq 0$ and $V \geq 0$.

(a) Sketch several vertical slices of the graph of g parallel to the pT-plane, that is, for several values $V = V_0$. Describe these slices.

(b) Sketch several vertical slices of the graph of g parallel to the VT-plane, that is, for several values $p = p_0$. Describe these slices. (The relationship of the temperature and volume of a fixed quantity of a gas is a form of **Charles' Law.**)

(c) Sketch a contour plot for g in the first quadrant. (The relationship of the pressure and volume of a fixed quantity of gas on a level set of g is a form of **Boyle's Law.**)

(d) Use the information from your plots to sketch the graph of g in the first quadrant. Indicate the horizontal slices of the graph that correspond to the level sets you sketched in part (c).

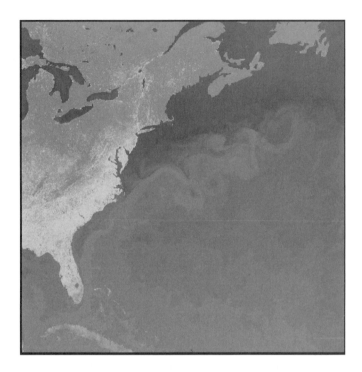

Figure 3.11 A gray scale rendering of an infrared satellite image of the North Atlantic showing the Gulf Stream. Image courtesy of the National Oceanic and Atmospheric Administration. (See Exercise 4.)

4. **The Gulf Stream.** The Gulf Stream is a warm-water current that moves up the east coast of North America. Generally, the water to the north of the Gulf Stream is cooler than the Gulf Stream, and the water in the middle of the Atlantic—the Sargasso Sea—is warmer than the Gulf Stream, as is indicated by Figure 3.11. As the Gulf Stream moves to the north and east, it appears to wander or oscillate on a very large scale. These oscillations are known as "meanders." On occasion, these meanders become large and they "pinch off" to form sizable eddies of water the temperature of the Gulf Stream. Describe in words what you would expect to see in a density plot of the ocean surface temperature of the North Atlantic when a Gulf Stream meander pinches off (i) to the north of the Gulf Stream and (ii) to the south of the Gulf Stream. Explain your answers.

5. **A Density Plot**

   (a) Sketch a density plot of the data set in Figure 3.12 on the square $[0, 6] \times [0, 6]$ using a grid with (i) 6×6 pixels, (ii) 3×3 pixels, (iii) 4×4 pixels, and (iv) 8×8 pixels.

6	0	0	0	0	0	0
5	0	0.5	0.5	0	0.5	0.5
4	1	1	1.5	1.5	1.5	0.5
3	1	1	1.5	2	1.5	1
2	1.5	1.5	2	3	2	0.5
1	1	0.5	0	0	0.5	0.5
	1	2	3	4	5	6

Figure 3.12 Data for Exercise 5.

Be sure to provide a key to explain the correspondence between values and shading.
(b) Explain your scheme for selecting values to use in shading the plots. In particular, explain why your plots are or are not identical.

6. **Density Plots.** For each of the following functions given in symbolic form, construct a density plot on the domain $[-2, 2] \times [-2, 2]$ using 8×8 pixels and 4 shades of gray. Be sure to provide a key to explain the correspondence between values and shading.

(a) $f(x, y) = y$.
(b) $f(x, y) = x + y$.

(c) $f(x, y) = |x + y|$.
(d) $f(x, y) = y - x^2$.

7. **Functions of One Variable**

(a) Suppose that $f(x, y) = g(x)$. Without choosing a particular function g, describe in as much detail as possible the contour plot of f, the vertical slices of the graph of f parallel to the xz-plane, the vertical slices of the graph of f parallel to the yz-plane, and the graph of f.
(b) Repeat part (a) for a function f of the form $f(x, y) = g(y)$.

8. **Extreme Points and Saddles.** For each of the following functions, use the accompanying contour plot to determine the approximate locations of the local extrema, the global extrema, and the saddle points of the function on the domain represented by the contour plot. For each of these points, give the xy-coordinates and the value of the function at the point.

(a) $f(x, y) = x^3 - x - y^2$. (See Figure 3.13(a).)
(b) $f(x, y) = \cos(x) + \sin(y)$. (See Figure 3.13(b).)
(c) $f(x, y) = x^3 - 3xy^2$. (See Figure 3.13(c).)

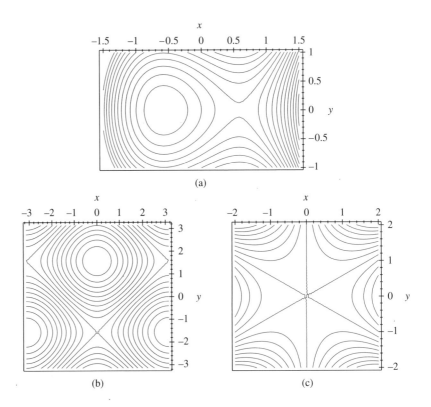

Figure 3.13   (a) A contour plot for $f(x,y) = x^3 - x - y^2$. (b) A contour plot for $f(x,y) = \cos(x) + \sin(y)$. (c) A contour plot for $f(x,y) = x^3 - 3xy^2$. (See Exercise 8(a,b,c).)

9. **Contour Plots.** For each of the following functions given in symbolic form, construct a contour plot on the domain $[-2, 2] \times [-2, 2]$. Be sure to label your contours.

(a) $f(x,y) = x - 2y$.

(b) $f(x,y) = x^2 + 2y^2$.

(c) $f(x,y) = -x^2 + 4y$.

(d) $f(x,y) = 2x^2 - y^2$.

10. **Comparing Plots.** Figure 3.14 contains a contour plot and a plot of the graph of $f(x,y) = 0.5x^2 + x^3 - x^4 - 0.5y^2$ on the domain $[-1, 1.5] \times [-2, 2]$. What features of this function are visible on one of these plots but not the other? Explain your answer.

11. **A Vertical Line Test?** In one-variable calculus, a curve represents the graph of a function if it passes the vertical line test. Is there a vertical line test for surfaces in space? Explain. Give examples of surfaces that pass the vertical line test and others that fail the vertical line test.

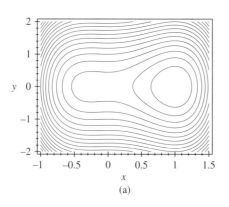 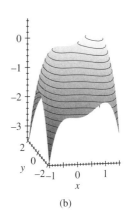

(a)   (b)

Figure 3.14 (a) The contour plot of $f(x, y) = 0.5x^2 + x^3 - x^4 - 0.5y^2$ on the domain $[-1, 1.5] \times [-2, 2]$. (b) The graph of $f(x, y) = 0.5x^2 + x^3 - x^4 - 0.5y^2$ on the domain $[-1, 1.5] \times [-2, 2]$. (See Exercise 10.)

12. **The Method of Slices.** For each of the following functions given in symbolic form: (i) describe and plot the vertical slices of the graph of f that lie over the x- and y-axes; (ii) describe and plot several level sets of f near the origin; and (iii) use these vertical slices and level sets to sketch the graph of f.

   (a) $f(x, y) = x^2 + 2y^2$.
   (b) $f(x, y) = -x^2 + 3y$.

   (c) $f(x, y) = \sqrt{x^2 + y^2}$.
   (d) $f(x, y) = x^2 + y^3$.

13. **Standing Waves.** In Example 3.5, we considered a function that describes a standing wave motion of a taut string. For each of the following functions, describe the corresponding standing wave motion of a string modeled by the interval $[0, L]$ for the given L.

   (a) $u(x, t) = \sin(x)\cos(2t)$, $L = \pi$.
   (b) $u(x, t) = 2\sin(3x)\cos(t)$, $L = \pi$.

   (c) $u(x, t) = \sin(4x)\cos(2t)$, $L = 2\pi$.
   (d) $u(x, t) = -\sin(2x)\cos(3t)$, $L = 3\pi$.

14. **Contour Surfaces.** For each of the following functions given in symbolic form, sketch and describe the contour surfaces for the given contour values on the domain $[-2, 2] \times [-2, 2] \times [-2, 2]$.

   (a) $f(x, y, z) = x - 2y + z$, $c = -2, -1, 0, 1, 2$.
   (b) $f(x, y, z) = x^2 + y^2 + z^2$, $c = 0, 1, 2, 3$.
   (c) $f(x, y, z) = x^2 + y^2 - z^2$, $c = -2, -1, 0, 1, 2$.
   (d) $f(x, y, z) = 4x^2 + 2y^2 + z^2$, $c = 0, 1, 2, 3$.

15. Acoustic Monopole

(a) Consider the monopole source of sound described in Example 3.6. How does changing the value of A change the collection of isobars? Explain your answer.

(b) Does this mathematical model give a reasonable description of isolated sources of sound in the physical world? Explain your answer.

16. Sound Pressure Levels. The expression for loudness or intensity of sound that is commonly used is a function of the rms pressure. If p_a is the rms pressure of an acoustic source thought of as an average pressure difference or variation from the air pressure p_0 in the absence of sound, then the **sound pressure level**, SPL, is defined by

$$ SPL = 20 \log_{10} \frac{p_a + p_0}{p_0}. $$

The unit of sound pressure is the **decibel**. The definition is based on the empirical fact that humans are sensitive to logarithmic changes in the rms pressure rather than linear changes. With normal hearing, humans can detect sounds as soft as a few decibels and feel discomfort with sounds of approximately 120 dB. The normal range for music listening is between 40 and 100 dB.[1] In the following, units of rms pressure are given in micropascals (µPa), with $p_0 = 20$ µPa, and the rms pressure for the monopole is given by $p_a(\mathbf{x}) = A/r$, where A is the amplitude of the source.

(a) Plot the SPL function for the monopole source as a function of r for several values of A. (This is a plot of curves in the plane whose coordinates are r and SPL.)

(b) How do the plots depend on A? How must A change in order to produce a 10 dB increase in the sound pressure level? Explain your answer in terms of the formula for the function SPL.

(c) For what range of values of A will the sound pressure level be in the range 40–100 dB at a distance of 2 m from the source?

(d) If the sound pressure level is 120 dB at a distance of 2 m from the source, how far must you be from the source for the sound pressure level to be 80 dB?

■ 3.2 Directional and Partial Derivatives

The **rate of change of a function at a point** plays a central role in our understanding of functions of one variable, and the same will be true for functions of two and three variables. Indeed, we have already seen in Section 3.1 that it is natural to ask questions

---

[1] *Physics of Musical Instruments*, Neville H. Fletcher and Thomas D. Rossing, Springer-Verlag, 1991.

about the rate of change of a function of two variables, especially when the function represents a physical quantity. Thus in Example 3.3B, we were concerned with the steepness of paths in the region represented by Figure 3.4, that is, with the rate of change of an altitude function in a particular direction. Here are two more physical examples.

Example 3.7	**Rates of Change**

A. Temperature. Figure 3.15 contains a density plot of the temperature T of a thin metal plate. Darker regions represent regions of lower temperature, and lighter regions represent regions of higher temperature. Given a point in the plate, we would like to know how the temperature changes as we move away from the point in a given direction. In particular, we would like to know in which direction the temperature is increasing fastest and in which direction the temperature is decreasing fastest. Or we might ask if there are directions in which the temperature is not changing. For example, at the point P in Figure 3.15, T appears to be increasing if we move in the direction $(-1, -1)$ and decreasing if we move in the direction $(1, 1)$.

B. Loudness. In Example 3.6, we considered the rms pressure of a monopole source of sound, which produces sound at a single frequency. If there are several monopole sources, then as we move around the monopoles, the rms pressure will vary. At a given location, we can ask, for example, in which directions the rms pressure is increasing and in which directions it is decreasing. In particular, we can ask in which direction the rms pressure is increasing most rapidly. With a single source of sound and no other factors to take into account—for example, objects that might reflect or absorb sound—it is intuitively clear

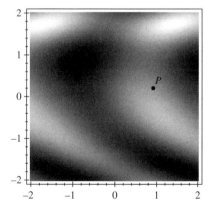

Figure 3.15 A density plot of a temperature function on a domain in the plane that represents the heat in a thin metal plate. Darker regions represent regions of lower temperature and lighter regions represent regions of higher temperature. (See Example 3.7A.)

that the rms pressure increases most rapidly in the direction of the source and decreases most rapidly in the direction opposite that of the source. However, if the source of sound is more complicated, or if other factors affect the rms pressure, these directions are no longer obvious.

We would now like to define the instantaneous rate of change of a function at a point. This will depend on the point and the direction in which the rate is to be calculated. Following the model of one-variable calculus, we will begin by defining an average rate of change and then use a limiting process to define the instantaneous rate of change.

Suppose $\mathbf{x}_0 = (x_0, y_0)$ is in the domain of f and $\mathbf{u} = (u_1, u_2)$ is a unit vector. For any $h \in \mathbb{R}$, $\mathbf{x}_0$ and $\mathbf{x}_0 + h\mathbf{u}$ are the endpoints of a line segment with direction $\mathbf{u}$. The average rate of change of f along this line segment is the quotient of the change in f by the length of the segment. We have the following definition:

Definition 3.3 The *average rate of change of f on the line segment with endpoints* $\mathbf{x}_0$ *and* $\mathbf{x}_0 + h\mathbf{u}$ is the *difference quotient*

$$\frac{f(\mathbf{x}_0 + h\mathbf{u}) - f(\mathbf{x}_0)}{h}. \quad \blacklozenge$$

Intuitively, by choosing smaller values of h, the average rate of change from P to Q will better approximate the behavior of f at P in the direction $\mathbf{u}$. More rigorously, we want to take the limit of average values as h approaches 0.

Definition 3.4 Suppose f is defined near $\mathbf{x}_0 = (x_0, y_0)$. Let $\mathbf{u}$ be a unit vector in $\mathbb{R}^2$. The *instantaneous rate of change* or *directional derivative of f at the point* $\mathbf{x}_0$ *in the direction* $\mathbf{u}$ is

$$\lim_{h \to 0} \frac{f(\mathbf{x}_0 + h\mathbf{u}) - f(\mathbf{x}_0)}{h},$$

if this limit exists. We denote this limit by $D_{\mathbf{u}} f(\mathbf{x}_0)$. $\quad \blacklozenge$

We can, in fact, use the ideas of one-variable calculus to evaluate directional derivatives. Define a function of one variable by $t \to f(\mathbf{x}_0 + t\mathbf{u}) = f(x_0 + tu_1, y_0 + tu_2)$. Using one-variable calculus, the derivative of this function at $t = 0$ is the limit of the difference quotient

$$\lim_{h \to 0} \frac{f(\mathbf{x}_0 + (0 + h)\mathbf{u}) - f(\mathbf{x}_0 + 0 \cdot \mathbf{u})}{h} = \lim_{h \to 0} \frac{f(\mathbf{x}_0 + h\mathbf{u}) - f(\mathbf{x}_0)}{h} = D_{\mathbf{u}} f(\mathbf{x}_0).$$

It follows that the directional derivative $D_{\mathbf{u}} f(\mathbf{x}_0)$ is the derivative of $t \to f(\mathbf{x}_0 + t\mathbf{u})$ at $t = 0$. The following example illustrates how to apply this fact.

Example 3.8

Directional Derivatives. Suppose $f(x, y) = x^2 y$ and $\mathbf{u} = (\frac{1}{\sqrt{5}}, \frac{2}{\sqrt{5}})$. To calculate the directional derivative of f at $\mathbf{x}_0$ in the direction $\mathbf{u}$, we should calculate the derivative of $t \to f(\mathbf{x}_0 + t\mathbf{u}) = (x_0 + \frac{1}{\sqrt{5}}t)^2(y_0 + \frac{2}{\sqrt{5}}t)$. Differentiating this expression with respect to t, we obtain $\frac{2}{\sqrt{5}}(x_0 + \frac{1}{\sqrt{5}}t)(y_0 + \frac{2}{\sqrt{5}}t) + (x_0 + \frac{1}{\sqrt{5}}t)^2 \cdot \frac{2}{\sqrt{5}}$. Evaluating at $t = 0$, we obtain $\frac{2}{\sqrt{5}}x_0 y_0 + \frac{2}{\sqrt{5}}x_0^2$. We conclude that $D_{(1,2)}f(\mathbf{x}_0) = \frac{2}{\sqrt{5}}x_0 y_0 + \frac{2}{\sqrt{5}}x_0^2$.

If we fix our attention on a particular point and allow the direction $\mathbf{u}$ to vary, we can determine in which directions a function increases or decreases by finding the directions for which $D_\mathbf{u} f > 0$ or $D_\mathbf{u} f < 0$. The following example carries out such a calculation.

Example 3.9

Directions of Increase and of Decrease. Once again, let $f(x, y) = x^2 y$ and set $\mathbf{x}_0 = (1, -1)$. The directional derivative of f in the direction $\mathbf{u} = (u_1, u_2)$ is the derivative of the function $t \to (1 + tu_1)^2(-1 + tu_2)$ at $t = 0$. A calculation then shows that $D_\mathbf{u} f(1, -1) = -2u_1 + u_2$. Thus f is increasing in the direction $\mathbf{u}$ if $D_\mathbf{u} f(1, -1) > 0$, that is, if $u_2 - 2u_1 > 0$ or $u_2 > 2u_1$. Similarly, f is decreasing in the direction $\mathbf{u}$ if $D_\mathbf{u} f(1, -1) < 0$, that is, if $u_2 - 2u_1 < 0$ or $u_2 < 2u_1$. The function is neither increasing nor decreasing if $D_\mathbf{u} f(1, -1) = 0$ or $u_2 = 2u_1$.

Let us interpret these calculations geometrically. Since $\mathbf{u}$ is a unit vector, $u_1^2 + u_2^2 = 1$. There are only two solutions to this equation and the equation $u_2 = 2u_1$: $\mathbf{u} = (\frac{1}{\sqrt{5}}, \frac{2}{\sqrt{5}})$ and its negative $-\mathbf{u} = (\frac{-1}{\sqrt{5}}, \frac{-2}{\sqrt{5}})$. These are the directions in which the instantaneous rate of change is zero. These vectors lie in the same line through $(-1, 1)$. This line divides $\mathbb{R}^2$ into two half-planes. (See Figure 3.16.) The half-plane above and to the left of the line corresponds to the directions $\mathbf{u}$ for which $u_2 - 2u_1 > 0$, that is, the directions in which f is increasing at $\mathbf{x}_0$. The half-plane below and to the right of the line corresponds to

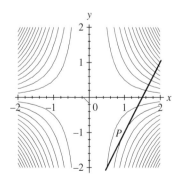

y

Figure 3.16 A contour plot of $f(x, y) = x^2 y$ and line through $\mathbf{x}_0 = (1, -1)$ with direction $\mathbf{u} = (\frac{1}{\sqrt{5}}, \frac{2}{\sqrt{5}})$. The function is increasing in directions from P that point to the left of this line and is decreasing in directions from P that point to the right of this line. (See Example 3.9.)

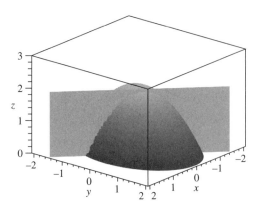

Figure 3.17   The intersection of the graph of a function f with the vertical plane that passes through $(x_0, y_0, 0)$ and contains the horizontal vector $(u_1, u_2, 0)$. The intersection is the graph of the function $t \to f(\mathbf{x} + t\mathbf{u})$.

the directions $\mathbf{u}$ for which $u_2 - 2u_1 < 0$, that is, the directions in which f is decreasing at $\mathbf{x}_0$.

We can identify the directional derivative with the slope of a curve lying on the graph of f. Let L be the line in the xy-plane through $(x_0, y_0, 0)$ in the direction $(u_1, u_2, 0)$. The function $t \to f(\mathbf{x}_0 + t\mathbf{u})$ is the composition $f \circ \alpha$ of f with α where $\alpha(t) = \mathbf{x}_0 + t\mathbf{u}$. The graph of $f \circ \alpha(t)$ can be identified with the curve on the graph of f "lying above" L. That is, it is the curve of intersection of the graph of f with the vertical plane containing L. Then $D_{\mathbf{u}}f(\mathbf{x}_0)$ is the slope of this curve at $(x_0, y_0, f(x_0, y_0))$. See Figure 3.17 for a plot of the intersection. This interpretation will prove particularly useful later in this section, when we introduce partial derivatives.

We can easily generalize the definition of a directional derivative to functions $f : \mathbb{R}^3 \to \mathbb{R}$.

Definition 3.5   Suppose f is defined near the point $\mathbf{x}_0 = (x_0, y_0, z_0)$. Let $\mathbf{u}$ be a unit vector in $\mathbb{R}^3$. Then the ***instantaneous rate of change*** or ***directional derivative*** of f at the point $\mathbf{x}_0$ in the direction $\mathbf{u}$ is

$$\lim_{h \to 0} \frac{f(\mathbf{x}_0 + h\mathbf{u}) - f(\mathbf{x}_0)}{h},$$

if this limit exists. We denote this limit by $D_{\mathbf{u}}f(\mathbf{x}_0)$. ◆

In the following example, we carry out the calculation of the directional derivative of a function of three variables.

Example 3.10 **Directional Derivatives in $\mathbb{R}^3$.** Suppose $f(x, y, z) = y\sqrt{x} + 1/z$ and $\mathbf{x}_0 = (x_0, y_0, z_0)$ is a point in the domain of f, thus $x_0 \geq 0$ and $z_0 \neq 0$. Let us compute the directional derivative of f in the direction $(1, 0, 0)$, $D_{(1,0,0)}f(\mathbf{x}_0)$. This is equal to the derivative of the function $t \to f(x_0 + t, y_0, z_0) = y_0\sqrt{x_0 + t} + 1z_0$ at $t = 0$. Completing this calculation, we have that $D_{(1,0,0)}f(\mathbf{x}_0) = y_0/(2\sqrt{x_0})$.

Like the ordinary derivative of functions of one variable, the directional derivative of a sum of functions is the sum of the directional derivatives, and the directional derivative of the product of a constant and a function is the product of the constant and the directional derivative. We state this as a proposition. It holds in the plane and in space. We leave the proofs to the exercises. (See Exercise 13.)

Proposition 3.1 Let f and g be defined on a domain $\mathcal{D} \subset \mathbb{R}^2$ or $\mathbb{R}^3$, and let $a \in \mathbb{R}$. If the directional derivatives $D_{\mathbf{u}}(f)(\mathbf{x}_0)$ and $D_{\mathbf{u}}(g)(\mathbf{x}_0)$ exist, then

1. $D_{\mathbf{u}}(f + g)(\mathbf{x}_0) = D_{\mathbf{u}}(f)(\mathbf{x}_0) + D_{\mathbf{u}}(g)(\mathbf{x}_0)$.

2. $D_{\mathbf{u}}(af)(\mathbf{x}_0) = aD_{\mathbf{u}}(f)(\mathbf{x}_0)$. ♦

Application: Directionality of Sound

In Example 3.7B, we informally discussed the directionality of sound. We are now in a position to make this rigorous for an acoustic monopole, which we introduced in Example 3.6.

Example 3.11 **The Acoustic Monopole Revisited**

A. The Directional Derivative. The rms pressure at $\mathbf{x}$ of a monopole source of sound located at the origin is given by

$$p_a(\mathbf{x}) = \frac{A}{\|\mathbf{x}\|}.$$

To compute the directional derivative of p_a at $\mathbf{x}_0$ in the direction $\mathbf{u}$, we must compute the ordinary derivative of the function

$$t \to p_A(\mathbf{x}_0 + t\mathbf{u}) = \frac{A}{\sqrt{(x_0 + tu_1)^2 + (y_0 + tu_2)^2 + (z_0 + tu_3)^2}}$$

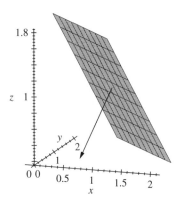

Figure 3.18   The plane passing through $\mathbf{x}_0 = (1,1,1)$ for which $D_\mathbf{u}(p_a) = 0$ in Example 3.11B. Also shown is a unit vector with initial point $\mathbf{x}_0$ for which $D_\mathbf{u}(p_a) > 0$. It points toward the half-space containing the origin, the source of the sound.

at $t = 0$. Carrying out this calculation, we see that

$$D_\mathbf{u} p_A(\mathbf{x}_0) = -\frac{A}{\|\mathbf{x}_0\|^3}(x_0 u_1 + y_0 u_2 + z_0 u_3).$$

B. Which Direction? Let us examine this expression for $D_\mathbf{u} p_a(\mathbf{x}_0)$ further. The coefficient $-A/\|\mathbf{x}_0\|^3$ depends on the point $\mathbf{x}_0$ and not the direction $\mathbf{u}$. It is not defined at the origin and is negative for $\mathbf{x}_0 \neq \mathbf{0}$. The pressure p_a is increasing in the direction $\mathbf{u}$ if $D_\mathbf{u} p_a(\mathbf{x}_0) > 0$, which occurs when $x_0 u_1 + y_0 u_2 + z_0 u_3 < 0$. It is decreasing if $D_\mathbf{u} p_a(\mathbf{x}_0) < 0$, which occurs when $x_0 u_1 + y_0 u_2 + z_0 u_3 > 0$. It is neither increasing nor decreasing when $D_\mathbf{u} p_a(\mathbf{x}_0) = 0$, which occurs when $x_0 u_1 + y_0 u_2 + z_0 u_3 = 0$.

Let us interpret this geometrically. Since $\mathbf{x}_0$ is fixed and nonzero, the set of $\mathbf{u}$ satisfying $x_0 u_1 + y_0 u_2 + z_0 u_3 = 0$ is a plane perpendicular to the vector $\mathbf{x}_0$. (See Figure 3.18.) This plane divides $\mathbb{R}^3$ into two half-spaces, one of which contains the origin. If $x_0 u_1 + y_0 u_2 + z_0 u_3 < 0$, $\mathbf{u}$ points into the half-space containing the origin. Since $D_\mathbf{u} p_a(\mathbf{x}_0) > 0$ in these directions, we see that p_a is increasing in directions pointing from $\mathbf{x}_0$ toward the origin. This coincides with our understanding that as we move toward an acoustic source, the sound should become louder. If $x_0 u_1 + y_0 u_2 + z_0 u_3 > 0$, $\mathbf{u}$ points away from the half-space containing the origin. Since $D_\mathbf{u} p_a(\mathbf{x}_0) < 0$ in these directions, we see that p_a is decreasing in directions pointing from $\mathbf{x}_0$ away from the origin. This coincides with our understanding that as we move away from an acoustic source, the sound should become less loud.

Partial Derivatives

The directional derivatives $D_{(1,0)}f$ and $D_{(0,1)}f$ in the directions of the coordinate axes are of particular importance. These derivatives are called partial derivatives. The following definition introduces the standard notation for partial derivatives in $\mathbb{R}^2$. A similar definition applies in $\mathbb{R}^3$.

Definition 3.6 Let $f : \mathbb{R}^2 \to \mathbb{R}$ be defined in a region containing the point $\mathbf{x}_0 = (x_0, y_0)$.

The **partial derivative of f with respect to** x, denoted $\frac{\partial f}{\partial x}$, is the function whose value at $\mathbf{x}_0$ is the limit

$$\frac{\partial f}{\partial x}(\mathbf{x}_0) = \lim_{h \to 0} \frac{f(x_0 + h, y_0) - f(x_0, y_0)}{h}.$$

The domain of $\frac{\partial f}{\partial x}$ is the set of $\mathbf{x}$ in the domain of f for which this limit exists.

The **partial derivative of f with respect to** y, denoted $\frac{\partial f}{\partial y}$, is the function whose value at $\mathbf{x}_0$ is the limit

$$\frac{\partial f}{\partial y}(\mathbf{x}_0) = \lim_{h \to 0} \frac{f(x_0, y_0 + h) - f(x_0, y_0)}{h}.$$

The domain of $\frac{\partial f}{\partial y}$ is the set of $\mathbf{x}$ in the domain of f for which this limit exists. ◆

At any point $\mathbf{x}_0$ in the domain of $\frac{\partial f}{\partial x}$, $\frac{\partial f}{\partial x}(\mathbf{x}_0) = D_{(1,0)}f(\mathbf{x}_0)$. Similarly, for any point $\mathbf{x}_0$ in the domain of $\frac{\partial f}{\partial y}$, $\frac{\partial f}{\partial y}(\mathbf{x}_0) = D_{(0,1)}f(\mathbf{x}_0)$. The partial derivatives of f at $\mathbf{x}_0$ are the slopes of the vertical slices of the graph of f parallel to the xz- and yz-coordinate planes. The partial derivative $\frac{\partial f}{\partial x}(\mathbf{x}_0)$ is the slope of the curve obtained by intersecting the graph of f with the plane $y = y_0$. That is, it is the slope of the curve obtained by moving through the point $(x_0, y_0, f(x_0, y_0))$ along the graph of f by changing x and leaving $y = y_0$ fixed. (See Figure 3.19(a).) Similarly, $\frac{\partial f}{\partial y}(\mathbf{x}_0)$ is the slope of the curve obtained by intersection of the graph of f with the plane $x = x_0$. It is the slope of the curve obtained by moving through the point $(x_0, y_0, f(x_0, y_0))$ along the graph of f by changing y and leaving $x = x_0$ fixed. (See Figure 3.19(b).)

The limit definition of partial derivatives will lead us to a convenient method of calculation. The partial derivative of f with respect to x is given by

$$\frac{\partial f}{\partial x}(\mathbf{x}_0) = \lim_{h \to 0} \frac{f(x_0 + h, y_0) - f(x_0, y_0)}{h}.$$

This is the derivative at x_0 of the function of x given by $x \to f(x, y_0)$ obtained by holding $y = y_0$ constant and allowing x to vary. Thus, if we have a formula for f, we can

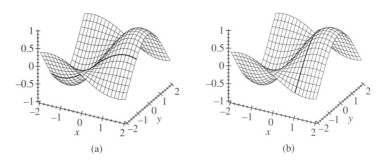

(a)   (b)

Figure 3.19   (a) Graph of $f(x, y)$ with vertical slice at $y = y_0$. (b) Graph of $f(x, y)$ with vertical slice at $x = x_0$.

compute $\frac{\partial f}{\partial x}$ by symbolically differentiating the formula for f with respect to x holding y constant. Similarly, we can compute $\frac{\partial f}{\partial y}$ by symbolically differentiating the formula for f with respect to y holding x constant. Here are several examples of these calculations.

Example 3.12   **Partial Derivatives**

A. Suppose $f(x, y) = x^2 y$. In computing $\frac{\partial f}{\partial x}$, we assume y is constant and differentiate x. This gives

$$\frac{\partial f}{\partial x}(x, y) = (2x)y = 2xy.$$

In computing $\frac{\partial f}{\partial y}$, we assume x, and thus x^2, is constant and differentiate with respect to y. This gives

$$\frac{\partial f}{\partial y}(x, y) = (x^2)(1) = x^2.$$

B. Suppose $f(x, y) = xe^{xy^2}$. Again, to compute $\frac{\partial f}{\partial x}$, we assume y is constant and differentiate x. In this case, we must use the product rule and the chain rule to differentiate this function with respect to x. This gives

$$\frac{\partial f}{\partial x}(x, y) = (x)((y^2)e^{xy^2}) + (1)e^{xy^2} = e^{xy^2}(xy^2 + 1).$$

In computing $\frac{\partial f}{\partial y}$, we hold x constant. In this case, we use the chain rule to differentiate this function as a function of y. This gives

$$\frac{\partial f}{\partial y}(x, y) = (x)e^{xy^2}(2xy) = 2x^2 ye^{xy^2}.$$

C. Suppose

$$f(x,y) = \frac{\sin(x^2 + y^2)}{x^2 + y^2}$$

for $(x,y) \neq (0,0)$ and $f(0,0) = 1$. For all points $(x,y) \neq (0,0)$, we can differentiate this function symbolically. Using the chain rule and the quotient rule, we obtain

$$\frac{\partial f}{\partial x}(x,y) = \frac{(x^2 + y^2)\cos(x^2 + y^2)(2x) - 2x\sin(x^2 + y^2)}{(x^2 + y^2)^2}$$

$$= \frac{2x((x^2 + y^2)\cos(x^2 + y^2) - \sin(x^2 + y^2))}{(x^2 + y^2)^2},$$

for all points $(x,y) \neq (0,0)$.

To compute the partial derivatives at $(0,0)$, we must use the definitions of the partial derivatives; that is, we must compute the limit of the appropriate difference quotients. Thus we have

$$\frac{\partial f}{\partial x}(0,0) = \lim_{h \to 0} \frac{f(0+h,0) - f(0,0)}{h}$$

$$= \lim_{h \to 0} \frac{\frac{\sin(h^2)}{h^2} - 1}{h}$$

$$= \lim_{h \to 0} \frac{\sin(h^2) - h^2}{h^3}.$$

This limit can be evaluated using techniques from one-variable calculus, for example, l'Hôpital's rule. This gives a limit of 0; thus $\frac{\partial f}{\partial x}(0,0) = 0$.

Similarly, for $(x,y) \neq (0,0)$,

$$\frac{\partial f}{\partial y}(x,y) = \frac{2y((x^2 + y^2)\cos(x^2 + y^2) - \sin(x^2 + y^2))}{(x^2 + y^2)^2},$$

and by evaluating a limit, $\frac{\partial f}{\partial y}(0,0) = 0$.

Application: Time-Dependent Functions

In many applications, partial derivatives take on physical meaning. Here are two examples involving functions for which time is one of the independent variables.

Example 3.13 **A Vibrating String.** In Example 3.5, we considered the motion of a vibrating string given by the function $u(x,t) = \sin(2x)\cos(3t)$. The position of the string at time t_0 is given by the portion of the graph of the function $x \to u(x, t_0)$ in the xz-plane extending from $x = 0$ to $x = \pi$. The partial derivative of u with respect to x at time $t = t_0$,

$$\frac{\partial u}{\partial x}(x, t_0) = 2\cos(2x)\cos(3t_0),$$

is the slope of this graph, that is, the slope of the string at position x at time $t = t_0$. On the other hand, if we fix a position x_0 on the string, the partial derivative of u with respect to t,

$$\frac{\partial u}{\partial t}(x_0, t) = -3\sin(2x_0)\sin(3t),$$

is the vertical velocity of the string at that point.

In the following example, we consider a function that describes the blood pressure inside a capillary as a function of position in the capillary and time in the heart cycle. Since there is no symbolic expression for the pressure, we will obtain information about the partial derivatives by analyzing the slices of the plot of the function.

Example 3.14 | **Capillary Pressure.** Figure 3.20 contains a plot of a function that represents the blood pressure $p = p(x, t)$ in a coronary capillary. The plot was constructed from empirical data and is defined on the rectangle $0 \le x \le L$ and $0 \le t \le T$ where L is the length of

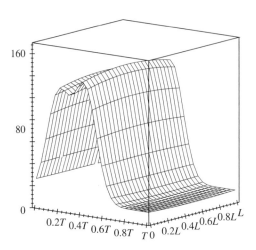

Figure 3.20 A plot of the pressure in a capillary as a function of time and position along the capillary. (See Example 3.14.)

the capillary and T is the length of a complete heart cycle. The heart cycle consists of two phases, **diastole**, when the heart is relaxed, and **systole**, when the heart is exerting pressure. Blood enters the cavities of the heart during diastole and is expelled during systole. In this plot, the beginning of systole occurs at $t = 0$ and the beginning of diastole occurs approximately at $t = 0.6T$.

With the above notation, $\frac{\partial p}{\partial t}(x_0, t)$ is the slope of a vertical slice of this surface parallel to the t-axis. Let us fix $x_0 = 0$ and consider $\frac{\partial p}{\partial t}(0, t)$ as t increases (note the intervals are only approximate):

$$0 < t < 0.2T : p \text{ increases rapidly, } \frac{\partial p}{\partial t}(0, t) \text{ is large and positive.}$$

$$0.2T < t < 0.3T : p \text{ decreases slowly, } \frac{\partial p}{\partial t}(0, t) \text{ is small and negative.}$$

$$0.3T < t < 0.5T : p \text{ increases slowly, } \frac{\partial p}{\partial t}(0, t) \text{ is small and positive.}$$

$$0.5T < t < 0.6T : p \text{ decreases rapidly, } \frac{\partial p}{\partial t}(0, t) \text{ is large and negative.}$$

$$0.6T < t < T : p \text{ decreases slowly, } \frac{\partial p}{\partial t}(0, t) \text{ is small and negative.}$$

The large positive values in the rate of change of pressure correspond to the beginning of systole, when the heart contracts. The large negative values correspond to the beginning of diastole, when the heart relaxes.

Fixing, for example, $t_0 = 0.5T$, we can analyze $\frac{\partial p}{\partial x}(x, 0.5T)$. The behavior is simpler than that for $\frac{\partial p}{\partial t}(x_0, t)$. For $0 < x < 0.5L$, p increases slowly, so that $\frac{\partial p}{\partial x}(x, 0.5T)$ is small and positive. For $0.5L < x < L$, p decreases slowly, so that $\frac{\partial p}{\partial x}(x, 0.5T)$ is small and negative.

Higher Order Derivatives

Since the partial derivatives $\frac{\partial f}{\partial x}$ and $\frac{\partial f}{\partial y}$ are themselves real-valued functions of two variables, we can also consider their partial derivatives. These second order partial derivatives are also real-valued functions of two variables, so we can compute their partial derivatives as well. This process can, of course, continue indefinitely in any region where the partial derivatives of each order are defined. In practice, we will only be using the first and second order partial derivatives of a function.

If f is a function of two variables, it has two first partial derivatives, $\frac{\partial f}{\partial x}$ and $\frac{\partial f}{\partial y}$. To compute the second partial derivatives, we can differentiate each of these with respect to x and y. Thus f has four second partial derivatives. We have the following definition of the second partial derivatives of f.

Definition 3.7 Suppose $f : \mathbb{R}^2 \to \mathbb{R}$ *second order partial derivatives* of f are the functions defined by

$$\tfrac{\partial^2 f}{\partial x^2}(x,y) = \tfrac{\partial}{\partial x}\left(\tfrac{\partial f}{\partial x}(x,y)\right), \qquad \tfrac{\partial^2 f}{\partial y \partial x}(x,y) = \tfrac{\partial}{\partial y}\left(\tfrac{\partial f}{\partial x}(x,y)\right),$$

$$\tfrac{\partial^2 f}{\partial x \partial y}(x,y) = \tfrac{\partial}{\partial x}\left(\tfrac{\partial f}{\partial y}(x,y)\right), \qquad \tfrac{\partial^2 f}{\partial y^2}(x,y) = \tfrac{\partial}{\partial y}\left(\tfrac{\partial f}{\partial y}(x,y)\right). \quad \blacklozenge$$

Example 3.15   **Second Order Partial Derivatives**

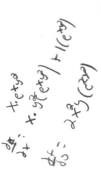

A. Suppose $f(x,y) = x^2 y$. We saw in Example 3.12A that $\frac{\partial f}{\partial x}(x,y) = 2xy$ and $\frac{\partial f}{\partial y}(x,y) = x^2$. In order to compute the second partial derivatives, we must differentiate each of these functions with respect to both x and y. This gives

$$\frac{\partial^2 f}{\partial x^2}(x,y) = \frac{\partial}{\partial x}(2xy) = 2y, \qquad \frac{\partial^2 f}{\partial y \partial x}(x,y) = \frac{\partial}{\partial y}(2xy) = 2x,$$

$$\frac{\partial^2 f}{\partial x \partial y}(x,y) = \frac{\partial}{\partial x}(x^2) = 2x, \qquad \frac{\partial^2 f}{\partial y^2}(x,y) = \frac{\partial}{\partial y}(x^2) = 0.$$

B. Suppose $f(x,y) = xe^{xy^2}$. In Example 3.12B, we computed the partial derivatives of this function to be $\frac{\partial f}{\partial x}(x,y) = e^{xy^2}(xy^2 + 1)$ and $\frac{\partial f}{\partial y}(x,y) = 2x^2 y e^{xy^2}$. The second order partial derivatives $\frac{\partial^2 f}{\partial x^2}$ and $\frac{\partial^2 f}{\partial y \partial x}$ are

$$\frac{\partial^2 f}{\partial x^2}(x,y) = \frac{\partial}{\partial x}(e^{xy^2}(xy^2+1)) = e^{xy^2}(y^2) + e^{xy^2}(y^2)(xy^2+1) = e^{xy^2}(xy^4 + 2y^2).$$

$$\frac{\partial^2 f}{\partial y \partial x}(x,y) = \frac{\partial}{\partial y}(e^{xy^2}(xy^2+1)) = e^{xy^2}(2xy) + e^{xy^2}(2xy)(xy^2+1) = e^{xy^2}(4xy + 2x^2 y^3).$$

In Exercise 11, you are asked to verify that

$$\frac{\partial^2 f}{\partial x \partial y}(x,y) = e^{xy^2}(4xy + 2x^2 y^3) \quad \text{and} \quad \frac{\partial^2 f}{\partial y^2}(x,y) = e^{xy^2}(4x^3 y^2 + 2x^2).$$

In both examples above, the $\frac{\partial^2 f}{\partial x \partial y}(x,y) = \frac{\partial^2 f}{\partial y \partial x}(x,y)$ for all x and y. This will be the case for almost all functions we will be using. The precise result is given in the following theorem.

Theorem 3.1 Let $f : \mathcal{D} \subset \mathbb{R}^2 \to \mathbb{R}$. If the second partial derivatives of f exist and are continuous, then the mixed partial derivatives, $\frac{\partial^2 f}{\partial x \partial y}$ and $\frac{\partial^2 f}{\partial y \partial x}$, are equal. $\quad \blacklozenge$

We will not give a proof of this theorem in this text. However, we will discuss continuous functions in the next section. Intuitively, a function is continuous if the graph of the function has no holes or breaks in it. As was the case in one-variable calculus, most of the functions we will be working with are continuous everywhere in their domains. The requirement that the second partial derivatives exist and are continuous is stronger than requiring that the function be continuous, but will still be true for most of the functions we will be working with.

The definitions of the first and second partial derivatives of a function are easily generalized to functions $f : \mathbb{R}^3 \to \mathbb{R}$.

Definition 3.8 Let f be a real-valued function, $f : \mathbb{R}^3 \to \mathbb{R}$. Then for any point $\mathbf{x}_0 = (x_0, y_0, z_0)$ in the domain of f, the **partial derivatives of f with respect to** x, y, and z are the functions $\frac{\partial f}{\partial x}$, $\frac{\partial f}{\partial y}$, and $\frac{\partial f}{\partial z}$ from $\mathbb{R}^3$ to $\mathbb{R}$ whose values at $\mathbf{x}$ are given by the following limits:

$$\frac{\partial f}{\partial x}(\mathbf{x}_0) = \lim_{h \to 0} \frac{f(x_0 + h, y_0, z_0) - f(x_0, y_0, z_0)}{h}$$

$$\frac{\partial f}{\partial y}(\mathbf{x}_0) = \lim_{h \to 0} \frac{f(x_0, y_0 + h, z_0) - f(x_0, y_0, z_0)}{h}$$

$$\frac{\partial f}{\partial z}(\mathbf{x}_0) = \lim_{h \to 0} \frac{f(x_0, y_0, z_0 + h) - f(x_0, y_0, z_0)}{h}.$$

The domains of $\frac{\partial f}{\partial x}$, $\frac{\partial f}{\partial y}$, and $\frac{\partial f}{\partial z}$ are the sets of all points $\mathbf{x}$ in the domain of f for which the corresponding limit exists. ◆

Analogous to the two variable case, each of the first partial derivatives is again a real-valued function of three variables. The partial derivative $\frac{\partial f}{\partial x}(\mathbf{x})$ gives the rate of change of f with respect to x at the point $\mathbf{x}$ holding y and z constant, $\frac{\partial f}{\partial y}(\mathbf{x})$ gives the rate of change of f with respect to y at the point $\mathbf{x}$ holding x and z constant, and $\frac{\partial f}{\partial z}(\mathbf{x})$ gives the rate of change of f with respect to z at the point $\mathbf{x}$ holding x and y constant.

Each of these first partial derivatives can be differentiated with respect to each of the three variables, giving nine second partial derivatives. Again in this case, if the second partial derivatives are continuous, the mixed partial derivatives will be equal. Thus, for example, the second order partial derivative with respect to x twice and the second order partial derivative with respect to y and x are

$$\frac{\partial^2 f}{\partial x^2}(\mathbf{x}) = \frac{\partial}{\partial x}\left(\frac{\partial f}{\partial x}(\mathbf{x})\right) \text{ and } \frac{\partial^2 f}{\partial x \partial y}(\mathbf{x}) = \frac{\partial}{\partial x}\left(\frac{\partial f}{\partial y}(\mathbf{x})\right) = \frac{\partial}{\partial y}\left(\frac{\partial f}{\partial x}(\mathbf{x})\right).$$

The other second order partials are written in a similar manner.

Summary

In this section, we were concerned with the ***rate of change*** of a function ***at a point***. We defined the ***average rate of change*** of f on the line segment with endpoints $\mathbf{x}_0$ and $\mathbf{x}_0 + h\mathbf{u}$ in the direction $\mathbf{u}$ to be the difference quotient

$$\frac{f(\mathbf{x}_0 + h\mathbf{u}) - f(\mathbf{x}_0)}{h},$$

where $\mathbf{u}$ is a ***unit vector***. The ***instantaneous rate of change at $\mathbf{x}_0$ in the direction $\mathbf{u}$*** is the limit of this difference quotient at $h \to 0$, which we called the ***directional derivative of f at $\mathbf{x}_0$ in the direction $\mathbf{u}$***,

$$D_{\mathbf{u}}f(\mathbf{x}_0) = \lim_{h \to 0} \frac{f(\mathbf{x}_0 + h\mathbf{u}) - f(\mathbf{x}_0)}{h}.$$

Similar definitions and formulas hold for functions defined on $\mathbb{R}^3$. We also applied these constructions to particular examples to find the rates of change of several quantities and to determine directions in which these quantities are ***increasing*** and ***decreasing***.

We paid particular attention to the directional derivatives of f in the direction parallel to the coordinate axes. These are the ***partial derivatives of f***. The ***second order*** partial derivatives of f, $\frac{\partial^2 f}{\partial x^2}$, $\frac{\partial^2 f}{\partial x \partial y}$, and $\frac{\partial^2 f}{\partial y^2}$, are defined by taking the respective partial derivatives in succession. These definitions extend to functions of three variables.

Section 3.2 Exercises

1. **Temperature in a Plate.** Figure 3.21(a) is density plot of a temperature distribution in a metal plate four units on a side. Lighter shadings correspond to higher temperatures and darker shadings correspond to lower temperatures. Based on the plot, estimate the directions in which the temperature is increasing most rapidly at each of the points $(1, 1)$, $(1, -1)$, $(-1, 1)$, and $(-1, -1)$. Specify the directions by giving a unit vector.

2. **Rates of Change by Isotherms.** Figure 3.21(b) is a contour plot of a temperature function defined on the square $[-5, 5] \times [-5, 5]$. The contour curves of temperature are called ***isotherms***. Based on the plot, estimate the directions in which the temperature is increasing most rapidly at each of the points $(2, 2)$, $(2, -2)$, $(-2, 2)$, and $(-2, -2)$. Specify the directions by giving a unit vector. The temperature difference between consecutive isotherms is 1 deg.

3. **Temperature in a Room.** Given the thermal properties of an object or space, it is possible to find a symbolic expression for temperature. For example, assume a rectangular room is 10 ft square and 8 ft high. Further, assume that one of the walls of the room is an external wall. If the temperature along the internal wall opposite the external wall is constant at 70°F, the temperature of the external wall is constant at 55°F, and the

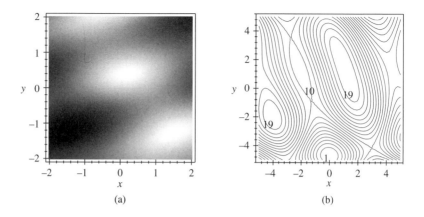

Figure 3.21   (a) The density plot for Exercise 1. (b) The contour plot for Exercise 2. The interval between contour values is 1 deg.

temperature is not changing in time, then temperature in the room is given by

$$T(x, y, z) = 70 - 1.5x,$$

where the room is represented by the set

$$\{(x, y, z) : \ 0 \le x \le 10, \ 0 \le y \le 10, \ 0 \le z \le 8\},$$

with $x = 10$ representing the external wall.

(a) What are the level sets in $\mathbb{R}^3$ of the function T?
(b) Compute the directional derivative of T at the point $\mathbf{x}_0$ in the direction $\mathbf{u}$.
(c) In what directions is $D_{\mathbf{u}}(T)(\mathbf{x}_0)$ positive and what directions is it negative? Does this correspond to your intuition about temperature? Explain.

4. **Mt. Moosilauke.** Figure 3.22 is a geographic contour plot of Mt. Moosilauke in New Hampshire. Note that the adjacent contours represent a change in altitude of 40 feet.

(a) Given a point P on the plot, how would you use the plot to determine in which direction the altitude function is increasing most rapidly? Explain your answer.
(b) Use your answer to part (a) to determine the direction in which the altitude function is increasing most rapidly at the point where Beaver Brook Trail crosses the 4600 ft contour curve to the northeast of the peak of Mt. Moosilauke. Specify the directions by giving a unit vector.
(c) Does your answer to (a) apply at all points on the plot? If so, explain why. If not, give an example of a point where it would not apply. Explain why it does not apply to the second point.

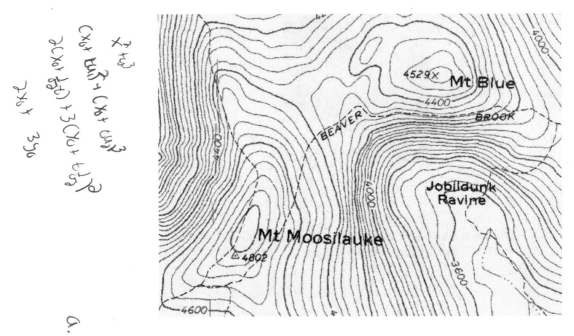

Figure 3.22   A contour plot of Mt. Moosilauke in New Hampshire. The contour curves are for 40-ft intervals of altitude. Courtesy of the USGS. (See Exercise 4.)

5. **Rates of Change at a Peak.** Consider the geographic contour plot in Figure 3.22. Note that the adjacent contours represent a change in altitude of 40 ft. The altitude function has a local maximum of 4529 ft at the peak of Mt. Blue. What are the values of the directional derivative of the altitude function at the maximum? Explain your answer.

6. **Directional Derivative Calculations.** For each of the following functions f, compute $D_{(1/\sqrt{2},1/\sqrt{2})}f(\mathbf{x}_0)$ and $D_{(0,1)}f(\mathbf{x}_0)$.

   (a) $f(x,y) = xy$.
   (b) $f(x,y) = y/x$.
   (c) $f(x,y) = x^2 + y^3$.
   (d) $f(x,y) = \sqrt{x}$.

7. **Directional Derivatives in Space.** For each of the following functions f and unit vectors $\mathbf{u}$, compute $D_{\mathbf{u}}f(\mathbf{x}_0)$.

   (a) $f(x,y,z) = x + y^2 z$ and $\mathbf{u} = (1/\sqrt{2}, 0, -1/\sqrt{2})$.
   (b) $f(x,y,z) = xyz$ and $\mathbf{u} = (0, 1/\sqrt{5}, 2/\sqrt{5})$.
   (c) $f(x,y,z) = \sqrt{yz}$ and $\mathbf{u} = (0, -1, 0)$.
   (d) $f(x,y,z) = z^3$ and $\mathbf{u} = (1/\sqrt{14}, -3/\sqrt{14}, 2/\sqrt{14})$.

8. **Directions of Increase and Decrease in $\mathbb{R}^2$.** For each of the following functions f, compute $D_{\mathbf{u}}f$ at the given point $\mathbf{x}_0$ for $\mathbf{u} = (u_1, u_2)$. Find the directions in which the

rate of change of f is 0, the directions in which f is increasing, and the directions in which f is decreasing at $\mathbf{x}_0$. Sketch a contour plot of f near $\mathbf{x}_0$, shading the half-plane representing directions in which f is increasing at $\mathbf{x}_0$.

(a) $f(x,y) = x + y^2$, $\mathbf{x}_0 = (2,1)$.
(b) $f(x,y) = xy$, $\mathbf{x}_0 = (1,-1)$.
(c) $f(x,y) = 3x^2 + y^2$, $\mathbf{x}_0 = (1,0)$.

9. **Directions of Increase and Decrease in $\mathbb{R}^3$.** For each of the following functions f, compute $D_{\mathbf{u}}f$ at the given point for $\mathbf{u} = (u_1, u_2, u_3)$. Find the directions in which the rate of change of f is 0, the directions in which f is increasing, and the directions in which f is decreasing from this point.

(a) $f(x,y,z) = x^2 + y^2 + z^2$ at the point $(1,2,1)$.
(b) $f(x,y,z) = xy + z$ at the point $(1,1,0)$.

10. **Partial Derivatives.** For each function f below, compute all the first and second order partial derivatives.

(a) $f(x,y) = x^2y + y^3$.
(b) $f(x,y) = e^{xy}$.

(c) $f(x,y) = \sin(x)\cos(y)$.
(d) $f(x,y) = (x^2 + y)^{-1}$.

11. **Verifications.** Verify the partial derivatives $\frac{\partial^2 f}{\partial x \partial y}$ and $\frac{\partial^2 f}{\partial y^2}$ from Example 3.15B.

12. **Partial Derivatives in Space.** For each function f below, compute all the first and second order partial derivatives.

(a) $f(x,y,z) = xyz + x^2$.

(b) $f(x,y,z) = \sqrt{x^2 + y^2 + z^2}$.

13. **Properties of $D_{\mathbf{u}}$.** Assume that f and g are defined on a domain $\mathcal{D} \subset \mathbb{R}^2$ or $\mathbb{R}^3$, and let a, $b \in \mathbb{R}$. Assume the directional derivatives $D_{\mathbf{u}}(f)(\mathbf{x}_0)$ and $D_{\mathbf{u}}(g)(\mathbf{x}_0)$ exist. Use the definition of the directional derivative and properties of limits to prove:

(a) $D_{\mathbf{u}}(f + g)(\mathbf{x}_0) = D_{\mathbf{u}}(f)(\mathbf{x}_0) + D_{\mathbf{u}}(g)(\mathbf{x}_0)$.
(b) $D_{\mathbf{u}}(af)(\mathbf{x}_0) = aD_{\mathbf{u}}(f)(\mathbf{x}_0)$.

14. **A Vibrating String.** To model the sound produced by a vibrating string, it is useful to think of the string as being a straight line. The rms pressure level at a point is then inversely proportional to the distance from the point to the string. Thus the pressure is a function p_a of the form

$$p_a(x,y,z) = \frac{A}{\sqrt{x^2 + y^2}},$$

where A is a positive constant.

(a) Describe the level sets of p_a.

(b) Let $\mathbf{x}_0 = (x_0, y_0, z_0)$ be a point not on the z-axis. Intuitively, in which direction would you expect the rms pressure to be increasing most rapidly and in which direction would you expect it to be decreasing most rapidly?

(c) Compute the directional derivative of p_a in the direction $\mathbf{u}$.

(d) Use your answer to (c) to determine the directions in which p_a is increasing and the directions in which it is decreasing at $\mathbf{x}_0$. Is your answer consistent with your answer to (b)? Explain.

15. **Using the Definition.** Let $f(x, y) = \frac{xy}{x^2+y^2}$ for $(x, y) \neq (0, 0)$ and $f(0, 0) = 0$.

(a) Compute the first partial derivatives of f, $\frac{\partial f}{\partial x}$ and $\frac{\partial f}{\partial y}$, for all points $(x, y) \neq (0, 0)$.

(b) Use the definitions of $\frac{\partial f}{\partial x}$ and $\frac{\partial f}{\partial y}$ to compute these partial derivatives at the point $(0, 0)$.

16. **The Perfect Gas Law.** (See Exercise 3 of Section 3.1.) If we rewrite the **perfect gas law** to express pressure as a function of temperature and volume, we see that the temperature, pressure, and volume of an ideal gas are related in the following way:

$$P = \frac{nRT}{V}.$$

Here n is the number of moles of gas present and R is Boltzmann's constant.

(a) Compute $\frac{\partial P}{\partial T}$ and $\frac{\partial P}{\partial V}$.

(b) Use the partial derivatives to describe the effect on pressure of a unit change in temperature or a unit change in volume. (It is not necessary to work with particular units.)

17. **van der Waals Equation.** The perfect gas law is only an approximation to the behavior of real gases since the derivation of the law assumes that the gas molecules are noninteracting point particles. A more realistic description of the relationship between pressure, temperature, and volume is given by the **van der Waals equation**, which states that

$$P = \left(\frac{nRT}{V - nb}\right) - \frac{n^2 a}{V^2},$$

where a and b are constants that depend on the particular gas, n is the number of moles of gas present, and R is **Boltzmann's constant**.

(a) Compute $\frac{\partial P}{\partial T}$ and $\frac{\partial P}{\partial V}$.

(b) For nitrogen, the constants a and b have values $a = 0.14$ and $b = 3.91 \times 10^{-5}$. Use these constants and the partial derivatives from part (a) to describe the effect on pressure of a unit change in temperature or a unit change in volume. (It is not necessary to work with particular units.)

(c) How does your answer to part (b) above compare to your answers to Exercise 16(b)?

■ 3.3 Limits and Continuous Functions

At this point, we want to pause in our discussion of differentiation to introduce limits and continuous functions. Although both continuity and differentiability can be explained intuitively without the concept of a limit, the careful mathematical definitions rely on limits. The limits in the plane and in space that we introduce here are not the same as the one-variable limit that we used in Section 3.2 to define directional and partial derivatives. Here, the variable of the limit will be a point (x, y) in the plane, and it will approach its limiting value through points in the plane rather than through real numbers.

Open and Closed Sets

An important concept when thinking about limits is what it means for points to be "near" or "close." This can be phrased in terms of *open sets*, which are defined in two steps.

Definition 3.9

1. An *open ball* $\mathcal{B}$ of radius $r > 0$ centered at a point $\mathbf{x}_0 = (x_0, y_0)$ is the set of points $\mathbf{x} = (x, y)$ of distance less than r from $\mathbf{x}_0$. Symbolically, this is given by

$$\mathcal{B} = \{(x, y) : (x - x_0)^2 + (y - y_0)^2 < r^2\}.$$

2. A set $\mathcal{O} \subset \mathbb{R}^2$ is called an *open set* if for every $\mathbf{x}_0 \in \mathcal{O}$, there is an open ball centered at $\mathbf{x}_0$ that is contained in $\mathcal{O}$. ◆

The open ball of radius r centered at $\mathbf{x}_0$ is the set of points inside a circle of radius r centered at $\mathbf{x}_0$. An open set has no particular shape or size except that it contains a nonempty open ball centered at each point. That is, if $\mathcal{O}$ is an open set that contains $\mathbf{x}_0$, there is an $r > 0$ so that

$$\{(x, y) : (x - x_0)^2 + (y - y_0)^2 < r^2\} \subset \mathcal{O}.$$

The choice of r that works for $\mathcal{O}$ and $\mathbf{x}_0$ depends on both $\mathcal{O}$ and $\mathbf{x}_0$. For different points in $\mathcal{O}$ it might be necessary to use balls of different radii. Also, an open ball centered at $\mathbf{x}_0$ that is contained in an open set $\mathcal{O}_1$ may not be contained in another open set $\mathcal{O}_2$ containing $\mathbf{x}_0$. (See Exercise 2.)

Intuitively, a set is an open set if it is defined by strict inequalities, that is by $<$ or $>$, and it is *not* an open set if it is defined by $=$, $\leq$, or $\geq$. More rigorously, in order to show that a set is an open set, we must show that it contains open balls centered at *each* of its points. An open ball is open by this intuitive criterion since it is defined by a strict inequality. Exercise 11 of Section 3.6 asks for a rigorous proof of this fact. To show that a set is not an open set, we need only produce one point for which there is *no* open ball centered at the point and contained in the set.

Example 3.16   **Open Sets**

A. Let $\mathcal{O} = \{(x, y) : x > 0\}$, the right half-plane. (See Figure 3.23(a).) According to our intuition, since this is defined by a strict inequality, it is an open set. For example, the point $(2, 1) \in \mathcal{O}$ and the open ball of radius 1 centered at $(2, 1)$, $\{(x, y) : (x - 2)^2 + (y - 1)^2 < 1\}$, is contained in $\mathcal{O}$. More generally, suppose $\mathbf{x}_0 = (x_0, y_0)$ is a point in $\mathcal{O}$. This means that $x_0 > 0$. An open ball of radius r centered at $\mathbf{x}_0$ will be contained in $\mathcal{O}$ if $r < x_0$. In particular, an open ball of radius $r = \frac{1}{2}x_0$ centered at $\mathbf{x}_0$ is contained in $\mathcal{O}$. Since this works for every point in $\mathcal{O}$, we have shown that $\mathcal{O}$ is an open set. This set $\mathcal{O}$ is called an open half-plane.

B. Let $\mathcal{O} = \{(x, y) : 0 < x < 1 \text{ and } 0 < y < 1\}$, a square of 1 unit on a side. (See Figure 3.23(b).) Once again, our intuition says that $\mathcal{O}$ is an open set. More carefully, if $(x_0, y_0) \in \mathcal{O}$, then we know that both $0 < x_0 < 1$ and $0 < y_0 < 1$. This means that an open ball centered at $\mathcal{O}$ must have radius r that is smaller than the distances from $\mathbf{x}_0$ to each of the four edges of the square: $r < x_0$, $r < 1 - x_0$, $r < y_0$, and $r < 1 - y_0$. Choosing any positive number r that satisfies these four strict inequalities will give us an open ball that is contained in $\mathcal{O}$. Since this was a general argument that can be applied to any point in $\mathcal{O}$, we have shown that $\mathcal{O}$ is an open set.

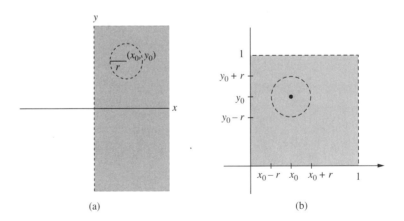

(a)   (b)

Figure 3.23   (a) An open ball contained in the set $\mathcal{O}$ of Example 3.16A. (b) An open ball contained in the set $\mathcal{O}$ of Example 3.16B.

Contrasting with the idea of an open set is that of a closed set. Intuitively, a closed set is defined by equalities, that is by $=$, $\leq$, or $\geq$. For example, the set $\{(x, y) : x \geq 0\}$ is a closed half-plane. The precise definition of a closed set involves the concept of the boundary of a set. We have:

Definition 3.10 Let $\mathcal{S}$ be a subset of $\mathbb{R}^2$.

1. A point $\mathbf{x} \in \mathbb{R}^2$ is a ***boundary point*** of S if every open ball centered at $\mathbf{x}$ contains points that are in S and points that are not in S. The set of boundary points of S is called the ***boundary*** of S. If a point of S is not on the boundary, it is called an ***interior*** point of S.
2. If S contains its boundary, it is called a ***closed*** set. ◆

The following example illustrates both the intuitive and the rigorous definitions of closed sets.

Example 3.17 **A Closed Set.** Let $\mathcal{S} = \{(x, y) : 0 \leq x \leq 1 \text{ and } 0 \leq y \leq 1\}$. (See Figure 3.24.) By our intuitive definition, $\mathcal{S}$ is a closed set since it is defined by equations involving $\leq$ and $\geq$. The boundary of $\mathcal{S}$ consists of the four line segments that make up its edges. However, to properly use the definition, we must analyze open balls centered at points that are in $\mathcal{S}$ and points that are not in $\mathcal{S}$.

From Example 3.16B, we know for every point $\mathbf{x_0} = (x_0, y_0)$ with $0 < x_0 < 1$ and $0 < y_0 < 1$, there is an open ball $\mathcal{B}$ contained in $\mathcal{S}$. This means that the open ball is

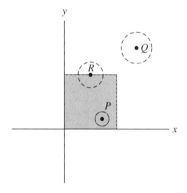

Figure 3.24 $\mathcal{S} = \{(x, y) : 0 \leq x \leq 1 \text{ and } 0 \leq y \leq 1\}$. There is an open ball centered at P that is entirely contained in $\mathcal{S}$, so it is not in the boundary of $\mathcal{S}$. An open ball centered at a point Q contains no points of $\mathcal{S}$, so Q is not in the boundary of $\mathcal{S}$. Every open ball centered at R contains points in $\mathcal{S}$ and points not in $\mathcal{S}$. The boundary of $\mathcal{S}$ consists of the four edges.

contained entirely in S. Thus $\mathbf{x}_0$ is in the interior of S. On the other hand, if a point $\mathbf{x}_0$ lies outside of S, so that $x_0 < 0$, $x_0 > 1$, $y_0 < 0$, or $y_0 > 1$, an argument of the type we made in Example 3.16 shows that there is an open ball $\mathcal{B}$ centered at $\mathbf{x}_0$ that lies entirely outside of S. This means that $\mathbf{x}_0$ is not a boundary point of S. We have yet to consider points lying on the four line segments that make up the edge of S, that is, points $\mathbf{x}_0$ that satisfy $x_0 = 0$, $x_0 = 1$, $y_0 = 0$, or $y_0 = 1$. Let us work with a particular point, say $\mathbf{x}_0 = (\frac{1}{2}, 1)$. Any open ball centered at $\mathbf{x}_0$ contains points $(\frac{1}{2}, y)$ where $y < 1$, which lie inside S, and points $(\frac{1}{2}, y)$ where $y > 1$, which lie outside S. This says that $(\frac{1}{2}, 1)$ is a boundary point of S. We could have made a similar argument for any point in S with $x_0 = 0$, $x_0 = 1$, $y_0 = 0$, or $y_0 = 1$.

To conclude, we have shown that the boundary of S consists of the points inside S that satisfy $x_0 = 0$, $x_0 = 1$, $y_0 = 0$, or $y_0 = 1$. Since S contains its boundary, it is a closed set.

It is worth noting that there are sets that are neither open nor closed. For example, consider $S = \{(x, y) : 0 \le x < 1,\ 0 \le y < 1\}$. The set S is not open since it contains points of the form $(0, y)$, which are on its boundary, and it is not closed because it does not contain, for example, $(\frac{1}{2}, 1)$, which, as we have seen, is a boundary point.

Let us return to the idea of "near" or "close" to a point. When we say that a property holds for points Q that are close to P, we mean that there is an open set $\mathcal{O}$ containing P so that the property holds for every $Q \in \mathcal{O}$. Such an open set is often called a ***neighborhood*** of P. For example, in Section 3.1, we said that a point P is a local maximum of a function f if for all nearby points Q, $f(P) \ge f(Q)$. We can now give this a precise mathematical definition. The point P is a local maximum of f if there exists an open set $\mathcal{O}$ containing P with the property that P is a maximum for f on $\mathcal{O}$, that is, $f(P) \ge f(Q)$ for all $Q \in \mathcal{O}$. In practice, the set $\mathcal{O}$ might be an open ball of a small radius centered at P, so that near means the condition holds for points a small distance from P.

Limits and Continuous Functions

Let us start with an intuitive version of the definition of limit. We suppose that $f = f(x, y)$ is defined on an open set $\mathcal{O}$ in the plane. The limit of f at a point $\mathbf{x}_0 \in \mathcal{O}$ will tell us how the values of f near $\mathbf{x}_0$, which means inside small open balls centered at $\mathbf{x}_0$, relate to the value of f at $\mathbf{x}_0$. We will say that the number L is the limit of f at $\mathbf{x}_0$ if, whenever we want the values of f to be close to L, say within some small positive number ϵ of L, we can choose an open ball $\mathcal{B}$ centered at $\mathbf{x}_0$, with the property that if $\mathbf{x} \in \mathcal{B}$, $f(\mathbf{x})$ is within ϵ of L. Different numbers ϵ may require different open balls $\mathcal{B}$, but in order for us to say the limit is L, we must be able to find a $\mathcal{B}$ for every ϵ. More concisely, we have the following definition of limit:

Definition 3.11   Let f be defined for all points $\mathbf{x} \neq \mathbf{x}_0$ in an open set $\mathcal{O}$ containing $\mathbf{x}_0$. We say L is the *__limit__* of f as $\mathbf{x}$ approaches $\mathbf{x}_0$ if for every $\epsilon > 0$, there is an open ball $\mathcal{B}$ centered at $\mathbf{x}_0$ with the property that for all $\mathbf{x} \in \mathcal{B}$, $\mathbf{x} \neq \mathbf{x}_0$, $|f(\mathbf{x}) - L| < \epsilon$. Symbolically, we write

$$\lim_{\mathbf{x} \to \mathbf{x}_0} f(\mathbf{x}) = L. \quad \blacklozenge$$

Notice that f must be defined in an open set containing $\mathbf{x}_0$, but it need not be defined at the point $\mathbf{x}_0$ itself. The reason for this distinction is that the limit is intended to give us information about the function as we approach the point, but not at the point.

It is a consequence of the definition that *__limits are unique__*. That is, if there are two values K and L that satisfy the ϵ condition of the definition of a limit for all $\epsilon > 0$, then $K = L$. To see this, we can proceed by contradiction. Suppose that $K \neq L$ are both limits of f at $\mathbf{x}_0$. Thus $\epsilon_0 = \frac{1}{2}|K - L| \neq 0$. By the definition of limit, we can find a ball of radius r_1 so that $|f(\mathbf{x}) - K| < \epsilon_0$ for $\|\mathbf{x} - \mathbf{x}_0\| < r_1$ and, similarly, a ball of radius r_2 so that $|f(\mathbf{x}) - K| < \epsilon_0$ for $\|\mathbf{x} - \mathbf{x}_0\| < r_1$. For a point $\mathbf{x}$ in the smaller of the two balls, $f(\mathbf{x})$ is within distance $\frac{1}{2}|K - L|$ from both K and L. This is impossible; thus it must be the case that $K = L$.

In computing the instantaneous rate of change of f at $\mathbf{x}_0$ in Section 3.2, we computed the limit as points $\mathbf{x}$ approach $\mathbf{x}_0$ in one particular direction $\mathbf{u}$. Here, $\mathbf{x}$ is not constrained to approach $\mathbf{x}_0$ from a particular direction. By using the language of open balls, we are allowing $\mathbf{x}$ to approach $\mathbf{x}_0$ from all directions. Intuitively, it follows that if $\lim_{\mathbf{x} \to \mathbf{x}_0} f(\mathbf{x}) = L$, we also have that the limit equals L for any direction $\mathbf{u}$. What is less intuitive is that the converse is false. We will explore this in Exercise 9.

It is tempting to say that the limit of a function f at a point $\mathbf{x}_0$ should be $f(\mathbf{x}_0)$, since this is the case for many functions. However, we know this is not true in one-variable calculus. For example, the function f defined by

$$f(x) = \begin{cases} 0 & \text{if } x \neq 0 \\ 1 & \text{if } x = 0 \end{cases}$$

has $\lim_{x \to 0} f(x) = 0$, but $f(0) = 1$. As was the case in one-variable calculus, we will avoid problems like this by focusing on functions for which $\lim_{\mathbf{x} \to \mathbf{x}_0} f(\mathbf{x}) = f(\mathbf{x}_0)$, that is, on continuous functions. We formalize this in the following definition:

Definition 3.12   Let f be defined on an open set $\mathcal{O}$ containing $\mathbf{x}_0 = (x_0, y_0)$. We say f is *__continuous__* at $\mathbf{x}_0$ if

$$\lim_{\mathbf{x} \to \mathbf{x}_0} f(\mathbf{x}) = f(\mathbf{x}_0).$$

If f is continuous at every point $\mathbf{x} \in \mathcal{O}$, we say that f is *__continuous__* on $\mathcal{O}$. $\blacklozenge$

Thus if a function f is continuous, we know immediately that the limit of f as $\mathbf{x}$ approaches $\mathbf{x}_0$ is just $f(\mathbf{x}_0)$. In order for this to truly simplify matters, we must have some way to identify continuous functions without evaluating limits. We do, because if f is built up from continuous functions by algebraic operations—addition, subtraction, multiplication, division—and by composition, then f is continuous. This only requires that we know that a few basic functions are continuous, in particular, that $f(x, y) = x$, $f(x, y) = y$, and the constant function $f(x, y) = c$ are continuous functions. Exercise 18 of Section 3.6 steps through these arguments. The following example illustrates how we might construct complicated continuous functions from simple continuous functions.

Example 3.18   **Continuous Functions**

A. Polynomials. A polynomial in two variables x and y is a sum of products of powers of x and y. For example, since $f(x, y) = x$ and $f(x, y) = y$ are continuous functions, their product, $f(x, y) = xy$ is also continuous. Similarly, $f(x, y) = 3y^2$ is the product of two copies of $f(x, y) = y$ and the continuous function $f(x, y) = 3$, so it too is continuous. Then $f(x, y) = xy + 3y^2$ is the sum of continuous functions and is therefore continuous.

B. Exponentials. Since $g(z) = e^z$ is a continuous function of one variable and $f(x, y) = xy + 3y^2$ is a continuous function of two variables, the composition, $(g \circ f)(x, y) = e^{xy + y^2}$ is also continuous.

Since we already have a large collection of continuous functions of one variable, we can see from Example 3.18B that there is a similarly large collection of continuous functions of two variables. For the most part, we will not make explicit reference to how a complicated continuous function is obtained from simpler ones; we will simply observe that the function under consideration is continuous.

On occasion, we will also want to consider functions that fail to be continuous either at points or on larger subsets, that is, that have *discontinuities*. The intuitive criteria for recognizing that f is discontinuous at $\mathbf{x}_0$ is that the graph of f has a hole or jump at $\mathbf{x}_0$ or is unbounded at $\mathbf{x}_0$. Conversely, we might say that f is continuous at $\mathbf{x}_0$ if the graph of f has *no* hole or jump at $\mathbf{x}_0$. More rigorously, we would say that f is discontinuous at $\mathbf{x}_0$ if the limit of f as $\mathbf{x}$ approaches $\mathbf{x}_0$ is not equal to $f(\mathbf{x}_0)$. Here are several examples of functions that are discontinuous at the origin.

Example 3.19   **Discontinuities**

A. A Hole. Let $f : \mathbb{R}^2 \to \mathbb{R}$ be defined by

$$f(x, y) = \begin{cases} 0 & \text{if } (x, y) \neq (0, 0) \\ 1 & \text{if } (x, y) = (0, 0). \end{cases}$$

The graph of f consists of the xy-plane with the origin removed and the single point $(0, 0, 1)$. The graph has a hole at the origin. Since $f(x, y) = 0$ for all points not equal to $(0, 0)$, $\lim_{\mathbf{x} \to (0,0)} f(x, y) = 0$. Since $0 \neq f(0, 0)$, f is not continuous at the origin. Of course, if the value of f at the origin were to be redefined to be 0, f would be continuous at the origin.

B. A Jump. Let $f : \mathbb{R}^2 \to \mathbb{R}$ be defined by

$$f(x, y) = \begin{cases} 0 & \text{if } x \leq 0 \\ 1 & \text{if } x > 0. \end{cases}$$

The graph of f consists of two half-planes, one at height 0 for points with $x \leq 0$ and one at height 1 for points with $x > 0$. The graph of f has a jump at the origin and all along the y-axis. In this case, $\lim_{\mathbf{x} \to (0,0)} f(x, y)$ does not exist, and it is impossible to redefine f at the origin to make it continuous there.

C. An Unbounded Function. Let $f : \mathbb{R}^2 \to \mathbb{R}$ be defined by

$$f(x, y) = \begin{cases} \frac{1}{x^2 + y^2} & \text{if } (x, y) \neq (0, 0) \\ 0 & \text{if } (x, y) = (0, 0). \end{cases}$$

Although f is defined for all $\mathbf{x}$, we can see that as $\mathbf{x}$ approaches $(0, 0)$, the value of f increases without bound. That is, $\lim_{\mathbf{x} \to (0,0)} f(x, y) = \infty$. Although this limit exists, it is impossible to redefine f to have a finite value at the origin to make f continuous there.

The definition of the limit of a function and the definition of continuity for functions of two variables can be extended to functions of three variables. The important difference between the two definitions is that the open sets $\mathcal{O}$ are subsets of $\mathbb{R}^3$. Unfortunately, since the graph of a function of three variables is a subset of $\mathbb{R}^4$, the intuitive graphical tests for continuity are unavailable, and it is necessary to work with the symbolic expression for a function.

Application: Potential Energy

When constructing a mathematical model of a chemical reaction, it is often useful to construct a model for the interaction of individual molecules. In particular, this involves developing a symbolic expression for the total energy of the system, which is the sum of the kinetic and potential energies of the molecules. The kinetic energy of a particle is one-half the product of its mass and the square of the length of its velocity. A more interesting problem is to determine the potential energy, which requires knowing the forces that act on the particles and the nature of the particles themselves. Here we will work with the potential energy function U for the interaction of two molecules with opposite electrical charges, so that there is an attractive force between them. Initially we will assume that the molecules are points that have no dimension, and then we will assume that they are three-dimensional "hard spheres."

Example 3.20   **Potential Energy**

A. Molecules as Points. To simplify our formulas, assume that one molecule, M_1 of charge $q_1 > 0$, is located at the origin in space and that a second molecule, M_2 of charge $q_2 < 0$, is located at $\mathbf{x} = (x, y, z)$, so that $\|\mathbf{x}\|$ is the distance between the molecules. The force $\mathbf{F}$ acting on M_2 is given by

$$\mathbf{F} = \frac{q_1 q_2}{4\pi\epsilon_0 \|\mathbf{x}\|^3} \mathbf{x}.$$

(See Example 1.7 of Section 1.2.) Since $q_1 q_2 < 0$, $\mathbf{F}$ points toward the origin and the force is attractive. The potential energy U of the system can also be expressed in terms of $\|\mathbf{x}\|$, so it too is a function of the position of M_2,

$$U(\mathbf{x}) = \frac{q_1 q_2}{4\pi\epsilon_0} \frac{1}{\|\mathbf{x}\|}.$$

This function is continuous for all $\mathbf{x} \neq \mathbf{0}$ and is not continuous at the origin. Since $q_1 q_2 < 0$, we also have that $U(\mathbf{x}) < 0$, which reflects the fact that $\mathbf{F}$ is an attractive force. Note also that as M_2 moves away from M_1, the potential energy U approaches 0.

B. Molecules as Hard Spheres. A more realistic model assumes that the molecules are three-dimensional, that is, they have radii $R_1 > 0$ and $R_2 > 0$, respectively. One approach is to redefine U to have a large positive value for $\|\mathbf{x}\| \leq R_1 + R_2$. Intuitively, this corresponds to a large repulsive force between the molecules that keeps their centers $R_1 + R_2$ units apart, that is, that prevents the two molecules from occupying the same region of space. We will take U to be

$$U(\mathbf{x}) = \begin{cases} \frac{q_1 q_2}{4\pi\epsilon_0} \frac{1}{\|\mathbf{x}\|} & \text{if } \|\mathbf{x}\| > R_1 + R_2 \\ K & \text{if } \|\mathbf{x}\| \leq R_1 + R_2. \end{cases}$$

where K is a positive constant. In this case, U is continuous for all points $\mathbf{x}$ with $\|\mathbf{x}\| \neq R_1 + R_2$ and discontinuous for points with $\|\mathbf{x}\| = R_1 + R_2$. In other words, U is discontinuous on a sphere of radius $R_1 + R_2$ centered at the origin. As M_2 approaches M_1, the potential energy decreases (recall $q_1 q_2 < 0$) until $\|\mathbf{x}\| = R_1 + R_2$, at which point it jumps to K. Since the potential energy of M_2 is large and negative, it cannot make this jump to a large positive potential energy without the addition of energy from an external source. Thus the region for which $U(\mathbf{x}) = K$ forms a **potential barrier** that M_2 cannot cross. We interpret this as meaning that the molecules consist of hard spheres.

Summary   In this section, we developed the language that we will need in the next section to define differentiability. First, we defined **open sets** and **closed sets**, giving definitions in terms of **open balls** and intuitive definitions in terms of inequalities and equalities. The definition of a closed set required the notion of the **boundary** of a set. We interpreted "near to a point" as meaning **within an open set containing the point**.

We then we gave the **ε-definition** of **limit**. A limit of f at $\mathbf{x}$ tells us the behavior of f as we **approach** $\mathbf{x}$ **but not at** $\mathbf{x}$. If the value of f at $\mathbf{x}$ is equal to the limit as f approaches $\mathbf{x}$, then we say that f is **continuous at** $\mathbf{x}$. If a function is continuous at every point in its domain, we say that it is **continuous**.

Section 3.3 Exercises

1. **Constructing Open Balls.** The set $\mathcal{O} \subset \mathbb{R}^2$ defined by

$$\mathcal{O} = \{(x, y) : x > y\}.$$

is an open set. This question asks you for details showing this to be the case.

(a) Sketch the set $\mathcal{O}$ in $\mathbb{R}^2$.
(b) Let $P = (4, 1)$, $P \in \mathcal{O}$. Find an $r > 0$ so that the open ball of radius r centered at P is contained in $\mathcal{O}$.
(c) $P = (x_0, y_0) \in \mathcal{O}$. Find an $r > 0$ so that the open ball of radius r centered at P is contained in $\mathcal{O}$. (*Hint:* Your answer should be that r is less than one or more expressions in x_0 and y_0.)
(d) Based on your answer to (c), is $\mathcal{O}$ an open set? Explain why or why not.

2. **Intersecting Open Sets.** Let $\mathcal{O}_1$ and $\mathcal{O}_2$ be the subsets of $\mathbb{R}^2$ given by

$$\mathcal{O}_1 = \{(x, y) : \ 0 < x < 4\} \quad \text{and} \quad \mathcal{O}_2 = \{(x, y) : \ x > y \text{ and } y > 0\}.$$

According to our intuitive criteria, both $\mathcal{O}_1$ and $\mathcal{O}_2$ are open sets.

(a) Sketch $\mathcal{O}_1$ and $\mathcal{O}_2$ on the same coordinate axes.
(b) Let $P = (2, 1)$. Give an example of an open ball centered at P that is contained in $\mathcal{O}_1$ but not $\mathcal{O}_2$, or an open ball centered at P that is contained in $\mathcal{O}_2$ but not $\mathcal{O}_1$.
(c) What does part (b) say about open balls and open sets?

3. **Open Sets in $\mathbb{R}^2$.** Which of the following subsets of the plane are open sets, closed sets, or neither? Explain your answer.

(a) $\mathcal{M} = \{(x, y) : 0 < \sqrt{x^2 + y^2} \leq 1\}$.
(b) $\mathcal{M} = \{(x, y) : 0 \leq \sqrt{x^2 + y^2} \leq 1\}$.
(c) $\mathcal{M} = \{(x, y) : x > 0 \text{ and } y \geq 0\}$.

(d) $\mathcal{M} = \{(x, y) : -1 \leq x < 1\}$.
(e) $\mathcal{M} = \{(x, y) : 0 < y < 2$ and $-1 < x < 2\}$.

4. **The Boundary.** Sketch each of the following subsets of $\mathbb{R}^2$, shading points that are in $\mathcal{S}$. Label each piece of the boundary of $\mathcal{S}$ by the equality that defines it.

   (a) $\mathcal{S} = \{(x, y) : x \geq 2\}$.
   (b) $\mathcal{S} = \{(x, y) : x \geq 2$ and $y < 1\}$.
   (c) $\mathcal{S} = \{(x, y) : |x| \geq 1$ and $y \leq -2\}$.
   (d) $\mathcal{S} = \{(x, y) : y \geq x^2\}$.
   (e) $\mathcal{S} = \{(x, y) : x^2 + y^2 < 4\}$.
   (f) $\mathcal{S} = \{(x, y) : x^2 + y^2 \ngeq 4\}$.

5. **Closed Sets.** Which of the subsets given in Exercise 4 are closed subsets of the plane? Explain.

6. **Isolated Extrema.** Let f be defined on $\mathbb{R}^2$, and let P be a maximum value of f. We say that P is an *isolated maximum* of f if there is an open set $\mathcal{O}$ containing P so that $f(P) > f(Q)$ for all $Q \in \mathcal{O}$, $Q \neq P$. An *isolated minimum* is defined in a similar manner. For which of the following functions is the origin an isolated minimum or maximum? Explain your answer in each case.

   (a) $f(x, y) = x^2 + y^2$.
   (b) $f(x, y) = (x - y)^2$.
   (c) $f(x, y) = x^2 - y^4$.
   (d) $f(x, y) = e^{-(x^2 + y^2)}$.

7. **Open Sets in $\mathbb{R}^3$.** Which of the following subsets $\mathcal{N}$ of $\mathbb{R}^3$ is an open set, a closed set, or neither? Explain your answer.

   (a) $\mathcal{N} = \{(x, y, z) : x > 0, \ y > 0\}$.
   (b) $\mathcal{N} = \{(x, y, z) : -2 < z \leq 1\}$.
   (c) $\mathcal{N} = \{(x, y, z) : x^2 + y^2 \neq 0\}$.
   (d) $\mathcal{N} = \{(x, y, z) : 0 < x < 1, \ 0 < y < 1, \ 0 < z < 1\}$.
   (e) $\mathcal{N} = \{(x, y, z) : x = 0\}$.
   (f) $\mathcal{N} = \{(x, y, z) : x^2 + y^2 + z^2 \leq 1\}$.

8. **The Boundary of a Subset of $\mathbb{R}^3$.** The definition of the boundary of a subset of space is analogous to that given in the text for subsets of the plane. Describe the boundary of each of the following subsets of space in words *and* in set notation.

   (a) $\mathcal{S} = \{(x, y, z) : z \leq 1\}$.
   (b) $\mathcal{S} = \{(x, y, z) : z < 1\}$.
   (c) $\mathcal{S} = \{(x, y, z) : |x| \leq 1$ and $|y| \leq 3\}$.
   (d) $\mathcal{S} = \{(x, y, z) : z \geq x^2 + y^2\}$.

9. **A Limit That Does Not Exist.** Let $f : \mathbb{R}^2 \rightarrow \mathbb{R}$ be defined by

$$f(x, y) = \begin{cases} 1 & \text{if } xy \geq 0 \\ 0 & \text{if } xy < 0, \end{cases}$$

so that f takes the value 1 in the first and third quadrants and on the axes and takes the value 0 in the second and fourth quadrants.

(a) Let $\mathbf{u} = (\frac{1}{\sqrt{2}}, \frac{1}{\sqrt{2}})$. What is the value of $\lim_{h \to 0} f(h\mathbf{u})$?

(b) Let $\mathbf{u} = (-\frac{1}{\sqrt{2}}, \frac{1}{\sqrt{2}})$. What is the value of $\lim_{h \to 0} f(h\mathbf{u})$?

(c) What do your answers to (a) and (b) tell us about $\lim_{\mathbf{x} \to \mathbf{0}} f(\mathbf{x})$? Explain your answer.

10. **Continuous Functions.** Are the following functions continuous everywhere in $\mathbb{R}^2$? If so, why are they continuous? If not, where do they fail to be continuous, and why are they not continuous at those points? (Intuitive explanations are sufficient.)

(a) $f(x, y) = x^2 - xy^2 - 3$.
(b) $f(x, y) = 1/(x^2 + y^2)$.
(c) $f(x, y) = 1$ if $x^2 + y^2 \leq 1$ and $f(x, y) = 0$ if $x^2 + y^2 > 1$.

11. **Diatomic Molecules.** A *diatomic molecule* consists of two atoms, say A_1 and A_2. The potential energy U of the molecule can be expressed as a function of the distance between the atoms. Working in the plane rather than space, assume A_1 is located at the origin and A_2 is located at $\mathbf{x} = (x, y)$. As in Example 3.20A, U can be expressed as a function of $||\mathbf{x}||$, the distance between the atoms,

$$U(\mathbf{x}) = E_0(e^{-2a(||\mathbf{x}||-r_0)} - 2e^{-a(||\mathbf{x}||-r_0)}),$$

where r_0 is the distance between the atoms when the molecule is in equilibrium, E_0 is the equilibrium potential energy that occurs when $||\mathbf{x}|| = r_0$, and a is related to the frequency of vibration of the molecule. In order to simplify matters, let us assume that $E_0 = 1$, $a = 1$, and $r_0 = 1$, so that $U(\mathbf{x}) = e^{-2(||\mathbf{x}||-1)} - 2e^{-(||\mathbf{x}||-1)}$. (*Hint*: In (b) and (c), you may find it useful to use a computer to plot U as a function of x and y.)

(a) Where is U continuous and where is it discontinuous? Explain.
(b) Describe the graph of U as a function of x and y.
(c) Where do the maximum and minimum values of U occur? Are the extrema isolated? Explain.

■ 3.4 Differentiable Functions

With our work in Section 3.3 on limits and continuity, we are now in a position to address the question of differentiability for functions of two variables. Although we have already introduced directional and partial derivatives, the existence of these derivatives falls short of ensuring that a function of several variables has properties analogous to those of differentiable functions of one variable. We will need to phrase our definition in terms of the several variable limit that we introduced in the previous section.

Before we begin to discuss functions of several variables, let us recall two important properties of differentiable functions of one variable. We will take these two ideas as the starting point of our discussion.

If a function $f : \mathcal{D} \subset \mathbb{R} \to \mathbb{R}$ is differentiable at $x_0 \in \mathcal{D}$, the limit of difference quotients

$$\lim_{x \to x_0} \frac{f(x) - f(x_0)}{x - x_0}$$

exists. This limit is just the derivative of f at x_0 and is denoted by $f'(x_0)$ or $\frac{df}{dx}(x_0)$. Geometrically, we interpret the derivative to be the slope of the tangent line to the graph of f at the point $(x_0, f(x_0))$. Consequently, the **tangent line** is the graph of the linear function $l(x) = f'(x_0)(x - x_0) + f(x_0)$. Also, we called l the **linear approximation** to f at x_0. Intuitively, l approximates f because as $x \to x_0$, the values of l become close to the values of f. More precisely, the function f and the function l satisfy the following limit:

$$\lim_{x \to x_0} \frac{f(x) - l(x)}{x - x_0} = 0.$$

Since the limit is zero, it tells us that the numerator goes to zero as $x \to x_0$ and, since the denominator also goes to zero as $x \to x_0$, that the numerator goes to zero faster than the denominator does.

The limit satisfied by f and l will be the starting point for our discussion of differentiability in two variables. Since we do not yet have an expression for the tangent plane, we will use the limit to find the correct expression for the linear approximation l. First, let us formalize this in the following definition:

Definition 3.13   Suppose that f is defined on an open set $\mathcal{O}$ containing $\mathbf{x}_0$. A **linear function** l,

$$l(x, y) = m_1(x - x_0) + m_2(y - y_0) + b,$$

is a **linear approximation to f at $\mathbf{x}_0$** if

$$\lim_{\mathbf{x} \to \mathbf{x}_0} \frac{f(\mathbf{x}) - l(\mathbf{x})}{\|\mathbf{x} - \mathbf{x}_0\|} = 0.$$

If there is an l so that this limit exists and equals 0, we will say that f is **differentiable** at $\mathbf{x}_0$. If f is differentiable at every $\mathbf{x} \in \mathcal{O}$, we say that f is **differentiable** on $\mathcal{O}$.   ◆

Now let us use the limit to construct l. A linear function l that satisfies $l(\mathbf{x}_0) = f(\mathbf{x}_0)$ can be written in the form

$$l(x, y) = m_1(x - x_0) + m_2(y - y_0) + f(x_0, y_0),$$

where m_1 and m_2 are constants. If l is a linear approximation, it must satisfy

$$\lim_{(x,y) \to (x_0, y_0)} \frac{f(x, y) - l(x, y)}{\|(x, y) - (x_0, y_0)\|} = 0.$$

Since this is a two-variable limit, we know that it must hold no matter how $\mathbf{x}$ approaches $\mathbf{x}_0$. In particular, it must hold if $\mathbf{x}$ approaches $\mathbf{x}_0$ parallel to the x-axis or parallel to the y-axis. First, considering the direction of the x-axis, we must have

$$\lim_{x \to x_0} \frac{f(x, y_0) - (m_1(x - x_0) + m_2(y_0 - y_0) + f(x_0, y_0))}{\|(x, y_0) - (x_0, y_0)\|} = 0.$$

Simplifying, the limit reduces to

$$\lim_{x \to x_0} \frac{f(x, y_0) - (m_1(x - x_0) + f(x_0, y_0))}{|x - x_0|} = \lim_{x \to x_0} \frac{f(x, y_0) - f(x_0, y_0)}{|x - x_0|} - m_1 \frac{x - x_0}{|x - x_0|}.$$

This limit is equal to 0 if and only if the limit of the same expression without the absolute value signs is equal to zero. (This would not be true if the limit were not equal to zero.) That is, if and only if

$$\lim_{x \to x_0} \frac{f(x, y_0) - f(x_0, y_0)}{x - x_0} - m_1 \frac{x - x_0}{x - x_0} = \lim_{x \to x_0} \frac{f(x, y_0) - f(x_0, y_0)}{x - x_0} - m_1 = 0.$$

This limit exists if and only if

$$\lim_{x \to x_0} \frac{f(x, y_0) - f(x_0, y_0)}{x - x_0}$$

exists. But this limit is $\frac{\partial f}{\partial x}(x_0, y_0)$. (In Section 3.2 we used h for $x - x_0$ and let $h \to 0$.) Thus if f is differentiable at (x_0, y_0), $m_1 = \frac{\partial f}{\partial x}(x_0, y_0)$.

Similarly, it can be shown that the partial derivative of f with respect to y at $\mathbf{x}_0$ exists and is equal to m_2. Putting these results together, we have the following proposition, which is analogous to the result for functions of one variable.

Proposition 3.2 If f is differentiable at $\mathbf{x}_0$, the partial derivatives of f at $\mathbf{x}_0$ exist and the linear approximation to f at $\mathbf{x}_0$ is given by

$$l(x, y) = \frac{\partial f}{\partial x}(\mathbf{x}_0)(x - x_0) + \frac{\partial f}{\partial y}(\mathbf{x}_0)(y - y_0) + f(\mathbf{x}_0). \quad \blacklozenge$$

The linear approximation l is also called the **Taylor polynomial of degree one of f at $\mathbf{x}_0$**. Intuitively, the fact that a differentiable function has a linear approximation at $\mathbf{x}_0$ means that near $\mathbf{x}_0$ the graph of f looks like a plane. In fact, we can see this by "zooming in" on the graph of a function at a point. This is demonstrated in Figure 3.25. This interpretation is sometimes referred to by saying that f is "locally linear" at $\mathbf{x}_0$. Given this proposition, the question arises as to whether the converse is true: Does the existence of the partial derivatives of f at a point imply that f is differentiable at that point? In fact, this is not true, as the following example demonstrates.

Example 3.21   **Partial Derivatives and Differentiability.** Consider the function f defined by

$$f(x, y) = \begin{cases} \frac{xy}{x^2+y^2} & \text{if } (x,y) \neq (0,0) \\ 0 & \text{if } (x,y) = (0,0). \end{cases}$$

Away from the origin, f is defined by a quotient of polynomials whose derivative is nonzero, so the partial derivatives of f exist for $\mathbf{x} \neq (0,0)$ and are given by

$$\frac{\partial f}{\partial x}(x, y) = \frac{y(y^2 - x^2)}{x^2 + y^2} \quad \text{and} \quad \frac{\partial f}{\partial y}(x, y) = \frac{x(x^2 - y^2)}{x^2 + y^2}.$$

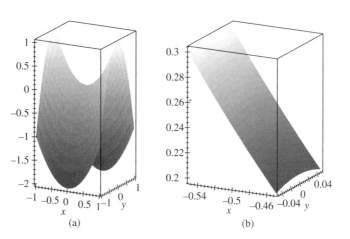

(a)   (b)

Figure 3.25   (a) The graph of $f(x, y) = x^2 - 2y^2$ on the domain $[-1, 1] \times [-1, 1]$. (b) "Zooming in" on the graph of $f(x, y)$ near the point $(-0.5, 0)$.

Since these formulas do not extend to the origin, we have to use the limit definition of the partial derivative at $(0,0)$. For $\frac{\partial f}{\partial x}(0,0)$, we have

$$\lim_{x \to 0} \frac{f(x,0) - f(0,0)}{x - 0} = \lim_{x \to 0} \frac{0 - 0}{x} = 0.$$

So, by a direct calculation, we see that $\frac{\partial f}{\partial x}(0,0) = 0$. Similarly, $\frac{\partial f}{\partial y}(0,0) = 0$. Thus the partial derivatives of f exist for all $\mathbf{x}$, including at the origin. Using the above formula, we define a linear function l by

$$l(\mathbf{x}) = \frac{\partial f}{\partial x}(0,0)x + \frac{\partial f}{\partial y}(0,0)y + f(0,0) = 0 \cdot x + 0 \cdot y + 0 = 0.$$

If f is to be differentiable at the origin, we must have

$$\lim_{\mathbf{x} \to (0,0)} \frac{f(\mathbf{x}) - l(\mathbf{x})}{\|\mathbf{x} - (0,0)\|} = 0.$$

Simplifying the numerator, we see that this limit is equal to

$$\lim_{\mathbf{x} \to (0,0)} \frac{\frac{xy}{x^2+y^2} - 0}{(x^2 + y^2)^{1/2}} = \lim_{\mathbf{x} \to (0,0)} \frac{xy}{(x^2 + y^2)^{3/2}}.$$

If $\mathbf{x} \to (0,0)$ along the line $y = x$, this limit is equal to

$$\lim_{x \to 0} \frac{x^2}{(2x^2)^{3/2}} = \lim_{x \to 0} \frac{x^2}{2^{3/2}|x|^3} = \lim_{x \to 0} \frac{1}{2^{3/2}|x|}.$$

But this limit is $+\infty$, not 0. Thus we conclude that f is not differentiable at the origin.

By an explicit calculation, the example shows that there are functions whose partial derivatives are defined but that are not differentiable. The difficulty here is that although the partial derivatives of f exist for all (x, y), the functions $\frac{\partial f}{\partial x}$ and $\frac{\partial f}{\partial y}$ are not continuous at the origin. Each partial derivative is zero at the origin, but the limit of each partial derivative as $\mathbf{x} \to (0,0)$ is undefined since, for example,

$$\lim_{(x,y) \to (0,0)} \frac{\partial f}{\partial x}(x, y) = \lim_{(x,y) \to (0,0)} \frac{y(y^2 - x^2)}{(x^2 + y^2)}$$

does not exist. It turns out that the continuity of the partial derivatives of f at $\mathbf{x}_0$ is the condition we need to guarantee that f is differentiable. The following proposition summarizes the relationship between the differentiability of a function and its partial derivatives.

Proposition 3.3 Let f be defined on an open set containing $\mathbf{x}_0$.

1. If f is differentiable at $\mathbf{x}_0$, then the partial derivatives of f exist at $\mathbf{x}_0$.
2. If the partial derivatives of f exist and are continuous at $\mathbf{x}_0$, then f is differentiable at $\mathbf{x}_0$. ◆

Application: A Nondifferentiable Function

Although our emphasis will generally be on differentiable functions, there are physical phenomena that display nondifferentiable behavior. Here is one such example from the field of cell biology where the important information about a function is when it fails to be differentiable.

Example 3.22

Signal Transduction in Cells. The concentration of calcium (Ca^{2+}) in a cell varies in response to external stimuli. For example, in higher organisms, the extracellular stimulus might be the binding of a hormone to the cell membrane.[2] The change in concentration is, in fact, a means of communicating information about the stimulus throughout the cell. Intuitively, we would expect that an increased stimulus should result in an increased concentration of calcium and that a decreased stimulus should result in a decreased concentration of calcium. However, in certain stimulus ranges this is not the case. Rather, the stimulus level is encoded by the frequency of a series of "spikes" in the concentration of calcium rather than by the concentration itself. (See Figure 3.26.) The spikes are places where the concentration, as a function of time and stimulus level, fails to be differentiable. Here we will look at a hypothetical graph of the concentration.

The graph of the concentration ρ of Ca^{2+} as a function of the extracellular hormone level h and time t is shown in Figure 3.26(a). It consists of a number of valleys separated by sharp ridges that extend in the positive h direction. For a constant hormone level, $h = h_0$, the graph of $t \to \rho(h_0, t)$ consists of a "train" of spikelike local maxima separated by local minima. (See Figure 3.26(b).) This graph is the vertical slice of the graph of ρ by the plane $h = h_0$. It indicates that the partial derivatives of ρ in the t-direction fail to exist at the local maxima. Thus ρ fails to be differentiable along the crests of the ridges. It is differentiable everywhere else. We can also see from the graph of ρ that the valleys become narrower as h increases, but that the concentrations at the local maxima do not change. That is, the **frequency** of the maxima, the number of maxima per unit time, increases with increasing stimulus, but the **amplitude** of the maxima, the value of ρ at the maxima, does not change.

The phenomena of calcium spiking and frequency encoding were first observed in liver cells in response to the binding of the hormone vasopressin, but it has since been observed

---

[2]See Hans G. Othmer, "Signal Transduction and Second Messenger Systems," *Mathematical Modeling: Ecology, Physiology, and Cell Biology*, Prentice-Hall, Englewood Cliffs, 1997, pp. 99-126.

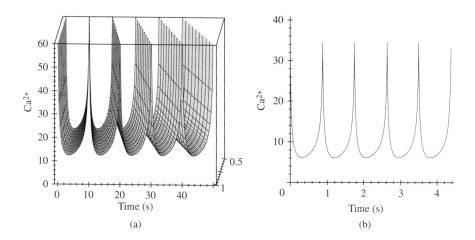

Figure 3.26   (a) A plot simulating the intracellular concentration ρ of calcium (Ca^{2+}) as a function of hormone concentration h and time t. (b) A vertical slice of the graph of ρ parallel to the $t\rho$-plane. (See Example 3.22.)

in many types of cells. The question of whether the spiking, that is, the nondifferentiable behavior of Ca^{2+} concentrations, plays a physiological role has not been resolved. However, there is speculation that the high transient concentrations of Ca^{2+} can be more easily differentiated from fluctuations in the concentration due to other cellular activity. Further, spiking allows cells to keep average concentrations of Ca^{2+} low, thus avoiding potential problems caused by high concentrations of Ca^{2+}, which are toxic.

The Tangent Plane

If f is differentiable at $\mathbf{x}_0$, we can use the linear approximation l to construct a plane that is tangent to the graph of f.

Definition 3.14   If f is differentiable at $\mathbf{x}_0$, the *tangent plane* to the graph of f at $\mathbf{x}_0$ is defined to be the graph of the linear approximation to f at $\mathbf{x}_0$. Thus the equation of the tangent plane is

$$l(x, y) = \frac{\partial f}{\partial x}(\mathbf{x}_0)(x - x_0) + \frac{\partial f}{\partial y}(\mathbf{x}_0)(y - y_0) + f(\mathbf{x}_0). \; \blacklozenge$$

The following example illustrates the calculation of a tangent plane.

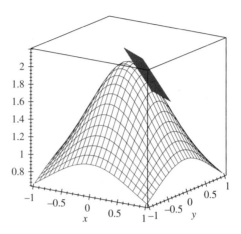

Figure 3.27   A plot of the graph of $f(x, y) = 2/(1 + x^2 + y^2)$ and the tangent plane to the graph of f at the point $(x_0, y_0) = (1/2, 1/4)$. (See Example 3.23.)

Example 3.23   **A Tangent Plane Calculation.** Let $f(x, y) = 2/(1 + x^2 + y^2)$ and $(x_0, y_0) = (1/2, 1/4)$. The partial derivatives of f are

$$\frac{\partial f}{\partial x}(x, y) = \frac{-4x}{(1 + x^2 + y^2)^2} \quad \text{and} \quad \frac{\partial f}{\partial y}(x, y) = \frac{-4y}{(1 + x^2 + y^2)^2}.$$

Since the partial derivatives of f exist and are continuous, f is differentiable. The tangent plane to the graph of f at $\mathbf{x}_0$ is given by

$$\begin{aligned} z &= \frac{\partial f}{\partial x}(1/2, 1/4)(x - 1/2) + \frac{\partial f}{\partial y}(1/2, 1/4)(y - 1/4) + f(1/2, 1/4) \\ &= -\frac{512}{441}(x - 1/2) - \frac{256}{441}(y - 1/4) + \frac{32}{21}. \end{aligned}$$

The graph of f and the tangent plane to the graph of f are shown in Figure 3.27.

Properties of Differentiable Functions

Differentiable functions of two variables satisfy a number of properties that are analogous to those of differentiable functions of one variable. First, there are the algebraic properties.

Proposition 3.4   Let f and g be differentiable functions on an open set $\mathcal{O}$ and let $c \in \mathbb{R}$.

1. cf is a differentiable function on $\mathcal{O}$.

2. $f + g$ and $f - g$ are differentiable functions on $\mathcal{O}$.
3. fg is a differentiable function on $\mathcal{O}$, and if $g(x, y) \neq 0$ on $\mathcal{O}$, f/g is a differentiable function on $\mathcal{O}$. ♦

Although we assumed that f was continuous in our construction of l, we did not assume it in the definition of differentiability. In fact, this is a consequence of differentiability.

Proposition 3.5   If f is differentiable at $\mathbf{x}_0$, f is continuous at $\mathbf{x}_0$. ♦

In subsequent sections, we will usually begin our discussion by setting out conditions on functions necessary for the exposition. Since differentiable functions are continuous, we will not need to specify that a function is continuous and differentiable. It suffices to specify that it is differentiable. However, as we have seen, differentiability of f does not guarantee continuity of its partial derivatives. Thus we will at least require that f be differentiable and that its partial derivatives be continuous.

Our development of the concept of differentiability for functions of two variables and the properties of differentiable functions extends to functions of three variables.

Definition 3.15   A function f is ***differentiable*** at $\mathbf{x}_0 = (x_0, y_0, z_0)$ if there exists a linear function

$$l(\mathbf{x}) = m_1(x - x_0) + m_2(y - y_0) + m_3(z - z_0) + f(\mathbf{x}_0)$$

so that

$$\lim_{\mathbf{x} \to \mathbf{x}_0} \frac{f(\mathbf{x}) - l(\mathbf{x})}{\|\mathbf{x} - \mathbf{x}_0\|} = 0,$$

where $\mathbf{x} = (x, y, z)$. If f is differentiable at $\mathbf{x}_0$, the ***linear approximation*** to f at $\mathbf{x}_0$ is given by

$$l(\mathbf{x}) = \frac{\partial f}{\partial x}(\mathbf{x}_0)(x - x_0) + \frac{\partial f}{\partial y}(\mathbf{x}_0)(y - y_0) + \frac{\partial f}{\partial z}(\mathbf{x}_0)(z - z_0) + f(\mathbf{x}_0). ♦$$

If the first order partial derivatives of f exist and are continuous at $\mathbf{x}_0$, f is differentiable at $\mathbf{x}_0$.

Summary   In this section, we defined differentiability for functions of two variables. A function is ***differentiable*** at $\mathbf{x}_0$ if there exists a linear function

$$l(x, y) = m_1(x - x_0) + m_2(y - y_0) + f(x_0, y_0)$$

that satisfies

$$\lim_{\mathbf{x} \to \mathbf{x}_0} \frac{f(\mathbf{x}) - l(\mathbf{x})}{\|\mathbf{x} - \mathbf{x}_0\|} = 0.$$

We showed that if f is differentiable at $\mathbf{x}_0$, the partial derivatives of f at $\mathbf{x}_0$ exist and are the coefficients of l:

$$l(x, y) = \frac{\partial f}{\partial x}(\mathbf{x}_0)(x - x_0) + \frac{\partial f}{\partial y}(\mathbf{x}_0)(y - y_0) + f(\mathbf{x}_0).$$

We stated the result that **if the partial derivatives of f exist and are continuous at $\mathbf{x}_0$, then f is differentiable at $\mathbf{x}_0$.**

We called this l the **linear approximation** to f at $\mathbf{x}_0$. Intuitively, if we **zoom in** on the graph of f near $\mathbf{x}_0$, it looks increasingly like the graph of l. The graph of l is the **tangent plane** to the graph of f at $\mathbf{x}_0$. Finally, we extended these ideas to functions of three variables.

Section 3.4 Exercises

1. **Differentiability.** In the text, we stated the result that if the partial derivatives of f exist and are continuous, then f is differentiable. Suppose we are given a symbolic expression for f on an open set $\mathcal{O}$.

   (a) Describe how you would use this result to show that f is differentiable.
   (b) Apply this argument to show that $f(x, y) = \sin(x + y)$ is differentiable.

2. **Linear Approximations in $\mathbb{R}^2$.** For each of the following functions $f : \mathbb{R}^2 \to \mathbb{R}$ and points $\mathbf{x}_0 = (x_0, y_0)$, find the linear approximation to f at $\mathbf{x}_0$.

   (a) $f(x, y) = 3x^2 - xy$, $\mathbf{x}_0 = (1, -1)$.
   (b) $f(x, y) = \sin(x + y)$, $\mathbf{x}_0 = (\pi/2, \pi/2)$.
   (c) $f(x, y) = ye^x$, $\mathbf{x}_0 = (0, 2)$.
   (d) $f(x, y) = \frac{1}{x^2 + y^2}$, $\mathbf{x}_0 = (-1, 2)$.

3. **Linear Approximations in $\mathbb{R}^3$.** For each of the following functions $f : \mathbb{R}^3 \to \mathbb{R}$ and points $\mathbf{x}_0 = (x_0, y_0, z_0)$, find the linear approximation to f at $\mathbf{x}_0$.

   (a) $f(x, y, z) = xyz^2$, $\mathbf{x}_0 = (0, 1, -1)$.
   (b) $f(x, y, z) = e^{-(x^2 + y^2 + z^2)}$, $\mathbf{x}_0 = (0, 0, 0)$.
   (c) $f(x, y, z) = \cos(x - y - z)$, $\mathbf{x}_0 = (\pi/4, \pi/4, -\pi/4)$.
   (d) $f(x, y, z) = \ln(\sqrt{x^2 + y^2 + z^2})$, $\mathbf{x}_0 = (0, 1, 0)$.

4. **Differentiable?** For each of the following functions f, determine if f is differentiable at the given $\mathbf{x}_0$. Justify your answer.

   (a) $f(x, y) = x^2 + 3xy + 2y^2$, $\mathbf{x}_0 = (-1, 2)$.
   (b) $f(x, y) = x^{1/3}y^{1/3}$, $\mathbf{x}_0 = (0, 0)$.
   (c) $f(x, y) = \sin(xy)/(xy)$, $\mathbf{x}_0 = (0, 0)$.
   (d) $f(x, y) = |xy|$, $\mathbf{x}_0 = (0, 0)$.

5. **Tangent Planes.** For each of the following functions f and points $\mathbf{x}_0$, find the tangent plane to the graph of f at $\mathbf{x}_0$.

   (a) $f(x, y) = x^2 - 2y^2$, $\mathbf{x}_0 = (1, 1)$.
   (b) $f(x, y) = \cos(x + y) - \sin(x - y)$, $\mathbf{x}_0 = (\pi/2, 0)$.
   (c) $f(x, y) = e^{-x^2 - y^2}$, $\mathbf{x}_0 = (0, 0)$.
   (d) $f(x, y) = 1/(x^2 + y^2)$, $\mathbf{x}_0 = (1, 2)$.

6. **rms Pressure of a Monopole.** In Example 3.6A of Section 3.1, we introduced the function

$$p_a(\mathbf{x}) = \frac{A}{\|\mathbf{x}\|}$$

   to model the rms pressure at $\mathbf{x} = (x, y, z)$ of a monopole source of sound located at the origin, where A is a positive constant that depends on the source. Note that $\|\mathbf{x}\|$ is the distance of the point from the source.

   (a) Where is p_a continuous and where is it not continuous? Explain.
   (b) How might we interpret the continuity of p_a in physical terms? Explain.
   (c) Where is p_a differentiable and where is it not differentiable? Explain.
   (d) How might we interpret the differentiability of p_a in physical terms? Explain.

7. **Points of Nondifferentiability.** Each of the following functions fails to be differentiable at the origin. Give an intuitive explanation for why this is the case. Are the functions continuous at the origin?

   (a) $f(x, y) = |x|$.
   (b) $f(x, y) = \sqrt{x^2 + y^2}$.
   (c) $f(x, y) = (x^2 + y^2)^{1/3}$.
   (d) $f(x, y) = \frac{1}{x^2 + y^2}$.

■ 3.5 The Chain Rule and the Gradient

As in one-variable calculus, the chain rule is an important tool for working with the composition of functions. In its most general form, it says that the composition of differentiable functions is differentiable, and it gives an expression for the derivative of the composition in terms of their derivatives. In calculus, we used the chain rule to compute the derivative of the composition $f \circ g$ of two functions $f : \mathbb{R} \to \mathbb{R}$ and $g : \mathbb{R} \to \mathbb{R}$. In this case, $(f \circ g)'(x) = f'(g(x))g'(x)$. In Section 2.2, we presented the chain rule for the composition $\alpha \circ g$ of a function $\alpha : \mathbb{R} \to \mathbb{R}^2$ with a function $g : \mathbb{R} \to \mathbb{R}$. In this case, $(\alpha \circ g)'(t) = \alpha'(g(t))g'(t)$. In this section, we present the chain rule for the composition $f \circ \alpha$ of a function $f : \mathbb{R}^2 \to \mathbb{R}$ with a function $\alpha : \mathbb{R} \to \mathbb{R}^2$.

Assume that $\alpha : (a, b) \to \mathbb{R}^2$ with $\alpha(t) = (x(t), y(t))$, $f : \mathcal{O} \subset \mathbb{R}^2 \to \mathbb{R}$, and the image of α is contained in $\mathcal{O}$. Then the composition $f \circ \alpha$ is a real-valued function with domain (a, b). Thus, $(f \circ \alpha)(t) = f(\alpha(t)) = f(x(t), y(t))$. We will not prove that the composition is differentiable if f and α are differentiable, but we will determine a formula for the derivative of the composition. Suppose that t_0 is a point in (a, b) and that $\alpha(t_0) = (x(t_0), y(t_0)) = (x_0, y_0)$, so that $(f \circ \alpha)(t_0) = f(x_0, y_0)$.

First, we know that the derivative of the composition at t_0 is given by

$$(f \circ \alpha)'(t_0) = \lim_{t \to t_0} \frac{(f \circ \alpha)(t) - (f \circ \alpha)(t_0)}{t - t_0} = \lim_{t \to t_0} \frac{f(x(t), y(t)) - f(x_0, y_0)}{t - t_0}.$$

We will use the linear approximation $l(x, y)$ of f at (x_0, y_0) in the numerator of the difference quotient. For (x, y) close to (x_0, y_0),

$$f(x, y) \approx l(x, y) = \frac{\partial f}{\partial x}(x_0, y_0)(x - x_0) + \frac{\partial f}{\partial y}(x_0, y_0)(y - y_0) + f(x_0, y_0).$$

Since α is continuous, if t is close to t_0, $\alpha(t) = (x(t), y(t))$ is close to $\alpha(t_0) = (x_0, y_0)$, and we have

$$f(x(t), y(t)) - f(x_0, y_0) \approx l(x(t), y(t)) - f(x_0, y_0)$$
$$= \frac{\partial f}{\partial x}(x_0, y_0)(x(t) - x_0) + \frac{\partial f}{\partial y}(x_0, y_0)(y(t) - y_0).$$

Dividing this last expression by $t - t_0$ and simplifying, we can approximate the above difference quotient by

$$\frac{\partial f}{\partial x}(x_0, y_0)\frac{x(t) - x_0}{t - t_0} + \frac{\partial f}{\partial y}(x_0, y_0)\frac{y(t) - y_0}{t - t_0}.$$

If we take the limit of this expression as $t \to t_0$, we obtain the following formula for the limit; hence for the derivative of the composition,

$$\frac{\partial f}{\partial x}(x_0, y_0)x'(t_0) + \frac{\partial f}{\partial y}(x_0, y_0)y'(t_0).$$

Replacing x_0 and y_0 by $x(t_0)$ and $y(t_0)$, we have the formula for the chain rule. We summarize this in the following proposition.

Proposition 3.6 **The Chain Rule.** Let f be a differentiable function on the open set $\mathcal{O} \subset \mathbb{R}^2$, and let $\alpha : (a, b) \to \mathbb{R}^2$ be a differentiable function whose image is contained in $\mathcal{O}$.

    1. The composition $f \circ \alpha$ is differentiable.
    2. The derivative of $f \circ \alpha$ at $t_0 \in (a, b)$ is given by

$$(f \circ \alpha)'(t_0) = \frac{\partial f}{\partial x}(x(t_0), y(t_0))x'(t_0) + \frac{\partial f}{\partial y}(x(t_0), y(t_0))y'(t_0). \quad \blacklozenge$$

We can also interpret the composition of f and α and its derivative geometrically. The image of α is a curve in $\mathbb{R}^2$ parametrized by α, which is contained in the domain of f. The composition $f \circ \alpha$ gives the values of the function f along this curve, and the derivative of the composition, $(f \circ \alpha)'$, is the instantaneous rate of change of f along this curve. In the following example, we use the chain rule to compute the instantaneous rate of change of a function f on the unit circle parametrized by $\alpha(t) = (\cos t, \sin t)$.

Example 3.24

The Chain Rule. Let $f(x, y) = x^2 + xy^3$ and $\alpha(t) = (\cos t, \sin t)$. The derivative of the composition of f with α is given by

$$(f \circ \alpha)'(t_0) = \frac{\partial f}{\partial x}(x(t_0), y(t_0))x'(t_0) + \frac{\partial f}{\partial y}(x(t_0), y(t_0))y'(t_0)$$

$$= (2\cos(t_0) + \sin^3(t_0))(-\sin(t_0)) + (3\cos(t_0)\sin^2(t_0))\cos(t_0).$$

For example, if $t_0 = \pi/4$, then

$$(f \circ \alpha)'(\pi/4) = (2\frac{\sqrt{2}}{2} + (\frac{\sqrt{2}}{2})^3)(-\frac{\sqrt{2}}{2}) + (3\frac{\sqrt{2}}{2}(\frac{\sqrt{2}}{2})^2)\frac{\sqrt{2}}{2} = -\frac{1}{2}.$$

This calculation shows that the instantaneous rate of change of f on the unit circle parametrized by α at $t_0 = \pi/4$ is $-1/2$.

An immediate consequence of the chain rule is an expression for the directional derivative of a function in terms of its partial derivatives. In Section 3.2, we said that the directional derivative $D_{\mathbf{u}}f(\mathbf{x}_0)$ was equal to the derivative of the composition of f and the parametrization $\alpha(t) = \mathbf{x}_0 + t\mathbf{u}$ of the line through $\mathbf{x}_0$ with direction vector $\mathbf{u} = (u_1, u_2)$, where $\mathbf{u}$ is a unit vector. Notice $\alpha'(t) = \mathbf{u}$. Applying the chain rule to this composition, we have the following result:

Proposition 3.7 If f is a differentiable function on an open set $\mathcal{O}$ containing $\mathbf{x}_0 = (x_0, y_0)$ and $\mathbf{u}$ is a unit vector, then the directional derivative of f at $\mathbf{x}_0$ in the direction $\mathbf{u}$ is given by

$$D_{\mathbf{u}}f(\mathbf{x}_0) = \frac{\partial f}{\partial x}(\mathbf{x}_0)u_1 + \frac{\partial f}{\partial y}(\mathbf{x}_0)u_2. \; \blacklozenge$$

The Gradient

The previous results can be simplified and explored more readily if we express them in the language of vectors.

Definition 3.16 If f is differentiable at $\mathbf{x}_0$, we define the **gradient vector** of f at $\mathbf{x}_0$ to be the vector whose entries are the partial derivatives of f at $\mathbf{x}_0$. We denote this vector by $\nabla f(\mathbf{x}_0)$. Thus

$$\nabla f(\mathbf{x}_0) = (\frac{\partial f}{\partial x}(\mathbf{x}_0), \frac{\partial f}{\partial y}(\mathbf{x}_0)). \; \blacklozenge$$

If f is differentiable on the open set $\mathcal{O}$, the partial derivatives of f are also defined on all of $\mathcal{O}$, and this formula defines a vector field ∇f on $\mathcal{O}$, which we call the **gradient vector field** of f on $\mathcal{O}$. The coordinate functions of ∇f are the partial derivatives of f:

$$\nabla f(\mathbf{x}) = (\frac{\partial f}{\partial x}(\mathbf{x}), \frac{\partial f}{\partial y}(\mathbf{x})).$$

If f is differentiable on the open set $\mathcal{O}$ containing $\mathbf{x}_0$, and the image of $\alpha : [a, b] \to \mathbb{R}^2$ is contained in $\mathcal{O}$, the instantaneous rate of change of f along the image of α is given by

$$(f \circ \alpha)'(t) = \frac{\partial f}{\partial x}(x(t), y(t))x'(t) + \frac{\partial f}{\partial y}(x(t), y(t))y'(t)$$
$$= \nabla f(\alpha(t)) \cdot \alpha'(t).$$

In addition, if $\mathbf{u} = (u_1, u_2)$ is a unit vector, the directional derivative of f at a point $\mathbf{x} = (x, y)$, $D_{\mathbf{u}}f(\mathbf{x})$, is given by

$$D_{\mathbf{u}}f(\mathbf{x}) = \frac{\partial f}{\partial x}(\mathbf{x})u_1 + \frac{\partial f}{\partial y}(\mathbf{x})u_2$$
$$= \nabla f(\mathbf{x}) \cdot \mathbf{u}.$$

Since these two formulas involving the gradient will be used regularly, let us summarize them here.

∇f *Formulas*

1. If f is differentiable and $\alpha : [a, b] \to \mathbb{R}^2$ is a differentiable function, then

$$(f \circ \alpha)'(t) = \nabla f(\alpha(t)) \cdot \alpha'(t).$$

2. If f is differentiable and $\mathbf{u}$ is unit vector, then

$$D_{\mathbf{u}}f(\mathbf{x}) = \nabla f(\mathbf{x}) \cdot \mathbf{u}.$$

Expressed in this way, these two results provide information about the behavior of f at $\mathbf{x}_0$. First, let us investigate the instantaneous rate of change of f along a level curve of f at a point where $\nabla f(\mathbf{x}_0) \neq \mathbf{0}$. Suppose that α is a parametrization of the level curve through $\mathbf{x}_0$ that satisfies $\alpha(0) = \mathbf{x}_0$ and $\alpha'(0) \neq \mathbf{0}$. Since the image of α is contained in a level curve, f is constant on the image of α. Thus the rate of change of f along the curve parametrized by α is 0. Therefore, $(f \circ \alpha)'(0) = 0$, which gives

$$\nabla f(\alpha(0)) \cdot \alpha'(0) = 0.$$

That is, $\nabla f(\mathbf{x}_0)$ is orthogonal to α' at $\mathbf{x}_0 = \alpha(0)$. In particular, since α' is tangent to the level curve, $\nabla f(\mathbf{x}_0)$ *is orthogonal to the level curve of* f, and the direction of $\nabla f(\mathbf{x}_0)$ is one of the two directions orthogonal to the level curve of f through $\mathbf{x}_0$ if $\alpha'(0) \neq \mathbf{0}$.

In order to determine which orthogonal direction is the direction of $\nabla f(\mathbf{x}_0)$, we must examine the directional derivatives of f. We have that

$$\begin{aligned} D_{\mathbf{u}}f(\mathbf{x}_0) &= \nabla f(\mathbf{x}_0) \cdot \mathbf{u} \\ &= \|\nabla f(\mathbf{x}_0)\| \, \|\mathbf{u}\| \cos \theta, \end{aligned}$$

where θ is the angle between $\nabla f(\mathbf{x}_0)$ and $\mathbf{u}$. Since $\mathbf{u}$ is a unit vector,

$$D_{\mathbf{u}}f(\mathbf{x}_0) = \|\nabla f(\mathbf{x}_0)\| \cos \theta.$$

It follows that $D_{\mathbf{u}}f$ attains a maximum value when $\theta = 0$, that is, when the direction of $\mathbf{u}$ is that of $\nabla f(\mathbf{x}_0)$. Thus *the direction of* $\nabla f(\mathbf{x}_0)$ *is the direction of the maximum increase of* f at a point where $\nabla f \neq \mathbf{0}$. Moreover, the rate of change of f at the point $\mathbf{x}_0$ in the direction of $\nabla f(\mathbf{x}_0)$ is equal to $\|\nabla f(\mathbf{x}_0)\|$. Thus we have proven the following proposition.

Proposition 3.8 Let f be a differentiable function on an open set $\mathcal{O} \subset \mathbb{R}^2$, and let $\mathbf{x}_0 \in \mathcal{O}$ with $\nabla f(\mathbf{x}_0) \neq \mathbf{0}$.

1. $\nabla f(\mathbf{x}_0)$ is orthogonal to the level set of f at $\mathbf{x}_0$, and it points in the direction of maximum increase of f at $\mathbf{x}_0$.
2. The maximum rate of increase of f at $\mathbf{x}_0$ is $\|\nabla f(\mathbf{x}_0)\|$, the length of $\nabla f(\mathbf{x}_0)$.   ◆

If we change our point of view to focus on the gradient vector field of f, we obtain additional information about the behavior of f. A flow line of ∇f is a function $\alpha : \mathbb{R} \to \mathbb{R}^2$ with the property that $\alpha'(t) = \nabla f(\alpha(t))$. Since the tangent vector to α is a gradient vector, we know from above that it must be orthogonal to the level set of f. Further, we know that it points in the direction of maximum increase of f at the point, which is always positive. It follows that f is increasing on each flow line as t increases. Rephrasing the above results, we have the following proposition about the gradient vector field.

Proposition 3.9 Let f be a differentiable function on the open set $\mathcal{O} \subset \mathbb{R}^2$ and let $\alpha : \mathbb{R} \to \mathcal{O}$ be a flow line of ∇f. At every point for which $\alpha'(t) \neq \mathbf{0}$, α is orthogonal to the level set of f and points in the direction of maximum increase of f. Further, f is an increasing function on each flow line of ∇f on which $\nabla f \neq \mathbf{0}$.   ◆

Functions of Three Variables

Our previous results can be extended to functions of three variables. If $f : \mathcal{O} \subset \mathbb{R}^3 \to \mathbb{R}$ is a differentiable function and $\mathbf{x}_0 \in \mathcal{O}$, we define the **gradient vector of f at $\mathbf{x}_0$** by

$$\nabla f(\mathbf{x}_0) = \left(\frac{\partial f}{\partial x}(\mathbf{x}_0), \frac{\partial f}{\partial y}(\mathbf{x}_0), \frac{\partial f}{\partial z}(\mathbf{x}_0) \right).$$

Let $\mathbf{x}_0$ be a point in the level set of f for the value c, that is $f(\mathbf{x}_0) = c$, which satisfies $\nabla f(\mathbf{x}_0) \neq \mathbf{0}$. Suppose that α is a parametrization of a curve that lies in the level set of f for the value c and that satisfies $\alpha(0) = \mathbf{x}_0$ and $\alpha'(0) \neq \mathbf{0}$. Then, as above, we have

$$\nabla f(\alpha(0)) \cdot \alpha'(0) = 0.$$

It follows that $\nabla f(\mathbf{x}_0)$ is orthogonal to $\alpha'(0)$. Thus for any curve through the point $\mathbf{x}_0$ that is contained in the level set of f, $\nabla f(\mathbf{x}_0)$ is orthogonal to the tangent vector of the curve at this point. Since $\nabla f(\mathbf{x}_0) \neq \mathbf{0}$, this is equivalent to saying that these tangent vectors are contained in the plane orthogonal to $\nabla f(\mathbf{x}_0)$ that passes through the point $\mathbf{x}_0$.

We will use this construction to define the tangent plane to a surface that is represented as the level surface of a function of three variables.

Definition 3.17   If a surface $\mathcal{S}$ is the level set of a function of three variables for the value c, that is, $\mathcal{S} = \{(x, y, z) : f(x, y, z) = c\}$, then at a point $\mathbf{x}_0 = (x_0, y_0, z_0) \in \mathcal{S}$ with $\nabla f(\mathbf{x}_0) \neq \mathbf{0}$, we define the ***tangent plane*** to $\mathcal{S}$ at $(\mathbf{x}_0)$ to be the plane

$$\{(x, y, z) : \nabla f(\mathbf{x}_0) \cdot (x - x_0, y - y_0, z - z_0) = 0\}. \quad \blacklozenge$$

Earlier in this section, we defined the tangent plane to the graph of a function. If a surface can be represented both as the level set of a function of three variables and as the graph of a function of two variables, these two definitions coincide. We demonstrate this in the following example.

Example 3.25   **Tangent Planes**

A. Let $f(x, y, z) = x^2 + y^2 + z^2$ and let $(x_0, y_0, z_0) = (1, 1, 2)$. Then $f(1, 1, 2) = 6$, and the level set of f passing through $(1, 1, 2)$ is a sphere of radius $\sqrt{6}$ centered at the origin. We have $\nabla f(1, 1, 2) = (2, 2, 4)$, and the tangent plane to the level set of f at $(1, 1, 2)$ is the plane given by

$$\begin{aligned}
0 &= \nabla f(x_0, y_0, z_0) \cdot (x - x_0, y - y_0, z - z_0) \\
&= (2, 2, 4) \cdot (x - 1, y - 1, z - 2) \\
&= 2(x - 1) + 2(y - 1) + 4(z - 2).
\end{aligned}$$

(See Figure 3.28.)

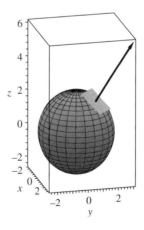

Figure 3.28   The tangent plane and gradient vector to the level surface $f(x, y, z) = x^2 + y^2 + z^2 = 6$ at the point $(1, 1, 2)$. (See Example 3.25A.)

B. The upper hemisphere of the level set of part A is the graph of the function $g(x, y) = \sqrt{6 - x^2 - y^2}$. In Section 3.4, we defined the tangent plane to the graph of g to be the graph of the linear function $z = l(x, y)$ given by

$$z = \frac{\partial g}{\partial x}(x_0, y_0)(x - x_0) + \frac{\partial g}{\partial y}(x_0, y_0)(y - y_0) + g(x_0, y_0)$$

$$= \frac{\partial g}{\partial x}(1, 1)(x - 1) + \frac{\partial g}{\partial y}(1, 1)(y - 1) + 2$$

$$= (-1/2)(x - 1) + (-1/2)(y - 1) + 2.$$

Rearranging this formula yields the formula in part A. This shows that the implicit (part A) and explicit (part B) derivations of the tangent plane to this surface are equivalent.

Application: Conservative Forces

The relationship between a function and its gradient is useful in understanding the relationship between a force and the corresponding potential energy when the force is **conservative**. The following example presents the general relationship and applies it to the model for charged particles that we considered in Example 3.20A of Section 3.3.

Example 3.26   Conservative Forces

A. Potential Energy and the Gradient. Suppose that a particle of mass m moves through space and is subject only to the force $\mathbf{F}$ once it is set in motion. Let $\alpha(t) = (x(t), y(t), z(t))$ be a parametrization of the motion of the particle, and let $\mathbf{v}(t) = \alpha'(t)$ denote the velocity of the particle and $\mathbf{a}(t) = \alpha''(t)$ denote the acceleration of the particle. The **kinetic energy** of the particle is the quantity $\frac{1}{2}m\mathbf{v} \cdot \mathbf{v}$. The **potential energy** U of the particle depends on the force and the particle. The total energy E of the system is the sum of the kinetic energy and the potential energy. In our notation,

$$E(\alpha(t)) = \frac{1}{2}m\mathbf{v}(t) \cdot \mathbf{v}(t) + U(\alpha(t)).$$

Many force fields, including electric fields, are **conservative**. One way to express this is that **the total energy remains constant over time when the particle is put in**

motion. If we assume that $\mathbf{F}$ is conservative, so that E is constant, then $(E \circ \alpha)'(t_0) = 0$ for t_0 in the domain of α. Substituting the expression for E, we have

$$\frac{d}{dt}\left(\frac{1}{2}m\mathbf{v}(t) \cdot \mathbf{v}(t) + U(\alpha(t))\right) = m\mathbf{v}(t) \cdot \mathbf{a}(t) + \nabla U(\alpha(t)) \cdot \mathbf{v}(t)$$

$$= \mathbf{v}(t) \cdot (m\mathbf{a}(t) + \nabla U(\alpha(t))),$$

where we differentiated the dot product using Exercise 3(c) of Section 2.2. Since, according to ***Newton's second law***, $\mathbf{F} = m\mathbf{a}$, we can substitute into the last equation to arrive at

$$\mathbf{v}(t) \cdot (\mathbf{F}(\alpha(t)) + \nabla U(\alpha(t))) = 0.$$

Now let us focus our attention on the force at a particular point $\mathbf{x}_0$, and let us suppose that $\alpha(t_0) = \mathbf{x}_0$. The result of the derivative calculation can be rewritten

$$\mathbf{v}(t_0) \cdot (\mathbf{F}(\mathbf{x}_0) + \nabla U(\mathbf{x}_0)) = 0.$$

This equation holds regardless of the direction that the particle takes in passing through $\mathbf{x}_0$, that is, regardless of the vector $\mathbf{v}(t_0)$. Since this dot product is zero for all possible vectors $\mathbf{v}$, a calculation shows that the second vector in the dot product must be the zero vector. That is,

$$\mathbf{F}(\mathbf{x}_0) + \nabla U(\mathbf{x}_0) = \mathbf{0}.$$

(See Exercise 8.) Since these calculations apply at any point in the domain of $\mathbf{F}$, we have shown that for a conservative force field, ***the gradient of the potential energy is equal to the negative of the force field***, $\nabla U = -\mathbf{F}$. Thus for a conservative force it is possible to derive the force from the potential energy. In Section 6.2, we will show that it is also possible to recover the potential energy from the force field for a conservative force.

Further, since ∇U is orthogonal to the level set of U at every point where it is nonzero and the force vector of a conservative force equals $-\nabla U$, it is also orthogonal to the level set of the potential energy where it is nonzero. Since $\mathbf{F}$ points in the direction opposite to that of ∇U, $\mathbf{F}$ points in the direction of maximum decrease of U.

B. Two Charged Particles. In Example 3.20A, we introduced the potential energy function U corresponding to the electric force $\mathbf{F}$ exerted by one charged particle on another. If the first particle is fixed at the origin in space and the second is located at $\mathbf{x} = (x, y, z)$, then

$$\mathbf{F} = \frac{q_1 q_2}{4\pi\epsilon_0 \|\mathbf{x}\|^3}\mathbf{x} \quad \text{and} \quad U(\mathbf{x}) = \frac{q_1 q_2}{4\pi\epsilon_0}\frac{1}{\|\mathbf{x}\|}.$$

A calculation verifies that the gradient of the potential energy is equal to the negative of the force vector. (See Exercise 9.) One can show that this is equivalent to saying that $\mathbf{F}$ is a conservative force field. The level set of U at a point $\mathbf{x}_0$ is a sphere of radius r_0

centered at the origin. (See Exercise 23 of Section 3.6.) The force vector $\mathbf{F}(\mathbf{x}_0)$ points toward the origin, which is orthogonal to the sphere and in the direction of the maximum decrease of U.

The flow lines of a conservative force field turn out to have physical importance. Suppose we have a charged particle that is free to move in the conservative field of a collection of charged particles that are not free to move. If the charged particle is placed at $\mathbf{x}_0$ with no initial velocity, that is, with no initial kinetic energy, and is allowed to move according the force $\mathbf{F}$, then the particle moves in the direction $\mathbf{F} = -\nabla U$. Since the tangent vector to its motion is always a vector of the field, a parametrization of this motion must be a flow line of the vector field. That is, under these conditions, the flow lines of a force field are the paths of particles in the field. Let us see how this looks for a system consisting of three charged particles of identical charge q. In order to illustrate this relationship, we will do all the calculations in the plane.

Example 3.27

Flow Lines of a Force Field. Consider a system consisting of an electric dipole with particles located at $(1,0)$ and $(-1,0)$ and a third particle located at $\mathbf{x} = (x,y)$ that is free to move in the plane. The potential energy is given by

$$U(\mathbf{x}) = \frac{q^2}{4\pi\epsilon_0}\frac{1}{r_1(\mathbf{x})} + \frac{q^2}{4\pi\epsilon_0}\frac{1}{r_2(\mathbf{x})} + \frac{q^2}{4\pi\epsilon_0}\frac{1}{2},$$

where $r_1(\mathbf{x}) = \sqrt{(x-1)^2 + y^2}$ and $r_2(\mathbf{x}) = \sqrt{(x+1)^2 + y^2}$ are the distances of the free particle to the particles in the dipole. Note that the third term is constant.

Since the force exerted by the dipole on the third particle is a conservative force, $\mathbf{F} = -\nabla U$. Thus,

$$\mathbf{F}(\mathbf{x}) = -\nabla U(\mathbf{x})$$

$$= -\left(\frac{\partial U}{\partial x}(\mathbf{x}),\ \frac{\partial U}{\partial y}(\mathbf{x}) \right)$$

$$= -\frac{q^2}{4\pi\epsilon_0}\left(-\frac{x-1}{r_1^3} - \frac{x+1}{r_2^3},\ -\frac{y}{r_1^3} - \frac{y}{r_2^3} \right).$$

(See Exercise 26 of Section 3.6.)

Figure 3.29(a) contains a plot of $-\nabla U$ and several flow lines of $-\nabla U$ in a neighborhood of the origin. We can see from the plot, for example, that there are flow lines with initial points near the origin that flow toward the origin and flow lines that flow away from the origin. Figure 3.29(b) is a plot of U on a neighborhood of the origin that includes the points $(-1,0)$ and $(1,0)$, the locations of the particles of the dipole. Note that $U(x,y)$ approaches ∞ as (x,y) approaches either $(-1,0)$ or $(1,0)$. If, for example, a particle is placed at $(0.25, 0.05)$, which is closer to the end of the dipole located at $(1,0)$, then the

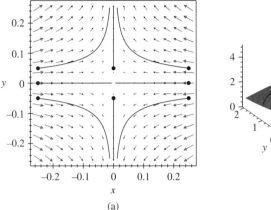 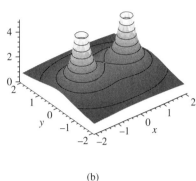

(a)   (b)

Figure 3.29   (a) The vector field $\mathbf{F} = -\nabla U$ and several flow lines of $\mathbf{F}$ on the domain $[-0.25, 0.25] \times [-0.25, 0.25]$. The initial points of the flow lines are marked by dots. (b) The graph of U on the domain $[-2, 2] \times [-2, 2]$. Notice that particles that follow the flow lines of $\mathbf{F}$ flow *down* the graph of U. (See Example 3.27.)

particle initially will move directly away from that end of the dipole. As the distances from the particle to each end of the dipole become close to each other, the repulsive force of the dipole causes the particle to move off in the positive y-direction. Notice that the particle moves in a direction to decrease U. It should be emphasized that this interpretation holds only if the initial kinetic energy of the particle is zero. If the particle is given a nonzero initial velocity, hence a nonzero initial kinetic energy, the particle will follow different paths depending on the initial velocity.

Summary

In this section, we introduced the ***chain rule*** for compositions $f \circ \alpha$ of differentiable functions $\alpha : (a, b) \to \mathbb{R}^2$ and $f : \mathbb{R}^2 \to \mathbb{R}$, and we expressed it in terms of the ***gradient vector field*** ∇f of f:

$$\nabla f(x, y) = (\frac{\partial f}{\partial x}(x, y), \frac{\partial f}{\partial y}(x, y)).$$

We observed that $(f \circ \alpha)'(t) = \nabla f(\alpha(t)) \cdot \alpha'(t)$. We also expressed the ***directional derivative*** of f in terms of ∇f,

$$D_{\mathbf{u}} f(\mathbf{x}) = \nabla f(\mathbf{x}) \cdot \mathbf{u},$$

where $\mathbf{u}$ is a unit vector.

We showed that ∇f is orthogonal to the level sets of f, points in the direction of maximum increase of f, and has magnitude equal to the maximum rate of increase of f at a point. Further, the flow lines of f are orthogonal to the level sets of f, and f is increasing on each flow line for which $\nabla f \neq \mathbf{0}$.

We extended these ideas to functions of three variables and showed how to use the gradient vector of $f = f(x, y, z)$ to construct a **tangent plane** to a level surface of f. Finally, we used the gradient vector to develop the idea of a **conservative force field**.

Section 3.5 Exercises

1. **Chain Rule Calculations.** Compute the derivative of the composition $(f \circ \alpha)(t)$ for the following functions f and α.

   (a) $f(x, y) = x^2 - xy^3$ and $\alpha(t) = (t, 4t)$.
   (b) $f(x, y) = \cos(x - y)$ and $\alpha(t) = (t - 1, t + 2)$.
   (c) $f(x, y, z) = x^2 - 3y^2 - z^2/4$ and $\alpha(t) = (t - 1, t + 2, 4t)$.
   (d) $f(x, y, z) = \sin(x - y)e^z$ and $\alpha(t) = (t, t^2, t^3)$.

2. **Gradient Calculations.** Compute $\nabla f(\mathbf{x}_0)$ for each of the following functions f and points $\mathbf{x}_0$.

   (a) $f(x, y) = 2x^2 - xy + y^2$ and $\mathbf{x}_0 = (2, -1)$.
   (b) $f(x, y) = \sqrt{x^2 + y^2}$ and $\mathbf{x}_0 = (1, 3)$.
   (c) $f(x, y, z) = x^2 - y^2 + yz$ and $\mathbf{x}_0 = (0, -1, -1)$.
   (d) $f(x, y, z) = e^{-x^2 - y^2 - z^2}$ and $\mathbf{x}_0 = (1, -1, -1)$.

3. **Directional Derivative Calculations.** Compute $D_\mathbf{u} f(\mathbf{x}_0)$ for the following functions f, points $\mathbf{x}_0$, and directions $\mathbf{u}$.

   (a) $f(x, y) = xy^2 - y$, $\mathbf{x}_0 = (1, 2)$, and $\mathbf{u} = (1/\sqrt{2}, 1/\sqrt{2})$.
   (b) $f(x, y) = \tan(xy)$, $\mathbf{x}_0 = (1, 0)$, and $\mathbf{u} = (1/\sqrt{3}, \sqrt{2}/\sqrt{3})$.
   (c) $f(x, y, z) = y^2 - z^2$, $\mathbf{x}_0 = (-2, 1, 2)$, and $\mathbf{u} = (1/\sqrt{2}, 0, 1/\sqrt{2})$.
   (d) $f(x, y, z) = \ln(x^2 + y^2 + z^2)$, $\mathbf{x}_0 = (0, 1, -1)$, and $\mathbf{u} = (1/\sqrt{3}, 1/\sqrt{3}, -1/\sqrt{3})$.

4. **Rates of Change.** For each of the functions in parts (a)–(d) of Exercise 3, identify those directions $\mathbf{u}$ for which the directional derivative is positive, zero, and negative.

5. **Horizontal Tangent Planes.** Let f be a differentiable function on an open set $\mathcal{O}$ containing (x_0, y_0). If the tangent plane to the graph of f is horizontal at the point $(x_0, y_0, f(x_0, y_0))$, what can we say about the directional derivatives of $f(x, y)$ at (x_0, y_0)? Explain your answer.

6. **Extrema of Rates of Change.** For each of the following functions f and points $\mathbf{x}_0$, find the direction of the maximum rate of increase of f and the minimum rate of increase of f at $\mathbf{x}_0$.

(a) $f(x, y) = x^2 - 2xy + y^2 + y$ and $\mathbf{x}_0 = (1, -1)$.
(b) $f(x, y) = \sin(xy) + \cos(x - y)$ and $\mathbf{x}_0 = (\pi/2, 0)$.
(c) $f(x, y, z) = x^3 - y^3 + z^3$ and $\mathbf{x}_0 = (0, 2, -1)$.
(d) $f(x, y, z) = \sqrt{x^2 + y^2 + z^2}$ and $\mathbf{x}_0 = (-2, 1, 2)$.

7. **Tangent Planes to Level Surfaces.** Find an equation for the tangent plane to the surface defined by the equation $f(x, y, z) = c$ at the given point $\mathbf{x}_0$.

(a) $x^2 + y^2 - z = 1$ and $\mathbf{x}_0 = (1, 1, 1)$.   (c) $e^{xy-z} = 1$ and $\mathbf{x}_0 = (2, 1, 2)$.
(b) $x^2 + y^2 - z^2 = 0$ and $\mathbf{x}_0 = (1, 0, 1)$.   (d) $xyz = 8$ and $\mathbf{x}_0 = (2, -2, -2)$.

8. **A Dot Product Calculation.** In this exercise, we want to verify the claim about dot products made in Example 3.26A. Suppose that $\mathbf{w}_0 = (x_0, y_0, z_0)$ is a fixed vector that has the property that for any vector $\mathbf{v} = (v_1, v_2, v_3)$, $\mathbf{v} \cdot \mathbf{w}_0 = 0$. Show that $\mathbf{w}_0 = (0, 0, 0)$. (*Hint:* Since the property holds for all vectors, it holds for the vectors $(1, 0, 0)$, $(0, 1, 0)$, and $(0, 0, 1)$.)

9. **Potential Energy.** Carry out the calculation of Example 3.26B to show that $-\nabla U = \mathbf{F}$ for the potential energy function U and vector field $\mathbf{F}$ given in the example.

10. **Potential Energy of a Diatomic Molecule.** In Exercise 11 of Section 3.3, we introduced the potential energy function U of a diatomic molecule. Suppose that the molecule is in equilibrium when the distance between the atoms is r_0 and that the equilibrium potential energy is E_0. If one atom is located at the origin and the second atom is located at $\mathbf{x} = (x, y, z)$, the potential energy U is given by

$$U(\mathbf{x}) = E_0(e^{-2a(\|\mathbf{x}\| - r_0)} - 2e^{-a(\|\mathbf{x}\| - r_0)}),$$

where a is related to the frequency of vibration of the molecule.

(a) Assuming that the force binding the atoms together is conservative, compute $\mathbf{F}$ given the above U.
(b) Describe the behavior of $\mathbf{F}$. In particular, where is it an attractive force?
(c) What are the implications of your answer to part (b) for the structure of the molecule? Explain.

■ 3.6 End of Chapter Exercises

1. **Contour Plots.** For each of the following functions given in symbolic form, construct a contour plot on the domain $[-2, 2] \times [-2, 2]$. Be sure to label your contours.

(a) $f(x, y) = -x^2/2 - y^2 + 4$.   (c) $f(x, y) = x^2 - 2xy + y^2$.
(b) $f(x, y) = xy$.   (d) $f(x, y) = x^2 + y^2 + 2y$.

2. **Composition of Functions**

   (a) Let $f(x, y) = x^2 + y^2$ and $g(z) = z^2$. What, if any, is the relationship between the level sets of $f(x, y)$ and the level sets of the composition $(g \circ f)(x, y) = g(f(x, y)) = (x^2 + y^2)^2$? Explain your answer.

   (b) Suppose that $g(z)$ is an arbitrary function of z. What, if anything, can you say about level sets of $(g \circ f)(x, y) = g(f(x, y)) = g(x^2 + y^2)$? Explain your answer.

   (c) Suppose that $f(x, y)$ is an arbitrary function of (x, y) and $g(z)$ is an arbitrary function of z. What, if anything, can you say about the level sets of $(g \circ f)(x, y) = g(f(x, y))$? Explain your answer.

   (d) For each of the following functions h, write h as a composition of a function $f = f(x, y)$ and a function $g = g(z)$. For each h, use this and part (c) to construct a contour plot on the domain $[-2, 2] \times [-2, 2]$. Be sure to label your contours.

   (i) $h(x, y) = (x - 2y)^3$.   (iii) $h(x, y) = e^{2x^2 - y^2}$.

   (ii) $h(x, y) = \sqrt{x^2 + 2y^2}$.   (iv) $h(x, y) = \ln(1 + x^2 + y^2)$.

3. **Method of Slices.** For each of the following functions given in symbolic form: (i) describe and plot the vertical slices of the graph of f that lie over the x- and y-axes; (ii) describe and plot several level sets of f near the origin; and (iii) use these vertical slices and level sets to sketch the graph of f.

   (a) $f(x, y) = -x^2 - y^2/2$.   (c) $f(x, y) = e^{-x^2 - y^2}$.

   (b) $f(x, y) = x^3 - x^2 + y^2$.   (d) $f(x, y) = x^2 + xy - y^2$.

4. **Computer Generated Plots.** Use a computer to produce contour plots and graphs for the following functions. Analyze the plots and graphs in order to determine local and global extrema and saddle points of the function.

   (a) $f(x, y) = x^3 - x + 2(y^4 - y)$.

   (b) $f(x, y) = e^{-2(x^2 - y^2)}$.

   (c) $f(x, y) = e^{-(x^2 - y^2)} + 2e^{-((x-1)^2 - y^2)}$.

   (d) $f(x, y) = \cos(x - y)$.

5. **Directional Derivative Calculations.** Compute the directional derivative $D_{\mathbf{u}}f(\mathbf{x}_0)$ for the following functions f, points $\mathbf{x}_0$, and directions $\mathbf{u}$.

   (a) $f(x, y) = 2x - 3yx$, $\mathbf{x}_0 = (1, -1)$, and $\mathbf{u} = (1/\sqrt{5}, 2/\sqrt{5})$.

   (b) $f(x, y) = e^{x^2 + y^2}$, $\mathbf{x}_0 = (0, 0)$, and $\mathbf{u} = (1/2, -\sqrt{3}/2)$.

   (c) $f(x, y, z) = 1/(1 + x^2 + y^2 + z^4)$, $\mathbf{x}_0 = (0, 0, 0)$, and $\mathbf{u} = (1/\sqrt{2}, -1/2, 1/2)$.

   (d) $f(x, y, z) = \sec(x - y + z)$, $\mathbf{x}_0 = (0, \pi/4, 0)$, and $\mathbf{u} = (1/\sqrt{3}, -1/\sqrt{3}, 1/\sqrt{3})$.

6. **Instantaneous Rates of Change.** Let f be a differentiable function in a neighborhood $\mathcal{O}$ of (x_0, y_0). Suppose that $\alpha(t) = (x(t), y(t))$ and $\beta(t) = (\overline{x}(t), \overline{y}(t))$ are differentiable parametrizations of curves with the property that $\alpha(0) = (x_0, y_0)$, $\beta(0) = (x_0, y_0)$, and $\alpha'(0) = \beta'(0)$. Show that the instantaneous rate of change of f along α at $t = 0$ is the same as the instantaneous rate of change of f along β at $t = 0$.

7. **Partial Derivatives.** For each function f below, compute all the first and second order partial derivatives.

   (a) $f(x, y) = x^3/(1 - y)$.
   (b) $f(x, y) = \ln(x^2 + y)$.

   (c) $f(x, y) = \sin(xe^y)$.
   (d) $f(x, y) = x^y$.

8. **Partial Derivatives in Space.** For each function f below, compute all the first and second order partial derivatives.

   (a) $f(x, y, z) = e^z \cos(xy)$.

   (b) $f(x, y, z) = \ln(xyz)$.

9. **Partial Derivatives and Slices.** Let $f : \mathcal{O} \subset \mathbb{R}^2 \to \mathbb{R}$.

   (a) Suppose $(x_0, y_0) \in \mathcal{O}$ and $\frac{\partial f}{\partial x}(x_0, y_0) = 0$. What are the possible shapes of the vertical slice of the graph of f through this point and parallel to the x-axis? Explain your answer.

   (b) Suppose $(x_0, y_0) \in \mathcal{O}$ and $\frac{\partial f}{\partial y}(x_0, y_0) = 0$. What are the possible shapes of the vertical slice of the graph of f through this point and parallel to the y-axis? Explain your answer.

   (c) Suppose $(x_0, y_0) \in \mathcal{O}$ and $\frac{\partial f}{\partial x}(x_0, y_0) = \frac{\partial f}{\partial y}(x_0, y_0) = 0$. What would the graph of f look like near this point? Explain your answer.

10. **The Chain Rule.** Suppose that the first partial derivatives of $g = g(s, t)$ exist and that $f = f(x)$ is a differentiable function of one variable. Define the function $h : \mathbb{R}^2 \to \mathbb{R}$ by $h = f \circ g$, so that $h(s, t) = f(g(s, t))$.

    (a) Construct an argument to show that

$$\frac{\partial h}{\partial s}(s, t) = f'(g(s, t))\frac{\partial g}{\partial s}(s, t) \quad \text{and} \quad \frac{\partial h}{\partial t}(s, t) = f'(g(s, t))\frac{\partial g}{\partial t}(s, t).$$

    (*Hint:* Use the chain rule for the composition of functions of one variable.)

    (b) Use the result of part (a) to show that $\frac{\partial}{\partial x}f(x - ct) = f'(x - ct)$ and $\frac{\partial}{\partial t}f(x - ct) = -cf'(x - ct)$.

11. **An Open Ball is an Open Set.** Let $\mathcal{B}$ be the open ball of radius r_0 centered at the origin. Show that $\mathcal{B}$ is an open set. That is, show that for every $\mathbf{x}_0 \in \mathcal{O}$, there is an $r > 0$ so that the open ball of radius r centered at $\mathbf{x}_0$ is contained in $\mathcal{O}$.

12. **The Chain Rule for $f \circ g$.** Suppose that $f : \mathbb{R} \to \mathbb{R}$ is a differentiable function of one variable and $g : \mathbb{R}^3 \to \mathbb{R}$ is a differentiable function of three variables. Show that $\nabla(f \circ g)(x, y, z) = f'(g(x, y, z))(\frac{\partial g}{\partial x}, \frac{\partial g}{\partial y}, \frac{\partial g}{\partial z})$.

13. **Closed Sets I.** Suppose that $\mathcal{A}$ is a subset $\mathbb{R}^2$ and $\mathcal{B}$ is its boundary. Show that $\mathcal{S} = \mathcal{A} \cup \mathcal{B}$ is a closed subset of $\mathbb{R}^2$. (*Hint:* You must show that $\mathcal{S}$ contains all of its boundary points. That is, if $\mathbf{x}_0$ is a boundary point of $\mathcal{S}$, then in fact $\mathbf{x}_0 \in \mathcal{B}$.)

14. **Closed Sets II.** Show that each of the following sets is closed and describe the boundary of each set in words and in symbols.

(a) $S = \{(x, y) : x - y \leq 1\}$.
(b) $S = \{(x, y) : x = 1 \}$.
(c) $S = \{(x, y) : x^2 + y^2 = 1 \}$.
(d) $S = \{(x, y) : x \geq 0, \ y \geq 0, \ \text{and} \ x + y \leq 1\}$.

15. **Complements.** The complement of a subset $S \subset \mathbb{R}^2$ is the set S^c consisting of all points not in S. Suppose that S is a closed set in the plane and that $\mathbf{x}_0 = (x_0, y_0) \in S^c$.

(a) Construct an argument to show there is an $r > 0$ so that the open ball of radius r centered at $\mathbf{x}_0$ is contained in S^c. (*Hint:* Think about the definition of a boundary point.)
(b) Since $\mathbf{x}_0$ was an arbitrary point in S^c, what does this tell us about the complement of a closed set? Explain.

16. **True or False?** For each of the following statements, find an open set containing P in which the condition holds, thus showing the claim is true, or show that the claim fails in every open ball about P, thus showing the claim is false.

(a) Near $P = (1, 2)$ the coordinates of every point are positive.
(b) Near $P = (0, 1)$ the y-coordinate of a point is greater than the x-coordinate.
(c) Near $P = (2, 2)$ the x- and y-coordinate of every point are equal.
(d) Near $P = (2, 1)$ the difference between the x- and y-coordinates of a point is not 0.

17. **Boundary Surfaces and Curves.** In this exercise, we are concerned with describing curves and surfaces that form the boundary of more complicated regions. For each of the following closed sets S:

(i) Describe each of the surfaces that make up the boundary $\mathcal{B}$ of S by writing it as the level set of a differentiable function g. Use set notation. (Each surface may have its own function g.)
(ii) Describe the curves of intersection of each pair of surfaces from (i) that make up $\mathcal{B}$. Use set notation. (Two surfaces may intersect in the empty set, in which case it is not necessary to provide a description.)

(a) S is the solid cylinder of height 3 and radius 1 centered on the z-axis with its base resting on the xy-plane.
(b) S is the solid spherical cap lying between spheres of radius 2 and 4 centered at the origin with a nonnegative z-coordinate.
(c) S is the set of points of distance less than or equal to 3 units from the origin and having a nonnegative y-coordinate and a nonpositive z-coordinate.
(d) S is the solid cube whose vertices are the points $(0, 0, 0)$, $(1, 0, 0)$, $(0, 1, 0)$, $(0, 0, 1)$, $(1, 1, 0)$, $(1, 0, 1)$, $(0, 1, 1)$, and $(1, 1, 1)$.
(e) S is the solid cone whose vertex is the origin and whose base is a circle of radius 1 centered on the z-axis and lying in the plane parallel to the xy-plane at height 1 on the z-axis.
(f) S is the solid tetrahedron with vertices $(0, 0, 0)$, $(1, 0, 0)$, $(0, 1, 0)$, and $(0, 0, 1)$.

18. **Continuous Functions Using Limits.** In this exercise, we want to use the definition of limit to show that constant functions and coordinate functions are continuous. To show that $\lim_{\mathbf{x}\to\mathbf{x}_0} f(\mathbf{x}) = f(\mathbf{x}_0)$, we must show that for every $\epsilon > 0$, there is an open ball $\mathcal{B}$ centered at $\mathbf{x}_0$ so that $|f(\mathbf{x}) - f(\mathbf{x}_0)| < \epsilon$ for $\mathbf{x} \in \mathcal{B}$.

    (a) Let us work with the particular constant function $f(x, y) = 3$. (The general case can be proven in a similar manner.)

      (i) Prove that f is continuous at $\mathbf{x}_0 = (1, 2)$. (You must show that for $\epsilon > 0$, there is a $\mathcal{B}$ so that $\mathbf{x} \in \mathcal{B}$ implies that $|f(\mathbf{x}) - f(1, 2)| < \epsilon$.)
      (ii) Modify your argument to show that f is continuous at every point $\mathbf{x}_0$.

    (b) Let $f(x, y) = x$. (The argument for $f(x, y) = y$ is similar.)

      (i) Show that f is continuous at $\mathbf{x}_0 = (3, 1)$. (*Hint:* Describe the set $\{(x, y) : 3 - \epsilon < f(x, y) < 3 + \epsilon\}$.)
      (ii) Modify your argument to show that f is continuous at every point $\mathbf{x}_0$.

19. **Continuous Functions.** Are the following functions continuous everywhere in $\mathbb{R}^2$? If so, why are they continuous? If not, where do they fail to be continuous and why are they not continuous at those points? (Intuitive explanations are sufficient.)

    (a) $f(x, y) = e^{x^2 + y^2}$.
    (b) $f(x, y) = \sqrt{x^2 + y^2}$.
    (c) $f(x, y) = e^{1/(x^2 + y^2)}$.

20. **Chain Rule Calculations.** Compute the derivative of the composition $(f \circ \alpha)(t)$ for the following functions f and α.

    (a) $f(x, y) = (x + y)^2$ and $\alpha(t) = (-2t, t^2)$.
    (b) $f(x, y) = e^{x+y}$ and $\alpha(t) = (t^3, t^2)$.
    (c) $f(x, y, z) = \ln(x^2 + y^2 + z^2)$ and $\alpha(t) = (\cos t, 0, \sin t)$.
    (d) $f(x, y, z) = \cos(xyz)$ and $\alpha(t) = (t, \sqrt{t}, t^2)$.

21. **Gradient Calculations.** Compute the $\nabla f(\mathbf{x}_0)$ for each of the following functions f and points $\mathbf{x}_0$.

    (a) $f(x, y) = xy^2 - 2x^2y$ and $\mathbf{x}_0 = (1, 3)$.
    (b) $f(x, y) = \cos(x^2 - y^2)$ and $\mathbf{x}_0 = (-\pi/4, \pi/2)$.
    (c) $f(x, y, z) = \sin(xy)\cos(z)$ and $\mathbf{x}_0 = (1, \pi/2, 0)$.
    (d) $f(x, y, z) = xze^y$ and $\mathbf{x}_0 = (1, 0, 2)$.

22. **Diffusion.** Diffusion is a time-dependent process in which molecules of one substance, the **solute**, disperse by random molecular motion through molecules of a second substance, the **solvent**. The dispersal of an airborne substance in the atmosphere or one liquid in another are examples of diffusion. In some cases, it is possible to give a simple mathematical description of diffusion. Let us consider a particular function that can be used to model an instantaneous release of a solute in a solvent in a pipe. We assume that the concentration is uniform within every cross section of the pipe and that the solute

spreads in both directions along the pipe by random molecular motion. Let us suppose that the pipe is aligned along the x-axis and an amount M of the solute is released at the origin. Then the concentration of the substance per unit volume at time $t > 0$ at position x in the pipe can be described by

$$c(x,t) = \frac{M}{\sqrt{4\pi Dt}} e^{-x^2/4Dt},$$

where D is a constant, called the **diffusivity**, that depends on the solute and the solvent. Initially, let us assume that $D = 1$.

(a) Let $x = 1$. Describe the behavior of $c(1,t)$ as t increases from 0.01 to ∞. That is, describe the vertical slice of the graph of c for $x = 1$ on the interval $[0.01, \infty)$.
(b) Describe the concentration c at $t = 1$ as a function of x. That is, describe the vertical slice of the graph of c for $t = 1$.
(c) How does the vertical slice in part (b) change as t increases?
(d) Suppose that $D = 2$. How does this change your answers to parts (b) and (c)?

23. **Level Sets of Potential Energy.** In this exercise, we consider the potential energy function U for two charged particles in space. If there are no other charges present, the potential energy is a function of the distance between the two particles. To simplify matters, let us assume that a particle of charge q_1 is located at the origin in space and that a particle of charge q_2 is located at $\mathbf{x} = (x, y, z)$, so that the distance r between them is given by $\|\mathbf{x}\| = \sqrt{x^2 + y^2 + z^2}$. Then U can be written in the form

$$U(x, y, z) = \frac{q_1 q_2}{4\pi\epsilon_0} \frac{1}{\|\mathbf{x}\|},$$

where the constant $\epsilon_0 > 0$ is the permittivity of a vacuum. (See Example 3.26B.)

(a) Describe the level set of U passing through the point (x_0, y_0, z_0) of distance r_0 from the origin.
(b) Describe the collection of level sets of U.
(c) The charges q_1 and q_2 may be positive or negative. How do the signs of q_1 and q_2 affect your answers to parts (a) and (b)? Explain.

24. **The Potential Energy of an Electric Dipole I.** In this exercise, we consider the potential energy function U of a system of three particles in space, two of which form an electric dipole. (See Example 3.27.) A dipole consists of two charged particles a fixed distance apart. If there are no other charges present, the potential energy is a function of the distance from the third particle to each particle of the dipole. To simplify matters, let us assume that the dipole consists of particles located at $(1, 0, 0)$ and $(-1, 0, 0)$. Assume these are of charge q_1 and q_2, respectively. From the superposition principle, the potential

energy of the system of three particles is the sum of the potential energies of the three pairs of particles (see Exercise 23). Thus if the third particle has charge q_3,

$$U(x, y, z) = \frac{q_3 q_1}{4\pi\epsilon_0} \frac{1}{r_1(\mathbf{x})} + \frac{q_3 q_2}{4\pi\epsilon_0} \frac{1}{r_2(\mathbf{x})} + \frac{q_1 q_2}{4\pi\epsilon_0} \frac{1}{2},$$

where $r_1(\mathbf{x}) = \sqrt{(x-1)^2 + y^2 + z^2}$ and $r_2(\mathbf{x}) = \sqrt{(x+1)^2 + y^2 + z^2}$ are the distances of the particle of charge q_3 to the particles in the dipole. Note that the third term is constant. (You may find it useful to use a computer plotting routine to answer the following questions.)

(a) Assume that all the charges are positive, so that values of U are always positive.

    (i) Describe the level sets of U. In particular, describe them for large values of U and for small values of U.

    (ii) As the values of U decrease, how do the corresponding level sets change? Explain.

(b) Assume that q_1 is negative and q_2 and q_3 are positive. Thus U may take both positive values and negative values.

    (i) Describe the level sets of U. In particular, describe them for large values of U and for small values of U.

    (ii) As the values of U decrease, how do the corresponding level sets change? Explain.

(c) Explain the differences, if any, between your answers to parts (a) and (b).

25. **The Potential Energy of an Electric Dipole II.** In this exercise, we investigate the continuity of the potential energy function U of Exercise 24.

(a) Assume that q_1, q_2, and q_3 are positive.

    (i) Where is U continuous and where is it discontinuous? Explain.

    (ii) As in Example 3.27, consider the two-dimensional version of U. Describe the graph of U as a function of x and y. In particular, describe the behavior of U where it is discontinuous. (You may find it useful to use a computer plotting routine to answer this question.)

(b) Assume that q_1 is negative and q_2 and q_3 are positive.

    (i) Where is U continuous and where is it discontinuous? Explain.

    (ii) As in Example 3.27, consider the two-dimensional version of U. Describe the graph of U as a function of x and y. In particular, describe the behavior of U where it is discontinuous. (You may find it useful to use a computer plotting routine to answer this question.)

(c) Explain the differences, if any, between your answers to parts (a) and (b).

26. **Potential Energy of an Electric Dipole III.** In this exercise, we investigate the force corresponding to the potential energy function U of Exercise 24.

    (a) Assuming that the electric force is conservative, compute the force vector $\mathbf{F}$ at $\mathbf{x} = (x, y, z)$.

    (b) Now let us focus our attention on points $\mathbf{x}$ that lie in the yz-plane, the plane bisecting the line between the particles of the dipole, for the situation in which the charges of the dipole have the same magnitude, $|q_1| = |q_2|$. Note that $x = 0$ implies that $r_1 = r_2$. Simplify your expression for $\mathbf{F}$.

    (c) How does the direction of $\mathbf{F}$ at a point of the form $(0, y, z)$ depend on the signs of the charges q_1, q_2, and q_3? Explain your answer.

    (d) Does your answer to part (c) correspond to your intuition about the forces exerted by the dipole on the particle of charge q_3? Explain.

Chapter

4

Critical Points and Optimization

In this chapter, we will continue our work on differentiable functions of two and three variables. Having developed the fundamental material on differentiability in Chapter 3, we are ready to apply these ideas to analyze the behavior of functions. Generally, we will be concerned with the problem of optimizing or locating maximum and minimum values of functions on both open and closed sets.

We will begin in Section 4.1 by developing two graphical techniques for locating and identifying critical points of a differentiable function defined on an open set in the plane. Analogous to the one variable case, a critical point of a function of several variables is a point where the first partial derivatives of the function are all zero. Expanding on ideas introduced in Section 3.1, we will show how to use the contour plot of a function to locate and identify critical points. In addition, we continue our analysis of the gradient vector field of a differentiable function. The critical points of the gradient vector field are also critical points of the function and vice versa. Further, we will be able to determine the type of the critical point of the function based on its classification as a critical point of the gradient vector field. Where possible, we will extend these ideas to differentiable functions of three variables. In Section 4.2, we will develop a symbolic second derivative test for classifying critical points of functions of two variables. This is based on an analysis of the critical points of quadratic polynomials in two variables.

In Section 4.3 and Section 4.4, we will consider the problem of finding the extreme values of a differentiable function on a closed set. We will develop a technique for finding extrema on a boundary known as the method of Lagrange multipliers. In Section 4.3, we consider the planar case, where the boundary is a union of curves, and in Section 4.4, we consider subsets of space, where the boundary is a union of curves and surfaces.

A Collaborative Exercise—Critical Points

Section 3.5 developed fundamental properties of the direction and length of the gradient vector field of a differentiable function at points where the gradient is nonzero. Here we want to explore the behavior of f near points where the gradient is zero, that is, at critical points of ∇f. It should be no surprise that the condition for being a critical point of the vector field ∇f, $\nabla f(x_0, y_0) = (0, 0)$, amounts to having a critical point of f. As in one-variable calculus, a point should be a critical point of f if the first derivatives of f are zero at the point. Formalizing this, we have the following definition.

Definition 4.1 Let f be a differentiable function defined on an open set $\mathcal{O} \subset \mathbb{R}^2$. A point $\mathbf{x}_0 = (x_0, y_0) \in \mathcal{O}$ is a ***critical point*** of f if the first partial derivatives of f at $\mathbf{x}_0$ are zero. That is, $\mathbf{x}_0$ is a critical point of f if

$$\frac{\partial f}{\partial x}(\mathbf{x}_0) = 0 \ \text{ and } \ \frac{\partial f}{\partial y}(\mathbf{x}_0) = 0. \ \blacklozenge$$

Given that critical points of ∇f are critical points of f, we want to use the properties of ∇f to draw a connection between the type of a critical point of ∇f and its type as a critical point of f. That is, if $\mathbf{x}_0 = (x_0, y_0)$ satisfies $\nabla f(\mathbf{x}_0) = \mathbf{0}$, knowing that $\mathbf{x}_0$ is a sink, center, source, or saddle for ∇f should tell us whether $\mathbf{x}_0$ is a minimum, maximum, or saddle of f. We will make this connection by examining Figure 4.1, which contains a plot of ∇f overlaid on a contour plot of f for a sufficiently complicated function. Since we are making the connection on an intuitive level, we do not need a symbolic expression for f or ∇f. (You may want to make a copy of the figure.)

1. Locate the critical points of ∇f, sketch flow lines of ∇f near the critical points, and identify the type of each these as a critical point of ∇f. (Note that the vectors are scaled.)
2. Using the contour plot, identify the type of each critical point of f. (Note that the values of f on contours are given on the plot.)
3. Based on your answers to questions 1 and 2, what is the correspondence relating the types of critical points of ∇f (sink, center, source, or saddle) with the types of critical points of f (minimum, maximum, or saddle)?
4. Describe the behavior of flow lines of ∇f near a minimum, a maximum, and a saddle of f.

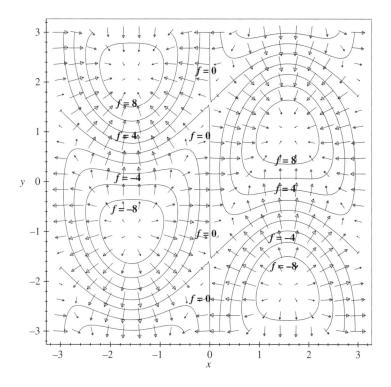

Figure 4.1   Labeled contours and gradient vector field for a function of two variables.

■ 4.1 Graphical Analysis of Critical Points

In Section 3.1, we introduced *maxima, minima,* and *saddle points* for functions of two variables, and in Example 3.3A of that section, we saw how to identify them on the contour plot of a function. Maxima, minima, and saddles are all examples of *critical points* of functions of several variables. As in one-variable calculus, the definition of a critical point for a differentiable function of several variables is given in terms of the derivative, in this case the first partial derivatives, of the function.

Also as in one-variable calculus, there is a geometric interpretation for critical points of functions of two variables. In Section 3.4, we saw that if f is differentiable at $\mathbf{x}_0$, then the tangent plane to the graph of f at $\mathbf{x}_0$ is given by

$$z = f(\mathbf{x}_0) + \frac{\partial f}{\partial x}(\mathbf{x}_0)(x - x_0) + \frac{\partial f}{\partial y}(\mathbf{x}_0)(y - y_0).$$

If $\mathbf{x}_0$ is a critical point of f, then $\frac{\partial f}{\partial x}(\mathbf{x}_0) = \frac{\partial f}{\partial y}(\mathbf{x}_0) = 0$ so that

$$z = f(\mathbf{x}_0) + 0 \cdot (x - x_0) + 0 \cdot (y - y_0)$$
$$= f(\mathbf{x}_0).$$

Thus the tangent plane to the graph of f at the critical point $\mathbf{x}_0$ is the horizontal plane $z = f(\mathbf{x}_0)$. In sum, *for differentiable functions of two variables, critical points are points where the tangent plane to the graph is horizontal.*

Since the directional derivatives of f can be expressed in terms of the partial derivatives of f, the fact that $\mathbf{x}_0$ is a critical point of f also provides information about $D_{\mathbf{u}}(f)(\mathbf{x}_0)$. Recall from Section 3.5 that if f is a differentiable function in an open set containing $\mathbf{x}_0$ and $\mathbf{u} = (u_1, u_2)$ is a unit vector, then

$$D_{\mathbf{u}}(f)(\mathbf{x}_0) = \frac{\partial f}{\partial x}(\mathbf{x}_0)u_1 + \frac{\partial f}{\partial y}(\mathbf{x}_0)u_2.$$

It follows that if $\mathbf{x}_0$ is a critical point of f, $D_{\mathbf{u}}(f)(\mathbf{x}_0) = 0$. That is, *the instantaneous rate of change of f is zero in every direction at a critical point*. There is also a relationship between critical points of f and the gradient vector of f, which we will explore later in this section.

The following example demonstrates how to find critical points of a function of two variables using the techniques we developed in Section 3.2. Notice that since critical points satisfy two equations, $\frac{\partial f}{\partial x}(x, y) = 0$ and $\frac{\partial f}{\partial y}(x, y) = 0$, finding critical points algebraically amounts to finding the solutions to a system of two equations in two unknowns.

Example 4.1   **Finding Critical Points**

A. Let $f(x, y) = -x^4 + 2x^2 - \frac{1}{2}x - 1 - y^2$. The critical points of f are the points that satisfy the system of equations

$$\frac{\partial f}{\partial x}(x, y) = -4x^3 + 4x - \frac{1}{2} = 0$$

$$\frac{\partial f}{\partial y}(x, y) = -2y = 0.$$

Since each of these equations involves only one of the variables, the equations can be solved separately. The solutions x to the first equation can be found by Newton's method for functions of one variable and are approximately equal to -1.06, 0.13, and 0.93. Clearly, $y = 0$ is the only solution to the second equation. Thus the critical points of f are approximately located at $(-1.06, 0)$, $(0.13, 0)$, and $(0.93, 0)$. (See Example 3.4 of Section 3.1.)

B. Let $f(x, y) = x^3 - 3x^2y^2$. The critical points of f are the points that satisfy the system of equations

$$\frac{\partial f}{\partial x}(x, y) = 3x^2 - 6xy^2 = 0$$

$$\frac{\partial f}{\partial y}(x, y) = -6x^2y = 0.$$

Since these equations involve both variables, it is not possible to solve them separately, as was the case in part A. The second equation is satisfied by $y = 0$ or $x = 0$. If we substitute $y = 0$ into the first equation, we get $x = 0$ as well. Thus $(0, 0)$ is a critical point. If we substitute $x = 0$ into the first equation, we see that this equation is solved for any value of y. Thus the critical points of f are the points $(0, y)$ for any $y \in \mathbb{R}$. That is, the set of critical points of f consists of the entire y-axis.

In both parts of Example 4.1, the function f had more than one critical point. However, the critical points of the function in Example 4.1A were "separated" by a positive distance. Since this case will be of the most interest for us, we make the following definition.

Definition 4.2 Let $\mathbf{x}_0 \in \mathcal{O}$ be a critical point of the differentiable function $f : \mathcal{O} \subset \mathbb{R}^2 \to \mathbb{R}$. We say that $\mathbf{x}_0$ is

1. *isolated* if there is an open ball $\mathcal{B}$ centered at $\mathbf{x}_0$ and contained in $\mathcal{O}$ with the property that for all $\mathbf{x} \in \mathcal{B}$, $\mathbf{x} \neq \mathbf{x}_0$, $\mathbf{x}$ is not a critical point of f. That is, $\mathbf{x}_0$ is the only critical point of f in $\mathcal{B}$.
2. *nonisolated* if it is not isolated. ◆

The critical points of the function f of Example 4.1A are isolated, and the critical points of the function f of Example 4.1B are nonisolated. From the definition, a critical point is either isolated or nonisolated; it cannot be both. However, it is possible for a function to have both isolated and nonisolated critical points. (See Exercise 1.)

As is the case for functions of one variable, we would like to classify the critical points of a differentiable function. First, however, we would like to restate the definitions of a maximum and of a minimum. Notice that these definitions do *not* require that the function be differentiable.

Definition 4.3 Suppose $f : \mathcal{O} \subset \mathbb{R}^2 \to \mathbb{R}$. A point $\mathbf{x}_0 \subset \mathcal{O}$ is a

1. *global maximum* of f on $\mathcal{O}$ if for all $\mathbf{x} \in \mathcal{O}$, $f(\mathbf{x}) \leq f(\mathbf{x}_0)$,

2. *local* or *relative maximum* of f if there is an open ball $\mathcal{B}$ centered at $\mathbf{x}_0$ and contained in $\mathcal{O}$ with the property that for all $\mathbf{x} \in \mathcal{B}$, $f(\mathbf{x}) \leq f(\mathbf{x}_0)$,
3. *global minimum* of f on $\mathcal{O}$ if for all $\mathbf{x} \in \mathcal{O}$, $f(\mathbf{x}) \geq f(\mathbf{x}_0)$, and a
4. *local or relative minimum* of f if there is an open ball $\mathcal{B}$ centered $\mathbf{x}_0$ and contained in $\mathcal{O}$ with the property that for all $\mathbf{x} \in \mathcal{B}$, $f(\mathbf{x}) \geq f(\mathbf{x}_0)$.

We say that a point is a *global extremum* of f if it is either a global maximum or minimum of f and a *local or relative extremum* of f if it is either a local maximum or a local minimum. ◆

Since the domain $\mathcal{O}$ of a differentiable function is an open set, it follows that global extrema of a differentiable function are always local extrema. The converse, of course, is not true; local extrema are not necessarily global extrema. (See Exercise 5.)

Now let us return to the question of finding extreme values. As is the case for functions of one variable, local extrema of a differentiable function f of two variables on an open set are also critical points of f. We state this in the form of a proposition.

Proposition 4.1 Suppose $f : \mathcal{O} \subset \mathbb{R}^2 \to \mathbb{R}$ is a differentiable function. If $\mathbf{x}_0$ is a local extremum of f, then it is a critical point of f. ◆

Proof: In order to show that $\mathbf{x}_0$ is a critical point, we must show that the first partial derivatives of f are zero at $\mathbf{x}_0$. We will outline the the proof for the case of a local minimum; the proof for a local maximum is analogous.

Suppose $\mathbf{x}_0$ is a local minimum of f, so that $f(\mathbf{x}_0) \leq f(\mathbf{x})$ for all points $\mathbf{x}$ in an open ball $\mathcal{B}$ centered at $\mathbf{x}_0$. If $\mathcal{C}_1$ is the vertical slice of the graph of f through $\mathbf{x}_0$ parallel to the xz-plane and $\mathcal{C}_2$ is the vertical slice of the graph of f through $\mathbf{x}_0$ parallel to the yz-plane, then $\mathbf{x}_0$ is a local minimum of f along the curves $\mathcal{C}_1$ and $\mathcal{C}_2$. It follows from one-variable calculus that the slope of these curves at $\mathbf{x}_0$ is 0. As we saw in Section 3.2, the slope of $\mathcal{C}_1$ at $\mathbf{x}_0$ is $\frac{\partial f}{\partial x}(\mathbf{x}_0)$ and the slope of $\mathcal{C}_2$ at $\mathbf{x}_0$ is $\frac{\partial f}{\partial y}(\mathbf{x}_0)$. Since $\frac{\partial f}{\partial x}(\mathbf{x}_0) = 0$ and $\frac{\partial f}{\partial y}(\mathbf{x}_0) = 0$, $\mathbf{x}_0$ is a critical point of f. ∎

While all extreme points of f are critical points, not all critical points are extreme points. We have the following definition.

Definition 4.4 If $\mathbf{x}_0$ is an isolated critical point of f that is not an extreme value of f, we will say that $\mathbf{x}_0$ is a *saddle point* of f. ◆

If $\mathbf{x}_0$ is a saddle point of a differentiable function f, then for every open ball $\mathcal{B}$ centered at $\mathbf{x}_0$, it must be the case that there are points $\mathbf{x} \in \mathcal{B}$ for which $f(\mathbf{x}) < f(\mathbf{x}_0)$ *and* points for which $f(\mathbf{x}) > f(\mathbf{x}_0)$. Since this holds for all $\mathcal{B}$, in particular, for arbitrarily small $\mathcal{B}$, this condition can be rephrased as follows: if $\mathbf{x}_0$ is a saddle point, there are points arbitrarily close to $\mathbf{x}_0$ with larger function values than at $\mathbf{x}_0$ and points arbitrarily close to $\mathbf{x}_0$ with smaller function values than at $\mathbf{x}_0$. The saddle point in Example 3.3A satisfies these conditions.

To this point we have characterized maxima, minima, and saddle points, but we have not provided techniques to identify the type of a critical point. In Example 3.3, we saw how to identify critical points using a contour plot of a function. Let us review this in the following example before we introduce a second graphical method for identifying critical points.

Example 4.2

Classifying Critical Points. Let $f(x, y) = e^{-x^2 - y^2} xy$. The critical points of f are the points that satisfy the system of equations

$$\frac{\partial f}{\partial x}(x, y) = e^{-x^2 - y^2}(-2x^2 y + y) = 0$$

$$\frac{\partial f}{\partial y}(x, y) = e^{-x^2 - y^2}(-2xy^2 + x) = 0.$$

Since $e^{-x^2 - y^2} \neq 0$, the solutions of this system are the same as the solutions of the system

$$y(-2x^2 + 1) = 0$$
$$x(-2y^2 + 1) = 0.$$

The first equation is satisfied if $y = 0$ or $x = \pm 1/\sqrt{2}$. If we substitute $y = 0$ into the second equation, we get $x = 0$. If we substitute $x = \pm 1/\sqrt{2}$ into the second equation, we get $y = \pm 1/\sqrt{2}$. Thus f has five critical points: $(0, 0)$, $(1/\sqrt{2}, 1/\sqrt{2})$, $(1/\sqrt{2}, -1/\sqrt{2})$, $(-1/\sqrt{2}, 1/\sqrt{2})$, and $(-1/\sqrt{2}, -1/\sqrt{2})$.

Figure 4.2(a) contains the contour plot of the function f. Let us focus our attention on the behavior of f near $\mathbf{x}_1 = (1/\sqrt{2}, 1/\sqrt{2})$. Notice that we can choose an open ball around $\mathbf{x}_1$ that contains none of the other critical points. That is, $\mathbf{x}_1$ is an isolated critical

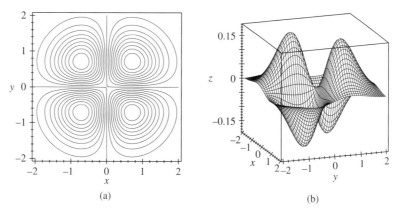

(a) (b)

Figure 4.2   (a) Contour plot of $f(x, y) = e^{-x^2 - y^2} xy$. (b) Graph of $f(x, y) = e^{-x^2 - y^2} xy$. (See Example 4.2.)

point. In the region around $\mathbf{x}_1$, the contour curves of f are closed curves that encircle $\mathbf{x}_1$. Thus it is a local maximum or a local minimum. Since it is isolated, it suffices to compare the value of f at the critical point to the value of f at a nearby point. (See Exercise 9.) For example, compare $f(\mathbf{x}_1) = \frac{1}{2e}$ to $f(1,1) = \frac{1}{e^2}$. Since $\frac{1}{2e} > \frac{1}{e^2}$, we conclude that $\mathbf{x}_1$ is a local maximum of f. In a similar manner, we can conclude that $(-1/\sqrt{2}, -1/\sqrt{2})$ is also a local maximum and that $(1/\sqrt{2}, -1/\sqrt{2})$ and $(-1/\sqrt{2}, 1/\sqrt{2})$ are local minima. Note that $f(-1/\sqrt{2}, -1/\sqrt{2}) = \frac{1}{2e}$ and $f(-1/\sqrt{2}, 1/\sqrt{2}) = f(1/\sqrt{2}, -1/\sqrt{2}) = \frac{-1}{2e}$. Further, since the values of f are small in absolute value outside the square $[-2, 2] \times [-2, 2]$ and approach 0 as $x^2 + y^2$ increases, the local maxima and minima are also global maxima and minima.

Since $f(0,0) = 0$ and f is positive in the first and third quadrants and is negative in the second and fourth quadrants, we conclude that $(0,0)$ is a saddle point of f. In addition, the contours in an open set of the origin are shaped like hyperbolas so that following Example 3.3A, we would also conclude that the origin is a saddle.

We can see from the example that starting with a symbolic expression for a function and then using the contour plot has the potential to provide significant information beyond what is available from the contour plot alone. In particular, it may be possible to solve for critical points of f symbolically. This is important because if the values of f at nearby critical points are close together, the critical points might be located between adjacent contours on a contour plot. Thus we could not identify the critical points from the plot. However, even if the locations of all the critical points of a function are known, it might be necessary to generate several contour plots using a computer before the type of the critical point can be identified with confidence.

The Gradient Vector Field

An alternative graphical method of classifying the isolated critical points of a function is provided by examining its gradient. Let us suppose that f is a differentiable function defined on the open set $\mathcal{O} \subset \mathbb{R}^2$. The gradient vector field of f, ∇f, is given by

$$\nabla f(\mathbf{x}) = (\frac{\partial f}{\partial x}(\mathbf{x}), \frac{\partial f}{\partial y}(\mathbf{x})).$$

It is immediate that ***the critical points of the gradient vector field of f are the same as the critical points of f***, since both are defined by the conditions $\frac{\partial f}{\partial x}(\mathbf{x}) = 0$ and $\frac{\partial f}{\partial y}(\mathbf{x}) = 0$. In Section 2.4, we identified four types of critical points of vector fields: sinks, centers, sources, and saddles. Sinks and centers are stable critical points, and sources and saddles are unstable critical points. In the collaborative exercise at the beginning of this chapter, we paired these types of critical points (except for centers) with minima, maxima, and saddles of f. Here we apply the results of Section 3.5 to give a more detailed explanation of the pairing. We consider each of the four possibilities separately.

If $\mathbf{x}_0$ is a *sink*, then there is a neighborhood $\mathcal{O}$ of $\mathbf{x}_0$ so that each flow line that begins at a point in $\mathcal{O}$ flows toward $\mathbf{x}_0$. From Section 3.5, it follows that the values of f increase everywhere along such flow lines as they "move" toward $\mathbf{x}_0$. Since f cannot decrease on these flow lines, we conclude that the value of f at $\mathbf{x}_0$ must be greater than the value of f at any other point in $\mathcal{O}$. That is, $\mathbf{x}_0$ is an *isolated local maximum* of f.

If $\mathbf{x}_0$ is a *source*, then there is a neighborhood $\mathcal{O}$ of $\mathbf{x}_0$ so each flow line that begins at a point in $\mathcal{O}$ flows away from $\mathbf{x}_0$. Thus the values of f increase as these flow lines move away from $\mathbf{x}_0$. We conclude that the value of f at $\mathbf{x}_0$ must be less than the value of f at any other point in $\mathcal{O}$. That is, $\mathbf{x}_0$ is an *isolated local minimum* of f.

If $\mathbf{x}_0$ is a *saddle*, then there are flow lines that begin near $\mathbf{x}_0$ that flow away from $\mathbf{x}_0$ and others that flow toward $\mathbf{x}_0$. Since these flow lines begin arbitrarily close to $\mathbf{x}_0$, the fact that flow lines flow away from $\mathbf{x}_0$ means that the value of f at $\mathbf{x}_0$ is less than those of f at some nearby points, as was the case for a source. The fact that flow lines flow toward $\mathbf{x}_0$ means that the value of f at $\mathbf{x}_0$ is greater than those of f at some nearby points, as was the case for a sink. We conclude that $\mathbf{x}_0$ is a *saddle point* of f.

Finally, we claim that a critical point of a gradient vector field cannot be a center. If $\mathbf{x}_0$ is a *center*, there is a neighborhood $\mathcal{O}$ of $\mathbf{x}_0$ so that the flow line α of ∇f through $\mathbf{x} \in \mathcal{O}$ is a closed curve. Thus there are points $t_1 < t_2$ in the domain of α with $\alpha(t_1) = \alpha(t_2) = \mathbf{x}$. Since f is increasing on α, it follows that $f(\mathbf{x}) = f(\alpha(t_1)) < f(\alpha(t_2)) = f(\mathbf{x})$, which is absurd. We conclude that ∇f *can have no centers*.

We summarize this analysis in the following proposition.

Proposition 4.2   Let $f : \mathcal{O} \subset \mathbb{R}^2 \to \mathbb{R}$ be a differentiable function. If $\mathbf{x}_0$ is a critical point of f, then it is also a critical point of ∇f. Further, if $\mathbf{x}_0$ is an isolated critical point of f, then:

1. if $\mathbf{x}_0$ is a sink of ∇f, it is a local maximum of f;
2. if $\mathbf{x}_0$ is a source of ∇f, it is a local minimum of f; and
3. if $\mathbf{x}_0$ is a saddle of ∇f, it is a saddle of f.

Finally, a critical point of ∇f cannot be a center.   ◆

We note that the converses are also true. For example, if $\mathbf{x}_0$ is local maximum of f, then $\mathbf{x}_0$ is a sink of ∇f. The following example illustrates the proposition.

Example 4.3   **Critical Points of f and ∇f.** Let $f(x,y) = -x^4 + 2x^2 - \frac{1}{2}x - 1 - y^2$. (We carried out a detailed graphical analysis of the behavior of f in Example 3.4.) The gradient vector field of f is given by

$$\nabla f(x,y) = (-4x^3 + 4x - \frac{1}{2}, -2y).$$

In Example 4.1A, we found that the critical points of f, hence the critical points of ∇f, are located approximately at $(-1.06, 0)$, $(0.13, 0)$, and $(0.93, 0)$. Figure 4.3(a) is a plot of

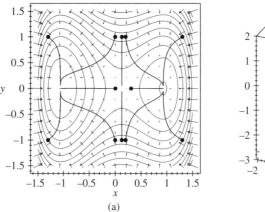
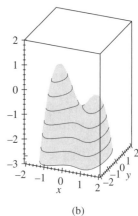

(a)   (b)

Figure 4.3   (a) The gradient vector field, flow lines of the gradient vector field, and level sets of $f(x, y) = -x^4 + 2x^2 - \frac{1}{2}x - 1 - y^2$. The initial points of the flow lines are marked by dots. (b) The graph of $f(x, y) = -x^4 + 2x^2 - \frac{1}{2}x - 1 - y^2$ showing horizontal slices of the graph of f. (See Example 4.3).

∇f, flow lines of ∇f, and level sets of f. Notice that the flow lines are orthogonal to the level sets. Now let us analyze the critical points of ∇f.

Based on the vector field and the flow lines with initial points $(-1.3, 1)$, $(0, 1)$, $(0, 0)$, $(0, -1)$, and $(-1.3, -1)$, which all flow toward $(-1.06, 0)$, we conclude that $(-1.06, 0)$ is a sink for ∇f, hence a local maximum of f. Similarly, we conclude that $(0.93, 0)$ is a sink for ∇f and a local maximum for f.

The flow lines with initial points $(0.13, 1)$ and $(0.13, 1)$ flow toward $(0.13, 0)$ while the flow lines with initial points $(0, 0)$ and $(0.2, 0)$ flow away from $(0.13, 0)$. We conclude that $(0.13, 0)$ is a saddle of ∇f and a saddle for f.

These conclusions agree with graph of f, which is shown in Figure 4.3(b). Note the two local maxima and the saddle point.

As was the case in Example 4.2, where we used a contour plot to classify the critical points of a function, we must employ caution when using a plot of the gradient of a function to classify its critical points. First, it is difficult to locate the critical points of a function from the plot of its gradient alone. However, given a symbolic expression for the function, we can solve for its critical points as we have done previously. Even so, it is usually necessary to plot several flow lines of ∇f for each critical point in order to identify the type of the critical point reliably.

Functions of Three Variables

The initial definitions and propositions of this section extend almost verbatim to functions of three variables. In particular, if $f : \mathcal{O} \subset \mathbb{R}^3 \to \mathbb{R}$ is a differentiable function on an open set $\mathcal{O}$, then $\mathbf{x}_0 \in \mathcal{O}$ is *a critical point of f if the partial derivatives of f at* $\mathbf{x}_0$ are 0,

$$\frac{\partial f}{\partial x}(\mathbf{x}_0) = 0, \ \frac{\partial f}{\partial y}(\mathbf{x}_0) = 0, \ \text{and} \ \frac{\partial f}{\partial z}(\mathbf{x}_0) = 0.$$

Thus in order to find the critical points of a function of three variables, it is necessary to solve a system of three equations in three variables. Extrema of differentiable functions are critical points, so the task of finding extrema again reduces to locating and classifying critical points of the function.

Generally, the graphical identification of the critical points of a function, either by analyzing its level sets or by analyzing a plot of its gradient vector field, is significantly more difficult for functions of three variables than for functions of two variables. It is still true, however, that the critical points of f and ∇f are the same and that sources of ∇f correspond to maximums of f and sinks of ∇f correspond to minimums of f.

Let us consider a simple example of a maximum.

Example 4.4   **A Maximum in $\mathbb{R}^3$.** Let $f(x, y, z) = e^{-x^2 - y^2 - z^2}$, which is defined and differentiable on all of space. The gradient vector field of f is given by

$$\nabla f(x, y, z) = (-2xe^{-x^2 - y^2 - z^2}, -2ye^{-x^2 - y^2 - z^2}, -2ze^{-x^2 - y^2 - z^2})$$

$$= e^{-x^2 - y^2 - z^2}(-2x, -2y, -2z).$$

Since the exponential factor is never 0, the only solution to the equation $\nabla f(\mathbf{x}) = \mathbf{0}$ is the origin, $(0, 0, 0)$. Further, since the exponent of e is nonpositive, the largest value of the exponential expression occurs when the exponent is 0, so the origin is a maximum point. Figure 4.4 contains a plot of ∇f near the origin. We can see from the plot that ∇f points toward the origin, so that the flow lines will also be directed toward the origin. That is, the origin is a sink for ∇f and we see the correspondence between sinks of ∇f and maxima of f.

Other graphical techniques for classifying the critical points of functions of three variables are more complicated, and we will not consider them.

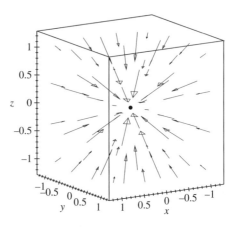

Figure 4.4   A plot of the gradient vector field of $e^{-x^2-y^2-z^2}$ in an open set of the origin. Notice ∇f points toward the origin at every point. (See Example 4.4.)

Summary

In this section, we expanded upon the ***graphical analysis of critical points*** that we introduced in Section 3.1. A point $\mathbf{x}_0$ is a ***critical point*** of the differentiable function f if the partial derivatives of f at $\mathbf{x}_0$ are zero. The tangent plane to the graph of f at a critical point is ***horizontal***.

We went on to define ***isolated*** and ***nonisolated*** critical points and to define ***local*** and ***global maxima*** and ***minima***. We showed that if f is differentiable, then every ***extreme point is a critical point***. Isolated critical points that are not extrema are called ***saddles***.

We established a ***correspondence between the classification of critical points of f and the critical points of*** ∇f: a ***sink*** of ∇f is a ***local maximum*** of f; a ***source*** of ∇f is a ***local minimum*** of f; and a ***saddle*** of ∇f is a ***saddle*** of f. A critical point of ∇f ***cannot be a center***. Finally, we indicated that these ideas extend to differentiable functions of three variables.

Section 4.1 Exercises

1. **Graphical Classification of Critical Points.** Determine the critical points of each of the following functions f. The contour plot of each function is given in Figure 4.5. Use

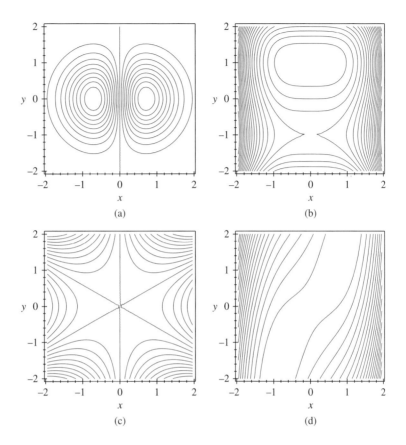

Figure 4.5 Plots for Exercise 1.

these contour plots to classify the critical points as local maxima, local minima, or saddle points. Determine if the local extrema are global extrema.

(a) $f(x,y) = xe^{-x^2-y^2}$.

(b) $f(x,y) = x^4 + y^3 - 3y$.

(c) $f(x,y) = x^3 - 3xy^2$.

(d) $f(x,y) = 4xy - 2x^4 - y^2$.

2. **Isolated Critical Points.** For each of the following functions $f : \mathbb{R}^2 \to \mathbb{R}$, find the critical points and determine if they are isolated or nonisolated. For each f, use a computer to generate a contour plot and use the plot to classify the critical points of f.

(a) $f(x,y) = 1/(1 + x^2 + y^2)$.

(b) $f(x,y) = x^2 - 4xy + 4y^2$.

(c) $f(x,y) = xy(1 - x - y)$.

(d) $f(x,y) = (2x^2 + 3y^2)e^{-x^2-y^2}$.

3. **Specified Critical Points.** Suppose that $f : \mathbb{R}^2 \to \mathbb{R}$ is a differentiable function on the entire plane. For each of the following descriptions, assume that f has critical points of the specified type with the given critical values at the given locations and only these critical points. Based on this information, sketch a possible plot of the contour curves of f on the square $[-2, 2] \times [-2, 2]$. Explain the features of the plot.

   (a) $f(0,0) = 3$ is an isolated global maximum.
   (b) $f(-1,0) = 0$ is an isolated saddle. $f(1,0) = 2$ is an isolated local maximum.
   (c) $f(0,1) = 0$ is an isolated local minimum. $f(0,0) = 2$ is an isolated saddle. $f(1,0) = -5$ is an isolated global minimum.
   (d) $f(1,1) = 1$ is an isolated global maximum. $f(-1,1) = -1$ is an isolated global minimum.

4. **Graphical Classification of Critical Points.** Determine the critical points of each of the following functions f. The plot of the gradient vector field ∇f for each function is given in Figure 4.6. Use these plots and the classification of critical points of vector fields

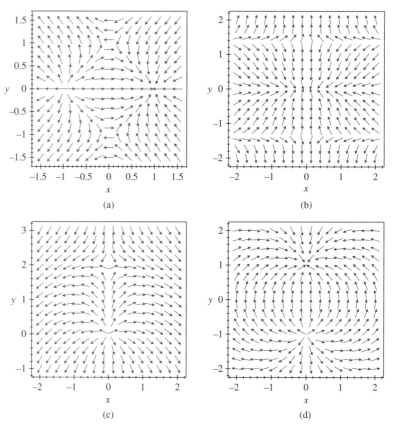

Figure 4.6   Plots for Exercise 4. The vectors in the plots have been rescaled to have the same length in order to better distinguish their directions when $\|\nabla f(\mathbf{x})\|$ is close to zero.

from Section 2.4 to classify the critical points of f as local maxima, local minima, or saddle points. Determine if the local extrema are global extrema.

(a) $f(x, y) = 3x - x^3 - 3xy^2$.

(b) $f(x, y) = y^4 - 4y^2 - x^2$.

(c) $f(x, y) = (x^2 + y^2)e^{-y}$.

(d) $f(x, y) = y/(1 + x^2 + y^2)$.

5. **Global vs. Local Extrema.** Let $f : \mathbb{R}^2 \to \mathbb{R}$ be a differentiable function on the entire plane.

   (a) Is it necessarily the case that f has global extrema? If so, explain why. If not, produce a counterexample.

   (b) Is it necessarily the case that f has local extrema? If so, explain why. If not, produce a counterexample.

6. **f and ∇.** In the text, we gave an intuitive argument for the identification of a critical point $\mathbf{x}_0$ of a differentiable function f based on knowing its type as a critical point of ∇f. Here you must reverse the argument. Using the facts that have been developed in this section about ∇f and its flow lines, construct a direct argument to show each of the following:

   (a) If $\mathbf{x}_0$ is a local isolated maximum of f, then $\mathbf{x}_0$ is a sink of ∇f.

   (b) If $\mathbf{x}_0$ is a local isolated minimum of f, then $\mathbf{x}_0$ is a source of ∇f.

   (c) If $\mathbf{x}_0$ is an isolated saddle point of f, then $\mathbf{x}_0$ is a saddle point of ∇f.

7. **f and $-f$.** Suppose that $f : \mathcal{O} \subset \mathbb{R}^2 \to \mathbb{R}$ is a differentiable function. An isolated critical point of f is also an isolated critical point of $-f$.

   (a) What, if any, is the relationship between the type of $\mathbf{x}_0$ as a critical point of f and the type of $\mathbf{x}_0$ as a critical point of $-f$? Explain your answer. (Be sure to consider all the possibilities for the type of $\mathbf{x}_0$.)

   (b) What, if any, is the relationship between the type of $\mathbf{x}_0$ as a critical point of ∇f and the type of $\mathbf{x}_0$ as a critical point of $\nabla(-f)$? Explain your answer. (Be sure to consider all the possibilities for the type of $\mathbf{x}_0$.)

8. **Isolated Minima.** Suppose that f is a differentiable function on the open subset $\mathcal{O} \subset \mathbb{R}^2$ and that $\mathbf{x}_0 \in \mathcal{O}$ is an isolated critical point of f. If $\mathbf{x}_0$ is a minimum for f, show that there is an open ball $\mathcal{B}$ centered at $\mathbf{x}_0$ with the property that $f(\mathbf{x}) > f(\mathbf{x}_0)$ for all $\mathbf{x} \in \mathcal{B}$, $\mathbf{x} \neq \mathbf{x}_0$. (*Hint*: Assume that for each open ball $\mathcal{B}$ of $\mathbf{x}_0$, there is an $\mathbf{x} \in \mathcal{B}$ with $\mathbf{x} \neq \mathbf{x}_0$ and $f(\mathbf{x}) = f(\mathbf{x}_0)$. Show that this leads to a contradiction.)

9. **Isolated Extrema.** Suppose that f is a differentiable function, $f : \mathcal{O} \subset \mathbb{R}^2 \to \mathbb{R}$, and that $\mathbf{x}_0 \in \mathcal{O}$ is a critical point of f.

   (a) In the text, it was claimed that if $\mathbf{x}_0$ is an isolated extrema of f, then in order to determine if $\mathbf{x}_0$ is a maximum or a minimum, it suffices to compare the value of f at $\mathbf{x}_0$ to the value of f at a nearby point. Explain why this claim is valid. How should the nearby point be chosen?

   (b) In part (a), is it necessary that $\mathbf{x}_0$ be isolated? Explain.

10. **Critical Points in $\mathbb{R}^3$.** For each of the following functions $f : \mathbb{R}^3 \to \mathbb{R}$, find the critical points of f and determine if they are isolated or nonisolated.

    (a) $f(x, y, z) = x^2 + y^2 + z^2 + xy.$

    (b) $f(x, y, z) = x^3 - x + y^2 + z^3 - z.$

    (c) $f(x, y, z) = e^{x^2 + y^2 + z^2}.$

    (d) $f(x, y, z) = x^2 + y^2.$

■ 4.2 Algebraic Classification of Critical Points

In Section 4.1, we developed two graphical techniques to classify the critical points of differentiable functions of two variables. In this section, we want to develop an **algebraic classification of critical points**. This classification is based on a quadratic approximation to the function at the critical point and is analogous to the second derivative test for relative extrema in one-variable calculus. In one variable, if x_0 is a critical point of the function f, we can determine if x_0 is a relative extrema by examining the sign of $f''(x_0)$. If $f''(x_0) > 0$, the graph of f is concave up in an open interval containing x_0, and we conclude that x_0 is a relative minimum. If $f'' < 0$, the graph of f is concave down in an open interval containing x_0, and we conclude that x_0 is a relative maximum. If $f''(x_0) = 0$, we cannot use this test to determine the type of critical point.

In two variables, we begin by considering quadratic polynomials of the form

$$f(x, y) = k + Ax^2 + 2Bxy + Cy^2,$$

where $k, A, B,$ and C are constants and at least one of A, B, and C is nonzero. Notice that f has no degree one terms. Each function of this form has a critical point at $(0, 0)$ with $f(0, 0) = k$. Because of the simple algebraic form of f, it is possible to determine the type of the critical point directly without differentiating f. First, we will rewrite the symbolic expression for f by completing the square. Assume that $A \neq 0$, which allows us to complete the square with respect to x. This involves adding and subtracting a constant term, which does not change the value, so that the new constant term and the existing x terms form a perfect square. We obtain

$$f(x, y) = k + A\left(x^2 + \frac{2B}{A}xy + \frac{B^2}{A^2}y^2\right) - \frac{B^2}{A}y^2 + Cy^2$$

$$= k + A\left(x + \frac{B}{A}y\right)^2 + \frac{1}{A}(AC - B^2)y^2.$$

Thus we have written the degree two term of f as a sum or difference of two squares, $(x + \frac{B}{A}y)^2$ with coefficient A and y^2 with coefficient $\frac{1}{A}(AC - B^2)$. We can determine the type of the critical point of f at $(0, 0)$ by analyzing the signs of these coefficients.

If both $AC - B^2 > 0$ and $A > 0$, the degree two term is a sum of two squares so that $f(x, y) > f(0, 0)$ for $(x, y) \neq (0, 0)$. It follows that $(0, 0)$ is a ***relative minimum*** of f.

If $AC - B^2 > 0$ and $A < 0$, the degree two term is the negative of a sum of squares so that $f(x, y) < f(0, 0)$ for $(x, y) \neq (0, 0)$. It follows that $(0, 0)$ is a ***relative maximum*** of f.

If $AC - B^2 < 0$, then regardless of the sign of A, the degree two term is a difference of squares. Let us assume that $A > 0$, so that $\frac{1}{A}(AC - B^2) < 0$. If $x \neq 0$ and $y = 0$,

$$f(x, 0) = k + \frac{A}{2}x^2 > f(0, 0).$$

Thus in any open ball centered at $(0, 0)$, there are points where the value of f is greater than $f(0, 0)$. On the other hand, if $x \neq 0$ and $y = -\frac{A}{B}x$,

$$f(x, -\frac{A}{B}x) = k + A(0)^2 + \frac{1}{A}(AC - B^2)\left(-\frac{A}{B}x\right)^2$$
$$= k + \frac{A}{B^2}(AC - B^2)x^2.$$

Since $AC - B^2 < 0$ and $A > 0$, this value is less than k. Thus in any open ball centered at $(0, 0)$, there are also points where the value of f is less than $f(0, 0)$. We conclude that $(0, 0)$ is a ***saddle point***. If $A < 0$, a similar analysis also shows that $(0, 0)$ is a saddle point. (See Exercise 1.)

If $AC - B^2 = 0$ and $A > 0$, $f(x, y) = k + Ax^2 \geq k$ and is equal to k for $x = 0$, so that $(0, 0)$ is a ***minimum*** for f that is ***not isolated***. If $A < 0$, $(0, 0)$ is a maximum for f that is not isolated.

This analysis was based on the assumption that $A \neq 0$. If $A = 0$ and $C \neq 0$, we could instead begin by completing the square with respect to y. Since this analysis is similar to the above case and the conclusions are equivalent, we will leave this argument as an exercise. (See Exercise 2(a).) If $A = C = 0$ and $B \neq 0$, it can be shown that $(0, 0)$ is a saddle point of f. We leave this argument for the exercises. (See Exercise 2(b).)

More generally, if g is a quadratic polynomial with a critical point at (x_0, y_0), g can be written

$$g(x, y) = k + A(x - x_0)^2 + 2B(x - x_0)(y - y_0) + C(y - y_0)^2,$$

where k, A, B, and C are constants and at least one of A, B, and C are nonzero. The function g can be derived from $f(x, y) = k + Ax^2 + 2Bxy + Cy^2$ by shifting the independent variables x and y by x_0 and y_0, respectively. Geometrically, we have seen that the graph of g is the graph of f shifted by the vector $\mathbf{v} = (x_0, y_0)$. Thus the critical point of f at $(0, 0)$ is shifted to a critical point of g at (x_0, y_0), but the type of the critical point remains the same. That is, an identical analysis would allow us to classify the critical point (x_0, y_0) for g based on the signs of A and $AC - B^2$.

We summarize these results in the following proposition.

Proposition 4.3   Let g be a quadratic polynomial given by

$$g(x,y) = k + A(x - x_0)^2 + 2B(x - x_0)(y - y_0) + C(y - y_0)^2,$$

where k, A, B, and C are constants and at least one of A, B, and C is nonzero. Then if $AC - B^2 \neq 0$, g has an isolated critical point at $\mathbf{x}_0 = (x_0, y_0)$ and if

1. $AC - B^2 > 0$ and $A > 0$, then $\mathbf{x}_0$ is an isolated local minimum;
2. $AC - B^2 > 0$ and $A < 0$, then $\mathbf{x}_0$ is an isolated local maximum; and
3. $AC - B^2 < 0$, then $\mathbf{x}_0$ is a saddle point.

If $AC - B^2 = 0$, g has a nonisolated critical point at $\mathbf{x}_0 = (x_0, y_0)$ and if

1. $A > 0$ or $C > 0$, then $\mathbf{x}_0$ is a nonisolated minimum; and
2. $A < 0$ or $C < 0$, then $\mathbf{x}_0$ is a nonisolated maximum.   ◆

We note that our analysis in fact shows that if f has a local extremum at $\mathbf{x}_0$, it is a global extremum.

Quadratic Approximations

In order to complete a similar analysis of a critical point $\mathbf{x}_0$ of an arbitrary differentiable function f, we will construct a quadratic polynomial, $q = q(x, y)$, called the **Taylor polynomial of degree two** or **quadratic approximation** of f at $\mathbf{x}_0$, which approximates f in a neighborhood of $\mathbf{x}_0$. This will be modeled on the construction of the linear approximation l to f given in Section 3.4. Recall that l is given by

$$l(x,y) = f(\mathbf{x}_0) + \frac{\partial f}{\partial x}(\mathbf{x}_0)(x - x_0) + \frac{\partial f}{\partial y}(\mathbf{x}_0)(y - y_0).$$

It has the properties that its value and the value of its first partial derivatives at $\mathbf{x}_0$ agree with those of f.

We will require that the values of q and its first and second partial derivatives at $\mathbf{x}_0$ agree with the corresponding values of f at $\mathbf{x}_0$. Let us write q in the form

$$q(x,y) = a_0 + b_1(x - x_0) + b_2(y - y_0) + c_1(x - x_0)^2 + c_2(x - x_0)(y - y_0) + c_3(y - y_0)^2.$$

Then $q(\mathbf{x}_0) = f(\mathbf{x}_0)$ implies that $a_0 = f(\mathbf{x}_0)$. Since $\frac{\partial q}{\partial x}(\mathbf{x}_0) = b_1$ and $\frac{\partial q}{\partial y}(\mathbf{x}_0) = b_2$, the requirement that the first partial derivatives agree implies that $b_1 = \frac{\partial f}{\partial x}(\mathbf{x}_0)$ and $b_2 = \frac{\partial f}{\partial y}(\mathbf{x}_0)$. Thus

$$q(x, y) = l(x, y) + c_1(x - x_0)^2 + c_2(x - x_0)(y - y_0) + c_3(y - y_0)^2.$$

The second partial derivatives of q are $\frac{\partial^2 q}{\partial x^2}(\mathbf{x}_0) = 2c_1$, $\frac{\partial^2 q}{\partial x \partial y}(\mathbf{x}_0) = c_2$, and $\frac{\partial^2 q}{\partial y^2}(\mathbf{x}_0) = 2c_3$. It follows that

$$c_1 = \frac{1}{2}\frac{\partial^2 f}{\partial x^2}(\mathbf{x}_0), \quad c_2 = \frac{\partial^2 f}{\partial x \partial y}(\mathbf{x}_0), \quad \text{and} \quad c_3 = \frac{1}{2}\frac{\partial^2 q}{\partial y^2}(\mathbf{x}_0).$$

Combining these statements, q is given by

$$q(x, y) = l(x, y) + \frac{1}{2}\left(\frac{\partial^2 f}{\partial x^2}(\mathbf{x}_0)(x - x_0)^2 + \right.$$
$$\left. 2\frac{\partial^2 f}{\partial x \partial y}(\mathbf{x}_0)(x - x_0)(y - y_0) + \frac{\partial^2 f}{\partial y^2}(\mathbf{x}_0)(y - y_0)^2\right),$$

where l is the linear approximation to f at $\mathbf{x}_0$.

As we did for l, we can express the fact that this polynomial approximates f near $\mathbf{x}_0$ in terms of a limit. We state this as a proposition.

Proposition 4.4 Let f be defined on a neighborhood of $\mathbf{x}_0$, and suppose that the first and second partial derivatives of f exist and are continuous at $\mathbf{x}_0$. The quadratic polynomial q defined above satisfies

$$\lim_{\mathbf{x} \to \mathbf{x}_0} \frac{f(\mathbf{x}) - q(\mathbf{x})}{\|\mathbf{x} - \mathbf{x}_0\|^2} = 0. \quad \blacklozenge$$

Since both the numerator and the denominator of the quotient in the limit approach zero, the fact that the limit is zero implies that as $\mathbf{x}$ approaches $\mathbf{x}_0$, the value of $f(\mathbf{x}) - q(\mathbf{x})$ approaches zero faster than $\|\mathbf{x} - \mathbf{x}_0\|^2$. That is, as $\mathbf{x}$ approaches $\mathbf{x}_0$, the value of $q(\mathbf{x})$ approaches that of $f(\mathbf{x})$ faster than the square of the distance from $\mathbf{x}$ to $\mathbf{x}_0$ approaches zero.

The converse of this proposition is also true.

Proposition 4.5 Let f be defined on a neighborhood $\mathcal{O}$ of $\mathbf{x}_0$. If there exists a quadratic polynomial q that satisfies

$$\lim_{\mathbf{x}\to\mathbf{x}_0} \frac{f(\mathbf{x}) - q(\mathbf{x})}{\|\mathbf{x} - \mathbf{x}_0\|^2} = 0,$$

then the first and second partial derivatives of f exist at $\mathbf{x}_0$, and q is given by

$$q(x, y) = f(\mathbf{x}_0) + \frac{\partial f}{\partial x}(\mathbf{x}_0)(x - x_0) + \frac{\partial f}{\partial y}(\mathbf{x}_0)(y - y_0)$$

$$+ \frac{1}{2}\left(\frac{\partial^2 f}{\partial x^2}(\mathbf{x}_0)(x - x_0)^2 + 2\frac{\partial^2 f}{\partial x \partial y}(\mathbf{x}_0)(x - x_0)(y - y_0) + \frac{\partial^2 f}{\partial y^2}(\mathbf{x}_0)(y - y_0)^2 \right). \blacklozenge$$

The following example illustrates the calculation of q.

Example 4.5

Second Degree Taylor Polynomials. In this example, we will consider the function $f(x, y) = y^3 - y + x^4 + x^2$.

A. Let us calculate the second degree Taylor polynomial or quadratic approximation q of f at $\mathbf{x}_0 = (0, 1.25)$. First, we must calculate the values of f and its partial derivatives at $(0, 1.25)$. The partial derivatives of f are

$$\frac{\partial f}{\partial x}(\mathbf{x}_0) = 4x^3 + 2x, \quad \frac{\partial f}{\partial y}(\mathbf{x}_0) = 3y^2 - 1,$$

$$\frac{\partial^2 f}{\partial x^2}(\mathbf{x}_0) = 12x^2 + 2, \quad \frac{\partial^2 f}{\partial y^2}(\mathbf{x}_0) = 6y, \quad \frac{\partial f}{\partial y \partial x}(\mathbf{x}_0) = 0.$$

Evaluating the function and these derivatives at $x = 0$ and $y = 1.25$, we have

$$f(0, 1.25) = 0.703125, \quad \frac{\partial f}{\partial x}(0, 1.25) = 0, \quad \frac{\partial f}{\partial y}(0, 1.25) = 3.6875,$$

$$\frac{\partial^2 f}{\partial x^2}(0, 1.25) = 2, \quad \frac{\partial^2 f}{\partial y^2}(\mathbf{x}_0) = 7.5, \quad \frac{\partial f}{\partial y \partial x}(\mathbf{x}_0) = 0.$$

We substitute these values into our formula for q to obtain

$$q(x, y) = 0.703125 + 0(x - 0) + 3.6875(y - 1.25)$$

$$+ \frac{1}{2}\left(2(x - 0)^2 + 2 \cdot 0(x - 0)(y - 1.25) + 7.5(y - 1.25)^2 \right)$$

$$= 0.703125 + 3.6875(y - 1.25) + \frac{1}{2}\left(2x^2 + 7.5(y - 1.25)^2 \right).$$

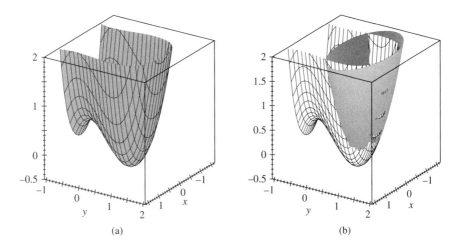

(a)   (b)

Figure 4.7 (a) The graph of $f(x,y) = y^3 - y + x^4 + x^2$. (b) The graph of $f(x,y)$ and the quadratic approximation to the graph of f at $\mathbf{x}_0 = (0, 1.25)$. (See Example 4.5A.)

A plot of the graph of f is shown in Figure 4.7(a), and plots of the graph of f and q are shown in Figure 4.7(b). Notice that the graphs of q and f are almost indistinguishable near the point $(0, 1.25)$ and that the graphs are not close for points (x, y) that are not close to $(0, 1.25)$.

B. Here we compute the second degree Taylor polynomial at a critical point of f. To find a critical point, we set the equations above for $\frac{\partial f}{\partial x}$ and $\frac{\partial f}{\partial y}$ equal to zero. Solving these equations for (x, y), we see that the critical points of f are $(0, -1/\sqrt{3})$ and $(0, 1/\sqrt{3})$. We will focus on $\mathbf{x}_0 = (0, -1/\sqrt{3})$. If we substitute $x = 0$ and $y = -1/\sqrt{3}$ into the formulas for f and its partial derivatives above, we have

$$f(0, -1/\sqrt{3}) = -2\sqrt{3}/9, \quad \frac{\partial f}{\partial x}(0, -1/\sqrt{3}) = 0, \quad \frac{\partial f}{\partial y}(0, -1/\sqrt{3}) = 0,$$

$$\frac{\partial^2 f}{\partial x^2}(0, -1/\sqrt{3}) = 2, \quad \frac{\partial^2 f}{\partial y^2}(0, -1/\sqrt{3}) = -2\sqrt{3}, \quad \frac{\partial^2 f}{\partial y \partial x}(0, -1/\sqrt{3}) = 0.$$

Substituting these values into our formula for the second Taylor polynomial, we obtain

$$
\begin{aligned}
q(x, y) &= -2\sqrt{3}/9 + 0(x - 0) + 0(y + 1/\sqrt{3}) \\
&\quad + \frac{1}{2}\left(2(x - 0)^2 + 0(x - 0)(y + 1/\sqrt{3}) - 2\sqrt{3}(y + 1/\sqrt{3})^2\right) \\
&= -2\sqrt{3}/9 + \frac{1}{2}\left(2x^2 - 2\sqrt{3}(y + 1/\sqrt{3})^2\right).
\end{aligned}
$$

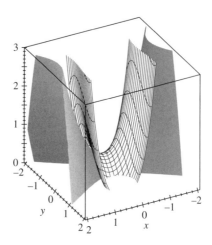

Figure 4.8   The graph of $f(x, y) = y^3 - y + x^4 + x^2$ and its quadratic approximation at the saddle point $(0, \frac{-1}{\sqrt{3}})$. (See Example 4.5B.)

The quadratic polynomial q also has a critical point at $(0, -1/\sqrt{3})$. Notice that the degree two term of q is a difference of two squares, so from our analysis above we conclude that $(0, -1/\sqrt{3})$ is a saddle point of q. Once again the graphs of f and q are indistinguishable near $(0, -1/\sqrt{3})$ so that we conclude that this point is a saddle point of f as well. (See Figure 4.8.)

The Second Derivative Test

If f has a critical point at $\mathbf{x}_0$, it follows that the quadratic approximation q to f at $\mathbf{x}_0$ also has a critical point at $\mathbf{x}_0$. This is immediate since $\frac{\partial q}{\partial x}(\mathbf{x}_0) = \frac{\partial f}{\partial x}(\mathbf{x}_0) = 0$ and $\frac{\partial q}{\partial y}(\mathbf{x}_0) = \frac{\partial f}{\partial y}(\mathbf{x}_0) = 0$. Thus the formula for q reduces to

$$q(x, y) = f(\mathbf{x}_0) + \frac{1}{2}\left(A(x - x_0)^2 + 2B(x - x_0)(y - y_0) + C(y - y_0)^2\right),$$

where $A = \frac{\partial^2 f}{\partial x^2}(\mathbf{x}_0)$, $B = \frac{\partial^2 f}{\partial x \partial y}(\mathbf{x}_0)$, and $C = \frac{\partial^2 f}{\partial y^2}(\mathbf{x}_0)$. It is evident that q is determined by f. Of course, this is the content of Proposition 4.5. Based on Proposition 4.4, it is tempting to conclude that f is determined by q near $\mathbf{x}_0$. The correct interpretation is that f is determined by q up to second order near $\mathbf{x}_0$. When $AC - B^2 = \frac{\partial^2 f}{\partial x^2}(\mathbf{x}_0)\frac{\partial^2 f}{\partial y^2}(\mathbf{x}_0) - \frac{\partial^2 f}{\partial x \partial y}(\mathbf{x}_0)^2 \neq 0$, this is sufficient to identify the type of the critical point of f at $\mathbf{x}_0$. That is, the algebraic criteria for determining the type of the critical point of q give identical algebraic criteria for determining the type of the critical point of f at $\mathbf{x}_0$. We obtain the following classification result for critical points of f.

Proposition 4.6 **The Second Derivative Test.** Let f be defined as an open set $\mathcal{O} < \mathbb{R}^2$, and suppose that the first and second partial derivatives of f exist and are continuous on $\mathcal{O}$. Let $\mathbf{x}_0$ be a critical point of f contained in O. Let $A = \frac{\partial^2 f}{\partial x^2}(\mathbf{x}_0)$, $B = \frac{\partial^2 f}{\partial x \partial y}(\mathbf{x}_0)$, and $C = \frac{\partial^2 f}{\partial y^2}(\mathbf{x}_0)$ with $AC - B^2 \neq 0$. If

1. $AC - B^2 > 0$ and $A > 0$, then $\mathbf{x}_0$ is a local minimum;
2. $AC - B^2 > 0$ and $A < 0$, then $\mathbf{x}_0$ is a local maximum; and
3. $AC - B^2 < 0$, then $\mathbf{x}_0$ is a saddle.

If $AC - B^2 = 0$, the type of the critical point cannot be determined from the values of the second partial derivatives of f at the point. ◆

As we noted above, if the quadratic function q has a minimum or maximum at $\mathbf{x}_0$, it must be a global minimum or maximum. However, since the behavior of q at $\mathbf{x}_0$ only reflects the behavior of f *near* $\mathbf{x}_0$, the most that we can say is that $\mathbf{x}_0$ is a local minimum or maximum for f. In some cases, it is possible to use the form of f or the geometry of the graph to determine if local extrema are global extrema.

In order to complete our analysis, we must examine the case $AC - B^2 = 0$. In Proposition 4.3, we saw that the quadratic polynomial can have a nonisolated extremum in this case. In the following example, we consider two functions that satisfy $AC - B^2 = 0$ and have the same quadratic approximation but exhibit different behavior at the critical point. The reason for this discrepancy will be that the higher order terms (degree greater than two) of the functions are different.

Example 4.6 **Critical Points of Higher Order Polynomials**

A. Let $f(x, y) = x^2 + y^4$. The origin is an isolated critical point for f. The quadratic approximation to f at the origin is $q(x, y) = \frac{1}{2}(2x^2)$, so that $A = 2$ and $B = C = 0$, making $AC - B^2 = 0$. We can see that q has a local minimum at $(0, 0)$, but it is not an isolated minimum. Returning to f, we see that the origin is a local (and global) minimum for f, since $f(x, y) \geq 0$ for all (x, y) and $f(x, y) = 0$ only at the origin.

B. Now let us consider $g(x, y) = x^2 - y^4$. The origin is an isolated critical point for g. The quadratic approximation to g at the origin is also $q(x, y) = \frac{1}{2}(2x^2)$. Thus we again have $A = 2$ and $B = C = 0$, so that $AC - B^2 = 0$, and we see that the origin is again a local minimum of q that is not isolated. However, returning to g, we see that the origin is a minimum for g if we restrict our attention to the x-axis and a maximum for g if we restrict our attention to the y-axis. Therefore the origin is a saddle point for g.

The example shows that if the coefficients of the quadratic approximation q to f at a critical point of f satisfy the condition $AC - B^2 = 0$, it is not possible to determine the type of the critical point of f based on an analysis of q alone. For this reason, we distinguish critical points that satisfy $AC - B^2 = 0$ from those that satisfy $AC - B^2 \neq 0$. We say that a critical point of f is **degenerate** if $AC - B^2 = 0$ and **nondegenerate** if $AC - B^2 \neq 0$.

Example 4.7   **Global Extrema**

A. Let $f(x, y) = x^4/4 - x^3/3 - x^2 + 1 + y^2$. Then $\nabla f(x, y) = (x^3 - x^2 - 2x, 2y)$, so that critical points of f satisfy $x^3 - x^2 - 2x = 0$ and $2y = 0$. Solving these equations for x and y, we obtain $x = -1$, 0, or 2 and $y = 0$. Thus the critical points for f are $(-1, 0)$, $(0, 0)$, and $(2, 0)$. Now let us compute the second order partial derivatives at the critical points. These are

$$\frac{\partial^2 f}{\partial x^2}(x, y) = 3x^2 - 2x - 2, \quad \frac{\partial^2 f}{\partial x \partial y}(x, y) = 0, \quad \frac{\partial^2 f}{\partial y^2}(x, y) = 2.$$

For $(-1, 0)$ we have $A = 3$, $B = 0$, and $C = 2$, so that $AC - B^2 = 6 > 0$ and $A > 0$. Therefore, the critical point at $(-1, 0)$ is a local minimum. For $(0, 0)$, we have $A = -2$, $B = 0$, and $C = 2$, so that $AC - B^2 = -4 < 0$ and the critical point at $(0, 0)$ is a saddle. For $(2, 0)$, $A = 6$, $B = 0$, and $C = 2$, so that $AC - B^2 = 12 > 0$ and $A > 0$. Therefore, the critical point at $(2, 0)$ is a local minimum. Since f is the sum of a polynomial in x and a polynomial in y (there are no terms involving xy), and the terms of highest degree in the variable x and in the variable y are of even degree and have positive coefficients, f increases without bound as one or both of the variables x and y increase. Thus f has no absolute maximum, but must have an absolute minimum. Since $f(-1, 0) = 7/12$ and $f(2, 0) = -5/3$, we conclude that the local minimum at $(2, 0)$ is an absolute minimum. A plot of the graph of f is shown in Figure 4.9.

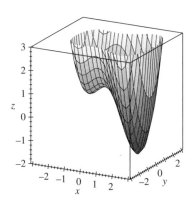

Figure 4.9   The graph of $f(x, y) = x^4/4 - x^3/3 - x^2 + 1 + y^2$ showing two local minimums and one saddle, all of which are nondegenerate. (See Example 4.7A.)

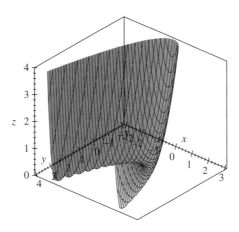

Figure 4.10   The graph of $f(x,y) = y^2 - 2yx^2 + x^4$ showing nonisolated minima along the curve $y = x^2$. Note the view is from below the xy-plane. (See Example 4.7B.)

B. Let $f(x,y) = y^2 - 2yx^2 + x^4$. Then $\nabla f(x,y) = (-4yx + 4x^3, 2y - 2x^2)$, so that the critical points of f satisfy $-4yx + 4x^3 = 0$ and $2y - 2x^2 = 0$. These equations can be rewritten as $-4x(y - x^2) = 0$ and $2(y - x^2) = 0$. We see that the set of critical points of f is the set of points satisfying $y = x^2$. Let us consider, for example, the critical point $(x,y) = (1,1)$. We have

$$\frac{\partial^2 f}{\partial x^2}(\mathbf{x}_0) = -4y + 12x^2, \quad \frac{\partial^2 f}{\partial x \partial y}(\mathbf{x}_0) = -4x, \quad \frac{\partial^2 f}{\partial y^2}(1,1) = 2.$$

Evaluating at the point $(1,1)$, we have $AC - B^2 = 8 \cdot 2 - (-4)^2 = 0$ and the second derivative test does not apply. In this case, we must resort to a more direct analysis of the function. Notice that $f(x,y) = y^2 - 2yx^2 + x^4 = (y - x^2)^2$. Thus $f(x,y) = 0$ for each critical point of f and $f(x,y) > 0$ for (x,y) not of the form $y = x^2$. We conclude that $(1,1)$ is a minimum of f. Since there are points (x,y) that are arbitrarily near to $(1,1)$ with $f(x,y) = f(1,1)$, the point $(1,1)$ is not an isolated minimum of f. See Figure 4.10 for a plot of the graph of f.

In the following example, we consider a function that models the displacement of points on a square drumhead when the drum is struck.

Example 4.8   **A Square Drum.** When a square drumhead is struck, the points in the drumhead are displaced vertically over time. This displacement is cyclical; that is, each point moves vertically in a positive or negative direction, returns to its initial position, moves vertically in the opposite direction, and then returns to its original position and repeats its cycle. We can, for example, represent the drumhead by the square $-\pi \leq x \leq \pi$ and $-\pi \leq y \leq \pi$ in the xy-plane. If the vertical displacement of the drumhead is small, we can model

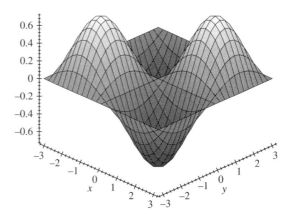

Figure 4.11 The graph of $f(x, y) = F(x, y, \pi/4) = \frac{1}{\sqrt{2}} \sin(x) \sin(y)$ modeling the position of a vibrating square drumhead. (See Example 4.8.)

the movement of the drumhead over time by a function of x, y, and t. One simple motion of the drumhead is modeled by $F(x, y, t) = \sin(t) \sin(x) \sin(y)$. At time $t = 0$, $F(x, y, 0) = 0$ and the drumhead is flat. As time increases, the points on the drumhead are set in motion. Here we would like to investigate the position of the drumhead for a fixed time t_0. Let us assume $t_0 = \pi/4$, so the position of drumhead is modeled by $f(x, y) = F(x, y, \pi/4) = \frac{1}{\sqrt{2}} \sin x \sin y$. (See Figure 4.11.) Since this is now a function of two variables, x and y, we can use the techniques of this section to analyze its critical points.

The first partial derivatives are

$$\frac{\partial f}{\partial x}(\mathbf{x}) = \frac{1}{\sqrt{2}} \cos x \sin y, \quad \frac{\partial f}{\partial y}(\mathbf{x}) = \frac{1}{\sqrt{2}} \sin x \cos y.$$

Setting these equal to zero, we see that this function has critical points at the five points $(0, 0)$, $(\pm\frac{\pi}{2}, \frac{\pi}{2})$, and $(\pm\frac{\pi}{2}, -\frac{\pi}{2})$. The second partial derivatives of this function are

$$\frac{\partial^2 f}{\partial x^2}(x, y) = -\frac{1}{\sqrt{2}} \sin x \sin y, \quad \frac{\partial^2 f}{\partial x \partial y}(x, y) = \frac{1}{\sqrt{2}} \cos x \cos y,$$

$$\frac{\partial^2 f}{\partial y^2}(x, y) = -\frac{1}{\sqrt{2}} \sin x \sin y.$$

Using the second derivative test, we see that $(0, 0)$ is a saddle point. The points $(\frac{\pi}{2}, \frac{\pi}{2})$ and $(-\frac{\pi}{2}, -\frac{\pi}{2})$ are relative maxima, and the points $(\frac{\pi}{2}, -\frac{\pi}{2})$ and $(\frac{\pi}{2}, -\frac{\pi}{2})$ are relative minima. The maximum position is $\frac{1}{\sqrt{2}}$; the minimum position is $-\frac{1}{\sqrt{2}}$.

In order to understand the movement of the drumhead more completely, it is necessary to investigate the position of the drumhead for different times and then to put these "snapshots" of the position of the drumhead together. (See Exercise 6.)

Summary

In this section, we developed an ***algebraic classification of critical points*** for differentiable functions to go along with the geometric classification that we developed in Section 4.1. By analyzing the second partial derivatives of quadratic polynomials at their critical points, we devised conditions on the values of these derivatives that told us whether the critical point was a maximum, a minimum, a saddle, or a nonisolated maximum or minimum. By applying these criteria to the ***second Taylor polynomial of a function at a critical point***, we obtained the ***second derivative test*** for classifying critical points of functions whose first and second partial derivatives exist and are continuous. Define

$$A = \frac{\partial^2 f}{\partial x^2}(\mathbf{x}_0), \quad B = \frac{\partial^2 f}{\partial x \partial y}(\mathbf{x}_0), \quad \text{and } C = \frac{\partial^2 f}{\partial y^2}(\mathbf{x}_0).$$

If $\mathbf{x}_0$ is a ***nondegenerate*** critical point of f, that is $AC - B^2 \neq 0$, then if
  1. $AC - B^2 > 0$ and $A > 0$, then $\mathbf{x}_0$ is a local minimum;
  2. $AC - B^2 > 0$ and $A < 0$, then $\mathbf{x}_0$ is a local maximum; and
  3. $AC - B^2 < 0$, then $\mathbf{x}_0$ is a saddle.

If $\mathbf{x}_0$ is a degenerate critical point, that is $AC - B^2 = 0$, the type of the critical point cannot be determined from the values of the second partial derivatives of f at the point.

Section 4.2 Exercises

1. **A Quadratic Saddle.** Suppose that $f(x, y) = k + Ax^2 + 2Bxy + Cy^2$. Verify that if $A < 0$ and $AC - B^2 < 0$, then $(0, 0)$ is a saddle point of f.

2. **Completing the Square**

   (a) Suppose that $f(x, y) = k + 2Bxy + Cy^2$, where $C \neq 0$. Rewrite the symbolic expression for f by completing the square with respect to y, and then use this expression to analyze the type of critical point of f at $(0, 0)$.

   (b) Show that if $f(x, y) = k + 2Bxy$, $B \neq 0$, then $(0, 0)$ is a saddle point for f.

3. **Taylor Polynomials.** For each of the following functions f of two variables, compute the first and second degree Taylor polynomials of f at the given point $\mathbf{x}_0$.

   (a) $f(x, y) = 3x^2 - xy + 5x - 2y + 3$, $\mathbf{x}_0 = (0, 0)$.
   (b) $f(x, y) = x^2y^2 - x^2y + xy$, $\mathbf{x}_0 = (-1, 3)$.
   (c) $f(x, y) = e^{x^2 + y^2}$, $\mathbf{x}_0 = (0, 0)$.
   (d) $f(x, y) = \cos(x - y)$, $\mathbf{x}_0 = (0, 0)$.

4. **Second Derivative Test Calculations.** For each of the following functions f, determine the critical points of f and use the second derivative test to identify the type of each of the critical points of f.

(a) $f(x, y) = 3x^2 - 2xy + y^2$.
(b) $f(x, y) = x^2 + 2xy - y^2 + 3x - 2y$.
(c) $f(x, y) = x^4/4 - x^3 + x^2 + 1 - y^2$.

(d) $f(x, y) = e^{-3x^2 - y^2}$.
(e) $f(x, y) = \frac{1}{x^2 + y^2 + 1}$.

5. **Degenerate Critical Points.** For each of the following functions f, determine the critical points of f and identify the type of each of the nondegenerate critical points of f using the second derivative test. If the critical point is degenerate, determine the type of the critical point by other means.

(a) $f(x, y) = x^4 - y^2$.
(b) $f(x, y) = x^3 - y^3$.

(c) $f(x, y) = x^2 - 2xy + y^2$.
(d) $f(x, y) = x^3 - 3xy^2$.

6. **The Square Drum.** In this exercise, we will explore the model of a vibrating square drumhead, which was introduced in Example 4.8. At a fixed time t_0, the position of the drumhead is modeled by the function $f(x, y) = \sin(t_0) \sin(x) \sin(y)$ for $-\pi \le x \le \pi$ and $-\pi \le y \le \pi$.

(a) Locate the critical points of this function. Does the location of the critical points depend on the time t_0? Explain your answer.
(b) Use the second derivative test to classify the critical points of this function. Does your classification depend on the time t_0? Explain your answer.
(c) Give a description of the motion of the vibrating drumhead by describing how the position changes for time from $t = 0$ to $t = 2\pi$.

7. **Quadratic Approximations**

(a) Let f be a twice differentiable function of two variables, and let c be a constant. Show that the quadratic approximation to cf at $\mathbf{x}_0$ is equal to cq where q is the quadratic approximation to f at $\mathbf{x}_0$.
(b) Let f and g be twice differentiable functions of two variables. Show that the quadratic approximation to the sum $f + g$ at $\mathbf{x}_0$ is equal to the sum of the quadratic approximation to f at $\mathbf{x}_0$ and the quadratic approximation to g at $\mathbf{x}_0$.

8. **Quadratic Polynomials.** Let $f(x, y) = ax^2 + cxy + dy^2 + mx + ny + l$ be a quadratic polynomial in two variables.

(a) Compute the linear and quadratic approximations to f at $(0, 0)$.
(b) Compute the linear and quadratic approximations to f at $\mathbf{x}_0$.
(c) What is the relationship between f and your answers to (a) and (b)? (*Hint:* Expand and simplify your answer to (b).)

9. **Cubic Polynomials.** A general polynomial of degree three in two variables contains all possible terms of total degree less than or equal to three. For example, the monomial $x^2 y$ has total degree $2 + 1 = 3$. Thus a general polynomial of degree three in two variables is

$$f(x,y) = a_1 x^3 + a_2 x^2 y + a_3 xy^2 + a_4 y^3$$
$$+ b_1 x^2 + b_2 xy + b_3 y^2 + c_1 x + c_2 y + d,$$

where a_i, b_j, c_k, and d are constants.

(a) Compute the linear and quadratic approximations to f at $(0,0)$.
(b) What is the relationship between f and your answers to (a)? Explain.

■ 4.3 Constrained Optimization in the Plane

In Sections 4.1 and 4.2, we were concerned with the problem of locating and identifying critical points of a differentiable function $f : \mathcal{O} \subset \mathbb{R}^2 \to \mathbb{R}$ on the open set $\mathcal{O}$. Here we are interested in the related question of finding the ***extrema of f on closed subsets of $\mathcal{O}$***. A simple physical example shows that it is not sufficient to find the critical points of f on $\mathcal{O}$ that happen to lie in the closed set.

Example 4.9

Extrema of Temperature. Suppose that a constant heat source is located at a point $\mathbf{x}_0$ so that the highest temperature occurs at $\mathbf{x}_0$ and the temperature decreases as we move away from $\mathbf{x}_0$. The temperature function has an isolated global maximum at $\mathbf{x}_0$ and has no other critical points. Suppose an object is located close to but does not contain the heat source. Intuitively, the point on the object closest $\mathbf{x}_0$ will be hottest, and the point on the object farthest from $\mathbf{x}_0$ will be coldest. That is, the extreme values of temperature on the object occur on the edge or boundary of the object. However, neither of these points is an extrema for the temperature function; thus the techniques that we have developed so far would fail to find extrema of temperature on the object.

The example leads us to conclude that if we restrict our attention to closed subsets of the domain, the extreme values of a function on the subset can occur at points other than critical points. Before we begin, it will be helpful to recall the analogous situation for functions of one variable. In $\mathbb{R}$, the closed subsets of interest are closed intervals $[a, b]$, which are the union of the open interval (a, b) and its boundary, which consists of the points a and b. The extrema of a differentiable function f on $[a, b]$ occur at the critical points of f on the open interval (a, b), or at the endpoints a or b. For example, consider the differentiable function $f(x) = x^3 - 3x$ on the interval $[0, 2]$: It has critical points at

$x = \pm 1$. The only critical point in the interval $(0, 2)$ is $x = 1$, which is a local minimum with $f(1) = -2$. Since $f(0) = 0$ and $f(2) = 2$, the minimum value of f on $[0, 2]$ occurs at the critical point $x = 1$ and the maximum value occurs at the endpoint $x = 2$.

Differentiable functions of two variables display similar behavior when we restrict our attention to closed subsets of the plane. Recall from Section 3.3 that a point $\mathbf{x}_0$ is in the **boundary** of a set $\mathcal{S}$ if every open ball centered at $\mathbf{x}_0$ contains points that are in $\mathcal{S}$ and points that are not in $\mathcal{S}$. A **closed set** is a set that contains all of its boundary points. Intuitively, closed sets are defined by inequalities that are not strict. For example, $\{(x, y) : x^2 + y^2 \leq 1\}$ is a closed set and its boundary is $\{(x, y) : x^2 + y^2 = 1\}$.

Closed sets play an important role in the theory of critical points because of the following theorem about extreme values of continuous functions, which we state without proof.

Theorem 4.1 **The Maximum Principle.** Let f be a continuous function defined on the open set $\mathcal{O} \subset \mathbb{R}^2$. If $\mathcal{S} \subset \mathcal{O}$ is a closed bounded subset of $\mathbb{R}^2$, then f has a maximum on $\mathcal{S}$. That is, there is a point $\mathbf{x}_0 \in \mathcal{S}$ so that $f(\mathbf{x}) \leq f(\mathbf{x}_0)$ for all points $\mathbf{x} \in \mathcal{S}$. ◆

Since a differentiable function is also continuous, the maximum principle holds for differentiable functions. The proof of the maximum principle is usually given in advanced courses in analysis. There is a corresponding **minimum principle** for continuous functions on closed sets. (See Exercise 11.) The requirements that the set be closed and that the function be continuous are essential. (See Exercise 9.)

Although the maximum principle and the minimum principle guarantee the existence of extrema, they do not provide a method for locating the extrema. While it is not possible to formulate such a method for continuous functions on arbitrary closed sets, it is possible to do so for differentiable functions when the closed set is the union of an open set and its boundary where the boundary is a curve or collection of curves.

Let us assume that $f : \mathcal{O} \rightarrow \mathbb{R}$ is a differentiable function and that the open set $\mathcal{N}$ and its boundary $\mathcal{B}$ are contained in $\mathcal{O}$. Then $\mathcal{S} = \mathcal{N} \cup \mathcal{B}$ is a closed set contained in $\mathcal{O}$. Further, we will assume that $\mathcal{B}$ is a curve. By the maximum principle and the minimum principle, we know that f has extrema on the closed set $\mathcal{S}$. The extrema can occur either on $\mathcal{N}$ or on $\mathcal{B}$. Since $\mathcal{N}$ is an open set, the extrema of f on $\mathcal{N}$ occur at the critical points of f on $\mathcal{N}$. If we have a symbolic expression for f, we can compute the partial derivatives of f, solve the equation $\nabla f(x, y) = (0, 0)$, and apply the second derivative test. If we do not have a symbolic expression for f, if we are unable to solve the equation $\nabla f(x, y) = (0, 0)$, or if the second derivative test fails, we can use graphical techniques, that is, we can analyze a contour plot, a density plot, plots of the slices of f, or a plot of the gradient vector field of f. It remains for us to develop symbolic and graphical techniques for finding the extrema of f on the curve $\mathcal{B}$.

Initially we will assume that the boundary $\mathcal{B}$ is a set of points satisfying a single equation of the form $g(x, y) = c$, where g is a differentiable function, and that $\nabla g(x, y) \neq \mathbf{0}$ for $(x, y) \in \mathcal{B}$. We will call a condition of this type a **constraint**, and we speak of *optimizing f subject to a constraint.* If we are given a contour plot of f showing the constraint curve $\mathcal{B}$, it is possible to use the information of the plot to find the approximate location of the extrema of f on $\mathcal{B}$ and to determine the type of the extrema. Given a symbolic expression for f, we will reduce the problem of locating extrema to solving a system of three equations, two involving the partial derivatives of f and g that must be solved subject to the third equation, the constraint equation $g(x, y) = c$.

The following example leads to the graphical approach for finding the extrema of f on $\mathcal{B}$.

Example 4.10

Finding Extrema Graphically. Let us consider the extrema of $f(x, y) = x^2/4 + y^2$ on the closed set $\mathcal{S} = \{(x, y) : x^2 + y^2/3 \leq 1 \}$. The boundary $\mathcal{B}$ of $\mathcal{S}$ is the ellipse given by $x^2 + y^2/3 = 1$. The function f has a single critical point, an absolute minimum, which is located at the origin. A contour plot of f and $\mathcal{B}$ is shown in Figure 4.12. We can see from the plot that the largest value that f attains on $\mathcal{B}$ is 3 and that this occurs at $(0, \pm\sqrt{3})$, and the smallest value that f attains on $\mathcal{B}$ is $1/4$ and that this occurs at $(\pm 1, 0)$. We conclude that the maximum value of f on $\mathcal{S}$ is 3 and the minimum value is 0, which occurs at the origin.

Examining the plot more closely, we can see that at each of the extrema of f on $\mathcal{B}$, the contour of f appears to be tangent to $\mathcal{B}$. This will the basis of the graphical approach.

In general, **the extrema of f on $\mathcal{B}$ will occur at points on $\mathcal{B}$ where the contour of f is tangent to $\mathcal{B}$ or at critical points of f that lie on $\mathcal{B}$.** We will verify this analytically in the process of deriving a symbolic approach to finding the extreme values of f on $\mathcal{B}$.

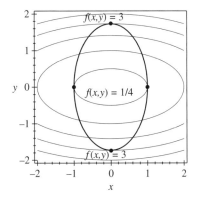

Figure 4.12   A plot of the contours of $f(x, y) = x^2/4 + y^2$ and the constraint curve $\mathcal{B}$ given by $g(x, y) = x^2 + y^2/3 = 1$. The contours of f are tangent to the constraint curve at the extrema of f on $\mathcal{B}$, which are marked on the plot. (See Example 4.10.)

In order to find the extrema of f on the constraint curve $\mathcal{B}$ given by $g(x, y) = c$, we must parametrize $\mathcal{B}$. Let us suppose that $\alpha(t) = (x(t), y(t))$ is a differentiable parametrization of $\mathcal{B}$ with nonzero derivative, that is, $\alpha : (a, b) \to \mathbb{R}^2$, $g(\alpha(t)) = c$, and $\alpha'(t) \neq \mathbf{0}$. Now let us suppose that $\mathbf{x}_0 = \alpha(t_0)$ is a local extremum of f on $\mathcal{B}$. Then the composition $f \circ \alpha$ has a local extremum at t_0. Since f and α are differentiable, the composition is also differentiable. From one-variable calculus, we know that an extreme value of a differentiable function is a critical point of the function, so we must have $(f \circ \alpha)'(t_0) = 0$. We can compute this derivative via the chain rule,

$$(f \circ \alpha)'(t_0) = \nabla f(\alpha(t_0)) \cdot \alpha'(t_0).$$

Since this derivative is zero and $\alpha'(t) \neq \mathbf{0}$, either $\nabla f(\alpha(t_0)) = \nabla f(\mathbf{x}_0) = \mathbf{0}$ or $\nabla f(\mathbf{x}_0)$ and $\alpha'(t_0)$ are orthogonal. In the first case, $\mathbf{x}_0$ is a critical point of f. In the second case, $\nabla f(\mathbf{x}_0)$ is orthogonal to $\mathcal{B}$. Since $\nabla f(\mathbf{x}_0)$ is also orthogonal to the contour curve of f passing through $\mathbf{x}_0$, the contour and $\mathcal{B}$ must be tangent at $\mathbf{x}_0$. This justifies the graphical analysis that follows Example 4.10.

To proceed with the symbolic argument, let us assume that $\nabla g(\mathbf{x}_0) \neq \mathbf{0}$. Since the constraint curve $\mathcal{B}$ is a level set of g, ∇g is orthogonal to the constraint curve and, therefore, to the tangent vector of the parametrization α, $\alpha'(t_0)$. Since $\nabla g(\mathbf{x}_0)$ and $\nabla f(\mathbf{x}_0)$ are orthogonal to the same vector, they are parallel. It follows that at an extreme point $\mathbf{x}_0$ of f on the constraint curve $\mathcal{B}$, $\nabla f(\mathbf{x}_0)$ is a scalar multiple of $\nabla g(\mathbf{x}_0)$. That is, there is a nonzero constant λ so that

$$\nabla f(\mathbf{x}_0) = \lambda \nabla g(\mathbf{x}_0).$$

This equation is called the **multiplier equation** and this method of finding critical points of f on the constraint curve is called the **method of Lagrange[1] multipliers**. The equation $g(x, y) = c$ that defines $\mathcal{B}$ is called the **constraint equation**. The following statement summarizes this method.

Lagrange Multipliers in the Plane	Let f and g be differentiable functions of two variables, and let $\mathcal{B} = \{(x, y) : g(x, y) = c\}$ be a curve contained in the domain of f with the property that $\nabla g(x, y) \neq \mathbf{0}$ for $(x, y) \in \mathcal{B}$. Then the extreme values of f on $\mathcal{B}$ are solutions to the system of equations $$\nabla f(x, y) = \lambda \nabla g(x, y)$$ $$g(x, y) = c,$$ where λ is a nonzero scalar.

[1] J. L. Lagrange (1736–1813) was an Italian mathematician and astronomer known for his wide-ranging and significant contributions to mathematics.

The multiplier equation is a vector equation that is equivalent to the coordinate equations $\frac{\partial f}{\partial x}(x, y) = \lambda \frac{\partial g}{\partial x}(x, y)$ and $\frac{\partial f}{\partial y}(x, y) = \lambda \frac{\partial g}{\partial y}(x, y)$. Thus to apply the method of Lagrange multipliers to find the extrema of f on $\mathcal{B}$, we must solve the following three equations for x, y and λ.

$$\frac{\partial f}{\partial x}(x, y) = \lambda \frac{\partial g}{\partial x}(x, y)$$
$$\frac{\partial f}{\partial y}(x, y) = \lambda \frac{\partial g}{\partial y}(x, y)$$
$$g(x, y) = c.$$

Note that the proposition says that the extreme points of f on $\mathcal{B}$ are **contained** in the set of solutions to this system of equations. Therefore this system may have solutions that are not extrema for f. In order to determine which of the solutions are extrema, we must evaluate f on each solution to determine whether it is a maximum or minimum for f.

Now let us consider an example of finding the extreme value of a function on a closed set consisting of an open set and its boundary curve.

Example 4.11   **Solving the Lagrange Multiplier System.** Let us find the extreme values of $f(x, y) = x^2 - y^2$ on the set $\mathcal{S} = \{(x, y) : (x - 1)^2 + y^2/4 \leq 2\}$, which is the union of the open set $\mathcal{N} = \{(x, y) : (x - 1)^2 + y^2/4 < 2\}$ and its boundary $\mathcal{B} = \{(x, y) : (x - 1)^2 + y^2/4 = 2\}$. First, let us examine a contour plot of f showing $\mathcal{S}$ (see Figure 4.13). There are three points, roughly the two points where $\mathcal{B}$ intersects the line $x = 1$ and the right-hand intersection point of $\mathcal{B}$ with the x-axis, where the contours appear to be tangent to $\mathcal{B}$ and a fourth point, the left-hand intersection point of $\mathcal{B}$ with the x-axis where we can

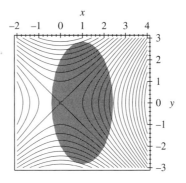

Figure 4.13   A contour plot of $f(x, y) = x^2 - y^2$ showing the closed set $\{(x, y) : (x - 1)^2 + y^2/4 \leq 2\}$. (See Example 4.11.)

infer the existence of a point of tangency. These are potentially extrema of f on $\mathcal{B}$. Now let us turn to the calculation of the extrema of f on $\mathcal{S}$.

To find the extrema of f in $\mathcal{N}$, we first find the critical points of f in $\mathcal{N}$ by solving $\nabla f(x, y) = (2x, -2y) = (0, 0)$. The origin, $(0, 0)$, is the only solution to this equation. Either by observing the contour plot of f or by using the second derivative test, we see that $(0, 0)$ is a saddle point of f. Since there are no local extrema of f in $\mathcal{N}$, we conclude that the extrema for f in $\mathcal{N} \cup \mathcal{B}$ must occur on $\mathcal{B}$.

The multiplier and constraint equations for f on the constraint curve $(x - 1)^2 + y^2/4 = 2$ are

$$2x = \lambda 2(x - 1)$$

$$-2y = \lambda y/2$$

$$(x - 1)^2 + y^2/4 = 2.$$

Since $\lambda \neq 0$, the first equation implies that $x \neq 0$. Further, if $x = 1$, the right side is zero, implying that $x = 0$. This is a contradiction, so we may assume that $x \neq 1$. This allows us to solve for λ in the first equation, $\lambda = x/(x - 1)$. Substituting into the second equation, we can eliminate λ, and the system of three equations in three unknowns reduces to a system of two equations in two unknowns,

$$-4y(x - 1) = xy$$

$$(x - 1)^2 + y^2/4 = 2.$$

The first equation can be rewritten in the form $y(4 - 5x) = 0$, which implies that $y = 0$ or $x = 4/5$. If $y = 0$, the constraint equation becomes $(x - 1)^2 = 2$, so that $x = 1 \pm \sqrt{2}$. If $x = 4/5$, the constraint equation becomes $(1/5)^2 + y^2/4 = 2$, so that $y = \pm 14/5$.

We have found four solutions to the Lagrange multiplier system for f on $\mathcal{B}$: $(1 + \sqrt{2}, 0)$, $(1 - \sqrt{2}, 0)$, $(4/5, 14/5)$, and $(4/5, -14/5)$. Evaluating f at each of the points, we have $f(1 + \sqrt{2}, 0) = (1 + \sqrt{2})^2 \approx 5.83$, $f(1 - \sqrt{2}, 0) = (1 - \sqrt{2})^2 \approx 0.17$, and $f(4/5, 14/5) = f(4/5, -14/5) = -7.2$. Combining these calculations with the earlier observation that f has no extrema in $\mathcal{N}$, we conclude that the minimum value for f on $\mathcal{S} = \mathcal{N} \cup \mathcal{B}$ is approximately -7.2, which occurs at $(4/5, \pm 14/5)$, and the maximum value is 5.83, which occurs at $(1 + \sqrt{2}, 0)$. Notice that the point $(1 - \sqrt{2}, 0)$ is a critical point for f on $\mathcal{B}$ but not an extreme point.

Although our derivation of the Lagrange multiplier method relied on the assumption that the constraint curve $\mathcal{B}$ is given by a single equation $g(x, y) = c$ and that $\nabla g \neq \mathbf{0}$ on $\mathcal{B}$, the technique applies more generally. If $\mathcal{B}$ consists of a union of curves, each of which is described by an equation of the form $g(x, y) = c$, then we can solve the multiplier and constraint equations on each curve to locate extrema on $\mathcal{B}$. In addition, we must also check the individual points on $\mathcal{B}$ where the different curves come together. We demonstrate this in the following example.

Example 4.12

Checking for Extrema on the Boundary. Let us find the extreme values of $f(x,y) = 5 - x^2 - (y-2)^2$ on the closed set $\mathcal{S} = \{(x,y) : x^2 + y^2 \leq 2 \text{ and } y \geq 0\}$, which is the union of the open set

$$\mathcal{N} = \{(x,y) : x^2 + y^2 < 2 \text{ and } y > 0\}$$

and its boundary $\mathcal{B}$, which is itself a union of the semicircle

$$\mathcal{B}_1 = \{(x,y) : x^2 + y^2 = 2 \text{ and } y \geq 0\}$$

and the diameter

$$\mathcal{B}_2 = \{(x,y) : -\sqrt{2} \leq x \leq \sqrt{2} \text{ and } y = 0\}.$$

We can see from a contour plot of f showing $\mathcal{S}$ (see Figure 4.14) that there is a point on $\mathcal{B}_1$ and a point on $\mathcal{B}_2$ where $\mathcal{B}$ appears to be tangent to level curves of f. These two points and the two points where $\mathcal{B}_1$ and $\mathcal{B}_2$ intersect are potentially extrema of f on $\mathcal{B}$.

The only critical point of f is located at $(0,2)$, which, by the second derivative test, is a local maximum for f. However, it lies outside of $\mathcal{S}$. Thus f has no critical points in $\mathcal{N}$ and the extrema of f on $\mathcal{S}$ must lie on $\mathcal{B}$.

Now let us apply the method of Lagrange multipliers to $\mathcal{B}_1$. The Lagrange multiplier system for the semicircle is

$$-2x = \lambda 2x$$

$$-2(y-2) = \lambda 2y$$

$$x^2 + y^2 = 2 \quad \text{and} \quad y \geq 0.$$

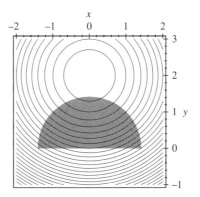

Figure 4.14 A plot of the contours of $f(x,y) = 5 - x^2 - (y-2)^2$ and the region $\{(x,y) : x^2 + y^2 \leq 2 \text{ and } y \geq 0\}$. (See Example 4.12.)

Rewriting the first equation in the form $2x(\lambda + 1) = 0$, we see that $x = 0$ or $\lambda = -1$. If $\lambda = -1$, the second equation reduces to $-y + 2 = -y$, which has no solution. If $x = 0$, then the constraint equation yields $y^2 = 2$, or $y = \pm\sqrt{2}$. Since we must have $y > 0$, the only solution to the system of equations is $(0, \sqrt{2})$. Notice that if $y = \sqrt{2}$, choosing $\lambda = (2 - \sqrt{2})/\sqrt{2}$ shows that this point also solves the second of the multiplier equations. Thus f has a critical point of f on $\mathcal{B}_1$ at $(0, \sqrt{2})$.

In applying the method of Lagrange multipliers to the diameter, we use the constraint equation $g(x, y) = y = 0$ subject to $-\sqrt{2} \le x \le \sqrt{2}$. The Lagrange multiplier system is then

$$-2x = \lambda 0$$
$$-2(y - 2) = \lambda 1$$
$$y = 0 \quad \text{and} \quad -\sqrt{2} \le x \le \sqrt{2}.$$

Once again we see that $x = 0$. From the constraint equation, we see that the only solution to the system of equations is $(0, 0)$.

It remains to check the values of f at the points $(0, \sqrt{2})$, $(0, 0)$, and the points of intersection of $\mathcal{B}_1$ and $\mathcal{B}_2$, $(\pm\sqrt{2}, 0)$. We see that

$$f(0, \sqrt{2}) = -1 + 4\sqrt{2} \approx 4.66$$
$$f(0, 0) = 1$$
$$f(\pm\sqrt{2}, 0) = -1.$$

Thus we conclude that on $\mathcal{S}$, f has a maximum value of $-1 + 4\sqrt{2}$, which occurs at $(0, \sqrt{2})$, and a minimum value of -1, which occurs at $(\pm\sqrt{2}, 0)$.

For most functions f and constraint curves $g(x, y) = c$, it is impossible to solve the multiplier equations by a simple sequence of symbolic manipulations as we have done in the previous two examples. Approximate information about the location of the extrema on the boundary can be derived from a contour plot of f showing the constraint curve. It is possible to use a computer to solve the Lagrange multiplier system numerically, but often these methods are cumbersome and may require additional information about the approximate location of solutions.

Summary

In this section, we saw how to **find the extreme values of f on a closed subset** of $\mathbb{R}^2$. The **maximum principle** and the **minimum principle** guarantee that continuous functions always have a maximum and a minimum on a closed set.

To find the extreme values of f on a set $\mathcal{S}$ consisting of an open set $\mathcal{N}$ and its boundary curve $\mathcal{B}$, $\mathcal{S} = \mathcal{N} \cup \mathcal{B}$, we first applied the graphical and symbolic techniques of Sections 4.1 and 4.2 to locate and identify extrema on $\mathcal{N}$. To find the extrema of f on $\mathcal{B}$, that is, **subject to the constraint** that the extrema lie on $\mathcal{B}$, we developed the **method of Lagrange multipliers**. Geometrically, **the extrema of f on $\mathcal{B}$ occur at points $\mathbf{x} \in \mathcal{B}$ where the level set of f is tangent to $\mathcal{B}$**. Analytically, if $\mathcal{B}$ is a level set of a differentiable function g, the extreme values of f on $\mathcal{B}$ are solutions to the system of equations

$$\nabla f(x, y) = \lambda \nabla g(x, y)$$
$$g(x, y) = c,$$

where $\lambda \neq 0$. We call the first equation the **multiplier equation** and the second equation the **constraint equation**. The multiplier equation is a vector equation that is equivalent to two scalar equations.

Finally, we applied these techniques to find the extreme values of differentiable functions on closed sets $\mathcal{S} = \mathcal{N} \cup \mathcal{B}$, including the case when $\mathcal{B}$ consisted of more than one curve.

Section 4.3 Exercises

1. **Finding Extrema.** Identify the extrema of each of the following functions on the given closed set $\mathcal{S}$ by elementary methods, that is, without using the method of Lagrange multipliers.

   (a) $f(x, y) = x + y$ on $\mathcal{S} = \{(x, y) : |x| \leq 1 \text{ and } |y| \leq 2\}$.
   (b) $f(x, y) = x^2 + y^2$ on $\mathcal{S} = \{(x, y) : 0 \leq x \leq 1 \text{ and } -1 \leq y \leq 0\}$.
   (c) $f(x, y) = e^{-x^2 - y^2}$ on $\mathcal{S} = \{(x, y) : x^2 + y^2 \leq 1\}$.
   (d) $f(x, y) = \sin(x) + \cos(y)$ on $\mathcal{S} = \{(x, y) : |x| \leq 2\pi \text{ and } |y| \leq 2\pi\}$.

2. **Lagrange Multipliers.** Use the method of Lagrange multipliers to find the extreme values of the following functions f on the constraint curve $g(x, y) = c$.

   (a) $f(x, y) = x^2 - 2y^2$ and $g(x, y) = x^2 + y^2 = 1$.
   (b) $f(x, y) = x^2 + y^2/2$ and $g(x, y) = x^2/2 + y^2 = 1$.
   (c) $f(x, y) = xy$ and $g(x, y) = 3x^2 - 2xy + y^2 = 4$.
   (d) $f(x, y) = (x - 1)^2 + 4y^2$ and $g(x, y) = 2x^2 + y^2 = 3$.

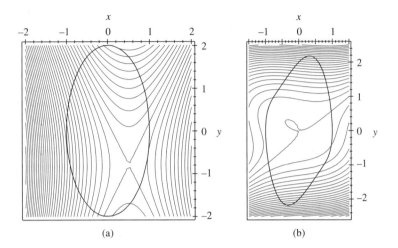

Figure 4.15 Plots for Exercise 4.

3. **Extrema on Closed Sets.** Find the extreme values of the following functions f on the closed sets $\mathcal{S}$.

   (a) $f(x, y) = x^2 + y^2 - 2x + 1$ and $\mathcal{S} = \{(x, y) : x^2 + y^2 \leq 4\}$.
   (b) $f(x, y) = e^{xy}$ and $\mathcal{S} = \{(x, y) : \frac{1}{4}(x - 1)^2 + y^2 \leq 1\}$.
   (c) $f(x, y) = x^2 - 3y^2$ and $\mathcal{S} = \{(x, y) : x^2 + \frac{1}{9}y^2 \leq 1 \text{ and } x \leq 0\}$.
   (d) $f(x, y) = x^2 y^2 - 2x - 4y + 5$ and $\mathcal{S} = \{(x, y) : x \geq 0, \ y \geq 0, \text{ and } x + y \leq 4\}$.

4. **Geometric Test for Extrema.** We have seen that the extrema of a differentiable function f on a constraint curve are located at points where the constraint curve is tangent to a contour of f. For each of the plots in Figure 4.15 of a function f and a constraint curve, use this technique to locate the extrema of f on the constraint curve.

5. **Using the Gradient.** It follows from the derivation of the method of Lagrange multipliers that the extrema of a differentiable function f on a constraint curve occur where ∇f is orthogonal to the constraint curve. For each of the plots in Figure 4.16 of the gradient vector field of a function f and a constraint curve, use this technique to locate the critical points of f on the constraint curve.

6. **Global vs. Constrained Extrema.** Suppose that $f : \mathcal{O} \subset \mathbb{R}^2 \to \mathbb{R}$ is a differentiable function on the open set $\mathcal{O}$. Let $\mathcal{S} \subset \mathcal{O}$ be a closed set where $\mathcal{S} = \mathcal{N} \cup \mathcal{B}$ is the union of an open set and its boundary.

   (a) If $\mathbf{x}_0 \in \mathcal{B}$ is an extreme value of f on $\mathcal{O}$, is $\mathbf{x}_0$ an extremum for f on $\mathcal{B}$? Explain your answer.
   (b) If $\mathbf{x}_0 \in \mathcal{B}$ is a saddle point of f on $\mathcal{O}$, is $\mathbf{x}_0$ an extremum for f on $\mathcal{B}$? Explain your answer.

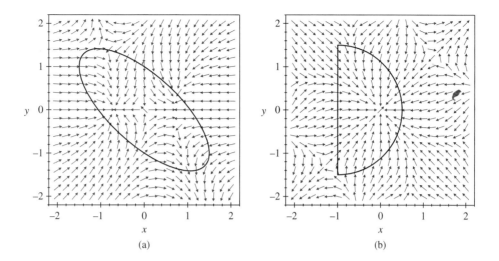

Figure 4.16   Plots for Exercise 5.

7. **Extreme Points on Curves**. Suppose that $f : \mathcal{O} \subset \mathbb{R}^2 \to \mathbb{R}$ is a differentiable function on the open set $\mathcal{O}$. Suppose that $\mathcal{B}$ is a curve defined by the constraint equation $g(x, y) = c$ for the differentiable function g and that α is a parametrization of $\mathcal{B}$ with $\alpha' \neq \mathbf{0}$. If $(f \circ \alpha)'(t_0) = 0$, is $\mathbf{x}_0 = \alpha(t_0)$ an extreme point of f on $\mathcal{B}$? Explain your answer.

8. **Extremes of Temperature.** Suppose that a rotationally symmetric distribution of heat in the plane is given by $T(x, y) = 30 + \frac{5}{\ln 10} \ln(x^2 + y^2)$. The temperature in a rectangular plate in the plane is also given by T where the plate is described by the equations $-3 \leq x \leq -2$ and $-1 \leq y \leq 2$. Find the locations of the extreme values of T on the plate.

9. **Hypotheses for the Maximum Principle I.** The maximum principle requires that the function f be continuous and the set $\mathcal{S}$ be closed. In this exercise, we show by example that the maximum principle fails if $\mathcal{S}$ is not closed. Let $f : \mathbb{R}^2 \to \mathbb{R}$ be the continuous function defined by $f(x, y) = 1 - x^2 - y^2$, and let $\mathcal{S} = \{(x, y) \ : \ 0 < x^2 + y^2 \leq 1\}$.

(a) Show that $\mathcal{S}$ is not closed.

(b) On $\mathbb{R}^2$, $f(x, y) \leq 1$. Show that the values of f can be made arbitrarily close to 1 but not equal to 1 for points in $\mathcal{S}$. Thus we conclude that f has no maximum on $\mathcal{S}$.

10. **Hypotheses for the Maximum Principle II.** In this exercise, we show by example that the maximum principle fails if f is not continuous. Let $\mathcal{S} = \{(x, y) : x^2 + y^2 \leq 1\}$, and let $f : \mathbb{R}^2 \to \mathbb{R}$ be defined by

$$ f(x, y) = \begin{cases} 1 - x^2 - y^2 & \text{if } (x, y) \neq (0, 0) \\ 0 & \text{if } (x, y) = (0, 0). \end{cases} $$

(a) Show that $\mathcal{S}$ is closed.

(b) Show that f is not continuous.

(c) On $\mathbb{R}^2$, $f(x, y) < 1$. Show that the values of f can be made arbitrarily close to 1 but not equal to 1 for points in $\mathcal{S}$. Thus we conclude that f has no maximum on $\mathcal{S}$.

11. **The Minimum Principle.** The minimum principle for continuous functions is in fact a consequence of the maximum principle. Show that the minimum principle for a continuous function f follows from the maximum principle for the function $-f$.

■ 4.4 Constrained Optimization in Space

In Section 4.3, we considered the problem of finding the extreme values of a differentiable function on a closed set $\mathcal{S}$ in the plane. Here we would like to extend those ideas to address the problem of finding extreme values of differentiable functions on closed sets in space. The primary difference between working in the plane and working in space is that the geometry of regions in space is more involved than the geometry of regions in the plane.

The maximum and minimum principles for continuous functions on closed sets in the plane extend to closed sets in space. Once again, since differentiable functions are continuous, if f is a differentiable function defined on an open set $\mathcal{O}$ and $\mathcal{S}$ is a closed subset of $\mathcal{O}$, we can conclude that f has a maximum value and a minimum value on $\mathcal{S}$.

Initially, we will consider closed sets $\mathcal{S}$ contained in the domain of f that are the union of an open set $\mathcal{N}$ and its boundary $\mathcal{B}$ where the boundary is given by a single equation of the form $g(x, y, z) = c$ for a differentiable function g. Here the boundary is a surface in space rather than a curve in the plane, as was the case for the two-variable version of the problem. We say that $\mathcal{B}$ is an *implicitly defined surface* or that $\mathcal{B}$ is *defined implicitly* by the equation $g(x, y, z) = c$. Further, we will assume that $\nabla g \neq \mathbf{0}$ on the boundary. As before, *the extreme values of f can occur at the critical points of f in $\mathcal{N}$ and the extreme values of f on $\mathcal{B}$.* We can locate the critical points of f on $\mathcal{N}$ by solving the equation $\nabla f(\mathbf{x}) = \mathbf{0}$ for $\mathbf{x} \in \mathcal{N}$. To locate the extreme values of f on $\mathcal{B}$ we will extend the technique of Lagrange multipliers to subsets of space. If f is a differentiable function and $\nabla g \neq \mathbf{0}$ on $\mathcal{B}$, *the extreme values of f on $\mathcal{B}$ occur where ∇f and ∇g are parallel.* That is, at points $\mathbf{x}$ where there is a scalar λ so that $\nabla f(\mathbf{x}) = \lambda \nabla g(\mathbf{x})$. This is identical to the condition that holds for the corresponding planar problem. This technique is summarized in the accompanying text box.

Lagrange Multipliers for a Surface in Space	Let f and g be differentiable functions of three variables, and let $\mathcal{B} = \{(x, y, z) : g(x, y, z) = c\}$ be a surface contained in the domain of f with the property that $\nabla g(x, y, z) \neq \mathbf{0}$ for $(x, y, z) \in \mathcal{B}$. Then the extreme values of f on $\mathcal{B}$ are solutions to the system of equations $$\nabla f(x, y, z) = \lambda \nabla g(x, y, z)$$ $$g(x, y, z) = c,$$ where λ is a nonzero scalar.

The multiplier equation can be replaced by three scalar equations to obtain a system of four equations in four unknowns, x, y, z, and λ.

$$\frac{\partial f}{\partial x}(x, y, z) = \lambda \frac{\partial g}{\partial x}(x, y, z)$$
$$\frac{\partial f}{\partial y}(x, y, z) = \lambda \frac{\partial g}{\partial y}(x, y, z)$$
$$\frac{\partial f}{\partial z}(x, y, z) = \lambda \frac{\partial g}{\partial z}(x, y, z)$$
$$g(x, y, z) = c.$$

Once again, this system of equations may have solutions that are not extrema for f. In order to determine which of the solutions are extrema, we must evaluate f on each solution to determine the maximum and the minimum for f.

The following example illustrates the geometric interpretation of the extreme points on a constraint surface.

Example 4.13

Extrema on a Closed Set I. Here we want to find the extreme values of $f(x, y, z) = x$ on the closed set $\mathcal{S}$ defined by $g(x, y, z) \leq 1$, where

$$g(x, y, z) = \frac{1}{2}(x - z)^2 + y^2 + \frac{1}{8}(x + z)^2.$$

The set $\mathcal{S}$ is a "solid" ellipsoid whose major axis is the line $z = x$ in the xz-plane. Its boundary $\mathcal{B}$ is the ellipsoid given by $g(x, y, z) = 1$. (See Figure 4.17.) We will carry out an intuitive geometric analysis and also solve the Lagrange multiplier system of equations.

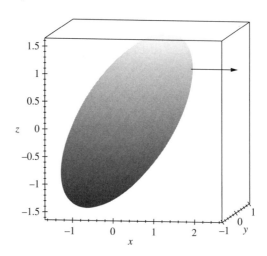

Figure 4.17   A plot of the constraint surface $g(x, y, z) = \frac{1}{2}(x - z)^2 + y^2 + \frac{1}{8}(x + z)^2 = 1$ of Example 4.13. The extreme points of $f(x, y, z) = x$ are the points on the constraint surface with the largest and smallest x-coordinates. The vector ∇f at the point with the largest x-coordinate is shown in the plot. It is orthogonal to the constraint surface.

A.   An Intuitive Solution. Since $f(x, y, z) = x$, the extreme values of f on $\mathcal{S}$ occur at those points in $\mathcal{S}$ with the largest x-coordinate and those points with the smallest x-coordinate. From the plot, we estimate that these points are located approximately at $(1.5, 0, 1)$ and $(-1.5, 0, -1)$. The gradient vector field of f, $\nabla f(x, y, z) = (1, 0, 0)$, always points in the direction of the positive x-axis and is orthogonal to the yz-plane. The geometry behind the method of Lagrange multipliers tells us that the extrema of f on $\mathcal{B}$ occur at points where ∇f is orthogonal $\mathcal{B}$. Examining the plot of $\mathcal{S}$, we can see that this is the case. At the point on $\mathcal{B}$ with the largest x-coordinate, $\nabla f = (1, 0, 0)$ appears to be orthogonal to $\mathcal{B}$.

B.   Solving the Equations. Since $\nabla f(x, y, z) = (1, 0, 0) \neq \mathbf{0}$, f has no critical points. Thus the extreme values of f on $\mathcal{S}$ must occur on $\mathcal{B}$, and we need to solve the Lagrange multiplier system of equations on $\mathcal{B}$. Since $\nabla g(x, y, z) = (\frac{5}{4}x - \frac{3}{4}z, \; 2y, \; -\frac{3}{4}x + \frac{5}{4}z)$, the equations are

$$1 = \lambda(\frac{5}{4}x - \frac{3}{4}z)$$

$$0 = \lambda \cdot 2y$$

$$0 = \lambda(-\frac{3}{4}x + \frac{5}{4}z)$$

$$1 = \frac{1}{2}(x - z)^2 + y^2 + \frac{1}{8}(x + z)^2.$$

Since we are looking for solutions where $\lambda \neq 0$, the second equation tells us that $y = 0$ and the third equation tells us that $-(3/4)x + (5/4)z = 0$ or $z = (3/5)x$. Substituting these values for y and z into the constraint equation and solving for x, we obtain $x = \pm 5/\sqrt{10} \approx \pm 1.5811$. It follows that the extreme values of f on $\mathcal{B}$ are located at $(5/\sqrt{10}, 0, 3/\sqrt{10})$ and $(-5/\sqrt{10}, 0, -3/\sqrt{10})$. Returning to the geometry of part A, notice, for example, that $\nabla g(5/\sqrt{10}, 0, 3/\sqrt{10}) = (4/\sqrt{10}, 0, 0)$, which is a scalar multiple of ∇f.

Let us consider one more example where the boundary of $\mathcal{S}$ is given by a single constraint. Although it is possible to solve the Lagrange multiplier system symbolically in this example, it is considerably more involved than the previous example.

Example 4.14

Extrema on a Closed Set II. Let us find the extreme values of the function $f(x, y, z) = x^2/4 - yz + z^2$ on the set $\mathcal{S}$ given by $g(x, y, z) \leq 2$, where $g(x, y, z) = x^2 + y^2 + 2z^2$. The set $\mathcal{S}$ is an ellipsoid with minor axis on the z-axis and circular cross sections orthogonal to the z-axis with the cross section at $z = 0$ having radius $\sqrt{2}$. The boundary $\mathcal{B}$ of $\mathcal{S}$ is given by the constraint equation $g(x, y, z) = 2$. To find the critical points of f on the interior of $\mathcal{S}$, we must solve $\nabla f(x, y, z) = \mathbf{0}$. We have

$$\nabla f(x, y, z) = (\tfrac{1}{2}x, -z, -y + 2z).$$

The only critical point occurs at the origin, where $f(0, 0, 0) = 0$. For points (x, y, z) near the origin with $y = 0$ and $x \neq 0$ or $z \neq 0$, $f(x, y, z) = x^2/4 + z^2 > 0$, so that the origin is not a local maximum. Similarly, for points near the origin with $x = 0$, $z > 0$, and $y > z$, $f(x, y, z) = -yz + z^2 < 0$, so that the origin is not a local minimum. It follows that f has no extrema in the interior of $\mathcal{S}$.

To find the extreme values of f on $\mathcal{B}$, we must solve the Lagrange multiplier system of equations:

$$\frac{1}{2}x = \lambda 2x$$
$$-z = \lambda 2y$$
$$-y + 2z = \lambda 4z$$
$$x^2 + y^2 + 2z^2 = 2.$$

The first equation is solved by $x = 0$ or $\lambda = 1/4$. Since $(0, 0, 0)$ does not solve the constraint equation, if $x = 0$ we cannot also have $y = 0$ and $z = 0$. Therefore if $x = 0$, we can solve for λ in the second equation and substitute this value into the third equation. Carrying out these calculations reduces the above system of four equations in four unknowns to the following system of two equations in two unknowns.

$$2z^2 + 2yz - y^2 = 0$$
$$y^2 + 2z^2 = 2.$$

We can solve the first equation for y using the quadratic formula to obtain $y = (1 \pm \sqrt{3})z$. Substituting these expressions into the second equation, we obtain a quadratic equation in z,

$$(2 + (1 \pm \sqrt{3})^2)z^2 = 2.$$

Solving for z, we obtain four solutions,

$$z_1 = \sqrt{\frac{1}{2} - \frac{1}{6}\sqrt{3}}, \ z_2 = -\sqrt{\frac{1}{2} - \frac{1}{6}\sqrt{3}}, \ z_3 = \sqrt{\frac{1}{2} + \frac{1}{6}\sqrt{3}}, \ z_4 = -\sqrt{\frac{1}{2} + \frac{1}{6}\sqrt{3}}.$$

Then $y_1 = (1 + \sqrt{3})z_1$, $y_2 = (1 + \sqrt{3})z_2$, $y_3 = (1 - \sqrt{3})z_3$, and $y_4 = (1 - \sqrt{3})z_4$. Putting this together, we obtain four solutions to the system of equations with $x = 0$.

$$(0, y_1, z_1) \approx (0, 1.26, 0.46)$$
$$(0, y_2, z_2) \approx (0, -1.26, -0.46)$$
$$(0, y_3, z_3) \approx (0, -0.65, 0.89)$$
$$(0, y_4, z_4) \approx (0, 0.65, -0.89)$$

Returning to the case $x \neq 0$ and $\lambda = 1/4$, we see from the second of our original equations, $-z = \lambda 2y$, that $z = -y/2$. From the third of our original equations, we have that $y = z$. Together, these equations imply that $y = z = 0$. Then the constraint equation implies that $x^2 = 2$. Thus we have two further solutions, $(\sqrt{2}, 0, 0)$ and $(-\sqrt{2}, 0, 0)$.

Evaluating f on each of the six solutions, we see that the maximum value for f is approximately 1.37, which occurs at $(0, y_3, z_3)$ and $(0, y_4, z_4)$, and the minimum value is approximately -0.37, which occurs at $(0, y_1, z_1)$ and $(0, y_2, z_2)$.

It is also possible to apply these techniques to a set $\mathcal{S} = \mathcal{N} \cup \mathcal{B}$ whose boundary $\mathcal{B}$ is a union of implicitly defined surfaces. For example, suppose $\mathcal{B} = \mathcal{B}_1 \cup \mathcal{B}_2$ with

$$\mathcal{B}_1 = \{(x, y, z) : g_1(x, y, z) = c_1\}$$
$$\mathcal{B}_2 = \{(x, y, z) : g_2(x, y, z) = c_2\},$$

where g_1 and g_2 are differentiable functions of three variables. In order to find the extreme values of f on $\mathcal{B}$, we must separately find the extreme values of f on $\mathcal{B}_1$ and on $\mathcal{B}_2$ using the technique of Lagrange multipliers as we did in Examples 4.13B and 4.14.

There is an additional difficulty in this case: **We must also find the extreme values of f on the curve of intersection of the two surfaces**, $\mathcal{B}_1 \cap \mathcal{B}_2$, which is the set of points that satisfy both constraint equations,

$$\mathcal{B}_1 \cap \mathcal{B}_2 = \{(x, y, z) : \; g_1(x, y, z) = c_1 \text{ and } \; g_2(x, y, z) = c_2\}.$$

If we assume that the curve $\mathcal{B}_1 \cap \mathcal{B}_2$ can be parametrized by a differentiable function α, then using the chain rule as we did in Section 4.3, it can be shown that the tangent vector to α is orthogonal to the gradient of f at the extreme points of f on $\mathcal{B}_1 \cap \mathcal{B}_2$. Since the curve is contained in a level surface for g_1, the tangent to the curve must be orthogonal to ∇g_1. Similarly, the tangent to the curve must be orthogonal to ∇g_2. If the vectors ∇g_1 and ∇g_2 are linearly independent at the extreme point, they determine a plane orthogonal to the tangent to the curve of intersection. Since ∇f is also orthogonal to the tangent to the curve, it must be contained in this plane. Thus we can write ∇f as a linear combination of ∇g_1 and ∇g_2 at an extreme point of f on the curve of intersection. That is, there are constants λ_1 and λ_2 so that

$$\nabla f(x, y, z) = \lambda_1 \nabla g_1(x, y, z) + \lambda_2 \nabla g_2(x, y, z)$$

at an extreme point of f on $\mathcal{B}_1 \cap \mathcal{B}_2$. If $\lambda_1 = \lambda_2 = 0$, then $\nabla f(x, y, z) = 0$ so solving the system amounts to finding a critical point of f. Thus, when applying this method, we look for solutions with $\lambda_1 \neq 0$ or $\lambda_2 \neq 0$. We summarize this in the following statement.

Lagrange Multipliers for Curves in Space

Let f, g_1, and g_2 be differentiable functions of three variables, and let $\mathcal{B} = \mathcal{B}_1 \cap \mathcal{B}_2$ be a curve contained in the domain of f, where

$$\mathcal{B}_1 = \{(x, y, z) : g_1(x, y, z) = c_1\}$$
$$\mathcal{B}_2 = \{(x, y, z) : g_2(x, y, z) = c_2\},$$

and ∇g_1 and ∇g_2 are linearly independent on $\mathcal{B}$. Then the extreme values of f on $\mathcal{B}$ are solutions to the system of equations

$$\nabla f(x, y, z) = \lambda_1 \nabla g_1(x, y, z) + \lambda_2 \nabla g_2(x, y, z)$$
$$g_1(x, y, z) = c_1$$
$$g_2(x, y, z) = c_2,$$

where $\lambda_1 \neq 0$ or $\lambda_2 \neq 0$.

As on the previous page, since the first equation is a vector equation, it is equivalent to three scalar equations. Thus the above system of equations is equivalent to the following system of five scalar equations in the five unknowns x, y, z, λ_1, and λ_2:

$$\frac{\partial f}{\partial x}(x, y, z) = \lambda_1 \frac{\partial g_1}{\partial x}(x, y, z) + \lambda_2 \frac{\partial g_2}{\partial x}(x, y, z)$$

$$\frac{\partial f}{\partial y}(x, y, z) = \lambda_1 \frac{\partial g_1}{\partial y}(x, y, z) + \lambda_2 \frac{\partial g_2}{\partial y}(x, y, z)$$

$$\frac{\partial f}{\partial z}(x, y, z) = \lambda_1 \frac{\partial g_1}{\partial z}(x, y, z) + \lambda_2 \frac{\partial g_2}{\partial z}(x, y, z)$$

$$g_1(x, y, z) = c_1$$

$$g_2(x, y, z) = c_2.$$

Example 4.15

Extrema on a Closed Set with Two Boundary Surfaces. Let us return to the functions of Example 4.14. We want to find the extreme values of the function $f(x, y, z) = x^2/4 - yz + z^2$ on the region

$$\{(x, y, z) : x^2 + y^2 + 2z^2 \leq 2 \text{ and } y \geq 0\}.$$

(See Figure 4.18.) This set is of the form $\mathcal{N} \cup \mathcal{B}$, $\mathcal{B} = \mathcal{B}_1 \cup \mathcal{B}_2$, where

$$\mathcal{N} = \{(x, y, z) : x^2 + y^2 + 2z^2 < 2 \text{ and } y > 0\}$$
$$\mathcal{B}_1 = \{(x, y, z) : x^2 + y^2 + 2z^2 = 2 \text{ and } y \geq 0\}$$
$$\mathcal{B}_2 = \{(x, y, z) : y = 0 \text{ and } x^2 + 2z^2 \leq 2\}.$$

We will let $g_1(x, y, z) = x^2 + y^2 + 2z^2$ and $g_2(x, y, z) = y$.

First, we must check for critical points of f that lie in $\mathcal{N}$; that is, we must find solutions to $\nabla f(\mathbf{x}) = \mathbf{0}$ in $\mathcal{N}$. We have

$$\nabla f(x, y, z) = (x/2, -z, 2z - y).$$

We have seen that the only solution to $\nabla f(x, y, z) = \mathbf{0}$ is the origin $(0, 0, 0)$, which lies in $\mathcal{B}_2$ and not $\mathcal{N}$.

From Example 4.14, we see that the solutions to the Lagrange multiplier equations for f, subject to the constraint $g_1(x, y, z) = 2$ with $y > 0$, are $(0, y_1, z_1) \approx (0, 1.26, 0.46)$ and $(0, y_4, z_4) \approx (0, 0.65, -0.89)$.

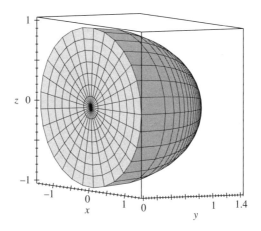

Figure 4.18   A plot of the constraint surface $\mathcal{B} = \mathcal{B}_1 \cup \mathcal{B}_2$ of Example 4.15, $\mathcal{B}_1 = \{(x, y, z) : x^2 + y^2 + 2z^2 = 2$ and $y \geq 0\}$ and $\mathcal{B}_2 = \{(x, y, z) : x^2 + 2z^2 \leq 2$ and $y = 0\}$.

Now we must solve the Lagrange multiplier equations for f subject to the constraint $g_2(x, y, z) = y = 0$ with $x^2 + 2z^2 < 2$. The equations are

$$\frac{1}{2}x = 0$$

$$-z = \lambda$$

$$-y + 2z = 0$$

$$y = 0, \ x^2 + 2z^2 < 2.$$

If $\lambda \neq 0$, there are no solutions to these equations. (Why?)

It remains to find the extreme values of f on $\mathcal{B}_1 \cap \mathcal{B}_2$ using the Lagrange multiplier equations for f subject to two constraints.

$$\frac{1}{2}x = \lambda_1 2x + \lambda_2 0$$

$$-z = \lambda_1 2y + \lambda_2$$

$$-y + 2z = \lambda_1 4z + \lambda_2 0$$

$$x^2 + y^2 + 2z^2 = 2$$

$$y = 0.$$

Since $y = 0$, the equations reduce to

$$\frac{1}{2}x = \lambda_1 2x$$

$$-z = \lambda_2$$

$$2z = \lambda_1 4z$$

$$x^2 + 2z^2 = 2.$$

If $x = 0$, the remaining constraint equation yields $z = \pm 1$. Then λ_1 and λ_2 can be chosen so that $(0, 0, \pm 1)$ is a solution to the system. If $x \neq 0$, $\lambda_1 = \frac{1}{4}$, $z = 0$, and we obtain the solutions $(\pm\sqrt{2}, 0, 0)$.

Now we must check the values of f on the points that we have found.

$$f(0, y_1, z_1) \approx -0.366$$

$$f(0, y_4, z_4) \approx 1.366$$

$$f(\pm\sqrt{2}, 0, 0) = 1/2$$

$$f(0, 0, \pm 1) = 1.$$

Thus the extreme values for f on $\mathcal{N} \cup \mathcal{B}$ are approximately 1.366 at $(0, y_4, z_4)$ and -0.366 at $(0, y_1, z_1)$.

While our focus so far has been on finding the extreme values of a differentiable function on a closed set consisting of an open set and its boundary, it is often the case that we want to find the extreme values of a function on a curve or surface without reference to an open set. The following physical example illustrates this point.

Example 4.16

Extrema of Potential Energy on a Curve. In previous examples, we have considered the potential energy function of a system of three particles consisting of an electric dipole fixed in space and a third particle free to move in space. Here we consider a dipole consisting of a particle of charge $+2$ located at the origin and a particle of charge -1 located at $(1, 0, 0)$, along with a third particle of charge $+1$ located at (x, y, z). The potential energy of the system is given by

$$U(\mathbf{x}) = \frac{2}{4\pi\epsilon_0} \frac{1}{r_1(\mathbf{x})} - \frac{1}{4\pi\epsilon_0} \frac{1}{r_2(\mathbf{x})} - \frac{2}{4\pi\epsilon_0},$$

where $r_1(\mathbf{x}) = \sqrt{x^2 + y^2 + z^2}$ and $r_2(\mathbf{x}) = \sqrt{(x-1)^2 + y^2 + z^2}$ are the distances from the particle of charge $+1$ to the particles of the dipole. We can see from the form of U that as the third particle moves toward the dipole and either r_1 or r_2 approaches 0, U takes arbitrarily large values. Thus U has no maximum value on any neighborhood that contains the dipole. On the other hand, if the third particle is constrained to lie on a

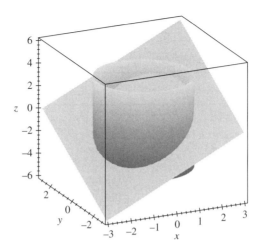

Figure 4.19 The surfaces of Example 4.16, a cylinder of radius 2 centered on the z-axis, $x^2 + y^2 = 4$, and the plane $x + y - z = 0$, intersect in an ellipse.

closed curve or surface that does not pass through the dipole, then U will have both a maximum and minimum.

Let us suppose that the third particle is constrained to lie on the curve of intersection of the cylinder of radius 2 centered on the z-axis and the plane given by $x + y - z = 0$. (See Figure 4.19.) The cylinder is the level set of the function $g_1(\mathbf{x}) = x^2 + y^2$ for the value 4, and the plane is the level set of the function $g_2(\mathbf{x}) = x + y - z$ for the value zero.

In order to find the extreme values of U on the curve of intersection, we must solve the Lagrange multiplier equations with two constraints for the function U and the constraint functions g_1 and g_2:

$$\nabla U(\mathbf{x}) = \lambda_1 \nabla g_1(\mathbf{x}) + \lambda_2 \nabla g_2(\mathbf{x})$$

$$g_1(\mathbf{x}) = 4$$

$$g_2(\mathbf{x}) = 0.$$

Because of the complicated form of U and its derivatives, we solved these equations using a numerical equation solver on a computer. We obtained the following six solutions and the corresponding values for the potential energy:

$$
\begin{aligned}
\mathbf{x}_1 &= (-1.466, 1.361, -0.105), & U(\mathbf{x}_1) &= -1.36/(4\pi\epsilon_0) \\
\mathbf{x}_2 &= (1.809, 0.853, 2.662), & U(\mathbf{x}_2) &= -1.74/(4\pi\epsilon_0) \\
\mathbf{x}_3 &= (-1.266, -1.548, -2.814), & U(\mathbf{x}_3) &= -1.68/(4\pi\epsilon_0) \\
\mathbf{x}_4 &= (1.791, -0.891, 0.900), & U(\mathbf{x}_4) &= -1.76/(4\pi\epsilon_0) \\
\mathbf{x}_5 &= (0.790, -1.837, -1.05), & U(\mathbf{x}_5) &= -1.59/(4\pi\epsilon_0) \\
\mathbf{x}_6 &= (2.000, 0.0, 2.061), & U(\mathbf{x}_6) &= -1.74/(4\pi\epsilon_0)
\end{aligned}
$$

The maximum value for U occurs at $\mathbf{x}_1$, and the minimum value occurs at $\mathbf{x}_4$. Note that we solved the multiplier system for $4\pi\epsilon_0 U$ and then divided the resulting function values by $4\pi\epsilon_0$ to obtain the values for U.

Summary

In this section, we extended our discussion of extrema of differentiable functions on closed sets to consider ***extrema on closed sets in space***. Once again, the discussion was set in motion by the ***maximum*** and ***minimum principles*** for continuous functions on closed sets, which guarantee the existence of extrema of continuous functions on closed sets.

To find the extreme values of f on a set $\mathcal{S}$ consisting of an open set $\mathcal{N}$ and its boundary surface $\mathcal{B}$, $\mathcal{S} = \mathcal{N} \cup \mathcal{B}$, we first locate the extrema of f on $\mathcal{N}$ by examining the behavior of f at its critical points in $\mathcal{N}$. To find the extrema of f on the surface $\mathcal{B}$, we again applied the ***method of Lagrange multipliers***. Geometrically, ***the extrema of f on $\mathcal{B}$ occur at points where ∇f and ∇g are parallel***, where $\mathcal{B}$ is defined implicitly by the ***constraint*** $g(x,y,z) = c$. Analytically, the extreme values of f on $\mathcal{B}$ are solutions to the system of equations

$$\nabla f(x,y,z) = \lambda \nabla g(x,y,z)$$
$$g(x,y,z) = c,$$

where $\lambda \neq 0$. The ***multiplier equation*** $\nabla f = \lambda \nabla g$ is a vector equation that is equivalent to three scalar equations.

We also extended these results to the case where the boundary $\mathcal{B}$ is a union of surfaces, $\mathcal{B} = \mathcal{B}_1 \cup \mathcal{B}_2$. Further, we saw how to apply the method of Lagrange multipliers to the ***curve of intersection*** $\mathcal{B}_1 \cap \mathcal{B}_2$. If $\mathcal{B}_1$ is given by $g_1(x,y,z) = c_1$, and $\mathcal{B}_2$ is given by $g_2(x,y,z) = c_2$, then extrema on $\mathcal{B}_1 \cap \mathcal{B}_2$ are solutions to the equations

$$\nabla f(x,y,z) = \lambda_1 \nabla g_1(x,y,z) + \lambda_2 \nabla g_2(x,y,z),$$
$$g_1(x,y,z) = c_1,$$
$$g_2(x,y,z) = c_2,$$

where $\lambda_1 \neq 0$ or $\lambda_2 \neq 0$.

Section 4.4 Exercises

1. **Extrema on a Constraint Surface.** Find the extreme values of the following functions f on the constraint surface $g(x,y,z) = c$.

   (a) $f(x,y,z) = x^2 - y^2 + 2z^2$ and $g(x,y,z) = x^2 + y^2 + z^2 = 1$.

    (b) $f(x, y, z) = x^2 - y^2 - z^2$ and $g(x, y, z) = 3x^2 + y^2 + 2z^2 = 1$.
    (c) $f(x, y, z) = xyz$ and $g(x, y, z) = (x - 1)^2 + y^2 + z^2 = 1$.
    (d) $f(x, y, z) = xy + yz$ and $g(x, y, z) = x^2/4 + y^2/9 + z^2 = 1$.

2. **Extrema on a Closed Set.** Find the extreme values of the following functions f on the closed set $\mathcal{S}$.

    (a) $f(x, y, z) = 2 - x^2 - y^2 - z^2/9$ and $\mathcal{S} = \{(x, y, z) : x^2 + y^2/4 + z^2 \le 1\}$.
    (b) $f(x, y, z) = x^2 + yz$ and $\mathcal{S} = \{(x, y, z) : x^2 + y^2 + z^2 \le 3\}$.

3. **Extrema on Closed Sets with Multiple Boundary Surfaces.** Find the extreme values of the following functions f on the closed set $\mathcal{S}$.

    (a) $f(x, y, z) = 4 - x^2 - y^2 + z^2$ and $\mathcal{S} = \{(x, y, z) : x^2 + y^2 \le 1 \text{ and } 0 \le z \le 1\}$.
    (b) $f(x, y, z) = x(y - 1)z$ and $\mathcal{S} = \{(x, y, z) : x^2 + y^2 + z^2 \le 1 \text{ and } z \ge 0\}$.

4. **Extrema on Curves.** Find the extreme values of the following functions f on the constraint curve $\mathcal{B} = \mathcal{B}_1 \cap \mathcal{B}_2$.

    (a) $f(x, y, z) = xy - yz$ and $\mathcal{B} = \{(x, y, z) : x^2 + y^2 = 1 \text{ and } z = 1\}$.
    (b) $f(x, y, z) = (x + y)^2$ and $\mathcal{B} = \{(x, y, z) : x^2 - y^2 - z^2 = 1 \text{ and } x = 2\}$.
    (c) $f(x, y, z) = x^2 - y^2 - z^2$ and $\mathcal{B} = \{(x, y, z) : x^2 + y^2 + z^2 = 1 \text{ and } x + y + z = 1\}$.

5. **Potential Energy on a Curve.** In Example 4.16, we found the extreme values for the potential energy function of a system of three particles consisting of a dipole and a particle constrained to lie on the curve of intersection $\mathcal{C}$ of a cylinder and a plane, which is in fact an ellipse.

    (a) Find a parametrization α of $\mathcal{C}$ that is defined on the interval $[0, 2\pi]$.
    (b) In what order does α pass through the points $\mathbf{x}_1, \dots, \mathbf{x}_6$?
    (c) Describe the behavior of the composition $U \circ \alpha : [0, 2\pi] \to \mathbb{R}$. (Where is it increasing, decreasing, etc.?)

6. **Coloring a Surface.** An intuitive method for finding extreme points of a function f on a constraint surface $g = c$ is to *shade* or *color* the surface according to the function f. For example, we might shade the surface using gray levels ranging from black through lighter shades of gray to white to represent increasing values of f on the surface, so that the shading is darkest near the minimum value of f and lightest near the maximum value of f. For this exercise, let $\mathcal{B}$ be the sphere of radius 1 centered at the origin in space. For each of the following functions f, shade a sketch of $\mathcal{B}$ according to the above prescription for the given f. Include boxed axes in your sketch to indicate directions. You may need to provide more than one sketch to indicate the behavior of f on the entire sphere.

    (a) $f(x, y, z) = x$.
    (b) $f(x, y, z) = x + y$.
    (c) $f(x, y, z) = x^2 + y^2$.

    (d) $f(x, y, z) = x^2 - y^2$.
    (e) $f(x, y, z) = x^2 - y^2 + z^2$.

7. **Extrema by Color.** Suppose a constraint surface $\mathcal{B}$ is shaded according to a function f using the prescription of Exercise 6.

   (a) Describe the shading that occurs around the minimum for f on $\mathcal{B}$. How does this compare to the shading around a local minimum of f on $\mathcal{B}$? Explain your answer.
   (b) Describe the shading that occurs around the maximum for f on $\mathcal{B}$. How does this compare to the shading around a local maximum of f on $\mathcal{B}$? Explain your answer.
   (c) How do you think $\mathcal{B}$ might be shaded near a point that is a solution to the Lagrange multiplier system for f on $\mathcal{B}$ that is not a local maximum or minimum? Explain your answer.

8. **Distance from the Origin on a Surface.** In this exercise, we want to analyze the problem of finding the extreme values of the function $f(x, y, z) = x^2 + y^2 + z^2$, which is the square of the distance from $\mathbf{x} = (x, y, z)$ to the origin, on surfaces in space. Suppose that the surface $\mathcal{B}$ is the level set of the differentiable function g for the value c and that $\nabla g \neq \mathbf{0}$ on $\mathcal{B}$. We will use $\mathbf{x}_0$ and $\mathbf{x}_1$ to refer to solutions of the Lagrange multiplier system of equations for f on the constraint surface $\mathcal{B}$.

   (a) Suppose that the maximum value for f on $\mathcal{B}$ is r_0^2 and it occurs only at $\mathbf{x}_0$, so that $f(\mathbf{x}_0) = r_0^2$. It follows that $\mathbf{x}_0$ is a point of intersection of $\mathcal{B}$ and $\mathcal{B}_0$, the level surface of f for the value r_0^2. Give a prose description of the plots of $\mathcal{B}$ and $\mathcal{B}_0$ near $\mathbf{x}_0$. (*Hints:* Think about the surface $\mathcal{B}_0$ and the interpretation of f. Also, you may find it helpful to sketch a generic picture of these plots near $\mathbf{x}_0$.)
   (b) Suppose the minimum value for f on $\mathcal{B}$ is r_1^2 and it occurs only at $\mathbf{x}_1$, so that $f(\mathbf{x}_1) = x_1^2 + y_1^2 + z_1^2 = r_1^2$. It follows that $\mathbf{x}_1$ is a point of intersection of $\mathcal{B}$ and $\mathcal{B}_1$, the level surface of f for the value r_1^2. Give a prose description of the plots of $\mathcal{B}$ and $\mathcal{B}_1$ near $\mathbf{x}_1$. (*Hints:* Think about the surface $\mathcal{B}_1$ and the interpretation of f. Also, you may find it helpful to sketch a generic picture of these plots near $\mathbf{x}_1$.)
   (c) If the hypothesis of parts (a) and (b) both apply, what is the relationship between $\mathcal{B}$ and $\mathcal{B}_0$ and $\mathcal{B}_1$? Explain your answer.

9. **Charge on a Wire.** In previous examples and exercises, we have considered the potential energy function of a system of charged particles. Here we consider the potential energy function of charge distribution that is evenly distributed along a straight wire. Suppose that a wire of length 2 lies along the x-axis in space from $(-1, 0, 0)$ to $(1, 0, 0)$ and has constant charge distribution C. Then the potential energy of the system consisting of the charged wire and a particle of charge 1 located at $\mathbf{x}$ is given by

$$U(\mathbf{x}) = C \left(\ln \left(1 - x + \sqrt{(x-1)^2 + y^2 + z^2} \right) - \ln \left(-1 - x + \sqrt{(x+1)^2 + y^2 + z^2} \right) \right).$$

Notice that U depends on $y^2 + z^2$ so that it has symmetry about the x-axis, which we would expect from the physical setup. Use a computer to answer the following questions.

   (a) Describe the behavior of the potential energy of the system when the free particle is constrained to lie on a sphere of radius 2 centered at the origin. (*Hint:* Examine a contour plot of U in the xy-plane holding $z = 0$. Why can we use this to understand the behavior in all of $\mathbb{R}^3$?)

(b) Describe the behavior of the potential energy of the system when the free particle is constrained to lie on the curve of intersection of the sphere of radius 2 and the plane given by $x + z = 0$. (*Hint:* Parametrize the curve of intersection; then plot and analyze the behavior of $t \to (U \circ \alpha)(t)$.)

■ 4.5 End of Chapter Exercises

1. **Isolated Critical Points.** For each of the following functions $f : \mathbb{R}^2 \to \mathbb{R}$, find the critical points of f and determine if they are isolated or nonisolated.

   (a) $f(x, y) = x^2 + y^2 + 2xy$.
   (b) $f(x, y) = \cos(x^2 + y^2)$.
   (c) $f(x, y) = x^4 + \frac{1}{2}y^4 - 4xy^2 + 2x^2 + 2y^2 + 1$.

2. **Taylor Polynomials.** For each of the following functions f of two variables, compute the first and second degree Taylor polynomials of f at the given point $\mathbf{x}_0$. That is, compute the linear and quadratic approximations to f at $\mathbf{x}_0$.

   (a) $f(x, y) = x^2 - y^2 - 2x - y$, $\mathbf{x}_0 = (1, -2)$.
   (b) $f(x, y) = x^3 - xy + y^3$, $\mathbf{x}_0 = (0, 0)$.
   (c) $f(x, y) = \cos(x) \sin(y)$, $\mathbf{x}_0 = (0, 0)$.
   (d) $f(x, y) = \ln(x + y)$, $\mathbf{x}_0 = (1, 1)$.

3. **Second Derivative Test Calculations.** For each of the following functions f, determine the critical points of f and use the second derivative test to identify the type of each of the critical points of f. Determine which of the critical points are global minima or global maxima.

   (a) $f(x, y) = x^2 + 2xy + y^2 + 3x - 2y$.   (c) $f(x, y) = e^{-3x^2 + y^2}$.
   (b) $f(x, y) = y^3 + x^2 - y - 1$.   (d) $f(x, y) = \frac{xy}{x^2 + y^2 + 1}$.

4. **Critical Points of $e^{f(x,y)}$.** Suppose $\mathbf{x}_0 = (x_0, y_0)$ is a critical point of the differentiable function $f : \mathcal{O} \subset \mathbb{R}^2 \to \mathbb{R}$.

   (a) Show that $g(x, y) = e^{f(x,y)}$ also has a critical point at $\mathbf{x}_0$ and that if $\mathbf{x}_0$ is an isolated critical point of f, it is also an isolated critical point of g.
   (b) If $\mathbf{x}_0$ is an isolated critical point of f, what is the relationship, if any, between the type of the critical point f at $\mathbf{x}_0$ and the type of the critical point of g at $\mathbf{x}_0$? Explain your answer. (Be sure to consider all the possibilities for the type of the isolated critical point of f at $\mathbf{x}_0$.)

5. **Composition of Functions.** Let $f = f(u)$ be a twice differentiable function of one variable and let $g = g(x, y)$ be a twice differentiable function of two variables. Let $\mathbf{x}_0 = (x_0, y_0)$ be a critical point of g.

   (a) Show that $f \circ g$ also has a critical point at $\mathbf{x}_0$.
   (b) For an arbitrary twice differentiable function f, is the critical point of $f \circ g$ at $\mathbf{x}_0$ the same type as the critical point of g at (x_0, y_0)? Explain. (*Hint:* See Exercise 4.)

6. **Poles.** If f is defined and differentiable on all of $\mathbb{R}^2$ except at the origin and $\lim_{(x,y)\to(0,0)} f(x, y) = \infty$, we say that f has a *pole* at the origin. The function $f(x, y) = \frac{1}{x^2+y^2}$ has a pole at the origin.

   (a) Describe the level sets of f in a neighborhood of the origin. How do they differ from the level sets near an isolated maximum of a differentiable function? Explain.
   (b) Describe ∇f in a neighborhood of the origin. How does ∇f near the origin differ from the gradient vector field of a differentiable function near an isolated maximum? Explain.

7. **An Electric Tripole.** Consider a system of four charged particles, each of charge q, that lie in the plane where three of the particles form an electric tripole located at $(1, 0)$, $(-1, 0)$, and $(0, 3)$. The fourth particle is free to move in the plane, and its location is given by $\mathbf{x} = (x, y)$. The potential energy is given by

$$U(\mathbf{x}) = \frac{q^2}{4\pi\epsilon_0} \frac{1}{r_1(\mathbf{x})} + \frac{q^2}{4\pi\epsilon_0} \frac{1}{r_2(\mathbf{x})} + \frac{q^2}{4\pi\epsilon_0} \frac{1}{r_3(\mathbf{x})} + \frac{q^2}{4\pi\epsilon_0} \frac{1}{2},$$

where $r_1(\mathbf{x}) = \sqrt{(x-1)^2 + y^2}$, $r_2(\mathbf{x}) = \sqrt{(x+1)^2 + y^2}$, and $r_3(\mathbf{x}) = \sqrt{x^2 + (y-3)^2}$ are the distances of the free particle to the particles in the tripole. Note that the fourth term is constant.

   (a) Use a computer to generate a contour plot of $-U$. (*Hint:* Use the values $q = 1$ and $\epsilon_0 = 1$ in U.) Locate and describe the critical points of $-U$. (Note that you may have to generate several plots to observe all the critical points of $-U$.)
   (b) Use a computer to generate a plot of $-\nabla U$ and flow lines of $-\nabla U$ in order to describe the behavior of the free particle near the critical points of $-\nabla U$ when it is released in the plane with no initial velocity. (Note that you may have to generate several plots in order to answer this question.)

8. **Extrema on Closed Sets.** Find the extreme values of the following function f on the closed set $\mathcal{S}$.

   (a) $f(x, y) = y^3 - x^2 - y$ and $\mathcal{S} = [-1, 1] \times [-1, 1]$.
   (b) $f(x, y) = \cos(x - y)$ and $\mathcal{S} = \{(x, y) : x^2 + y^2 \le 1\}$.
   (c) $f(x, y) = x^4 - x^2 - y^2$ and $\mathcal{S} = \{(x, y) : y \le 4 - |x| \text{ and } y \ge -1\}$.

(d) $f(x, y) = 2 + (x - 1)y$ and $S = \{(x, y) : x^2 + y^2 \leq 4\}$. (*Hint:* In this case, the Lagrange equations are difficult to solve. Try constructing a parametrization α of the boundary of S and find the extrema of $f \circ \alpha$.)

9. **The Geometric Test.** Suppose $\mathbf{x}_0$ is a point of tangency of a level set of the differentiable function f and the constraint curve $\mathcal{B}$. Is it possible to determine whether $\mathbf{x}_0$ is a local extrema of f on $\mathcal{B}$ from the geometry of the level sets of f near $\mathbf{x}_0$? Explain your answer. (You may use schematic diagrams of level sets and $\mathcal{B}$ to illustrate your answer.)

10. **What Happens on the Boundary?** Suppose that $f : \mathbb{R}^2 \to \mathbb{R}$ is a differentiable function and that the origin is the only critical point of f on the disk of radius 2 centered at the origin. If the type of the critical point of f at the origin is known, what, if anything, can be said about the extreme values of f on the circle of radius 1 centered at the origin? (*Hint:* You may find it useful to think about contour plots near an isolated critical point.)

11. **Extrema of Potential Energy.** In this exercise, we want to find the extreme values of the potential energy of a system of three particles consisting of an electric dipole and a free particle when the free particle is constrained to lie in a region in the plane. Assume the system consists of three charged particles that lie in the plane where two of the particles form an electric dipole located at $(2, 0)$ and $(-2, 0)$, the first having charge q and the second having charge $2q$. The third particle, also of charge q, is free to move in the plane, and its location is given by $\mathbf{x} = (x, y)$. The potential energy is given by

$$U(\mathbf{x}) = \frac{q^2}{4\pi\epsilon_0} \frac{1}{r_1(\mathbf{x})} + \frac{2q^2}{4\pi\epsilon_0} \frac{1}{r_2(\mathbf{x})} + \frac{q^2}{4\pi\epsilon_0} \frac{1}{2}$$

$$= \frac{q^2}{4\pi\epsilon_0} \left(\frac{1}{r_1(\mathbf{x})} + \frac{2}{r_2(\mathbf{x})} + \frac{1}{2} \right),$$

where $r_1(\mathbf{x}) = \sqrt{(x - 2)^2 + y^2}$ and $r_2(\mathbf{x}) = \sqrt{(x + 2)^2 + y^2}$ are the distances of the free particle q_3 to the particles in the dipole.

Let S be the region defined by $-2 \leq x \leq 2$ and $2 \leq y \leq 4$. Find the extreme values of the potential energy function when the free particle is constrained to lie in S. (*Hints:* (a) Since U is a scalar multiple of $V(x, y) = \frac{1}{r_1(\mathbf{x})} + \frac{2}{r_2(\mathbf{x})} + \frac{1}{2}$, it suffices to find the extreme values of V. (b) Use a computer to carry out your analysis of the behavior of V.)

12. **Extrema on a Closed Set in $\mathbb{R}^3$.** Find the extreme values of the following functions f on the closed set S.

(a) $f(x, y, z) = 1 - x^2 + y^2 - z^2$ and $S = \{(x, y, z) : (x - y)^2 + 4(x + y)^2 + z^2 \leq 1\}$.
(b) $f(x, y, z) = x^3 + y^3 + z^3 - x - y - z$ and $S = [-2, 2] \times [-2, 2] \times [-2, 2]$.

13. **Distance from the Origin on a Curve.** In this exercise, we want to analyze the problem of finding the extreme values of the function $f(x, y, z) = x^2 + y^2 + z^2$, which is the square of the distance from $\mathbf{x} = (x, y, z)$ to the origin, on curves in space. Suppose that the curve $\mathcal{C}$ is the intersection of the level sets of the differentiable function g_1 for the value c_1 and of the differentiable function g_2 for the value c_2 and that ∇g_1 and ∇g_2 are linearly

independent on $\mathcal{B}$. We will use $\mathbf{x}_0$ and $\mathbf{x}_1$ to refer to solutions of the Lagrange multiplier system of equations for f on the curve $\mathcal{C}$.

(a) Suppose that the maximum value for f on $\mathcal{C}$ is r_0^2 and it occurs only at $\mathbf{x}_0$, so that $f(\mathbf{x}_0) = r_0^2$. It follows that $\mathbf{x}_0$ is a point of intersection of $\mathcal{C}$ and $\mathcal{B}_0$, the level surface of f for the value r_0^2. Give a prose description of the plots of $\mathcal{C}$ and $\mathcal{B}_0$ near $\mathbf{x}_0$. (*Hints:* Think about the surface $\mathcal{B}_0$ and the interpretation of f. Also, you may find it helpful to sketch a generic picture of these plots near $\mathbf{x}_0$.)

(b) Suppose that the minimum value for f on $\mathcal{C}$ is r_1^2 and it occurs only at $\mathbf{x}_1$, so that $f(\mathbf{x}_1) = x_1^2 + y_1^2 + z_1^2 = r_1^2$. It follows that $\mathbf{x}_1$ is a point of intersection of $\mathcal{C}$ and $\mathcal{B}_1$, the level surface of f for the value r_1^2. Give a prose description of the plots of $\mathcal{C}$ and $\mathcal{B}_1$ near $\mathbf{x}_1$. (*Hints:* Think about the surface $\mathcal{B}_1$ and the interpretation of f. Also, you may find it helpful to sketch a generic picture of these plots near $\mathbf{x}_1$.)

(c) If the hypotheses of parts (a) and (b) both apply, what is the relationship between $\mathcal{C}$ and $\mathcal{B}_0$ and $\mathcal{B}_1$? Explain your answer.

(d) Suppose that $\mathbf{x}_2$ is a solution of the Lagrange multiplier system for f on that constraint curve $\mathcal{C}$, but that it is neither a minimum nor a maximum for f on $\mathcal{C}$. Give a prose description of the plots of $\mathcal{C}$ and $\mathcal{B}_2$, the level set of f passing through $\mathbf{x}_2$, near $\mathbf{x}_2$.

14. **Gypsy Moths on the Loose.** The spread of an aerosol in the atmosphere by a combination of diffusion and wind can be modeled by a function of three variables. If the coordinates are chosen so that the aerosol is released from the origin and the wind is blowing in the positive x-direction with velocity v_0, after a period of time, the concentration of the aerosol reaches a steady state independent of time at a point $\mathbf{x}$ that is given by

$$S(\mathbf{x}) = \frac{Q}{2\pi v_0 \sigma_y \sigma_z x^{2-n}} e^{-(y^2/\sigma_y^2 + z^2/\sigma_z^2)/(2x^{2-n})},$$

where Q is the rate of release of the chemical and σ_y, σ_z, and n are empirically determined constants. Distances are measured in centimeters, velocity in centimeters per second, and concentration in parts per cubic centimeter. The values $\sigma_y = 0.4$, $\sigma_z = 0.2$, and $n = 0.25$ give good results for wind velocities less than 5 m/s (= 500 cm/s). Among other applications, this has been used to model the diffusion of insect pheromones.[2] Pheromones are "odor" chemicals that are released by animals for chemical communication within a species. For example, for a single female gypsy moth, Q is on the order of 3×10^{13}

---

[2] See *Differential and Ecological Problems: Modern Perspectives*, second ed. Akira Okubo and Simon A. Levin. New York: Springer-Verlag, 2001, p. 113.

particles per cm^3. Given that male gypsy moths can detect as few as 100 particles per cm^3, let us investigate the range over which a male gypsy moth can detect a female gypsy moth.

(a) How would you use the method of Lagrange multipliers to detect the maximum distance downwind from a female gypsy moth over which a male gypsy moth can detect a female gypsy moth?

(b) Find the maximum distance downwind over which a male gypsy moth can detect a female when the wind is blowing at 100 cm/s, 300 cm/s, and 500 cm/s. (Use the computer if necessary.)

(c) What is the relationship between your three answers to (b)? Is this to be expected based upon the formula for the concentration of pheromone S? Explain.

Chapter

5

Integration

In this chapter, we will begin our discussion of integration in the plane and in space. We will be motivated by a number of applications where we want to compute the total accumulation of a function on a region. In each case, we will first approximate the total accumulation and then refine our approximation to obtain an exact value.

In Section 5.1, we provide an intuitive introduction to the idea of a Riemann sum approximation to the total accumulation of a function. The intuition will come from three examples that illustrate the construction of Riemann sums in different settings. The first, a calculation of average rainfall in Fort Collins, Colorado, illustrates the use of Cartesian coordinates in the plane; the second, a calculation of the total intensity of solar radiation at one wavelength, illustrates how we might take advantage of rotational symmetry in the plane; and the third, a calculation of the average temperature in a room, illustrates the use of Cartesian coordinates in space.

In Section 5.2, we consider Riemann sums and integration in the plane using the Cartesian coordinate system. We will show how to move from a sequence of Riemann sum approximations to the total accumulation of a function on a region to an exact value, or integral, for its total accumulation. We will then see how to express this integral in terms of iterated one-variable integrals, thus allowing us to use symbolic techniques from one-variable calculus. In Section 5.3, we introduce the polar coordinate system in the plane. This will augment the consideration of the total accumulation of functions with rotational symmetry on subsets of a disk, which arise in a variety of applications.

In Section 5.4, we consider Riemann sums and integration in space using Cartesian coordinates. Our approach here is modeled on the one we used in Section 5.2, the essential difference being the possible difficulties that arise from the more complicated geometry of regions in space than in the plane. In Sections 5.5 and 5.6, we introduce the cylindrical and spherical coordinates systems in space, which are both extensions of the polar coordinates system in the plane. We will use these to consider the total accumulation of functions with rotational or spherical symmetry in space.

A Collaborative Exercise—Riemann Sums

In February 1978, a blizzard hit the east coast of the United States. Using data from the National Climatic Data Center, we would like to determine the total volume of snow covering a region in the northeast consisting of the states of Pennsylvania, New York, New Jersey, Connecticut, Massachusetts, and Rhode Island.

Figure 5.1 contains the readings for the depth of snow cover. The locations of the data collection sites are given in latitude and longitude, and the snow measurements are given in inches. Figure 5.2 is a density plot for the snowfall data. (See Example 3.2 for more information about this data.)

lat.	long.	in.	lat.	long.	in.	lat.	long.	in.	lat.	long.	in.
42.22	−75.98	12	41.80	−78.63	32	41.18	−78.90	30	40.77	−73.90	14
42.93	−78.73	23	40.70	−74.17	17	42.75	−73.80	16	40.30	−78.32	12
40.65	−75.43	17	42.37	−71.03	29	41.93	−72.68	21	44.47	−73.15	14
43.20	−71.50	16	43.35	−73.62	14	40.22	−76.85	17	42.22	−71.12	33
44.27	−71.30	21	40.82	−82.52	16	41.63	−73.88	19	40.90	−78.08	34
43.65	−70.32	16	41.73	−71.43	27	43.12	−77.67	30	43.12	−76.12	11
40.22	−74.77	12	41.33	−75.73	12	41.25	−76.92	18	40.03	−74.35	18
42.15	−70.93	31	40.20	−75.15	19	41.42	−81.87	17	40.00	−82.88	7
42.42	−83.02	14	41.02	−83.67	17	42.97	−83.75	14	42.27	−84.47	26
42.78	−84.60	22	43.53	−84.08	20	41.25	−80.67	8	42.08	−80.18	13
44.20	−72.57	23	44.93	−74.85	24	40.78	−73.97	18	42.27	−71.87	30
43.63	−72.32	41	40.65	−73.78	14	44.00	−76.02	19	41.17	−71.58	5
43.15	−75.38	11	44.37	−84.68	20	40.50	−80.22	7	41.60	−83.80	18

Figure 5.1   Data for snow depth in the northeastern United States on February 8, 1978, the day after the blizzard of February 1978. The snow depth is given as a function of latitude and longitude. Note that west longitude is written as a negative number. Data is courtesy of the National Climatic Data Center.

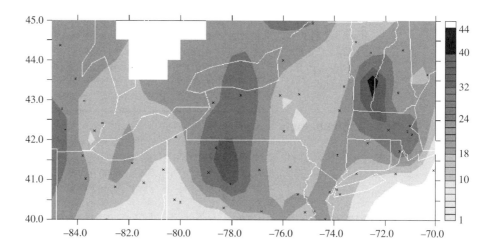

Figure 5.2   A density plot of the snow cover on February 8, 1978, in the northeastern United States the day after the blizzard of February 1978. The depth is measured in inches. Map courtesy of the National Climatic Data Center.

1. Suppose the snow depth in a region is constant. What is the total volume of snow that fell on that region? (Use units of feet × square miles.)
2. If the snow depth in a region is not constant (which, of course, it never is), but you have readings of snow depth at several different locations in the region, how might you approximate the total volume of snow that fell on that region? Would this approximation involve any choices on your part? If so, what are they?
3. Based on the data in Figure 5.1 for the region including the states of Pennsylvania, New York, New Jersey, Connecticut, Massachusetts, and Rhode Island, how would you proceed to approximate the total volume of snow that fell during the blizzard of 1978? (This region extends from approximately -73.5 to $-70.5°$ longitude and from 41.25 to $45.0°$ latitude. In this region, one degree of latitude is approximately 69 miles and one degree of longitude is approximately 51 miles.)
4. Carry out the approximation you describe in question 3.
5. How would you improve your method to produce a better approximation? Would you need more information than is provided in Figure 5.1?

■ 5.1 Riemann Sums—An Intuitive Introduction

In this section, we will begin our investigation of the problem of finding the ***total accumulation of a function of two or three variables***, that is, of summing the values of a function on a subset of its domain. Initially our goal is to develop a method to approximate the total accumulation of the function. In subsequent sections, we will develop a technique for improving upon a given approximation. In principle, we will then be able to compute an exact value for the total accumulation of the function by considering the limiting value of these successive approximations.

As motivation for our work, we consider several examples where the total accumulation of a function has a natural interpretation.

Example 5.1   **Total Accumulation**

A. **Volume.** Suppose that an object has a flat base and lies entirely over its base so that the height of the object over each point of the base is a function defined on the base. The total accumulation of the height function of the object is the *volume* of the object.

B. **Average Value.** The *average value* of a function defined on a region in the plane is the quotient of the total accumulation of the function on the region and the area of the region. If the function is defined on a region in space, then the average value is the quotient of the total accumulation of the function on the region and the volume of the region.

C. **Total Mass.** Suppose that a box is filled with a material of varying density. The density can be thought of as a function on a region corresponding to the interior of the box. The total accumulation of the density function of the box is the *total mass* of the material.

D. **Total Intensity.** An infrared photograph is a visual representation of the heat given off by an object. Calibrated appropriately, it is a density plot of the heat intensity of the object. Its total accumulation is the total heat of the object. Generally, when considering a function that represents the intensity of some form of radiation, the total accumulation of this function is the total intensity of the radiant object.

In the remainder of this section, we will consider three examples of naturally occurring functions and develop a method to approximate their total accumulation. In the first example, we will consider a contour plot of rainfall levels in Fort Collins, Colorado, and use the data from this plot to approximate the average rainfall in the region. In the second example, we will use a data set that gives intensity of the sun at a wavelength of 500 nm. We will use this data to approximate the total energy output of the sun at that frequency. Finally, in the third example, we will work with a data set that gives the temperature at different points in a room and approximate the average temperature in the room.

Rainfall in Fort Collins

Figure 5.3 is a contour plot indicating the rainfall totals in Fort Collins, Colorado, after severe storms on July 27 and 28, 1997. The region plotted is approximately 9 mi east to west by 16 mi north to south. The rainfall totals recorded ranged from a low of 2 in.

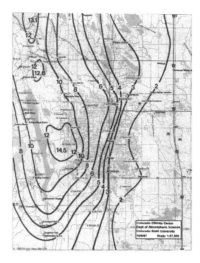

Figure 5.3 A contour plot indicating rainfall totals in Fort Collins, Colorado, July 28, 1997. Reproduced from Doesken, Nolan J. and Thomas B. McKee, *An Analysis of Rainfall for the July 28, 1997 Flood in Fort Collins, Colorado.* Climatology Report 98-1, Colorado Climate Center, Atmospheric Science Department, Colorado State University, 1998: pp. 28. Used with permission. (See Example 5.2.)

in the southeast corner of the region to a high of 14.5 in. in southwest Fort Collins just east of the Horsetooth Reservoir. We can see from the contour plot that there were three local maxima, 14.5 in. near Horsetooth Reservoir, 12.6 in. west of Fort Collins near Claymore lake, and 13.1 in. in the northwest corner of the plot.

We would like to use this plot to compute the average rainfall in the region. The average rainfall in the region is the ratio of the total accumulation of rainfall and the area of the region.

Example 5.2

Computing Average Rainfall. First, we will approximate the total accumulation of rainfall. In a region where the rainfall level is constant, the total accumulation of rainfall is the product of the rainfall level and the area of the region. Thus, if the rainfall level is close to the value t_0 everywhere, then we can approximate the total accumulation of rainfall on the region by the product of t_0 and the area of the region. In order to use this idea to approximate the total accumulation, we will partition the region into smaller regions or **cells** so that the rainfall level can be approximated by a constant in each cell. Then we can approximate the total accumulation of the rainfall on a cell by the product of a rainfall level in the cell and the area of the cell. We will then sum these values to obtain an approximation to the total accumulation of rainfall in the region.

We will partition the region into rectangles. First we will divide the region into three columns, each of width approximately 3 mi. We will then divide the column into four

4	12	5	3
3	12	6	4
2	12	8	3
1	10	6	2
	1	2	3

Figure 5.4 The sampled rainfall levels for a 3×4 partition of the region around Fort Collins. The values are given in inches.

cells of height approximately 4 mi. The dimensions of a cell are $\Delta x \times \Delta y$, where $\Delta x = 3$ mi and $\Delta y = 4$ mi.

In order to compute the total accumulation of rainfall, we will sample the rainfall level in each cell of the partition. In each cell, we use the rainfall value on the contour nearest the center of the cell for the sample rainfall value in the cell. This array of sampled values is shown in Figure 5.4. In cases where there was no contour curve through the center of the cell, we used the nearest contour value. The total accumulation of rainfall in each cell is approximated by the product of the sampled rainfall value and the area of the cell. Thus, for example, the accumulation of rainfall in the first cell in the first column is approximately $10 \times 3 \times 4$. The total accumulation in the region is then approximated by the sum of these values. Symbolically, if t_{ij} is the sampled rainfall value in the ij^{th} cell of the partition, the total accumulation of rainfall in the region is approximated by the sum,

$$\sum_{i,j=1}^{3,4} t_{ij} \Delta x \Delta y.$$

Note that we have used the Greek capital letter sigma to represent the sum. In this case, this denotes the summation of the quantities $t_{ij}\Delta x \Delta y$ for all possible combinations of i and j as i runs from 1 to 3 and j runs from 1 to 4.

If we substitute the rainfall levels shown in Figure 5.4 and the values given above for Δx and Δy into this sum, we obtain a value of 996 for the total accumulation. The total area of the region is approximately 144 sq mi. If we divide the total accumulation of rainfall by the area, we obtain an average rainfall of 6.92 in.

Before beginning the next example, let us summarize the technique we have developed. Given a function f defined on a region in the plane, in order to approximate the total accumulation of f on the region, we began by partitioning the region into $M \times N$

rectangular cells. We then chose a point in each cell and evaluated the function at that point. The product of the function value and the area of the cell gave an approximation for the total accumulation of the function on the cell. We then summed these values over all cells to obtain an approximation to the total accumulation of f on the region.

Let us label or index a cell by an ordered pair ij with $1 \leq i \leq M$ and $1 \leq j \leq N$, and let ΔA_{ij} denote the area of the cell labeled ij. If P_{ij} is the sample point chosen in the ij^{th} cell, then the total accumulation of f in the ij^{th} cell is approximated by $f(P_{ij})\Delta A_{ij}$. Thus the total accumulation of f on the region is approximated by the sum

$$\sum_{i,j=1}^{M,N} f(P_{ij})\Delta A_{ij}.$$

Again we remark that this notation indicates the summation of the quantities $f(P_{ij})\Delta A_{ij}$ for all possible combinations of i and j as i runs from 1 to M and j runs from 1 to N. This sum is called a **_Riemann_**[1] **_sum_** for the function f for this partition.

Solar Intensity

If we neglect sun spots, the intensity of an image of the sun at a point on the solar disk decreases as the point moves toward the edge of the disk, and this decrease is uniform in all directions from the center of the solar disk. This phenomenon is known as **_limb darkening_**. Thus the intensity of the sun can be modeled by a function I that depends only on the distance of a point on the solar disk from the center of the solar disk. Equivalently, we could say the function is **_rotationally symmetric_**; that is, if we rotate the image of the sun about its center, the values of I remain the same. Figure 5.5 contains a grayscale image of solar intensity at a wavelength of 500 nm. In order to determine the total energy output at a given frequency, we will approximate the total accumulation of I over the solar disk.

Figure 5.5 A density plot of the intensity of the sun at 500 nm.

---

[1]G. F. B. Riemann (1826–1866) was a German mathematician responsible for fundamental developments in several areas of mathematics.

We will use the technique we developed above to approximate the total accumulation of the intensity function. Thus we must partition the region into cells, and then approximate the total accumulation of I in each cell by taking the product of a sampled intensity value in the cell and the area of the cell. The total accumulation of I is then approximated by summing these values over all cells.

Example 5.3

Total Intensity. The accompanying table gives values for the intensity of the sun at a wavelength of 500 nm. The distances r are reported in terms of the solar radius $R_\odot$ so that $r = 0$ corresponds to the center of the solar disk and $r = R_\odot$ corresponds to the edge of the solar disk. Similarly, the intensity values are given as a fraction of I_0, where I_0 is the intensity at the center of the solar disk.

r	0	$0.6000R_\odot$	$0.8000R_\odot$	$0.8660R_\odot$	$0.9165R_\odot$
$I(r)$	$1.000I_0$	$0.877I_0$	$0.744I_0$	$0.675I_0$	$0.599I_0$
r	$0.9539R_\odot$	$0.9798R_\odot$	$0.9950R_\odot$	$0.9987R_\odot$	$0.9998R_\odot$
$I(r)$	$0.513I_0$	$0.425I_0$	$0.323I_0$	$0.260I_0$	$0.190I_0$

Since the intensity value $I(r)$ at a point depends only on the distance from the center of the solar disk, the intensity is constant on a circle of fixed radius r. Rather than partition the solar disk into rectangles, *it is more natural to partition it into cells that take advantage of the rotational symmetry of the disk and of the intensity* I. Consequently, we will choose each cell to be the region between two circles, an annulus. In order to partition the region so that each cell contains a sample point from the data set, we will choose values for the radii of these circles as follows:

$$r_1 = 0, \ r_2 = 0.5R_\odot, \ r_3 = 0.7R_\odot, \ r_4 = 0.85R_\odot, \ r_5 = 0.9R_\odot, \ r_6 = 0.94R_\odot,$$

$$r_7 = 0.96R_\odot, \ r_8 = 0.98R_\odot, \ r_9 = 0.996R_\odot, \ r_{10} = 0.999R_\odot, \ \text{and} \ r_{11} = R_\odot.$$

The ith cell of the partition is the region between the circle of radius r_i and the circle of radius r_{i+1}. The area of the ith cell is

$$\Delta A_i = \pi(r_{i+1}^2 - r_i^2).$$

We will denote the values for the radius given in the table above by r_i^* for $i = 1, \dots, 10$, and let $I(r_i^*)$ be the corresponding intensity value. Using these values, a calculation shows that the total accumulation of I is approximated by

$$\sum_{i=1}^{10} I(r_i^*)\Delta A_i \approx \pi(0.7901)R_\odot^2 I_0.$$

This gives a value for the total intensity at 500 nm of approximately $2.4822R_\odot^2 I_0$.

56.3	57.5	57.9
66.4	69.3	70.9
68.3	69.4	71.3
70.1	71.3	71.9
71.0	70.6	72.0

(a)

66.7	67.6	69.1
70.0	70.3	71.3
70.3	71.2	72.9
71.8	72.2	71.8
72.2	71.8	71.4

(b)

73.4	80.1	79.6
73.3	76.2	81.4
72.8	74.2	76.0
72.6	73.5	74.3
72.5	73.1	73.4

(c)

Figure 5.6   The sampled temperature values in an office: (a) along the floor, (b) at a height of 4.5 ft, and (c) along the ceiling. The values are given in degrees Fahrenheit. In each table, the first row contains the temperature readings along the outside wall and the last row contains the temperature readings along the wall with the door.

Average Temperature in Space

In this example, we would like to extend the techniques we developed above to *approximate the average temperature of a region in space.* The tables in Figure 5.6 give measurements of the temperature at a $5 \times 3 \times 3$ array of points in an office. The office measures 18 ft 9 in. long by 10 ft 6 in. wide by 8 ft 9 in. high. There are two windows along the western wall of the office with a door leading to a hallway on the opposite wall. The western wall is the only outside wall. In addition, there is a radiator that extends along the western wall and is located approximately 4 ft off the floor.

Example 5.4

Average Temperature in Space. The sampling points are arranged in fifteen columns with three points in each column. In Figure 5.6, Table (a) gives the fifteen temperature readings taken at a 5×3 array of points on the floor of the office, Table (b) gives the fifteen temperature readings taken at points 4.5 ft off the floor, and Table (c) gives the fifteen temperature readings taken at points on the ceiling. In each table, the first row contains the temperature readings along the outside wall, and the last row contains the temperature readings along the wall with the door. We can see from the readings that the temperature values in the office vary between 56.3 and 80.1°F.

Applying the average value construction to the temperature function, *the average temperature in a region in space is the quotient of the total accumulation of temperature and the volume of the region.* The volume of the region is approximately 1723 cu ft. In order to compute the total accumulation of temperature in the office, we will partition the office into cells, determine the temperature at a sample point in each cell, approximate the total accumulation of temperature in the cell by the product of the temperature and the volume of the cell, and then sum these values to approximate the total temperature in the office.

We will partition the room into $5 \times 3 \times 3$ rectangular boxes so that one sampled temperature value from the tables is located in each box. In order to do this, we will arrange fifteen columns of rectangular boxes in a 5×3 rectangular array. Each column will be divided evenly into three boxes. The dimensions of each box or cell in the partition are $\Delta x = 3$ ft 9 in., $\Delta y = 3$ ft 6 in., and $\Delta z = 2$ ft 11 in. The volume of each cell is approximately 38.3 cu ft. The total accumulation of temperature in each cell is approximated by the product of the temperature value and the volume of the box. Thus, for example, the total accumulation in the first cell is approximately $56.3 \times 38.3 = 2156.29$. If t_{ijk} is the sampled temperature value in the ijk^{th} cell, the total accumulation of temperature in the office is approximated by the Riemann sum

$$\sum_{i,j,k=1}^{5,3,3} t_{ijk} \Delta x \Delta y \Delta z \approx 122605.96.$$

Dividing this value by 1723, we obtain an approximate value of $71.16°$F.

Summary	The procedure we established in this section to **approximate the total accumulation of a function on a region in the plane or in space** requires that we **partition** the region into cells, **sample** the function in each cell of the partition, **scale** the sampled value by multiplying the sampled function value by the area (or volume) of the cell, and then **sum these values** to approximate the total accumulation on the region. Sums of this form are called **Riemann sum** approximations to the total accumulation of the function on the region.
	We applied these techniques to three examples: a calculation of **average rainfall** in a planar region, a calculation of **total intensity** in a circularly symmetric region, and a calculation of **average temperature** in space.

Section 5.1 Exercises

1. **Total Heat.** A heated rectangular aluminum plate is 4 cm by 5 cm by 1 cm thick. The temperature T at a point in the plate does not depend on the depth of the point in the plate (so we may think of it as a function of two variables). Suppose that the temperature at the corners of the plate is $20°$C, the temperature at the midpoints of the four sides is $17°$C, and the temperature in the middle of the plate is $15°$C. The density of the aluminum is 2.7 g/cm^3 and the specific heat of aluminum is 0.215 cal/g C$°$. The total heat in the plate is the accumulation of the function obtained by multiplying the temperature function by the product of the thickness of the plate, the density of the material, and the specific heat of the material. Use a Riemann sum to approximate the total heat contained in the aluminum plate.

2. **Average Temperature I.** The map in Figure 5.8 shows the maximum temperature in the Southwest United States on November 16, 2010. The table in Figure 5.7 gives the

City	lat.	long.	T	City	lat.	long.	T
Durango	37.16	107.52	46	Phoenix	33.27	112.4	76
Elko	40.50	115.45	48	Redding	40.34	122.22	73
Ely	39.15	114.52	50	Reno	39.31	119.49	59
Flagstaff	35.11	111.37	55	Salt Lake City	40.45	111.53	48
Fresno	36.44	119.46	72	San Diego	32.42	117.09	86
Grand Junction	39.04	108.34	46	San Francisco	37.46	122.25	70
Las Vegas	36.10	115.08	71	Tuscon	32.13	110.55	74
Los Angeles	34.03	118.15	89	Yuma	32.41	114.36	81

Figure 5.7   The daily maximum temperature in the Southwest United States on November 16, 2010, at cities marked on the density plot in Figure 5.8 and their latitudes and longitudes. (See Exercise 2).

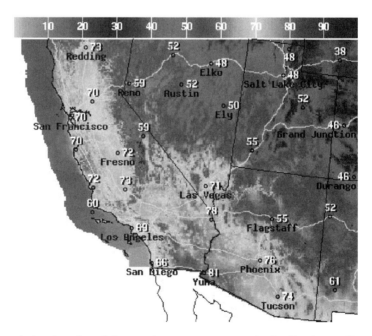

Figure 5.8   A density plot of the maximum temperature in the Southwest United States on November 16, 2010. Courtesy of the NWS/NOAA National Digital Forecast Database.

temperature readings at selected cities in the region. Use a Riemann sum to approximate the total accumulation of the temperature based on this map and the data set. Carefully explain the partition you used and your choice of sample points. Use your approximation to estimate the average temperature in this region. (The region extends from approximately 107° to 122° west latitude and from approximately 32° to 40° north longitude. In this region, one degree of latitude is approximately 69 miles and one degree of longitude is approximately 51 miles.)

3. **Refining an Approximation.** Consider the contour plot in Figure 5.3 for the total rainfall in Fort Collins. We used a partition of the region into 3 × 4 cells to approximate the average rainfall.

   (a) Construct a partition of the region into 6 × 8 cells. Describe the partition and use it to construct a Riemann sum to approximate the average rainfall in the region.
   (b) How does the value computed in part (a) compare to the value computed in the text? Which is a better approximation to the true average value? Why?
   (c) How would you improve the approximation further?

4. **Upper and Lower Sums.** In computing the average rainfall in Fort Collins, we partitioned the region into 3 × 4 cells and sampled the rainfall level at points on a contour lying closest to the center of each cell.

   (a) Use a Riemann sum to approximate the average rainfall by choosing the maximum rainfall level in each cell.
   (b) Use a Riemann sum to approximate the average rainfall by choosing the minimum rainfall level in each cell.
   (c) Based on your answers to (a) and (b), what can you say about the true average value?

5. **Volume.** An object has a square base that is 10 cm on a side. The height of the object at the corners is 1 cm, at the center is 5 cm, and at the points that trisect the sides is 2 cm. Use a Riemann sum to approximate the volume of the object.

6. **Average Temperature II.** A hot cup is placed on a thin square plate with sides of length 4. After 1 minute the temperature of the plate is given by $t(x, y) = 75 + 50e^{-x^2 - y^2}$ in °F, when the center of the plate is located at the origin and the sides are parallel to the axes. Compute a Riemann sum to approximate the total accumulation of the temperature of the plate. Use your approximation to estimate the average temperature of the plate.

7. **Average Temperature III.** A circular aluminum griddle is 20 cm in diameter and 1 cm thick. The temperature at the center of the griddle is 150°C and at each of eight equally spaced points on its circumference is 100°C. Use a Riemann sum to approximate the average temperature of the griddle.

8. **High and Low Approximations.** Consider the density plot of the snow cover on February 8, 1978, in the northeastern United States given in Figure 5.2. Let us focus on the region of the plot from the western border of Pennsylvania, approximately −80.5° longitude, to the eastern border of New Hampshire, approximately −70.5° longitude, and

between lower edge of the plot and the northern borders of Pennsylvania, New York, Vermont, and New Hampshire. Partition this region into 4×4 rectangular cells.

(a) Use this partition to construct a Riemann sum that gives an approximation to the total volume of snow on the ground that is less than the actual volume.
(b) Use this partition to construct a Riemann sum that gives an approximation to the total volume of snow on the ground that is greater than the actual volume.
(c) Explain why you believe the actual volume of snow is between the two approximations you calculated in parts (a) and (b).

9. **Average Value I.** Figure 5.9(a) is a density plot of the function $f(x, y) = \sin(x^2 + y^2)$ on the region $[0, 3] \times [0, 3]$. Use a Riemann sum to approximate the average value of f over this region. Describe the partition you used and indicate the sample points you chose in constructing your Riemann sum.

10. **Average Value II.** Figure 5.9(b) is a density plot of the function $f(x, y) = xy^2$ on the region $[-1, 1] \times [0, 1]$. The contours of f have been superimposed on this plot. Use a Riemann sum to approximate the average value of f over this region. Describe the partition you used and indicate the sample points you chose in constructing your Riemann sum.

11. **Total Mass.** A solid cube of edge length l consists of material of varying density. The density is 1 at each of the vertices of the cube, 2 at the center of each face of the cube, and 3 at the center of the cube. Construct a Riemann sum to approximate the total mass of the cube.

12. **Average Temperature in Space.** The tables in Figure 5.10 give sampled temperature values in a room that is approximately 10 ft wide by 12 ft long by 8 ft high. The sampling points are arranged in nine columns with three points in each column. Table (a) gives the nine temperature readings taken at a 3×3 array of points on the floor of the office,

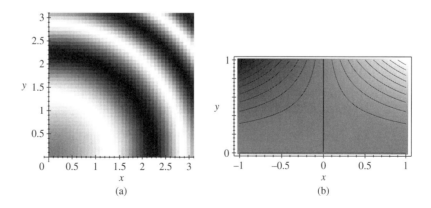

Figure 5.9 (a) Density plot of $f(x, y) = \sin(x^2 + y^2)$ for Exercise 9. (b) Density plot of $f(x, y) = xy^2$ for Exercise 10.

71.4	69.3	62.9
72.3	69.4	58.3
72.1	71.3	64.9

(a)

72.3	71.2	68.9
72.8	70.2	65.8
72.2	71.8	68.4

(b)

74.8	74.2	72.8
75.6	73.5	70.3
75.5	73.9	72.4

(c)

Figure 5.10   The sampled temperature values in a room (a) along the floor, (b) at a height of 4 ft off the floor, and (c) along the ceiling. The values are given in degrees Fahrenheit. In each table, the last column contains the temperature readings along the outside wall. (See Exercise 12.)

Table (b) gives the nine temperature readings taken at points 4 ft off the floor, and Table (c) gives the nine temperature readings taken at points on the ceiling. Use a Riemann sum to approximate the average temperature in the room.

■ 5.2 Integration of Functions of Two Variables

In Section 5.1, we saw how to approximate the total accumulation of a function on a region in the plane. We partitioned the region, sampled the function in each cell of the partition, scaled the sampled values by the area of each cell, and summed the scaled values. We called a sum of this type a Riemann sum. Intuitively, we would expect that we could obtain a more accurate approximation by choosing a larger number of sample points and smaller cells, and that where we have the ability to continue this process indefinitely, we could obtain an exact value for the total accumulation by taking a limit of Riemann sums as the number of sample points becomes infinite and the area of the cells approach zero. In this section, we want to make these ideas precise, and we want to develop a method for evaluating the total accumulation in the case where the function is defined symbolically. First, however, let us review the corresponding results for functions of one variable.

Integration in One Variable

In one-variable calculus, if f is a continuous function on an open interval containing $[a, b]$, the ***total accumulation*** of f is defined to be a ***limit of Riemann sums***. More precisely, if P is a partition of the interval $[a, b]$ into M subintervals, Δx_i is the length of the ith subinterval, and x_i^* is a sample point in the ith subinterval, then the sum

$$R(f, P) = \sum_{i=1}^{M} f(x_i^*)\Delta x_i$$

is called a *Riemann sum* for f on $[a, b]$ for the partition P. The *mesh* of the partition is defined to be the maximum of the values of Δx_i. An exact value for the total accumulation is obtained by taking a limit of Riemann sums for partitions P as the mesh of the partitions approaches 0. One can show that if f is continuous, then the limit of the Riemann sums exists and is independent of the choices of partitions and sampling points.

This value is called the definite integral of f from a to b, and we use the familiar integral notation for this limit:

$$\int_a^b f(x)dx = \lim_{\text{mesh}(P) \to 0} R(f, P).$$

The *fundamental theorem of calculus* provides a symbolic method for evaluating integrals.

Theorem 5.1 The Fundamental Theorem of Calculus. If f is continuous and F is an antiderivative of f, that is, $F'(x) = f(x)$, then

$$\int_a^b f(x)dx = F(b) - F(a). \ \blacklozenge$$

The techniques that are developed in one-variable calculus to evaluate integrals are rules for producing an antiderivative F for f by manipulating a symbolic expression for f.

Integration in Two Variables

Returning to functions of two variables, we want to take care in establishing our notation in order to utilize the results from one-variable calculus. Let us assume that $\mathcal{R}$ is a subset of the domain of f and that f is *bounded* on $\mathcal{R}$. Initially, we will assume that $\mathcal{R}$ *is the region between the graphs of the continuous functions g and h of one variable with x-coordinate between a and b*. That is,

$$\mathcal{R} = \{(x, y) \ : \ a \leq x \leq b \text{ and } g(x) \leq y \leq h(x)\}.$$

(See Figure 5.11(a).) Later in the section, we will consider more general regions.

We want to construct a *partition* P of the region $\mathcal{R}$ consisting of rectangular cells arranged in columns with the same number of cells in each column. Let M denote the number of columns and N the number of cells in a column. We say that the partition consists of $M \times N$ cells. This can be accomplished by choosing $M + 1$ values x_i, $i = 0, \ldots, M$, with

$$a = x_0 < x_1 < x_2 < \ldots < x_{M-1} < x_M = b.$$

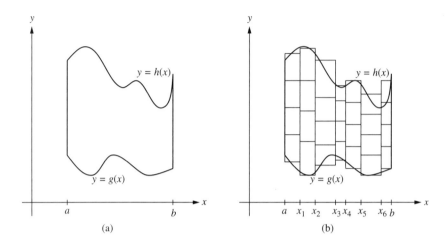

Figure 5.11 (a) A typical region of integration of the form $\mathcal{R} = \{(x, y) : a \leq x \leq b$ and $g(x) \leq y \leq h(x)\}$. (b) A partition of this region into $M = 7$ columns with $N = 4$ cells in each column. Note the columns have different widths and the cells have different heights.

This partitions $\mathcal{R}$ into M columns, where the ith column is

$$\{(x, y) : x_{i-1} \leq x \leq x_i \text{ and } g(x) \leq y \leq h(x)\}.$$

In order to partition each column into N rectangular cells, we must choose a height for the column. We select a point x_i^* with $x_{i-1} \leq x_i^* \leq x_i$ so that the ith column has minimum value $g(x_i^*)$ and maximum value $h(x_i^*)$. Then for each column, that is, for $i = 1, \ldots, M$, we choose $N + 1$ values y_{ij} with

$$g(x_i^*) = y_{i0} < y_{i1} < y_{i2} < \ldots < y_{iN} = h(x_i^*).$$

Then the ij^{th} rectangle of the partition, $\mathcal{R}_{ij}$, is defined by

$$\mathcal{R}_{ij} = \{(x, y) : x_{i-1} \leq x \leq x_i \text{ and } y_{i(j-1)} \leq y \leq y_{ij}\},$$

where $1 \leq i \leq M$ and $1 \leq j \leq N$. (See Figure 5.11(b).) We will often use the product notation $\mathcal{R}_{ij} = [x_{i-1}, x_i] \times [y_{i(j-1)}, y_{ij}]$ to describe a cell. Notice that the cells within a column have the same width, but that the columns may have different widths, and that the cells within a column may have different heights. The area of $\mathcal{R}_{ij}$ is Area($\mathcal{R}_{ij}$) $= \Delta x_i \Delta y_{ij}$, where $\Delta x_i = x_i - x_{i-1}$ and $\Delta y_{ij} = y_{ij} - y_{i(j-1)}$. The **mesh** of the partition P is the largest of the values Δx_i and Δy_{ij}, and we denote it by mesh(P).

We sample f on $\mathcal{R}$ by choosing one point P_{ij} from each cell $\mathcal{R}_{ij}$ of P. That is, we choose points $P_{ij} = (x_{ij}^*, y_{ij}^*)$, where

$$x_{i-1} \leq x_{ij}^* \leq x_i \quad \text{and} \quad y_{i(j-1)} \leq y_{ij}^* \leq y_{ij}.$$

We approximate the accumulation of f on $\mathcal{R}_{ij}$ by the scaled value

$$f(P_{ij})\text{Area}(\mathcal{R}_{ij}) = f(x_{ij}^*, y_{ij}^*)\Delta x_i \Delta y_{ij}.$$

Then we approximate the total accumulation of f on $\mathcal{R}$ by the sum

$$R(f, P) = \sum_{i,j=1}^{M,N} f(P_{ij})\text{Area}(\mathcal{R}_{ij}) = \sum_{i,j=1}^{M,N} f(x_{ij}^*, y_{ij}^*)\Delta x_i \Delta y_{ij}.$$

As before, this sum is called a **_Riemann sum_** for f on $\mathcal{R}$ for the partition P. Keep in mind that a summation expression of this form represents the sum of the values $f(P_{ij})\text{Area}(\mathcal{R}_{ij})$ for all the cells in the partition P, that is, for all combinations of the indices i and j where i runs from 1 to M and j runs from 1 to N. Since addition can be carried out in any order, it is not necessary to specify the order of the summation.

We expect that we can improve upon the approximation to the total accumulation given by a particular $R(f, P)$ by sampling f at more points and by choosing partitions with a smaller mesh. If we think of this as an iterative process, then in the limit, as the mesh approaches zero, we should obtain an exact value for the total accumulation of f on $\mathcal{R}$. If f is continuous, then this will be the case.

Theorem 5.2 Let f be continuous on an open set containing the region $\mathcal{R}$ given by

$$\mathcal{R} = \{(x, y) \,:\, a \leq x \leq b \text{ and } g(x) \leq y \leq h(x)\},$$

where g and h are continuous. Then

$$\lim_{\text{mesh}(P) \to 0} R(f, P)$$

exists and is independent of any choices that are made in constructing this limit. ◆

If the limit exists, we denote it by the symbol

$$\int\int_{\mathcal{R}} f \, dA_{x,y}.$$

We call this value the *integral* or *double integral* of f on $\mathcal{R}$. Thus the integral of f is the limit of sums

$$R(f, P) = \sum_{i,j=1}^{M,N} f(x_{ij}^*, y_{ij}^*) \Delta x_i \Delta y_{ij}$$

as mesh$(P) \to 0$. The function f is called the *integrand* of the integral.

If we assume that f is continuous, the limit of the Riemann sums does not depend on the choice of partitions or the choice of sampling points. This will be a key point that will allow us to connect integrals of functions of two variables with integrals of functions of one variable. Our goal in the following paragraphs is to develop a method of evaluating double integrals.

Fubini's Theorem

We would like to be able to employ the symbolic techniques of one-variable calculus to evaluate the integral of a function of two variables. In order to do this, we will express a double integral as a pair of nested or iterated single integrals. We will carry this out in three steps: first, writing a Riemann sum as an iterated sum; second, writing the limit of Riemann sums as an iterated limit; and third, recognizing the iterated limits as limits of Riemann sums of functions of one variable. This result is known as *Fubini's* [2] *theorem.* Let us assume that f is continuous on an open set containing $\mathcal{R} = \{(x, y) : a \leq x \leq b$ and $g(x) \leq y \leq h(x)\}$, where g and h are continuous. Since f is continuous, the integral does not depend on the choices involved in constructing the partitions and the Riemann sums used to calculate the limit. In particular, we can choose the x-coordinate of the sample points to be the same for each column of the partition. That is, we sample f at the points (x_i^*, y_{ij}^*). Using the notation that we established above, a Riemann sum for f on the region $\mathcal{R}$ for a partition consisting of $M \times N$ cells is given by

$$R(f, P) = \sum_{i,j=1}^{M,N} f(x_i^*, y_{ij}^*) \Delta x_i \Delta y_{ij}.$$

We will find it advantageous to reorder the summands so that we first carry out the sums corresponding to the columns of the partition to obtain an approximation to

---

[2]G. Fubini (1879–1943) was an Italian mathematician known for his work in geometry and analysis.

the total accumulation of f in a column, and then sum the values for the columns. We represent this symbolically by an **_iterated sum_**,

$$R(f, P) = \sum_{i=1}^{M} \left(\sum_{j=1}^{N} f(x_i^*, y_{ij}^*) \Delta x_i \Delta y_{ij} \right).$$

This expression instructs us to do the following: For each i from 1 to M, sum the values of $f(x_i^*, y_{ij}^*) \Delta x_i \Delta y_{ij}$ with j ranging from 1 to N. This gives an approximation to the total accumulation in column i. Then sum the resulting totals with i ranging from 1 to M. Notice that since Δx_i is constant with respect to the inner sum, we will be able to factor it out of the inner sum.

The integral of f over $\mathcal{R}$ is the limit of Riemann sums as $\text{mesh}(P) \to 0$. This implies that $M, N \to \infty$. Keeping in mind that we still require $\text{mesh}(P) \to 0$, we will rewrite this limit as a double limit as M and N approach infinity. It is an analytic fact (that we will not prove here) that **_if the double limit exists, it is equal to an iterated limit where we separately allow M and N to approach infinity_**. Thus

$$\int\int_R f \, dA_{x,y} = \lim_{M \to \infty} \left(\lim_{N \to \infty} R(f, P) \right).$$

Substituting the iterated sum for $R(f, P)$, we see the integral is equal to

$$\lim_{M \to \infty} \left(\lim_{N \to \infty} \sum_{i=1}^{M} \left(\sum_{j=1}^{N} f(x_i^*, y_{ij}^*) \Delta y_{ij} \right) \Delta x_i \right).$$

Since the outer sum does not depend on N, we can interchange the order of the outer sum and inner limit. That is, a limit of a sum is equal to a sum of limits when the index for the limit is not the index for the sum. Then, since Δx_i does not depend on N, we can factor it out of the inner limit. Combining these two steps, the integral is equal to

$$\lim_{M \to \infty} \sum_{i=1}^{M} \left(\lim_{N \to \infty} \sum_{j=1}^{N} f(x_i^*, y_{ij}^*) \Delta y_{ij} \right) \Delta x_i.$$

Now let us focus our attention on the inner limit,

$$\lim_{N \to \infty} \sum_{j=1}^{N} f(x_i^*, y_{ij}^*) \Delta y_{ij}.$$

Since i is fixed, x_i^* is constant in this expression, and we observe that the sum is in the form of a Riemann sum for the function $\tilde{f}$ of one variable given by $\tilde{f}(y) = f(x_i^*, y)$. Since the y_{ij}^* satisfy $g(x_i^*) \leq y_{ij}^* \leq h(x_i^*)$, this is a Riemann sum for $\tilde{f}$ on the interval $[g(x_i^*), h(x_i^*)]$. Since f is a continuous function of x and y, it is a continuous function of y when x is held fixed. Thus $\tilde{f}$ is a continuous function of y. It follows that this limit exists and is equal to the integral of $\tilde{f}$ on $[g(x_i^*), h(x_i^*)]$:

$$\int_{g(x_i^*)}^{h(x_i^*)} \tilde{f}(y)\, dy = \int_{g(x_i^*)}^{h(x_i^*)} f(x_i^*, y)\, dy.$$

Note that this expression depends on x_i^*. Let us write this as a function G of x, so that

$$G(x_i^*) = \int_{g(x_i^*)}^{h(x_i^*)} f(x_i^*, y)\, dy.$$

Substituting this into the iterated limit we have

$$\lim_{M \to \infty} \sum_{i=1}^{M} G(x_i^*) \Delta x_i.$$

This is a limit of Riemann sums for the function G on the interval $[a, b]$. Since f is continuous on $\mathcal{R}$, and g and h are continuous functions of x, G is continuous on (a, b). Consequently, this limit is equal to the integral of G on $[a, b]$. Putting this together, we have that

$$\lim_{\text{mesh}(P) \to 0} R(f, P) = \int_a^b G(x)\, dx$$

$$= \int_a^b \left(\int_{g(x)}^{h(x)} f(x, y) dy \right) dx.$$

We call this last expression an ***iterated integral*** for f on $\mathcal{R}$.

If $\mathcal{R} = \{(x, y) : c \leq y \leq d \text{ and } g(y) \leq x \leq h(y)\}$, then we can interchange the roles of x and y. In this case, $\mathcal{R}$ should be partitioned into N rows of M cells. Following our argument, we would be able to express the integral of f over $\mathcal{R}$ as an iterated integral of f over $\mathcal{R}$ where we first integrate with respect to x and then integrate with respect to y. We summarize this in the following theorem.

Theorem 5.3 **Fubini's Theorem.** Let f be a continuous function on an open set containing the region $\mathcal{R}$, and let g and h be continuous functions.

1. If $\mathcal{R} = \{(x, y) \ : \ a \le x \le b \text{ and } g(x) \le y \le h(x)\}$, then

$$\int \int_{\mathcal{R}} f \, dA_{x,y} = \int_a^b \int_{g(x)}^{h(x)} f(x, y) \, dy \, dx.$$

2. If $\mathcal{R} = \{(x, y) : c \le y \le d \text{ and } g(y) \le x \le h(y) \}$, then

$$\int \int_{\mathcal{R}} f \, dA_{x,y} = \int_c^d \int_{g(y)}^{h(y)} f(x, y) \, dx \, dy. \ \blacklozenge$$

This theorem will be our primary tool for computing the double integral of a function by symbolic means. In principle, we can evaluate a double integral by applying the fundamental theorem of calculus to each of the iterated integrals, beginning with the inner integral. For example, to evaluate an integral of the form

$$\int_{g(x)}^{h(x)} f(x, y) \, dy,$$

we must find a function $F(x, y)$ so that

$$\frac{\partial F}{\partial y}(x, y) = f(x, y).$$

Then

$$\int_{g(x)}^{h(x)} f(x, y) \, dy = F(x, h(x)) - F(x, g(x)).$$

In our first example of the use of Fubini's theorem, we will consider rectangular regions, so that the functions g and h are constant functions.

Example 5.5   **Applying Fubini's Theorem**

A. A Rectangular Region. Let us compute the integral of $f(x, y) = x^2 + y$ over the rectangle $\mathcal{R} = [-1, 3] \times [0, 1]$. By Fubini's theorem, we have

$$\int \int_{\mathcal{R}} (x^2 + y) \, dA_{x,y} = \int_{-1}^3 \int_0^1 (x^2 + y) dy \, dx.$$

Since x^2 is constant with respect to y, we treat it as a constant when we apply the fundamental theorem of calculus to evaluate the inner integral. The function $F(x, y) =$

$x^2y + y^2/2$ satisfies $\frac{\partial F}{\partial y}(x,y) = x^2 + y$, so that

$$\int_0^1 (x^2 + y)\,dy = x^2y + y^2/2 \,\big|_0^1$$
$$= (x^2 + 1^2/2) - 0$$
$$= x^2 + 1/2.$$

We must now evaluate the integral of this function on $[-1, 3]$:

$$\int_{-1}^3 (x^2 + 1/2)\,dx = x^3/3 + x/2 \,\big|_{-1}^3$$
$$= (9 + 3/2) - (-1/3 - 1/2) = 34/3.$$

Since the region of integration is a rectangle, it is also possible to express the integral as an iterated integral in the reverse order, first with respect to x and then with respect to y. We leave this calculation to the reader.

B. Average Temperature. Here we compute the average temperature in a rectangular plate. In coordinates, suppose that the plate is represented by the set

$$S = \{(x, y) : 0 \le x \le 1 \text{ and } 0 \le y \le 2\}.$$

The temperature in the plate as a function of x and y is given by

$$T(x, y) = 20\sin(\pi x)\sin(\pi y/2).$$

The average temperature is the total accumulation of the temperature function in the plate divided by the area of the plate. Thus the average temperature is

$$\frac{1}{2}\int\int_R T\,dA_{x,y} = \frac{1}{2}\int_0^1\int_0^2 20\sin(\pi x)\sin(\pi y/2)\,dy\,dx$$
$$= 10\int_0^1 \sin(\pi x)\left[-\frac{2}{\pi}\cos\left(\frac{\pi}{2}x\right)\right]_0^2 dx$$
$$= 10\int_0^1 \frac{4}{\pi}\sin(\pi x)\,dx$$
$$= 10\frac{4}{\pi}\frac{2}{\pi} = \frac{80}{\pi^2} \approx 8.1°\text{C}.$$

Now let us consider two examples where the region of integration is not a rectangle.

Example 5.6   **Integrals over Nonrectangles**

A. A Calculation. Let us compute the double integral of $f(x, y) = xe^y$ over the region $\mathcal{R} = \{(x, y) : 0 \le x \le 1 \text{ and } x^2 - 1 \le y \le x\}$. (See Figure 5.12(a).)

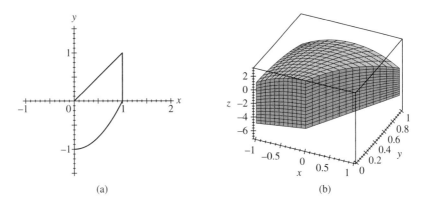

Figure 5.12 (a) The region of integration for Example 5.6A. (b) The region in space whose volume is computed in Example 5.6B.

We have

$$\int\int_{\mathcal{R}} f \, dA_{x,y} = \int_0^1 \int_{x^2-1}^x xe^y \, dy \, dx.$$

To evaluate the inner integral, we treat x as a constant and apply the fundamental theorem of calculus to xe^y as a function of y. In this case, we have

$$\int_{x^2-1}^x xe^y \, dy = xe^y \big|_{x^2-1}^x$$

$$= xe^x - xe^{x^2-1}.$$

Substituting this result into the iterated integral, we have

$$\int\int_{\mathcal{R}} f \, dA_{x,y} = \int_0^1 (xe^x - xe^{x^2-1}) \, dx$$

$$= (xe^x - e^x) - \frac{1}{2}e^{x^2-1} \bigg|_0^1$$

$$= \frac{1}{2} + \frac{1}{2e}.$$

B. Volume. Let us compute the volume of the region in space that lies over the trapezoid $\mathcal{R}$ defined by

$$\mathcal{R} = \{(x,y) : 0 \leq y \leq 1 \text{ and } -1 \leq x \leq y\}$$

and between the graphs of the functions $G(x,y) = x - 2y - 4$ and $H(x,y) = 3 - 2x^2 - 2y^2$. (See Figure 5.12(b).) The height of the region at the point (x,y) is $H(x,y) - G(x,y)$, so that the volume of the region is equal to $\int \int_R (H(x, y) - G(x, y)) \, dA_{x,y}$. In this case, it is natural to use the second formulation of Fubini's theorem, integrating first with respect to x and then with respect to y. We obtain the following:

$$\int_0^1 \int_{-1}^y (H(x, y) - G(x, y)) \, dx \, dy = \int_0^1 \int_{-1}^y ((3 - 2x^2 - 2y^2) - (x - 2y - 4)) \, dx \, dy$$

$$= \int_0^1 \left. -\frac{2}{3}x^3 - \frac{1}{2}x^2 + (7 + 2y - 2y^2)x \right|_{-1}^y \, dy$$

$$= \int_0^1 \left(-\frac{8}{3}y^3 - \frac{1}{2}y^2 + 9y + \frac{41}{6}\right) dy$$

$$= \left. -\frac{2}{3}y^4 - \frac{1}{6}y^3 + \frac{9}{2}y^2 + \frac{41}{6}y \right|_0^1$$

$$= \frac{21}{2}.$$

Notice in Example 5.6B that if the functions G and H were constant functions, say $G(x, y) = h_0$ and $H(x, y) = h_1$, where $h_0 < h_1$, then the region in space would be a slab of constant height $h_1 - h_0$. Its volume would be the height times the area of the base $\mathcal{R}$. In the general case, we can justify this by evaluating an iterated integral.

$$\int_a^b \int_{g(x)}^{h(x)} (H(x, y) - G(x, y)) \, dy \, dx = \int_a^b \int_{g(x)}^{h(x)} h_1 - h_0 \, dy \, dx$$

$$= (h_1 - h_0) \int_a^b \int_{g(x)}^{h(x)} 1 \, dy \, dx$$

$$= (h_1 - h_0) \int_a^b h(x) - g(x) \, dy.$$

The result is the height times the area of the region in the plane between the graphs of $y = g(x)$ and $y = h(x)$ and $x = a$ and $x = b$, that is, the area of the base of the slab. It follows that the area of a region in the plane is the double integral of the constant function $f(x, y) = 1$ over the region.

When evaluating iterated integrals, it might happen that the inner integral is straightforward to evaluate, whereas the resulting function might be difficult or even impossible to integrate symbolically, either by hand, or using tables, or using a symbolic integration routine on a computer. In this situation, we can apply a numerical routine on a computer. However, in some cases, "reversing" the order of integration can prove helpful. We demonstrate this in the following example.

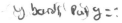

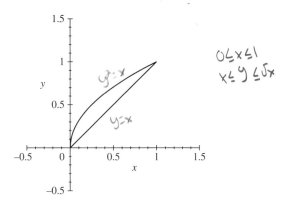

Figure 5.13   The region of integration for Example 5.7.

Example 5.7   **Reversing the Order of Integration**

A. Let $f(x, y) = \cos y$, and let $\mathcal{R}$ be the region in the first quadrant bounded by the line $y = x$ and the curve $y^2 = x$. (See Figure 5.13.) We describe the region as follows:

$$\mathcal{R} = \{(x, y) \: : \: 0 \le x \le 1, \, x \le y \le \sqrt{x} \, \}.$$

It is natural to compute the total accumulation of f as an iterated integral first with respect to y and then with respect to x. This integral is

$$\int\int_{\mathcal{R}} f \, dA_{x,y} = \int_0^1 \int_x^{\sqrt{x}} \cos y \, dy \, dx.$$

Evaluating the inner integral, we have

$$\int_x^{\sqrt{x}} \cos y \, dy = \sin(\sqrt{x}) - \sin x.$$

Substituting this result into the iterated integral we have,

$$\int\int_{\mathcal{R}} f \, dA_{x,y} = \int_0^1 \sin(\sqrt{x}) - \sin x \, dx.$$

The function $\sin(\sqrt{x})$ cannot be evaluated symbolically. However, using a numerical integration routine on a computer, we evaluated this integral to three decimal places to be 0.143.

B. If, however, we describe the region as

$$\mathcal{R} = \{(x, y) \: : \: 0 \le y \le 1, \, y^2 \le x \le y\},$$

it is more natural to represent the total accumulation as an iterated integral first with respect to x and then with respect to y. We then have

$$\int\int_{\mathcal{R}} f \, dA_{x,y} = \int_0^1 \int_{y^2}^{y} \cos y \, dx \, dy$$

$$= \int_0^1 x \cos y \big|_{y^2}^{y} \, dy$$

$$= \int_0^1 y \cos y - y^2 \cos y \, dy,$$

which can be evaluated using integration by parts. We obtain

$$(y \sin y + \cos y) - \left(y^2 \sin y + 2y \cos y - 2 \sin y\right)\big|_0^1 = 2\sin(1) - \cos(1) - 1 \approx 0.143.$$

Integration over More General Regions

There are regions in the plane that cannot be expressed in either form described in Fubini's theorem. That is, they cannot be written in the form

$$\{(x, y) \ : \ a \leq x \leq b \text{ and } g(x) \leq y \leq h(x)\}$$

or in the form

$$\{(x, y) \ : \ c \leq y \leq d \text{ and } g(y) \leq x \leq h(y)\}.$$

However, if the region $\mathcal{R}$ is not too complicated, it can be partitioned into a union of smaller regions that are of one of these two forms. The following theorem, which we will not prove, shows that we can then compute the total accumulation over $\mathcal{R}$ by summing the total accumulations over the subregions.

Theorem 5.4   Let f be a continuous function on an open set containing the region $\mathcal{R}$. If $\mathcal{R}_1, \mathcal{R}_2, \ldots, \mathcal{R}_n$ are nonoverlapping regions satisfying

$$\mathcal{R} = \mathcal{R}_1 \cup \mathcal{R}_2 \cup \ldots \cup \mathcal{R}_n,$$

then

$$\int\int_{\mathcal{R}} f \, dA_{x,y} = \int\int_{\mathcal{R}_1} f \, dA_{x,y} + \int\int_{\mathcal{R}_2} f \, dA_{x,y} + \cdots + \int\int_{\mathcal{R}_n} f \, dA_{x,y}. \; \blacklozenge$$

By nonoverlapping regions, we mean that two of the regions intersect at most along a common boundary curve. The following example should make this apparent.

Example 5.8

A Volume Calculation. Let us find the volume of the region in space that lies in the first octant between the cylinders of radius 1 and 2 and below the surface $z = 4 - x - y$. The volume of this region is the total accumulation of the height function $f(x, y) = 4 - x - y$ over the portion of the annulus $\mathcal{R} = \{(x, y) : 1 \leq x^2 + y^2 \leq 4\}$ lying in the first quadrant. (See Figure 5.14).

The region $\mathcal{R}$ cannot be written in either of the above forms, so we will subdivide it into two simpler domains as illustrated in Figure 5.14. We have

$$\mathcal{R}_1 = \{(x, y) \ : \ 0 \leq x \leq 1 \ , \ \sqrt{1 - x^2} \leq y \leq \sqrt{4 - x^2}\},$$
$$\mathcal{R}_2 = \{(x, y) \ : \ 1 \leq x \leq 2 \ , \ 0 \leq y \leq \sqrt{4 - x^2}\}.$$

Applying the previous theorem and writing the resulting integrals as iterated integrals, the volume of the region in space is equal to

$$\int_0^1 \int_{\sqrt{1-x^2}}^{\sqrt{4-x^2}} (4 - x - y) \, dy \, dx + \int_1^2 \int_0^{\sqrt{4-x^2}} (4 - x - y) \, dy \, dx.$$

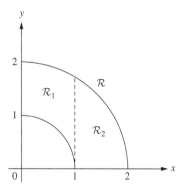

Figure 5.14 The domain of integration for Example 5.8 with $\mathcal{R} = \mathcal{R}_1 \cup \mathcal{R}_2$.

We will evaluate the second iterated integral and leave the first as an exercise.

$$\int_1^2 \int_0^{\sqrt{4-x^2}} (4 - x - y)\, dy\, dx = \int_1^2 4y - xy - y^2/2 \Big|_0^{\sqrt{4-x^2}} dx$$

$$= \int_1^2 \left((4-x)\sqrt{4-x^2} - (2 - x^2/2) \right) dx$$

$$= 2x\sqrt{4-x^2} + 8\arcsin(\frac{x}{2}) + \frac{(4-x^2)^{3/2}}{3} - 2x + \frac{x^3}{6} \Big|_1^2$$

$$= -\frac{5}{6} + \frac{8}{3}\pi - 3\sqrt{3},$$

where we used integration by parts to evaluate the first summand in the integral with respect to x. The value of the first integral is $-\frac{23}{6} + \frac{1}{3}\pi + 3\sqrt{3}$, so that the volume of the region in space is $3\pi - \frac{14}{3}$.

Summary

In this section, we began the formal presentation of *integration of functions of two variables*. We followed the corresponding ideas for integration in one variable. The goals were *to approximate the total accumulation of a function f on a region $\mathcal{R} \subset \mathbb{R}^2$, to produce an exact value for the total accumulation of f from these approximations*, and *to develop a method for computing this value*.

We started with regions $\mathcal{R}$ lying between the graphs of two functions $g = g(x)$ and $h = h(x)$ over an interval $a \le x \le b$. We *partitioned* $\mathcal{R}$ into M columns of N rectangular *cells*. We then *sampled* f in each cell and *scaled* the sampled value by multiplying by the area of its cell. We *summed* the scaled values to produce a *Riemann sum* approximation to the total accumulation of f on $\mathcal{R}$, which we denoted by $R(f, P)$. By taking a limit of Riemann sums as the *mesh* of the partition approached zero, we produced an exact value for the total accumulation, which is called the *integral* or *double integral* of f over $\mathcal{R}$. In symbols,

$$\lim_{\text{mesh}(P) \to 0} R(f, P) = \int\int_{\mathcal{R}} f\, dA_{x,y}.$$

We used **Fubini's theorem** to reduce the calculation of the double integral of f to an **iterated integral**:

$$\int\int_{\mathcal{R}} f \, dA_{x,y} = \int_a^b \left(\int_{g(x)}^{h(x)} f(x,y) \, dy \right) dx.$$

We proved Fubini's theorem by carefully rewriting the Riemann sum $R(f,P)$ as an **iterated sum** and applying ideas from one-variable integration. By interchanging the roles of x and y, we produced an iterated integral first with respect to x and then with respect to y. Finally, we showed how to extend these constructions to more general regions in the plane.

Section 5.2 Exercises

1. **Riemann Sum Calculations.** For each of the following functions f, regions $\mathcal{R}$, and values for M and N, (i) construct a partition $P_{M,N}$ of $\mathcal{R}$; (ii) sketch the domain, the partition, and a collection of sample points; and (iii) evaluate the corresponding Riemann sum approximation to the total accumulation of f on $\mathcal{R}$.

   (a) $f(x,y) = 1 + x - y$, $\mathcal{R} = [0,2] \times [0,1]$, and $M = 3$, $N = 4$.
   (b) $f(x,y) = x^2 - y^2$, $\mathcal{R} = \{(x,y) : -1 \leq x \leq 1 \text{ and } 0 \leq y \leq 1 - x^2\}$, and $M = 4$, $N = 4$.
   (c) $f(x,y) = \ln(y+1)$, $\mathcal{R} = \{(x,y) : 0 \leq y \leq 1 \text{ and } y^2 \leq x \leq y^2 + 1\}$, and $M = 5$, $N = 2$.

2. **Integrals over Rectangles.** Evaluate the following integrals.

   (a) $\int\int_{\mathcal{R}} x^3 y^2 \, dA_{x,y}$, where $\mathcal{R} = [-1,1] \times [0,5]$.
   (b) $\int\int_{\mathcal{R}} (x-y)^3 \, dA_{x,y}$, where $\mathcal{R} = [0,1] \times [1,2]$.
   (c) $\int\int_{\mathcal{R}} \cos(x+y) \, dA_{x,y}$, where $\mathcal{R} = [0,\pi/2] \times [0,\pi/2]$.
   (d) $\int\int_{\mathcal{R}} xy e^{x+y} \, dA_{x,y}$, where $\mathcal{R} = [0,2] \times [-2,0]$.

3. **Integrals over Nonrectangles.** Evaluate the following integrals.

   (a) $\int\int_{\mathcal{R}} (y-x) \, dA_{x,y}$, where $\mathcal{R} = \{(x,y) : 0 \leq x \leq 1, x^4 \leq y \leq x\}$.
   (b) $\int\int_{\mathcal{R}} (y+x) \, dA_{x,y}$, where $\mathcal{R} = \{(x,y) : 0 \leq y \leq 1, y \leq x \leq \sqrt{y}\}$.
   (c) $\int\int_{\mathcal{R}} e^{y/x} \, dA_{x,y}$, where $\mathcal{R} = \{(x,y) : 0 \leq x \leq 1, 0 \leq y \leq x^3\}$.
   (d) $\int\int_{\mathcal{R}} y \, dA_{x,y}$, where $\mathcal{R} = \{(x,y) : 0 \leq x \leq \pi, 0 \leq y \leq \sin(x)\}$.

4. **Iterated Integrals.** For each of the following iterated integrals, sketch the domain of integration and compute the integral.

(a) $\int_0^2 \int_{\sqrt{x}}^3 xy \, dy \, dx$.

(b) $\int_0^1 \int_1^{e^x} (x+y) \, dy \, dx$.

(c) $\int_0^\pi \int_0^{\sin(x)} y \sin x \, dy \, dx$.

(d) $\int_0^1 \int_0^y e^{y^2} \, dx \, dy$.

5. **Fubini's Theorem.** In the text, we indicated how to obtain the second form of Fubini's theorem, which holds for a region $\mathcal{R}$ of the form

$$\mathcal{R} = \{(x,y) \, : \, c \le y \le d \text{ and } g(y) \le x \le h(y)\}.$$

In this exercise, you are asked to fill in the steps of the derivation of Fubini's theorem for these regions. Assume f is continuous on an open set that contains $\mathcal{R}$.

(a) Sketch a representative domain of this form. Describe how you would partition this domain into $M \times N$ rectangles. What is the height and width of each cell in the partition?

(b) Write out the iterated Riemann sum to approximate the total accumulation of f on $\mathcal{R}$ that corresponds to this partition.

(c) Express the integral of f over $\mathcal{R}$ as an iterated limit of an iterated sum.

(d) Describe in detail how to derive an iterated integral for the total accumulation of f on $\mathcal{R}$ from the limit of part (b).

6. **Mesh(P).** In the discussion of Fubini's theorem, we claimed that if $\text{mesh}(P) \to 0$, then $M, N \to \infty$.

(a) Justify this claim.

(b) Show that the converse of this statement is false. That is, that $M, N \to \infty$ does not imply $\text{mesh}(P) \to 0$.

(c) If $\mathcal{R}$ is a rectangle, we say that a partition of P is regular if all of its cells have the same dimensions. Show that if the partitions P are regular, then $M, N \to \infty$ implies that $\text{mesh}(P) \to 0$.

7. **Product Functions.** Let $f(x,y) = g(x)h(y)$ where g and h are continuous functions of one variable. Let $\mathcal{R} = [a,b] \times [c,d]$.

(a) Why does the double integral of f exist on $\mathcal{R}$?

(b) Show that $\int \int_{\mathcal{R}} f \, dA_{x,y} = \left(\int_a^b g(x) \, dx \right) \left(\int_c^d h(y) \, dy \right)$.

8. **Volumes.** Compute the volume of each of the following regions of space.

(a) The region bounded above by the surface $z = 9 - x^2 - y^2$ and below by the xy-plane.

(b) The region bounded by the paraboloids $z = x^2 + 2y^2$ and $z = 12 - 2x^2 - y^2$.

(c) $\{(x,y,z) : \, 0 \le x \le \pi, \, -2 \le y \le 2, \text{ and } \sin(x)y^2 \le z \le xy^2\}$.

(d) $\{(x,y,z) : \, y + 2z - 4 \le x \le ye^{yz}, \, 0 \le y \le 2, \text{ and } 0 \le z \le 1\}$.

9. **Average Value.** The *average value* of a function on a region is defined to be the total accumulation of the function on the region divided by the area of the region. For each of the following functions f and regions $\mathcal{R}$, compute the average value of f over $\mathcal{R}$.

   (a) $f(x, y) = xe^y$, $\mathcal{R}$ is the region defined by $0 \le x \le 1$ and $0 \le y \le x$.
   (b) $f(x, y) = 3 - x - 2y$, $\mathcal{R}$ is the region defined by $0 \le y \le 2$ and $-y \le x \le y$.
   (c) $f(x, y) = 2xy$, $\mathcal{R}$ is the parallelogram with vertices $(0, 0)$, $(1, 1)$, $(0, 1)$, and $(1, 2)$.
   (d) $f(x, y) = x^2 - y^2$, $\mathcal{R}$ is the region defined by $-1 \le x \le 1$ and $x - 1 \le y \le x^2$.
   (e) $f(x, y) = \sqrt{1 + y^2}$, $\mathcal{R}$ is the region defined by $0 \le y \le 1$ and $0 \le x \le y$.

10. **Average Temperature.** Suppose that the region $\mathcal{R} = [0, L_1] \times [0, L_2] \times [0, \epsilon]$ represents a thin metal plate in space.

    (a) Find the average temperature of the plate if the temperature of the plate is given by

    $$T(x, y) = 30 \sin(\pi x / L_1) \sin(\pi y / L_2).$$

    (Notice that the temperature at a point does not depend on its z-coordinate.)
    (b) Find the average temperature of the plate if the temperature of the plate is given by

    $$T(x, y) = 30 \sin(\pi x / L_1) \sinh(\pi y / L_2),$$

    where sinh is the hyperbolic sine function, $\sinh(u) = \frac{1}{2}(e^u - e^{-u})$. (Notice that the temperature at a point does not depend on its z-coordinate.)

11. **Center of Mass.** The *center of mass* of an object or collection of objects plays an important role in determining the motion of the object. In this exercise, we will compute the center of mass for a plate or region in the plane. This definition will be given in terms of double integrals over the region. If a plate of varying density $\rho = \rho(x, y)$ is represented by a region $\mathcal{R}$ in the plane, the coordinates of the center of mass are given by

    $$\bar{x} = \frac{1}{M} \int \int_{\mathcal{R}} x\rho(x, y) \, dA_{x,y} \quad \text{and} \quad \bar{y} = \frac{1}{M} \int \int_{\mathcal{R}} y\rho(x, y) \, dA_{x,y},$$

    where $M = \int \int_{\mathcal{R}} \rho(x, y) dA_{x,y}$ is the total mass of the plate. These are "mass-weighted" averages of the x- and y-coordinates of points in $\mathcal{R}$. These expressions can be derived from the corresponding formulas for a finite collection of points in the plane. Find the center of mass of the following regions $\mathcal{R}$ with densities ρ.

    (a) $\mathcal{R}$ is the triangle with vertices $(1, 0)$, $(1, 1)$, and $(0, 1)$ and $\rho(x, y) = c$, a constant.
    (b) $\mathcal{R}$ is the square with vertices $(0, 0)$, $(1, 0)$, $(1, 1)$, and $(0, 1)$ and $\rho(x, y) = \cos x$.
    (c) $\mathcal{R}$ is half-disk of radius 1 centered at the origin with nonnegative x-coordinate and $\rho(x, y) = 2 - x^2 - y^2$.

■ 5.3 Integration in Polar Coordinates

In this section, we will adapt the techniques of Sections 5.1 and 5.2 to regions in the plane that have **rotational** or **circular symmetry**. In Section 5.1, we approximated the total intensity of the solar disk at the wavelength of 500 nm by partitioning the region into cells which were portions of disks and using this partition to construct a Riemann sum. In order to formalize partitions of this type and to be able to compute the limit of Riemann sums as we did in Section 5.2, we will introduce a new coordinate system in the plane: the **polar coordinate system**.

Polar Coordinates

The polar coordinate system provides an alternative method for representing a point in the plane by an ordered pair of real numbers. Let us begin with the usual Cartesian coordinate system in the plane that associates to each point P an ordered pair (x, y) that describes the location of the point relative to a horizontal and a vertical axis. Let θ denote the **angle that is measured counterclockwise from the positive x-axis to the line segment from the origin to** P. Recall that all angles are measured in radians. (See Figure 5.15(a).) If $x \neq 0$, $y/x = \tan\theta$ and $\theta = \arctan(y/x)$, where we must choose an appropriate branch of the arctan function. (See Exercise 1.) The **length of this line segment** is $\sqrt{x^2 + y^2}$, which we denote by r. We can express x and y in terms of r and θ,

$$x = r\cos\theta, \quad y = r\sin\theta.$$

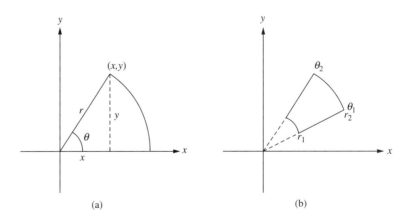

(a)   (b)

Figure 5.15   (a) The polar coordinate system. (b) The polar rectangle $\{(r, \theta) : r_1 \leq r \leq r_2 \text{ and } \theta_1 \leq \theta \leq \theta_2\}$.

The real numbers r and θ are called the **polar coordinates** of the point P, and we write them as an ordered pair, (r, θ). Note that since r is the distance from P to the origin, r is nonnegative, $r \geq 0$.

Conversely, this construction can be used to associate a point P in the plane to an ordered pair (r, θ), $r \geq 0$. The point P is the point whose rectangular coordinates are given by $(x, y) = (r \cos \theta, r \sin \theta)$. If $\theta > 0$, the angle θ is measured **counterclockwise** from the positive x-axis. If $\theta < 0$, the angle between the positive x-axis and the line segment from the origin to P has magnitude $|\theta|$ and is measured **clockwise** from the positive x-axis. As a consequence, the polar coordinates (r, θ) and $(r, \theta + 2k\pi)$ describe the same point in the plane for all integer values of k.

In the following example, we translate from the Cartesian coordinates of a point to the polar coordinates and vice-versa.

Example 5.9   **Polar Coordinates**

A.   Let $P = (4, 2)$. Then the polar coordinates of P are r and θ where $r = \sqrt{4^2 + 2^2} = 2\sqrt{5}$ and $\theta = \arctan(1/2)$. Since P lies in the first quadrant, we choose $\theta \approx 0.464$.

B.   Let $(r, \theta) = (2, -\pi/4)$. These are the polar coordinates of the point

$$P = (2 \cos(-\pi/4), 2 \sin(-\pi/4)) = (\sqrt{2}, -\sqrt{2}).$$

The sets that will be of most interest to us are **polar rectangles**, which are sets of the form

$$\{(r, \theta) : r_1 \leq r \leq r_2 \text{ and } \theta_1 \leq \theta \leq \theta_2\},$$

where r_1, r_2, θ_1, and θ_2 are constants. (See Figure 5.15(b).) The cells of our partition of the solar disk in Example 5.3 of Section 5.1 were of this form. In that case, the cells of the partition were $\{(r, \theta) : r_i \leq r \leq r_{i+1} \text{ and } 0 \leq \theta \leq 2\pi\}$. More generally, regions of this type will form the cells of our partitions in polar coordinates.

As we saw in the example in Section 5.1, we will need to calculate the area of a cell. Therefore, let us calculate the area of a polar rectangle $\{(r, \theta) : r_1 \leq r \leq r_2 \text{ and } \theta_1 \leq \theta \leq \theta_2\}$. Since the polar rectangle is a region between circles of radius r_1 and r_2 that spans an angle of $\theta_2 - \theta_1$, it has area equal to

$$(\pi r_2^2 - \pi r_1^2) \frac{(\theta_2 - \theta_1)}{2\pi} = (r_2 + r_1)(r_2 - r_1) \frac{(\theta_2 - \theta_1)}{2}.$$

We would like to rewrite this expression in a more convenient form. Let $\Delta\theta = \theta_2 - \theta_1$, $\Delta r = r_2 - r_1$, and $\bar{r} = (r_2 + r_1)/2$, the average of the radii r_1 and r_2. Then the **area of a polar rectangle** is equal to

$$\bar{r}\Delta r\Delta\theta.$$

Riemann Sums in Polar Coordinates

Let f be a function defined on a polar rectangle $\mathcal{R}$ of the form

$$\mathcal{R} = \{(r,\theta) : a \leq r \leq b \text{ and } \alpha \leq \theta \leq \beta\}.$$

Let us assume that the value for f at a point is expressed in terms of the polar coordinates of the point, $f = f(r,\theta)$. If f is given in Cartesian coordinates, we can change to polar coordinates by considering the function $f^*(r,\theta) = f(r\cos\theta, r\sin\theta)$. In order to approximate the total accumulation of f on $\mathcal{R}$, we will follow the procedure that we established in the previous sections. We will **partition** $\mathcal{R}$, **sample** the values of f in each cell of the partition, **scale** the sampled values by the areas of the cells, and then **sum** the scaled values over the partition. We observed in our earlier constructions that the resulting exact values for the total accumulation, that is, the corresponding integral, did not depend on any choices we made in the construction of the Riemann sums. Keeping this in mind, we will make our choices with an eye toward simplifying the algebra.

We will partition $\mathcal{R}$ into polar rectangles by choosing a collection of r values and a collection of θ values,

$$a = r_0 < r_1 < r_2 < \ldots < r_{M-1} < r_M = b,$$

and

$$\alpha = \theta_0 < \theta_1 < \theta_2 < \ldots < \theta_{N-1} < \theta_N = \beta.$$

Then we can construct a collection of $M \times N$ polar rectangles $\mathcal{R}_{ij}$, $1 \leq i \leq M$ and $1 \leq j \leq N$, of the form

$$\mathcal{R}_{ij} = \{(r,\theta) : r_{i-1} \leq r \leq r_i \text{ and } \theta_{j-1} \leq \theta \leq \theta_j\}.$$

(An example of such a partition is shown in Figure 5.16.) Let

$$\Delta r_i = r_i - r_{i-1}, \quad \Delta\theta_j = \theta_j - \theta_{j-1}, \text{ and } \bar{r}_i = (r_i + r_{i-1})/2,$$

so that the area of $\mathcal{R}_{ij}$ is equal to $\bar{r}_i\Delta r_i\Delta\theta_j$.

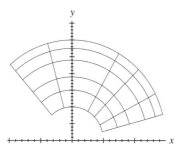

Figure 5.16   A partition of a polar rectangle into 5×5 cells, which are themselves polar rectangles.

We will choose sample points $P_{ij} = (r_i^*, \theta_j^*)$, $1 \le i \le M$ and $1 \le j \le N$, where

$$r_{i-1} \le r_i^* \le r_i \text{ and } \theta_{j-1} \le \theta_j^* \le \theta_j.$$

For convenience, we will choose $r_i^* = \overline{r}_i$. We can then approximate the accumulation of f on $\mathcal{R}_{ij}$ by the scaled value

$$f(P_{ij})\text{Area}(\mathcal{R}_{ij}) = f(r_i^*, \theta_j^*) r_i^* \Delta r_i \Delta \theta_j.$$

Summing over i and j, we obtain the following Riemann sum to approximate the total accumulation of f on $\mathcal{R}$:

$$R(f, P) = \sum_{i,j=1}^{M,N} f(r_i^*, \theta_j^*) r_i^* \Delta r_i \Delta \theta_j.$$

We can obtain an exact value for the total accumulation of f on $\mathcal{R}$ by taking the limit of $R(f, P)$ as the mesh of the partition approaches 0. In this case, the mesh of the partition is the maximum of the values of Δr_i and $\Delta \theta_j$. If this limit exists and is independent of the choices of partition and sample points, we call it the **_polar coordinate integral_** of f on $\mathcal{R}$. We denote this limit by

$$\int\int_{\mathcal{R}} f \, dA_{r,\theta}.$$

We shall not prove it here, but this value for the total accumulation of f on $\mathcal{R}$ is the same as the value we would obtain by using Euclidean coordinates to construct the Riemann sum.

Theorem 5.5 Let $f = f(x, y)$ be a continuous function on an open set containing the region $\mathcal{R}$ described in polar coordinates by

$$\mathcal{R} = \{(r, \theta) : a \leq r \leq b \text{ and } \alpha \leq \theta \leq \beta\}.$$

Let f^* be the function f described in polar coordinates, that is, $f^*(r, \theta) = f(r \cos \theta, r \sin \theta)$. Then

$$\int\int_{\mathcal{R}} f \, dA_{x,y} = \int\int_{\mathcal{R}} f^* \, dA_{r,\theta}. \quad \blacklozenge$$

Fubini's Theorem

As we did for double integrals in Section 5.2, we would like to equate the polar coordinate integral of f with an iterated integral of f with respect to r and θ. The Riemann sum that approximates the total accumulation of f is

$$R(f, P) = \sum_{i,j=1}^{M,N} f(r_i^*, \theta_j^*) r_i^* \Delta r \Delta \theta.$$

The total accumulation of f is the limit of this sum as $\text{mesh}(P)$ approaches 0. Thus the total accumulation of f is

$$\lim_{\text{mesh}(P) \to 0} \sum_{i,j=1}^{M,N} f(r_i^*, \theta_j^*) r_i^* \Delta r \Delta \theta.$$

This is the limit of a Riemann sum for the function $r f(r, \theta)$ on the region $[a, b] \times [\alpha, \beta]$ in the $r\theta$-coordinate plane. Since f is a continuous function of r and θ and r is continuous, $r f(r, \theta)$ is a continuous function. Thus we can apply Fubini's theorem for functions of two variables. This gives us the following version of Fubini's theorem.

Theorem 5.6 **Fubini's Theorem in Polar Coordinates.** Let f be a continuous function on an open set containing the region $\mathcal{R}$ described in polar coordinates by

$$\mathcal{R} = \{(r, \theta) : a \leq r \leq b \text{ and } \alpha \leq \theta \leq \beta\}.$$

Then the integral of f over $\mathcal{R}$ is equal to iterated integrals of f with respect to r and θ as follows:

$$\int\int_{\mathcal{R}} f \, dA_{r,\theta} = \int_{\alpha}^{\beta} \left(\int_{a}^{b} f(r, \theta) r \, dr \right) d\theta = \int_{a}^{b} \left(\int_{\alpha}^{\beta} f(r, \theta) r \, d\theta \right) dr. \quad \blacklozenge$$

In the following examples, we will use Fubini's theorem to compute the volume of a region in space and the total intensity of the sun at the wavelength 500 nm.

Example 5.10 **Applying Fubini's Theorem**

A. Volume. A region in space lies inside a cylinder of radius 2 centered on the z-axis, above the xy-plane, and under the graph of $f(x, y) = 3 - \sqrt{x^2 + y^2}$. (See Figure 5.17.) Because of the rotational symmetry of the region, it is advantageous to use polar coordinates to compute its volume. The base of the cylinder is a circle of radius 2 centered at the origin, which is the set

$$\{(r, \theta) : 0 \leq r \leq 2 \text{ and } 0 \leq \theta \leq 2\pi\}.$$

In order to carry out the calculation, we must express the height of the region over the base in terms of r and θ rather than x and y. Since $x = r \cos \theta$ and $y = r \sin \theta$, the height is given by

$$f(r \cos \theta, r \sin \theta) = 3 - \sqrt{(r \cos \theta)^2 + (r \sin \theta)^2}$$
$$= 3 - \sqrt{r^2}$$
$$= 3 - r,$$

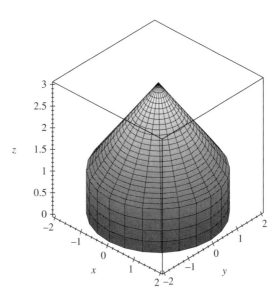

Figure 5.17 The solid described in Example 5.10A.

since $r \geq 0$. Thus the volume of the region is given by

$$\int_0^{2\pi} \left(\int_0^2 (3 - r)\, r\, dr \right) d\theta = \int_0^{2\pi} (10/3)\, d\theta = 20\pi/3.$$

B. Total Intensity. In Example 5.3 of Section 5.1, we used a Riemann sum to approximate the total intensity of the sun at the wavelength 500 nm. The intensity of the sun at a given wavelength can be modeled by the function I defined by

$$I(x, y) = I_0 \left[(1 - a - b) + a\sqrt{1 - \frac{x^2 + y^2}{R_\odot^2}} + b \left(1 - \frac{x^2 + y^2}{R_\odot^2} \right) \right],$$

where I_0 denotes the intensity at the center of the solar disk, $R_\odot$ denotes the radius of the sun, and a and b are constants that depend on the wavelength of the intensity function. At a wavelength of 500 nm, which is near the peak of the solar spectrum, $a = 0.97$ and $b = -0.22$. Using these values of a and b and $r^2 = x^2 + y^2$, we have

$$I(r, \theta) = I_0 \left(0.25 + 0.97\sqrt{1 - \frac{r^2}{R_\odot^2}} - 0.22 \left(1 - \frac{r^2}{R_\odot^2} \right) \right).$$

The solar disk can be described in polar coordinates as $\{(r, \theta) : 0 \leq r \leq R_\odot, 0 \leq \theta \leq 2\pi\}$. Thus the total accumulation of intensity at the wavelength 500 nm is

$$\int_0^{2\pi} \int_0^{R_\odot} I_0 \left(0.25 + 0.97\sqrt{1 - \frac{r^2}{R_\odot^2}} - 0.22 \left(1 - \frac{r^2}{R_\odot^2} \right) \right) r\, dr\, d\theta$$

$$= I_0 \int_0^{2\pi} \left(0.25\frac{r^2}{2} - \frac{0.97 R_\odot^2}{3} (1 - \frac{r^2}{R_\odot^2})^{3/2} + \frac{0.22 R_\odot^2}{4} (1 - \frac{r^2}{R_\odot^2})^2 \right) \Bigg|_0^{R_\odot} d\theta$$

$$= 2\pi I_0 R_\odot^2 (0.5033) \approx 3.163 I_0 R_\odot^2.$$

Summary

In this section, we introduced *polar coordinates* in the plane in order to take advantage of *circular* or *rotational symmetry* when evaluating planar integrals.

The *polar coordinates* (r, θ) and *Cartesian coordinates* of a point in the plane are related by the equations $x = r\cos\theta$ and $y = r\sin\theta$, where the *radial coordinate* r is the distance from (x, y) to the origin. The *angular coordinate* θ is the angle in radians measured counterclockwise from the positive x-axis to the radial segment from $(0, 0)$ to (x, y).

To approximate the total accumulation of a function $f = f(r, \theta)$ on a disk or a portion of a disk centered at the origin, it is natural to use polar coordinates. We *partitioned* these regions into *polar rectangles*, *sampled* f in each cell, *scaled* the sampled value by the area of its polar rectangle, and *summed* the scaled values to produce a *polar* Riemann sum. The limit of polar Riemann sums as the mesh of the partition approaches 0 is the *polar integral* of f over the region.

The polar integral is equal to the ordinary double integral over the region. Applying *Fubini's theorem* to the polar integral allowed us to rewrite the polar integral as an iterated integral in polar coordinates, that is,

$$\int\int_{\mathcal{R}} f\, dA_{r,\theta} = \int_{\alpha}^{\beta} \left(\int_{a}^{b} f(r, \theta)\, r\, dr \right) d\theta,$$

where $\mathcal{R}$ is defined by $\alpha \leq \theta \leq \beta$ and $a \leq r \leq b$. The iterated integral can be written in the reverse order. We applied polar integration to calculate the volume of a rotationally symmetric region and to calculate total solar intensity at fixed wavelength.

Section 5.3 Exercises

1. **Arctan.** In order to determine a value for $\arctan(y/x)$, it is necessary to specify a branch of the arctan function. The branches can be distinguished by their values at the origin. Which branches of the arctan function should be used to determine $\arctan(y/x)$ if we want arctan to take values in $[0, 2\pi)$? in $(-2\pi, 0]$? in $(-\pi, \pi]$? In particular, which branches do we use in each quadrant? Explain your answer.

2. **Polar Coordinates**

   (a) Find all possible polar coordinate representations of the points $(1, 1)$, $(0, -3)$, $(-2, 1)$, and $(-1, 5)$.

   (b) Find the Cartesian coordinates of the points given in polar coordinates by $(1, \pi/4)$, $(3, -5\pi/3)$, $(0, \pi/9)$, and $(2, 1)$.

3. **Polar Regions I.** Give a symbolic description for each of the following regions in the plane using polar coordinates and set notation.

   (a) The second quadrant including the positive y-axis and negative x-axis.
   (b) $\{(x, y) : x^2 + y^2 \geq 4\}$.
   (c) Inside a circle of radius 4 and outside a circle of radius 2.
   (d) Inside a circle of radius 3 in the first quadrant.
   (e) $\{(x, y) : y \geq |x|\}$.
   (f) $\{(x, y) : y \leq 2x \text{ and } x \leq 0\}$.

4. **Polar Regions II.** Describe in words each of the following regions in the plane.

   (a) $\{(r, \theta) : r \geq 2 \text{ and } 0 \leq \theta \leq \pi/2\}$.
   (b) $\{(r, \theta) : r \leq 1 \text{ and } -\pi/2 \leq \theta \leq \pi\}$.
   (c) $\{(r, \theta) : 1 \leq r \leq 5 \text{ and } |\theta| \leq \pi/2\}$.
   (d) $\{(r, \theta) : r = 4 \text{ and } \pi/4 \leq \theta \leq 5\pi/4\}$.
   (e) $\{(r, \theta) : \theta = \pi/6 \text{ or } \theta = 5\pi/6\}$.
   (f) $\{(r, \theta) : r = \theta \text{ and } \theta \geq 0\}$.

5. **Converting to Polar Coordinates.** Express each of the following functions f in terms of polar coordinates.

   (a) $f(x, y) = x^2 + y^2$.
   (b) $f(x, y) = \sqrt{x^2 + y^2}$.
   (c) $f(x, y) = e^{-x^2 - y^2}$.
   (d) $f(x, y) = y/x$.
   (e) $f(x, y) = y/(x^2 + y^2)$.
   (f) $f(x, y) = x/\sqrt{x^2 + y^2}$.

6. **Total Accumulation.** Find the total accumulation of the function f on the given region in the plane.

   (a) $f(x, y) = 1 - x^2 - y^2$ on a disk of radius 1 centered at the origin.
   (b) $f(x, y) = x$ in the first quadrant inside a circle of radius 2.
   (c) $f(x, y) = 1 + xy$ in the upper half-plane inside a circle of radius 1.
   (d) $f(x, y) = x^2 - y^2$ on the region $\{(x, y) : x \geq |y| \text{ and } x \leq 2\}$.

7. **Volumes.** Use a polar coordinate integral to find the volume of each of the following regions in space. (*Hint:* Position the region so as to make the integral as simple as possible.)

   (a) A hemisphere of radius of 2.
   (b) A quarter-cylinder of height 4 and radius 3.
   (c) The region inside a cone with base having radius 1 and with height 1.

8. **Polar vs. Rectangular Integrals.** For each of the following functions f and regions $\mathcal{R}$, compute the total accumulation of f on $\mathcal{R}$ using both Cartesian and polar coordinates.

   (a) $f(x, y) = y^2$, $\mathcal{R}$ is the top half of a disk of radius 2 centered at the origin.
   (b) $f(x, y) = y - x$, $\mathcal{R}$ is the region inside a circle of radius 3 and lying in the second and third quadrants.

9. **Fluid in a Pipe.** When a real fluid moves through a pipe, the fluid moves slower near the walls of the pipe than at the center of the pipe, and the layer of fluid in immediate contact with the walls of the pipe adheres to the pipe and has zero velocity. This is an example of what is known as a Poiseuille[3] flow. (See Example 2.22 of Section 2.4.) If the rate of flow of the fluid is constant, the speed of the fluid at a point in the pipe is in fact a function of the distance of the point from the center of the pipe. If the pipe has radius r_0, the velocity of the fluid at a distance r from the center of the pipe is given by

$$v(r) = v_0(1 - \frac{r^2}{r_0^2}),$$

where v_0 is the velocity of the fluid at the center of the pipe. Find the total volume of fluid that passes a given cross section of the pipe in one unit of time.

■ 5.4 Integration of Functions of Three Variables

In this section, we want to develop the concept of the integral for functions of three variables. That is, we want to develop a method for computing an exact value for the total accumulation of a function on a region of space. Our approach will be analogous to the one that we used in Section 5.2 for functions of two variables. To produce a Riemann sum approximation to the total accumulation, we will partition the region into three-dimensional cells, sample the function in each cell, scale the sampled values by the volume of the cells, and sum the scaled values. The integral of the function on the region will be the limit of Riemann sum approximations to the total accumulation. As in Section 5.2, the primary result of this section will be Fubini's theorem, which will allow us to express the integral of a function by a triple iterated integral. We begin by fixing the type of region we will examine and establishing our notation for a partition.

Let us assume that $\mathcal{R}$ is a subset of the domain of a function f and that f is continuous on $\mathcal{R}$. Initially, we will also assume that $\mathcal{R}$ is the region between the graphs of the continuous functions $h_1 = h_1(x, y)$ and $h_2 = h_2(x, y)$, and inside the vertical column whose horizontal cross section $\tilde{\mathcal{R}}$ in the xy-plane is a region of the type specified in Fubini's theorem for functions of two variables. We will use the shorthand terminology "$\mathcal{R}$ lies over $\tilde{\mathcal{R}}$" to describe regions of this type, although $\mathcal{R}$ might intersect or lie below the xy-plane. For now, let us assume that $\tilde{\mathcal{R}}$ is given by

$$\tilde{\mathcal{R}} = \{(x, y) : a \leq x \leq b \text{ and } g_1(x) \leq y \leq g_2(x)\},$$

---

[3]J. L. M. Poiseuille (1797–1869) was a French physiologist and physician who studied the flow of blood.

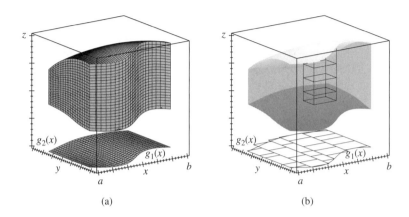

(a)   (b)

Figure 5.18   (a) A typical domain of integration of the form $\mathcal{R} = \{(x, y, z) : a \leq x \leq b,$ $g_1(x) \leq y \leq g_2(x),$ and $h_1(x, y) \leq z \leq h_2(x, y)\}$, which lies over the region $\tilde{\mathcal{R}} = \{(x, y) : a \leq x \leq b$ and $g_1(x) \leq y \leq g_2(x)\}$ in the xy-plane. (b) A cutaway view of a partition $\tilde{P}$ of $\tilde{\mathcal{R}}$ and one column of a partition P of $\mathcal{R}$. Notice that the vertical faces of $\mathcal{R}$ closest to the xz-plane and closest to the yz-plane are not shown.

where g_1 and g_2 are continuous functions of one variable. (See Figure 5.18(a).) As was the case in two variables, our construction will also apply to regions whose descriptions are obtained by interchanging the roles of x, y, and z.

We will partition $\mathcal{R}$ into $L \times M \times N$ cells, with $L \times M$ columns of cells with N cells in each column. The columns will lie over the $L \times M$ cells of a partition $\tilde{P}$ of $\tilde{\mathcal{R}}$ of the type we introduced in Section 5.2. (See Figure 5.18(b).) Thus we must first set up the partition $\tilde{P}$ of $\tilde{\mathcal{R}}$. We choose $L + 1$ values x_i, $i = 0, \ldots, L$, with

$$a = x_0 < x_1 < x_2 < \ldots < x_{L-1} < x_L = b,$$

and L points x_i^* with $x_{i-1} \leq x_i^* \leq x_i$. Then we choose $M + 1$ values y_{ij} with

$$g_1(x_i^*) = y_{i0} < y_{i1} < y_{i2} < \ldots < y_{iM} = g_2(x_i^*).$$

The ij^{th} cell of the partition $\tilde{P}$ of $\tilde{\mathcal{R}}$, $\tilde{\mathcal{R}}_{ij}$, is defined by

$$\tilde{\mathcal{R}}_{ij} = [x_{i-1}, x_i] \times [y_{i(j-1)}, y_{ij}],$$

where $1 \leq i \leq L$ and $1 \leq j \leq M$.

Each cell $\tilde{\mathcal{R}}_{ij}$ corresponds to a column of the partition P of $\mathcal{R}$, which we must partition into three-dimensional rectangular cells. For each cell $\tilde{\mathcal{R}}_{ij}$, we choose values y_{ij}^* so that

$(x_i^*, y_{ij}^*) \in \tilde{\mathcal{R}}_{ij}$. We use these points to determine the heights of the columns of P. For each column, we choose N points z_{ijk} so that

$$h_1(x_i^*, y_{ij}^*) = z_{ij0} < z_{ij1} < z_{ij2} < \ldots < z_{ijN} = h_2(x_i^*, y_{ij}^*).$$

We define the ijk^{th} cell of the partition P to be the rectangular box

$$\mathcal{R}_{ijk} = [x_{i-1}, x_i] \times [y_{i(j-1)}, y_{ij}] \times [z_{ij(k-1)}, z_{ijk}].$$

The volume of $\mathcal{R}_{ijk}$ is

$$\text{Vol}(\mathcal{R}_{ijk}) = \Delta x_i \Delta y_{ij} \Delta z_{ijk},$$

where $\Delta x_i = x_i - x_{i-1}$, $\Delta y_{ij} = y_{ij} - y_{i(j-1)}$, and $\Delta z_{ijk} = z_{ijk} - z_{ij(k-1)}$. The **mesh** of the partition P is the largest of the values Δx_i, Δy_{ij}, and Δz_{ijk}.

We sample f on $\mathcal{R}$ by choosing one point P_{ijk} from each cell $\mathcal{R}_{ijk}$ of P. That is, we choose points $P_{ijk} = (x_{ijk}^*, y_{ijk}^*, z_{ijk}^*)$, where

$$x_{i-1} \leq x_{ijk}^* \leq x_i, \quad y_{i(j-1)} \leq y_{ijk}^* \leq y_{ij}, \text{ and } z_{ij(k-1)} \leq z_{ijk}^* \leq z_{ijk}.$$

We approximate the accumulation of f on $\mathcal{R}_{ijk}$ by the scaled value

$$f(P_{ijk})\text{Vol}(\mathcal{R}_{ijk}) = f(x_{ijk}^*, y_{ijk}^*, z_{ijk}^*)\Delta x_i \Delta y_{ij} \Delta z_{ijk}.$$

We approximate the total accumulation of f on $\mathcal{R}$ by the Riemann sum

$$R(f, P) = \sum_{i,j,k=1}^{L,M,N} f(P_{ijk})\text{Vol}(\mathcal{R}_{ijk}) = \sum_{i,j,k=1}^{L,M,N} f(x_{ijk}^*, y_{ijk}^*, z_{ijk}^*)\Delta x_i \Delta y_{ij} \Delta z_{ijk}.$$

As before, this sum represents the sum of the values $f(P_{ijk})\text{Vol}(\mathcal{R}_{ijk})$ for all the cells in the partition P, that is, for all combinations of the indices i, j, and k where i runs from 1 to L, j runs from 1 to M, and k runs from 1 to N.

If, in the definition of $\mathcal{R}$, the roles of x, y, and z had been interchanged, the geometry of the partition must change accordingly. For example, if, instead of the above description, $\mathcal{R}$ were the region between the graphs of $h_1 = h_1(x, z)$ and $h_2 = h_2(x, z)$ "over" the region $\tilde{\mathcal{R}}$ in the xz-plane described by $a \leq z \leq b$ and $g_1(z) \leq x \leq g_2(z)$, the partition P would consist of $L \times M$ columns parallel to the y-axis and consisting of N cells. Since there are three ways to enclose $\mathcal{R}$ between the graphs of functions of two variables over a region $\tilde{\mathcal{R}}$ in a coordinate plane, and there are two ways to describe $\tilde{\mathcal{R}}$ as the region between the graphs of two functions of one variable, there are six possible descriptions of $\mathcal{R}$. We will use the same notation, $R(f, P)$, for the Riemann sum in each case.

As in Section 5.2, we obtain an exact value for the total accumulation of f on $\mathcal{R}$ by taking the limit of Riemann sums for partitions P as $\operatorname{mesh}(P) \to 0$. We want the limit to be independent of the choices we made in constructing the Riemann sums. It can be proven that if f is continuous, then this will be the case.

Theorem 5.7   Let f be continuous on an open set containing a region $\mathcal{R}$, which is one of the six types described above. Then

$$\lim_{\operatorname{mesh}(P) \to 0} R(f, P)$$

exists and is independent of any choices that are made in constructing this limit.   ◆

If the limit of the Riemann sums exists, we denote it by the symbol

$$\int \int \int_{\mathcal{R}} f \, dV_{x,y,z}.$$

We call this value the ***integral*** or ***triple integral*** of f on $\mathcal{R}$.

Fubini's Theorem

In space, Fubini's theorem expresses a triple integral as an iterated triple integral, which will allow us to employ symbolic techniques from one-variable calculus to evaluate triple integrals. We will employ the procedure that we used in Section 5.2. We will rewrite a Riemann sum as an iterated sum, reorder the limits and sums in the definition of the integral, identify the inner limit as a single integral, and then apply Fubini's theorem in the plane to the resulting expression.

Let us assume that f is continuous on an open set containing the region $\mathcal{R}$ given by

$$\mathcal{R} = \{(x, y, z) : a \le x \le b, \ g_1(x) \le y \le g_2(x), \ \text{and} \ h_1(x, y) \le z \le h_2(x, y)\},$$

where g_1 and g_2 are continuous functions of one variable and h_1 and h_2 are continuous functions of two variables. Since f is continuous on $\mathcal{R}$, the integral of f over $\mathcal{R}$ is independent of our choice of sampling scheme. We will take advantage of this fact in the following way: within each column of a partition P, we will sample f along a line parallel to the z-axis. That is, within the ij^{th} column, we will sample f at points with the same x- and y-coordinates, x_i^* and y_{ij}^*.

The resulting Riemann sum for f on the region $\mathcal{R}$ for a partition consisting of $L \times M \times N$ cells is given by

$$R(f, P) = \sum_{i,j,k=1}^{L,M,N} f(x_i^*, y_{ij}^*, z_{ijk}^*) \Delta x_i \Delta y_{ij} \Delta z_{ijk}.$$

We will rewrite this as an ***iterated sum*** where the inner sum is over the index k, which represents summing over the cells of a column of P to find an approximation to the total accumulation on a column,

$$R(f, P) = \sum_{i,j=1}^{L,M} \left(\sum_{k=1}^{N} f(x_i^*, y_{ij}^*, z_{ijk}^*) \Delta x_i \Delta y_{ij} \Delta z_{ijk} \right).$$

This expression instructs us to do the following: For all combinations of the indices i and j, sum the values of $f(x_i^*, y_{ij}^*, z_{ijk}^*) \Delta x_i \Delta y_{ij} \Delta z_{ijk}$ with k ranging from 1 to N, then sum the resulting totals with i ranging from 1 to L and j ranging from 1 to M. Note, since Δy_{ij} and Δx_i do not depend on k, we will be able to factor them out of the inner sum.

We will again rewrite the limit of $R(f, P)$ as mesh$(P) \to 0$ as an iterated limit. Here we need only separate the limit as $N \to \infty$. Keeping in mind that the mesh must still approach 0, we have

$$\int \int \int_{\mathcal{R}} f \, dV_{x,y,z} = \lim_{L,M,N \to \infty} R(f, P)$$

$$= \lim_{L,M \to \infty} \left(\lim_{N \to \infty} R(f, P) \right).$$

Substituting the iterated form of the Riemann sum into the iterated limit, we are able to interchange the order of the inner limit, which is taken with respect to N, and the outer sum, which is indexed by L and M:

$$\lim_{L,M \to \infty} \left(\lim_{N \to \infty} \sum_{i,j=1}^{L,M} \left(\sum_{k=1}^{N} f(x_i^*, y_{ij}^*, z_{ijk}^*) \Delta z_{ijk} \right) \Delta x_i \Delta y_{ij} \right)$$

$$= \lim_{L,M \to \infty} \sum_{i,j=1}^{L,M} \left(\lim_{N \to \infty} \sum_{k=1}^{N} f(x_i^*, y_{ij}^*, z_{ijk}^*) \Delta z_{ijk} \right) \Delta x_i \Delta y_{ij}.$$

Note that we also factored Δx_i and Δy_{ij} out of the inner limit since they do not depend on N.

As before, we recognize the inner limit as a limit of Riemann sums for a function $\tilde{f}$ of one variable. In this case, the function $\tilde{f} = \tilde{f}(z)$ is given by given by $\tilde{f}(z) = f(x_i^*, y_{ij}^*, z)$ and it is defined on the interval $[h_1(x_i^*, y_{ij}^*), h_2(x_i^*, y_{ij}^*)]$. Since f is a continuous function, it is a continuous function of z when x and y are held fixed. Thus $\tilde{f}$ is also a continuous function of z. It follows that the inner limit exists and is equal to the integral of $\tilde{f}$ on $[h_1(x_i^*, y_{ij}^*), h_2(x_i^*, y_{ij}^*)]$,

$$\lim_{N \to \infty} \sum_{k=1}^{N} f(x_i^*, y_{ij}^*, z_{ijk}^*) \Delta z_{ijk} = \int_{h_1(x_i^*, y_{ij}^*)}^{h_2(x_i^*, y_{ij}^*)} \tilde{f}(z) \, dz$$

$$= \int_{h_1(x_i^*, y_{ij}^*)}^{h_2(x_i^*, y_{ij}^*)} f(x_i^*, y_{ij}^*, z) \, dz.$$

Notice that this integral depends on x_i^* and y_{ij}^*. Thus we will write it as a function G of x and y, so that

$$G(x_i^*, y_{ij}^*) = \int_{h_1(x_i^*, y_{ij}^*)}^{h_2(x_i^*, y_{ij}^*)} f(x_i^*, y_{ij}^*, z) \, dz.$$

Since f, h_1, and h_2 are continuous, G is a continuous function of x and y.

Returning to our iterated limit, it now takes a form familiar to us from our work in Section 5.2:

$$\lim_{L, M \to \infty} \sum_{i,j=1}^{L,M} G(x_i^*, y_{ij}^*) \Delta x_i \Delta y_{ij}.$$

This a limit of Riemann sums for G on the region $\tilde{\mathcal{R}}$ in the xy-plane. Since G is continuous, this is equal to the integral of G over $\tilde{\mathcal{R}}$,

$$\iint_{\tilde{\mathcal{R}}} G \, dA_{x,y},$$

which, by Fubini's theorem for functions of two variables, is equal to the iterated integral

$$\int_a^b \left(\int_{g_1(x)}^{g_2(x)} G(x, y) dy \right) dx.$$

Combining this with the definition of G, we obtain a triple iterated integral for the integral of f on $\mathcal{R}$

$$\int\int\int_{\mathcal{R}} f\, dV_{x,y,z} = \int_a^b \left(\int_{g_1(x)}^{g_2(x)} G(x,y)dy \right) dx$$

$$= \int_a^b \left(\int_{g_1(x)}^{g_2(x)} \left(\int_{h_1(x,y)}^{h_2(x,y)} f(x,y,z)\, dz \right) dy \right) dx.$$

If the roles of x, y, and z in the description of $\mathcal{R}$ had been interchanged, we would obtain an iterated integral in which the order of integration and the limits of integration were also interchanged. For example, if $\mathcal{R}$ is the region between the graphs of $h_1 = h_1(x,z)$ and $h_2 = h_2(x,z)$ "over" the region $\tilde{\mathcal{R}}$ in the xz-plane described by $a \leq z \leq b$ and $g_1(z) \leq x \leq g_2$, we would have

$$\int\int\int_{\mathcal{R}} f\, dV_{x,y,z} = \int_a^b \left(\int_{g_1(z)}^{g_2(z)} \left(\int_{h_1(x,z)}^{h_2(x,z)} f(x,y,z)\, dy \right) dx \right) dz.$$

There are six possible orderings of the iterated integrals. Each ordering corresponds to one of the possible descriptions of $\mathcal{R}$ as the region between the graphs of continuous functions of two variables and lying over a Fubini-type region in the plane. These will be investigated in Exercise 5. Without elaborating on all the possible forms, let us state Fubini's theorem.

Theorem 5.8 **Fubini's Theorem in Three Variables.** Let f be a continuous function on an open set containing the region $\mathcal{R}$ in space. If $\mathcal{R}$ is the region lying between the graphs of continuous functions h_1 and h_2 of two variables and over the region $\tilde{\mathcal{R}}$ in coordinate plane of the type described in Fubini's theorem for functions of two variables, then the integral of f over $\mathcal{R}$ is equal to a triple iterated integral of f. For example, if

$$\mathcal{R} = \{(x,y,z) : a \leq x \leq b,\ g_1(x) \leq y \leq g_2(x),\ h_1(x,y) \leq z \leq h_2(x,y)\,\},$$

then

$$\int\int\int_{\mathcal{R}} f\, dV_{x,y,z} = \int_a^b \int_{g_1(x)}^{g_2(x)} \int_{h_1(x,y)}^{h_2(x,y)} f(x,y,z)\, dz\, dy\, dx. \quad \blacklozenge$$

Once again, this theorem will be our primary tool for computing the triple integral of a function by symbolic means. The calculations are analogous to those for functions of two variables and regions in the plane. However, we should keep in mind that the geometry of regions in space is more complicated than that of regions in the plane and that choosing the order of integration for the iterated integral is an essential part of evaluating an integral. The above construction gives a intuitive insight into this choice. In order to rewrite the Riemann sum as an iterated sum, it was important that we sampled the function along lines parallel to the z-axis. We chose the z-axis because the region $\mathcal{R}$ was described as the region between the graphs of two functions of x and y, the remaining variables. The parallel "sample" lines pass through each of these graphs. The inner variable of integration corresponds to the direction of these lines.

   In the following examples, we focus on describing the region of integration and setting up an appropriate iterated integral. We leave the evaluation of the iterated integrals to the reader.

Example 5.11

Setting Up a Triple Integral. Let us find the total accumulation of $f(x, y, z) = z$ on the region $\mathcal{R}$ bounded by a half-cylinder of radius r_0 lying in the upper half-space whose axis coincides with the x-axis and that extends from $x = -2$ to $x = 2$. (See Figure 5.19.) Thus $\mathcal{R}$ lies inside the cylinder $\{(x, y, z) : y^2 + z^2 = r_0^2 \}$ and

$$\mathcal{R} = \{(x, y, z) : y^2 + z^2 \le r_0^2, \ -2 \le x \le 2, \ \text{and} \ z \ge 0\}.$$

We will set up two different iterated integrals for the total accumulation of f on $\mathcal{R}$.

A. We can think of $\mathcal{R}$ as lying over the rectangle $\tilde{\mathcal{R}} = [-2, 2] \times [-r_0, r_0]$ in the xy-plane and between xy-plane and the curved surface of the cylinder. The z-coordinates of points in the region satisfy $0 \le z \le \sqrt{r_0^2 - y^2}$. Were we to partition the region and construct a Riemann sum, we would use columns of cells parallel to the z-axis and sample f on lines

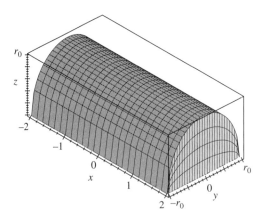

Figure 5.19   The region for Example 5.11, $\mathcal{R} = \{(x, y, z) : y^2 + z^2 \le r_0^2, \ -2 \le x \le 2, \ \text{and} \ z \ge 0\}$.

parallel to the z-axis. This description gives an iterated integral first with respect to z and then with respect to y and x.

$$\int_{-2}^{2} \int_{-r_0}^{r_0} \int_{0}^{\sqrt{r_0^2 - y^2}} z \, dz \, dy \, dx.$$

B. If we think of the region as lying over the half-disk in the yz-plane given by $y^2 + z^2 \le r_0^2$ and $z \ge 0$, and between the planes $x = -2$ and $x = 2$, we would sample f along lines parallel to the x-axis. In the corresponding iterated integral, we first integrate with respect to x. Since the values for z on the boundary of disk depend on y, we next integrate with respect to z. The iterated integral is

$$\int_{-r_0}^{r_0} \int_{0}^{\sqrt{r_0^2 - y^2}} \int_{-2}^{2} z \, dx \, dz \, dy.$$

Now let us consider an example where the sampling direction is evident, but the region in the coordinate plane is not.

Example 5.12   **Describing a Region.** Let us set up an iterated integral for the total accumulation of a function f on the region $\mathcal{R}$ that lies above the graph of the function $g(x,y) = x^2 + y^2$ and below the graph of the function $h(x,y) = 2 - x^2 + y^2$. (See Figure 5.20(a).) Since the region lies between the graphs of two functions of x and y, it is natural to think of

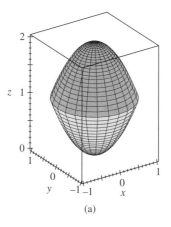
(a)

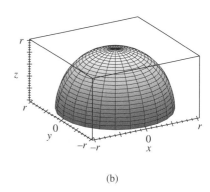
(b)

Figure 5.20   (a) The region $\mathcal{R}$ that lies above the graph of the function $g(x,y) = x^2 + y^2$ and below the graph of the function $h(x,y) = 2 - x^2 + y^2$. (b) The region $\mathcal{R}_0$ between the sphere $x^2 + y^2 + z^2 = r_0^2$ and the xy-plane.

sampling the region along lines parallel to the z-axis and to set up an iterated integral where the inner integral is with respect to z and with

$$x^2 + y^2 \leq z \leq 2 - x^2 - y^2.$$

The difficulty here is to determine the bounds in the xy-plane for the x- and y-coordinates of points in $\mathcal{R}$. We notice from the plot of the region (see Figure 5.20(a)) that the two graphs intersect in a curve. Since points on this curve must lie on both graphs, the z-coordinates of these points must equal both $x^2 + y^2$ and $2 - x^2 - y^2$. It follows that $x^2 + y^2 = 2 - x^2 - y^2$. Simplifying this expression, we see that $x^2 + y^2 = 1$, which describes a circle of radius 1 centered at the origin. Thus the x and y coordinates of points in $\mathcal{R}$ must satisfy $x^2 + y^2 \leq 1$. It follows that $\mathcal{R}$ is given by

$$\mathcal{R} = \{(x, y, z) : -1 \leq x \leq 1, \ -\sqrt{1 - x^2} \leq y \leq \sqrt{1 - x^2}, \ \text{and} \ x^2 + y^2 \leq z \leq 2 - x^2 - y^2\}.$$

The total accumulation of f on $\mathcal{R}$ is then given by the iterated integral

$$\int_{-1}^{1} \int_{-\sqrt{1-y^2}}^{\sqrt{1-y^2}} \int_{x^2+y^2}^{2-x^2-y^2} f(x, y, z) \, dz \, dy \, dx.$$

In many applications, the region over which we must integrate is a portion of a sphere. In the following example, we compute an average density of a substance in the atmosphere after it is released from a source on the ground.

Example 5.13 **Average Density.** If a substance is released into the atmosphere at an instant in time, it will, in the absence of wind, diffuse through the surrounding air in all directions. For example, we might think of the release of an airborne pollutant that is considered toxic above a certain threshold concentration. Intuitively, as the substance diffuses, the region of space where the concentration of the substance is above the threshold level will increase. Eventually, however, as the substance dissipates, the concentration will fall below the threshold level. If we take the point of release to be the origin, the concentration or density of the substance at a point $\mathbf{x}$ at time t can be modeled by the function

$$c(\mathbf{x}, t) = \frac{2M}{(4\pi\delta t)^{3/2}} e^{-(x^2+y^2+z^2)/(4\delta t)},$$

where M is the amount of chemical released and δ is the molecular diffusivity of the medium. For air, $\delta = 0.2 \ \text{cm}^2/\text{s}$.

Notice that since c is a function of the distance from $\mathbf{x}$ to the origin, it is spherically symmetric and the level sets of c for fixed values of t are hemispheres. (We are only concerned with $z \geq 0$.) Thus it is reasonable to determine the average concentration at a fixed time t_0 on the region $\mathcal{R}_0$ inside a hemisphere of radius r_0. (See Figure 5.20(b).)

The average concentration is the quotient of the total accumulation of c at time t_0 and the volume of the hemisphere $\mathcal{R}_0$, that is,

$$\frac{1}{\frac{2}{3}\pi r_0^3} \int \int \int_{\mathcal{R}_0} c(x, y, z, t_0)\, dV_{x,y,z}.$$

We want to set up an iterated integral for this integral. We can describe $\mathcal{R}_0$ as the region between the sphere $x^2 + y^2 + z^2 = r_0^2$ and the xy-plane. Thus $0 \le z \le \sqrt{r_0^2 - x^2 - y^2}$, and we think of sampling c on lines parallel to the z-axis. The hemisphere lies over the disk $x^2 + y^2 = r_0^2$ in the plane, which can be described by $-r_0 \le x \le r_0$ and $-\sqrt{r_0^2 - x^2} \le y \le \sqrt{r_0^2 - x^2}$. Thus the average concentration is equal to

$$\frac{1}{\frac{2}{3}\pi r_0^3} \int_{-r_0}^{r_0} \int_{-\sqrt{r_0^2 - x^2}}^{\sqrt{r_0^2 - x^2}} \int_0^{\sqrt{r_0^2 - x^2 - y^2}} \frac{2M}{(4\pi\delta t_0)^{3/2}} e^{-(x^2 + y^2 + z^2)/(4\delta t_0)}\, dz\, dy\, dx.$$

This integral cannot be evaluated by symbolic methods in Cartesian coordinates, but it can be evaluated by a numerical method on a computer if we assign numerical values to M, δ, and t_0. We explore this example further in Exercise 18(d) of Section 5.7.

Summary

In this section, we extended the concept of integration to functions of three variables. Once again our goals were to **define and compute the total accumulation** of a function $f = f(x, y, z)$ on a region $\mathcal{R} \subset \mathbb{R}^3$.

We started by considering regions $\mathcal{R}$ that lie **between the graphs of two functions** $h_1 = h_1(x, y)$ and $h_2 = h_2(x, y)$ that lie **over a region $\tilde{\mathcal{R}}$ in the xy-plane**. We assumed that $\tilde{\mathcal{R}}$ was a planar region lying between the graphs of two functions of one variable of the type we considered in Section 5.2. We **partitioned** $\mathcal{R}$ into $L \times M$ rectangular columns each containing N rectangular cells.

We **sampled** f in each cell of the partition, **scaled** the sampled values by the volume of the corresponding cells, and **summed** the scaled values to produce a **Riemann sum** approximation to the total accumulation of f on $\mathcal{R}$. An exact value for the total accumulation was obtained by taking the limit of Riemann sums as the mesh of the partition approaches 0. This limit is the **integral** of f over $\mathcal{R}$.

We evaluated the integral of f over $\mathcal{R}$ by applying Fubini's theorem in three variables, which replaces the triple integral by a ***triple iterated integral***. Depending on the geometry of $\mathcal{R}$, the six possible orderings of x, y, and z potentially give rise to six different iterated integrals for the integral of a function over a region. For example, if $\mathcal{R}$ is described by $a \leq x \leq b$, $g_1(x) \leq y \leq g_2(x)$, and $h_1(x,y) \leq z \leq h_2(x,y)$,

$$\int\int\int_{\mathcal{R}} f \, dV_{x,y,z} = \int_a^b \left(\int_{g_1(x)}^{g_2(x)} \left(\int_{h_1(x,y)}^{h_2(x,y)} f(x,y,z) \, dz \right) dy \right) dx.$$

Finally, we considered several examples of triple iterated integrals.

Section 5.4 Exercises

1. **Iterated Integrals on Boxes.** Evaluate the following iterated integrals.

   (a) $\int_{-1}^2 \int_{-1}^0 \int_0^1 (x^3 y - z) \, dz \, dy \, dx$.

   (b) $\int_{-1}^1 \int_{-1}^1 \int_0^1 x^5 y^6 z \, dz \, dy \, dx$.

   (c) $\int_1^3 \int_2^4 \int_{-1}^1 (xyz)^2 \, dx \, dy \, dz$.

   (d) $\int_0^2 \int_{-2}^0 \int_{-2}^0 x\sqrt{y + zxy} \, dz \, dx \, dy$.

2. **Integrals from the Text**

   (a) Evaluate the iterated integral that was set up in Example 5.11A.

   (b) Evaluate the iterated integral that was set up in Example 5.11B.

3. **Iterated Integrals on More General Regions.** For each of the following iterated integrals, sketch the region of integration and evaluate the iterated integral.

   (a) $\int_0^1 \int_0^1 \int_{x+y}^{3+2x-y} xy \, dz \, dy \, dx$.

   (b) $\int_{-1}^1 \int_{-1}^1 \int_{-y^2-z^2}^{1+y^2+z^2} z \, dx \, dz \, dy$.

   (c) $\int_0^1 \int_0^{x^2} \int_{-2}^{x+2y} z \, dz \, dy \, dx$.

   (d) $\int_{-1}^1 \int_{x^2-1}^{1-x^2} \int_{x^2+y^2-2}^3 x^2 \, dz \, dy \, dx$.

   (e) $\int_0^2 \int_y^{y+1} \int_{z+y}^{z+y+1} zy \, dx \, dz \, dy$.

   (f) $\int_{-1}^1 \int_{-z}^{z+2} \int_{x^2-2}^4 y \, dy \, dx \, dz$.

4. **Evaluating Triple Integrals.** For each of the following, sketch the region $\mathcal{R}$ and set up and evaluate an iterated integral for the total accumulation of f on $\mathcal{R}$.

   (a) $f(x,y,z) = x^2 - yz$ and $\mathcal{R} = [0,1] \times [-1,1] \times [0,2]$.

   (b) $f(x,y,z) = \sin(x)\cos(y)z$ and $\mathcal{R} = [0,\pi] \times [-\pi/2, \pi/2] \times [-1,0]$.

   (c) $f(x,y,z) = xy$ and $\mathcal{R} = \{(x,y,z) : -1 \leq x \leq 1, \ 0 \leq y \leq 1, \text{ and } y - 1 \leq z \leq x + 1 \}$.

   (d) $f(x,y,z) = x - y + z$ and $\mathcal{R} = \{(x,y,z) : -1 \leq x \leq 1, \ -x^2 - 1 \leq y \leq x^2 + 1, \text{ and } -1 \leq z \leq 1 \}$.

   (e) $f(x,y,z) = 3ze^x$ and $\mathcal{R} = \{(x,y,z) : -1 \leq x \leq 1, \ 0 \leq y \leq x + 2, \text{ and } x - y - 1 \leq z \leq x - y + 1 \}$.

5. **A Geometric Criterion.** In order to set up a Riemann sum for the total accumulation of a function on a region, we required that the region $\mathcal{R}$ be described as a region that

lies between the graphs of continuous functions h_1 and h_2 of two variables, and over the region $\tilde{\mathcal{R}}$ that is contained in a coordinate plane and lies between the graphs of continuous functions g_1 and g_2 of one variable.

(a) Give a geometric criterion that must be met in order for a region to be described in this way.
(b) Explain why the criterion guarantees that the region is of this form.
(c) Give an example of a region that fails to meet the criterion for a particular ordering of the variables.
(d) Give an example of a region that fails to meet the criterion for all orderings of the variables.

6. **Find an Iterated Integral.** For each of the following regions $\mathcal{R}$, find a single iterated integral equal to $\int \int \int_{\mathcal{R}} f dV_{x,y,z}$. (Do not evaluate any integrals in this problem.)

(a) $\mathcal{R}$ is the solid tetrahedron with vertices $(1,0,0)$, $(0,1,0)$, $(0,0,1)$, and $(0,0,0)$.
(b) $\mathcal{R}$ is the solid triangular prism with vertices $(\pm 1, 0, 0)$, $(\pm 1, 1, 0)$, $(0, 0, 1)$, and $(0, 1, 1)$.
(c) $\mathcal{R}$ is the region satisfying $y^2 + z^2 - x^2 \leq 1$ and $0 \leq x \leq 1$.
(d) $\mathcal{R}$ is the region satisfying $x^2 - y^2 + z^2 \leq 1$ and $-1 \leq y \leq 1$.

7. **More Complicated Regions.** As was the case in the plane, there are regions in space that cannot be represented in one of the six forms covered by Fubini's theorem in space. That is, there are regions in space for which we cannot express the total accumulation of a function on the region as an iterated integral. However, it will be possible to subdivide a region into regions of the type described in Fubini's theorem. If $\mathcal{R}$ is a union of nonoverlapping regions and f is continuous on $\mathcal{R}$, then one can prove that

$$\int \int \int_{\mathcal{R}} f \, dV_{x,y,z} = \int \int \int_{\mathcal{R}_1} f \, dV_{x,y,z} + \int \int \int_{\mathcal{R}_2} f \, dV_{x,y,z} + \cdots$$
$$+ \int \int \int_{\mathcal{R}_n} f \, dV_{x,y,z}.$$

Note that this is a statement about integrals, not iterated integrals. In choosing the regions $\mathcal{R}_i$, keep in mind that we want to evaluate the integrals on the right-hand side of the above equation by applying Fubini's theorem, but that we may need to use different orders of integration for these integrals.

For each of the following regions $\mathcal{R}$, explain why $\int \int \int_{\mathcal{R}} f dV_{x,y,z}$ cannot be written as a single iterated integral, write $\mathcal{R}$ as a union of regions $\mathcal{R}_i$ so that $\int \int \int_{\mathcal{R}_i} f dV_{x,y,z}$ can be written as an iterated integral, and write out these iterated integrals. (It may help to sketch the regions $\mathcal{R}$ and $\mathcal{R}_i$.)

(a) $\mathcal{R}$ is the solid tetrahedron with vertices $(1,1,1)$, $(-1,-1,1)$, $(1,-1,-1)$, and $(-1,1,-1)$.
(b) $\mathcal{R}$ is the solid octahedron with vertices $(\pm 1, 0, 0)$, $(0, \pm 1, 0)$, and $(0, 0, \pm 1)$.

(c) $\mathcal{R}$ is the region inside a sphere of radius 2 centered at the origin and outside a cylinder of radius 1 centered on the z-axis.

(d) $\mathcal{R}$ is the region between spheres of radius 1 and radius 2 centered at the origin.

8. **Center of Mass.** If an object in space is modeled by a density function on a region of space, the center of mass of the object can be expressed in terms of the integrals of its density function. (See Exercise 11 of Section 5.2.) If the object occupies the region $\mathcal{R}$ and density $\rho = \rho(x, y, z)$, the **center of mass** of the object is the point $\bar{\mathbf{x}} = (\bar{x}, \bar{y}, \bar{z})$ with coordinates

$$\bar{x} = \frac{1}{M} \int \int \int_{\mathcal{R}} x\rho(x, y, z) \, dV_{x,y,z}, \ \ \bar{y} = \frac{1}{M} \int \int \int_{\mathcal{R}} y\rho(x, y, z) \, dV_{x,y,z}, \ \text{and}$$

$$\bar{z} = \frac{1}{M} \int \int \int_{\mathcal{R}} z\rho(x, y, z) \, dV_{x,y,z},$$

where $M = \int \int \int_{\mathcal{R}} \rho(x, y, z) dV_{x,y}$ is the total mass of the object. These are "mass-weighted" averages of the coordinates of points in $\mathcal{R}$. These expressions can be derived from the corresponding formulas for a finite collection of points in the plane. Find the center of mass of each of the following regions in space.

(a) $\mathcal{R}$ is the region lying above the xy-plane and below the graph of $f(x, y) = 1 - x^2 - y^2$ with density $\rho(x, y, z) = z$.

(b) $\mathcal{R}$ is the cone with vertex $(0, 0, 2)$ and base the unit circle in the xy-plane with density $\rho(x, y, z) = 2$.

(c) $\mathcal{R}$ is the cube centered at the origin with edges of length 2 and faces aligned with the coordinate planes with density $\rho(x, y, z) = x + 1$.

(d) $\mathcal{R}$ is the tetrahedron with vertices $(0, 0, 0)$, $(1, 0, 0)$, $(0, 1, 0)$, and $(0, 0, 1)$ with density $\rho(x, y, z) = 1$.

9. **Average Temperature.** The temperature at time t in the rectangular region of space $\mathcal{R} = [0, L_1] \times [0, L_2] \times [0, L_3]$ is given by

$$T(x, y, z, t) = 20 + \sin(\pi x/L_1) \sin(\pi y/L_2) \sin(\pi z/L_3) e^{-\lambda K t},$$

where $\lambda = (\frac{\pi}{L_1})^2 + (\frac{\pi}{L_2})^2 + (\frac{\pi}{L_3})^2$ and K is a constant. Find the average temperature in the region when $t = 2$.

10. **Average Concentration.** In Example 5.13, we introduced a model for the diffusion of an airborne substance released from a point on the ground in the absence of wind. In the presence of wind, we will consider a different scenario and consequently a different

model. Here let us suppose that the source located at the origin emits a substance at a constant rate Q and that the wind is blowing in the positive x direction. Over time, the distribution of the substance in the atmosphere will reach a steady state, so that the concentration at a point will be independent of time. Under these circumstances, the concentration of the substance can be modeled by the function S,

$$c(\mathbf{x}) = \frac{2Q}{\pi v k_y k_z x^{7/4}} e^{-(1/x^{1/4})(y^2/k_y^2 + z^2/k_z^2)},$$

where k_y and k_z are empirically determined constants. In this case, the function c is not spherically symmetric. Suppose that we are concerned with the average concentration in the region of space $\mathcal{R}$ where the concentration is above a threshold, $c(\mathbf{x}) \geq c_0$.

(a) Use the equation $c(\mathbf{x}) = c_0$ to express the upper surface of the region $\mathcal{R}$ as the graph of a function of x and y.
(b) Where does the graph in part (a) intersect the xy-plane?
(c) Express the total accumulation of S over $\mathcal{R}$ as an iterated integral.
(d) Express the average concentration of S over $\mathcal{R}$ as a quotient of integrals. What is the integral in the denominator?

■ 5.5 Integration in Cylindrical Coordinates

In this section, we would like to introduce a non-Cartesian coordinate system in space that is an extension of the polar coordinate system in the plane. Using this coordinate system, we will be able to give simple symbolic descriptions of subsets of cylinders that will prove useful when evaluating integrals over these subsets. More importantly, this coordinate system is an essential tool in the construction and analysis of mathematical models of physical systems that display rotational symmetry. We say that a system has **rotational symmetry** if its behavior at a point in space depends on the distance of the point from a line or **axis of rotation**. Typically the coordinates are chosen so the axis of rotation is one of the Cartesian coordinate axes. Let us consider a physical example of this type of symmetry.

Example 5.14   **Potential Energy of a Dipole.** Previously we have considered the potential energy function U of a system of three charged particles consisting of a dipole fixed in space and a particle free to move in space. Intuitively, we do not expect the potential energy of the system to change if the free particle revolves about the axis of the dipole. We can verify this mathematically by choosing a coordinate system with the z-axis as the axis of the dipole and the origin midway between the particles of the dipole. Thus the dipole consists of a particle of charge q_1 located at $(0, 0, d/2)$ and a particle of charge q_2 located at $(0, 0, -d/2)$. If the free particle has charge q_3 and is located at $\mathbf{x}$, the potential energy is given by

$$U(\mathbf{x}) = \frac{q_1 q_3}{4\pi\epsilon_0 r_1(\mathbf{x})} + \frac{q_2 q_3}{4\pi\epsilon_0 r_2(\mathbf{x})} + \frac{q_1 q_2}{4\pi\epsilon_0 d},$$

where $r_1(\mathbf{x}) = \sqrt{x^2 + y^2 + (z - d/2)^2}$ is the distance between the free particle and the end of the dipole located at $(0, 0, d/2)$ and $r_2(\mathbf{x}) = \sqrt{x^2 + y^2 + (z + d/2)^2}$ is the distance between the free particle and the end of the dipole located at $(0, 0, -d/2)$. From the expressions for U, r_1, and r_2, we can see that the potential energy is unchanged if the free particle is revolved about the z-axis; that is, if the z-coordinate remains fixed and the x- and y-coordinates change so as to leave fixed the distance to the z-axis, $\sqrt{x^2 + y^2}$.

Cylindrical Coordinates

The coordinate system we will use to describe regions that are rotationally symmetric is the **cylindrical coordinate system**. In this coordinate system, we replace the xy-coordinates of a point by the corresponding polar coordinates. Here r is the distance from the point (x, y, z) to the z-axis, and θ is the angle between the positive xz-plane and the plane through the point (x, y, z) that contains the z-axis. Thus we have the following definition.

Definition 5.1 The **cylindrical coordinates** (r, θ, z) of a point (x, y, z) in space satisfy $x = r\cos\theta$, $y = r\sin\theta$, and $z = z$, where $r \geq 0$. ◆

It follows from this definition that $r = \sqrt{x^2 + y^2}$ and $\tan\theta = y/x$. Thus the coordinates r and θ form the polar representation of the point (x, y). The **radial coordinate** r is the perpendicular distance from the point (x, y, z) to the z-axis, that is, the distance from the point (x, y, z) to the point $(0, 0, z)$. The **angular coordinate** θ is the angle between the half-plane with $y = 0$ and $x \geq 0$ and the plane containing the z-axis and the point (x, y, z). (See Figure 5.21.)

As was the case for Cartesian coordinates, it will be important to understand sets that are described by fixing one or two cylindrical coordinates. We describe these in the following example.

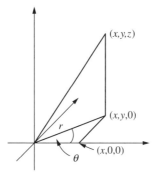

Figure 5.21   The cylindrical coordinates of the point (x, y, z) are (r, θ, z), where $x = r \cos \theta$ and $y = r \sin \theta$.

Example 5.15   **Subsets Described in Cylindrical Coordinates.** In each case that follows, k is constant.

A.   The set of points $\{(r, \theta, z) : r = k \text{ and } k > 0\}$ consists of points that are k units from the z-axis. This is a cylinder of radius k with central axis along the z-axis. (See Figure 5.22(a).)

B.   The set of points $\{(r, \theta, z) : \theta = k \}$ is a half-plane that contains the z-axis and forms an angle k with the half-plane, $y = 0$ and $x \geq 0$. (See Figure 5.22(b).)

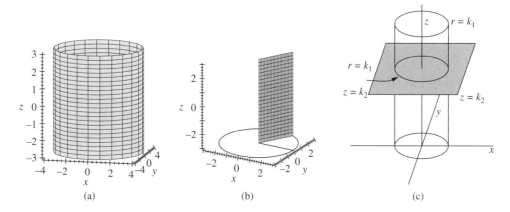

Figure 5.22   (a) The set of points satisfying $r = k$ in cylindrical coordinates. (b) The set of points satisfying $\theta = k$ in cylindrical coordinates. (c) The set of points $r = k_1$ and $z = k_2$ is the intersection of the cylinder $r = k_1$ with the horizontal plane $z = k_2$.

C. The set of points $\{(r, \theta, z) : z = k\}$ is a plane that is parallel to the xy-plane and contains the point $(0, 0, k)$. (See Figure 5.22(c).)

In using cylindrical coordinates for integration, we will be particularly interested in *cylindrical boxes*, which we will use to partition regions when constructing cylindrical Riemann sums.

Example 5.16   **Cylindrical Boxes**

A. Description of a Cylindrical Box. A *cylindrical box* is a region in space described in cylindrical coordinates by inequalities $0 \le r_1 \le r \le r_2$, $\theta_1 \le \theta \le \theta_2$, and $z_1 \le z \le z_2$, where r_1, r_2, θ_1, θ_2, z_1, and z_2 are constants. For example, the inequalities $1 \le r \le 2$, $0 \le \theta \le \pi/2$, and $0 \le z \le 3$ describe the region that lies in the first octant between the cylinder of radius 1 and the cylinder of radius 2 centered on the z-axis and below the plane $z = 3$. (See Figure 5.23.)

B. Volume of a Cylindrical Box. Since the cells of our cylindrical partitions will be cylindrical boxes, we need to know the volume of a cylindrical box. The volume of a cylindrical box is the product of the area of the base and the height. If the cylindrical box is given by

$$\{(r, \theta, z) : r_1 \le r \le r_2,\ \theta_1 \le \theta \le \theta_2,\ \text{and}\ z_1 \le z \le z_2\},$$

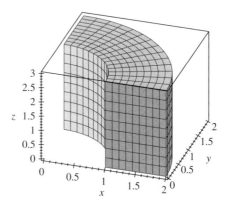

Figure 5.23   The region in the first octant between the cylinder of radius 1 and the cylinder of radius 2 and below the plane $z = 3$. See Example 5.16A.

the area of the base is $\frac{1}{2}(r_2^2 - r_1^2)(\theta_2 - \theta_1)$ and the ***volume of the cylindrical box*** is

$$\frac{1}{2}(r_2^2 - r_1^2)(\theta_2 - \theta_1)(z_2 - z_1) = \overline{r}\Delta r\Delta\theta\Delta z,$$

where $\overline{r} = (r_2 + r_1)/2$, $\Delta r = r_2 - r_1$, $\Delta\theta = \theta_2 - \theta_1$, and $\Delta z = z_2 - z_1$.

Riemann Sums in Cylindrical Coordinates

Let $\mathcal{R}$ be a region in space that lies over a polar rectangle $\tilde{\mathcal{R}}$. In cylindrical coordinates, $\mathcal{R} = \{(r, \theta, z) : a \leq r \leq b, \ \alpha \leq \theta \leq \beta, \ \text{and} \ h_1(r\cos\theta, r\sin\theta) \leq z \leq h_2(r\cos\theta, r\sin\theta)\}$. Let f be a function that is continuous on an open set containing $\mathcal{R}$. We will assume that f is expressed in cylindrical coordinates, $f = f(r, \theta, z)$. If f is expressed in Cartesian coordinates, the cylindrical coordinate expression for f is obtained from the Cartesian coordinate expression by $f^*(r, \theta, z) = f(r\cos\theta, r\sin\theta, z)$.

We will give the outline of how to construct a partition P of $\mathcal{R}$ consisting of $L \times M$ columns of N cylindrical boxes here. You are asked to complete the details in Exercise 6. The partition P consists of columns of cylindrical boxes that lie over a polar partition $\tilde{P}$ of the region $\tilde{\mathcal{R}}$ in the xy-plane defined by $\tilde{\mathcal{R}} = \{(r, \theta) : a \leq r \leq b, \ \alpha \leq \theta \leq \beta\}$. The partition $\tilde{P}$ will be constructed as we did in Section 5.3. From Example 5.16(b), we know that the volume of the ijk^{th} cell is

$$\text{Vol}(\mathcal{R}_{ijk}) = \overline{r}_i\Delta r_i\Delta\theta_j\Delta z_k,$$

where $\overline{r}_i$ is the average radius in the cell and Δr_i, $\Delta\theta_j$, and Δz_k are the change in radius, angle, and height in the cell. To approximate the accumulation of f in the cell $\mathcal{R}_{ijk}$, we will sample f at a point $P_{ijk} = (r_{ijk}^*, \theta_{ijk}^*, z_{ijk}^*)$. We have observed in our earlier constructions that the resulting exact value for the total accumulation, that is, the value of the integral, does not depend on any choices we made in the construction of the Riemann sums. With this in mind, we will sample f along lines parallel to the z-axis. We will choose $r_{ijk}^* = r_i^* = \overline{r}_i$, knowing that $\overline{r}_i$ occurs in the formula for the volume of $\mathcal{R}_{ijk}$, and $\theta_{ijk}^* = \theta_j^*$. Thus the sampled value for f in $\mathcal{R}_{ijk}$ is $f(r_i^*, \theta_j^*, z_{ijk}^*)$. Then we will approximate the accumulation of f in $\mathcal{R}_{ijk}$ by the scaled value

$$f(P_{ijk})\text{Vol}(\mathcal{R}_{ijk}) = f(r_i^*, \theta_j^*, z_{ijk}^*)\ r_i^*\Delta r_i\Delta\theta_j\Delta z_{ijk}.$$

Finally, the ***Riemann sum*** approximation for the total accumulation of f on $\mathcal{R}$ is the triply indexed sum

$$R(f, P) = \sum_{i,j,k=1}^{L,M,N} f(P_{ijk})\text{Vol}(\mathcal{R}_{ijk})$$

$$= \sum_{i,j,k=1}^{L,M,N} f(r_i^*, \theta_j^*, z_{ijk}^*)\ r_i^*\Delta r_i\Delta\theta_j\Delta z_{ijk}.$$

We can obtain an exact value for the total accumulation by taking the limit of $R(f, P)$ as the mesh of the partition approaches 0. If this limit exists and is independent of the choices of partition and sampling points, we call it the **cylindrical coordinate integral** of f over $\mathcal{R}$. We denote this limit by

$$\int \int \int_{\mathcal{R}} f \, dV_{r,\theta,z}.$$

The value for the total accumulation of f on $\mathcal{R}$ that results from the cylindrical coordinate integral is equal to the value we would obtain using Cartesian coordinates. We state this result below, but we will not prove it here.

Theorem 5.9 Let $f = f(x, y, z)$ be a continuous function on an open set containing the region $\mathcal{R}$ described in cylindrical coordinates by

$$\mathcal{R} = \{(r, \theta, z) : a \leq r \leq b, \ \alpha \leq \theta \leq \beta, \text{ and } h_1(r, \theta) \leq z \leq h_2(r, \theta) \}.$$

Let f^* be the cylindrical coordinate expression for f, that is, $f^*(r, \theta, z) = f(r \cos \theta, r \sin \theta, z)$. Then

$$\int \int \int_{\mathcal{R}} f^* \, dV_{r,\theta,z} = \int \int \int_{\mathcal{R}} f \, dV_{x,y,z}. \quad \blacklozenge$$

If we compare the above Riemann sum to the form of the Cartesian coordinate Riemann sums that we used previously, we see that it takes the form of a Riemann sum for the function $f(r, \theta, z) \cdot r$ on the region $\mathcal{R}$. Thus we are in a position to apply Fubini's theorem for functions of three variables. In this case, it says that the limit of cylindrical Riemann sums is equal to an iterated integral of $f(r, \theta, z) \cdot r$. We state this as a theorem.

Theorem 5.10 **Fubini's Theorem in Cylindrical Coordinates.** Let f be a continuous function on an open set containing the region $\mathcal{R} = \{(r, \theta, z) : a \leq r \leq b, \ \alpha \leq \theta \leq \beta, \text{ and } h_1(r, \theta) \leq z \leq h_2(r, \theta)\}$. Then

$$\int \int \int_{\mathcal{R}} f \, dV_{r,\theta,z} = \int_a^b \int_\alpha^\beta \int_{h_1(r,\theta)}^{h_2(r,\theta)} f(r, \theta, z) r \, dz \, d\theta \, dr. \quad \blacklozenge$$

The following example illustrates this result.

Example 5.17 **Cylindrical Integrals**

A. Let $f(x, y, z) = y$, and let $\mathcal{R}$ be the region bounded by the half-cylinder of radius r_0 lying over the first and second quadrants in the xy-plane, the xz-plane, and the

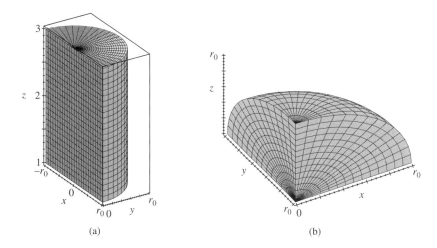

Figure 5.24 (a) The region $\mathcal{R}$ of Example 5.17A. (b) The region $\mathcal{R}$ of Example 5.17B.

planes $z = 1$ and $z = 3$. (See Figure 5.24(a).) In cylindrical coordinates, $f(r, \theta, z) = f(r\cos\theta, r\sin\theta, z) = r\sin\theta$, and

$$\mathcal{R} = \{(r, \theta, z) : 0 \le r \le r_0,\ 0 \le \theta \le \pi,\ \text{and } 1 \le z \le 3\ \}.$$

The total accumulation of f over $\mathcal{R}$ is

$$\int\int\int_{\mathcal{R}} f\, dV_{r,\theta,z} = \int_0^{r_0} \int_0^{\pi} \int_1^3 f(r, \theta, z)\, r\, dz\, d\theta\, dr$$

$$= \int_0^{r_0} \int_0^{\pi} \int_1^3 r\sin(\theta)\, r\, dz\, d\theta\, dr$$

$$= 2 \int_0^{r_0} \int_0^{\pi} r^2 \sin(\theta)\, d\theta\, dr$$

$$= 2(2) \int_0^{r_0} r^2 dr = \frac{4}{3}r_0^3.$$

B. Let us use cylindrical coordinates to set up and evaluate an iterated integral for the total accumulation of $f(x, y, z) = x^2 + y^2$ over the region $\mathcal{R}$ consisting of the portion of the ball of radius r_0 centered at the origin and lying in the first octant. (See Figure 5.24(b).) The sphere that bounds the ball is defined by $x^2 + y^2 + z^2 = r_0^2$ in Cartesian coordinates. In cylindrical coordinates this is given by $r^2 + z^2 = r_0^2$, so that $z = \sqrt{r_0^2 - r^2}$ in the upper half-space. Then $\mathcal{R}$ can be described in cylindrical coordinates by

$$\mathcal{R} = \{(r, \theta, z) : 0 \le r \le r_0,\ 0 \le \theta \le \pi/2,\ \text{and } 0 \le z \le \sqrt{r_0^2 - r^2}\ \}.$$

The total accumulation of f over $\mathcal{R}$ is

$$
\iiint_{\mathcal{R}} f \, dV_{r,\theta,z} = \int_0^{r_0} \int_0^{\pi/2} \int_0^{\sqrt{r_0^2 - r^2}} f(r,\theta,z) \, r \, dz \, d\theta \, dr
$$

$$
= \int_0^{r_0} \int_0^{\pi/2} \int_0^{\sqrt{r_0^2 - r^2}} r^2 \, r \, dz \, d\theta \, dr
$$

$$
= \int_0^{r_0} \int_0^{\pi/2} r^3 \sqrt{r_0^2 - r^2} \, d\theta \, dr
$$

$$
= \int_0^{r_0} \frac{\pi}{2} r^3 \sqrt{r_0^2 - r^2} \, dr = \frac{\pi}{15} r_0^5.
$$

The last integral was evaluated using a substitution.

Summary

In this section, we introduced the **cylindrical coordinate system** in space. The **cylindrical coordinates** (r, θ, z) of a point (x, y, z) satisfy $x = r \cos\theta$, $y = r \sin\theta$, and $z = z$, where $r \geq 0$.

A **cylindrical box** is a set defined by inequalities of the form $r_1 \leq r \leq r_2$, $\theta_1 \leq \theta \leq \theta_2$, and $z_1 \leq z \leq z_2$. The **volume** of a cylindrical box is $\bar{r} \Delta r \Delta \theta \Delta z$, where $\bar{r} = (r_2 + r_1)/2$, $\Delta r = r_2 - r_1$, $\Delta\theta = \theta_2 - \theta_1$, and $\Delta z = z_2 - z_1$.

We developed integration in cylindrical coordinates using a **Riemann sum** approximation to the total accumulation of a function f. We followed our usual procedure, **partitioning** the region $\mathcal{R}$, **sampling** f in each cell of the partition, **scaling** the sampled value by the volume of a cell, and **summing** the scaled values. In this case, the cells of the partition were cylindrical boxes. We then evaluated the limit of Riemann sums to obtain a **cylindrical integral** equal to the triple integral in rectangular coordinates of f over $\mathcal{R}$. Applying Fubini's theorem, we obtain a **triple iterated integral** in cylindrical coordinates,

$$
\iiint_{\mathcal{R}} f \, dV_{r,\theta,z} = \int_a^b \int_\alpha^\beta \int_{h_1(r,\theta)}^{h_2(r,\theta)} f(r,\theta,z) \, r \, dz \, d\theta \, dr.
$$

Section 5.5 Exercises

1. Coordinates of Points

(a) Find the cylindrical coordinates of the points $(1, -1, 2)$, $(2, 1, -2)$, $(\sqrt{3}, 1, 1)$, $(1, 1, \sqrt{2})$, and $(0, 0, 2)$.

(b) Find the Cartesian coordinates of the points with cylindrical coordinates $(2, \pi, 2)$, $(1, \pi/4, -1)$, $(\sqrt{3}, \pi/6, -3)$, $(3, 3\pi/4, 1)$, and $(1, -\pi/3, 2)$.

2. **Surfaces in Cylindrical Coordinates.** Express each of the following surfaces in terms of cylindrical coordinates.

   (a) The sphere of radius k centered at the origin, k a positive constant.
   (b) The half-plane $y = kx$, $y \geq 0$, k a constant.
   (c) The paraboloid $z = x^2 + y^2 + k$, k a constant.
   (d) The cone $z = k\sqrt{x^2 + y^2}$, k is a constant.

3. **Regions in Cylindrical Coordinates.** Sketch each of the following regions and express them in terms of cylindrical coordinates. All of the cylinders have central axis along the z-axis.

   (a) The half-space bounded by the xy-plane and containing the positive z-axis.
   (b) The half-space bounded by the xz-plane and containing the negative y-axis.
   (c) The region in the first octant bounded by the cylinder of radius 2 and the planes $z = 1$ and $z = 4$.
   (d) The region between the cylinder of radius 2 and the cylinder of radius 4, and the planes $z = -3$ and $z = 2$.
   (e) The region bounded below by the paraboloid $z = x^2 + y^2$ and above by the plane $z = 4$.
   (f) The region bounded below by the paraboloid $z = x^2 + y^2 - 2$ and above by the paraboloid $z = 4 - x^2 - y^2$.

4. **Sets Given in Cylindrical Coordinates.** Describe and sketch each of the following sets given in cylindrical coordinates.

   (a) $\{(r, \theta, z) : r = z \}$.
   (b) $\{(r, \theta, z) : \theta = \pi/4, \ r = z \}$.
   (c) $\{(r, \theta, z) : z = 4 - r^2 \}$.
   (d) $\{(r, \theta, z) : r = 2, \ \theta = \pi/3 \}$.
   (e) $\{(r, \theta, z) : r = 2, \ 0 < \theta < \pi/2, \ z = 1 \}$.

5. **The Solar Constant.** An important value in solar astronomy is the total solar irradiance, or **solar constant**. This is the total intensity of the sun taken over all wavelengths. Thus we can think of intensity as a function of three variables: two variables representing the location of a point on the solar disk and the third representing wavelength. Figure 5.25 contains values for the intensity of the sun at different radii r and different wavelengths λ. The radii are given as a fraction of the solar radius $R_\odot = 6.96 \times 10^{10}$ cm, so that a radius 0 corresponds to the center of the sun and a radius 1 corresponds to the edge of the solar disk. The wavelengths are given in nanometers (10^{-9} m). The intensity values are given as a fraction of the intensity I_0 at the wavelength 500 nm at the center of the sun, which is the maximum value for the intensity at a given wavelength.

   Describe how you would use this data to construct a cylindrical coordinate Riemann sum to approximate the solar constant.

λ \ r	0	0.6	0.8	0.866	0.92	0.95	0.98	0.99	1
200	0.06	0.05	0.04	0.04	0.04	0.03	0.03	0.00	0.00
220	0.11	0.06	0.04	0.03	0.02	0.02	0.01	0.00	0.00
245	0.20	0.14	0.10	0.08	0.07	0.06	0.05	0.00	0.00
265	0.29	0.20	0.12	0.09	0.07	0.05	0.04	0.00	0.00
280	0.36	0.26	0.17	0.14	0.10	0.08	0.06	0.00	0.00
300	0.46	0.36	0.26	0.22	0.18	0.14	0.10	0.06	0.00
320	0.56	0.45	0.35	0.30	0.25	0.20	0.15	0.10	0.00
350	0.70	0.59	0.47	0.41	0.34	0.28	0.21	0.15	0.00
370	0.78	0.66	0.54	0.47	0.40	0.33	0.26	0.18	0.15
380	0.81	0.68	0.54	0.47	0.39	0.32	0.24	0.18	0.15
400	0.87	0.73	0.58	0.51	0.43	0.35	0.27	0.19	0.16
450	0.97	0.84	0.69	0.62	0.54	0.45	0.37	0.27	0.20
500	1.00	0.88	0.74	0.67	0.60	0.51	0.42	0.32	0.26
550	0.98	0.87	0.75	0.69	0.62	0.55	0.46	0.36	0.30
600	0.93	0.84	0.73	0.68	0.62	0.55	0.47	0.38	0.33
800	0.64	0.59	0.54	0.51	0.48	0.44	0.39	0.34	0.30
1000	0.41	0.38	0.35	0.34	0.32	0.30	0.28	0.24	0.22
2000	0.06	0.06	0.05	0.05	0.05	0.05	0.04	0.04	0.03
3000	0.01	0.01	0.01	0.01	0.01	0.01	0.01	0.01	0.01

Figure 5.25 Values for the intensity of the sun at a given wavelength and radius. The wavelength is given in nanometers (10^{-9} m) and the value for intensity is given as a fraction of the intensity at the center of the sun at the wavelength 500 nm. The value for the radius is given as a fraction of the solar radius, $R_\odot$, which is 6.96×10^{10} cm. (See Exercise 5.)

6. **Riemann Sums in Cylindrical Coordinates.** Let the region $\mathcal{R}$ be given in cylindrical coordinates by

$$\mathcal{R} = \{(r, \theta, z) : a \le r \le b, \ \alpha \le \theta \le \beta, \ \text{and} \ h_1(r, \theta) \le z \le h_2(r, \theta) \},$$

where h_1 and h_2 are continuous functions. Let f be a continuous function on this region.

(a) Sketch a representative domain of this form and carefully describe how to partition the domain into $L \times M \times N$ cylindrical boxes.
(b) Write out the iterated Riemann sum to approximate the total accumulation of f on $\mathcal{R}$ that corresponds to this partition.
(c) Derive an iterated triple integral to compute the total accumulation of f on $\mathcal{R}$ using Riemann sums of this form.

7. **Cylindrical Integration.** For each of the following functions f and regions $\mathcal{R}$, sketch the region $\mathcal{R}$ and compute the total accumulation of f over $\mathcal{R}$ using cylindrical coordinates.

   (a) $f(x, y, z) = xyz$, $\mathcal{R}$ is the region bounded by the half-cylinder of radius 2 lying over the first and second quadrants in the xy-plane, the xy-plane, the xz-plane, and the plane $z = 3$.
   (b) $f(x, y, z) = z$, $\mathcal{R}$ is the region bounded below by the paraboloid $z = x^2 + y^2$ and above by the plane $z = 1$.
   (c) $f(x, y, z) = \sin(x^2 + y^2)$, $\mathcal{R}$ is the region between the cylinders of radius 2 and radius 4 centered on the z-axis and the planes $z = 1$ and $z = 3$.

8. **Set Up the Integral.** Set up but do not evaluate an iterated integral for the total accumulation of the function f on the region $\mathcal{R}$ using either Cartesian or cylindrical coordinates.

   (a) $f(x, y, z) = \cos(x^2)$ and $\mathcal{R}$ is a ball of radius 3 centered at the origin.
   (b) $f(x, y, z) = \sqrt{1 + x^3 y}$ and $\mathcal{R}$ is the region consisting of a half-ball of radius 4 centered at the origin and lying in the lower half-space.
   (c) $f(x, y, z) = xyz$ and $\mathcal{R}$ is an ellipsoid centered at the origin with semiminor axis of length 1 along the x-axis and semimajor axes of length 4 along the y- and z-axes.
   (d) $f(x, y, z) = x^2 + y^2 + z^2$ and $\mathcal{R}$ is a solid cone with height 1 and base a disk of diameter 2 resting on the xy-plane.

■ 5.6 Integration in Spherical Coordinates

In this section, we want to extend our work on integration in space using a new coordinate system: spherical coordinates. Spherical coordinates are an extension of the polar coordinate system in the plane and will allow us to give simple symbolic descriptions of subsets of spheres. In particular, we want to consider regions and functions that are spherically symmetric, that is, regions and functions that have the property that the behavior at a point in space depends on the distance of the point from a fixed point or center. We have already seen an example of such a function in considering the diffusion of an airborne substance in space. (See Example 5.13 of Section 5.4.) As was the case for cylindrical coordinates, the key difference between spherical constructions of Riemann sums and the Cartesian coordinate construction is that we will be using spherical boxes rather than rectangular boxes for the cells of our partitions. Consequently, when we construct a Riemann sum for the total accumulation of a function on a region, the scaling factor for each cell will have an additional factor that will also appear in the resulting iterated integral. Using the ideas of Section 5.4 on Riemann sums and integrals in space, we will be able to move directly from spherical Riemann sum to an iterated integral without the intervening steps involving the evaluation of limits.

Spherical Coordinates

The spherical coordinate system in space describes a point by giving the **distance ρ of the point from the origin** and two angles, θ measuring the **longitude** of the point and ϕ measuring the **latitude** of the point. Given (x, y, z) in $\mathbb{R}^3$, $\rho = \sqrt{x^2 + y^2 + z^2}$. The angle θ is the angle between the positive xz-plane and the plane through the point (x, y, z) that contains the z-axis, which is the same θ that is used in cylindrical coordinates. Once again, we must restrict θ to an interval of length 2π to obtain a unique longitude angle. The angle ϕ is the angle measured from the positive z-axis to the line through the origin and the point (x, y, z). We restrict ϕ to be in the interval $[0, \pi]$.

From Figure 5.26, we can see that $z/\rho = \cos\phi$, so that $z = \rho\cos\phi$. Since the radial distance r from $\mathbf{x}$ to the z-axis is $r = \rho\sin\phi$, and $x = r\cos\theta$ and $y = r\sin\theta$, we have that $x = \rho\cos\theta\sin\phi$ and $y = \rho\sin\theta\sin\phi$. We summarize these relationships in the following definition.

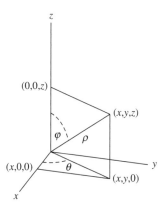

Figure 5.26   The spherical coordinates of (x, y, z) satisfy $x = \rho\cos\theta\sin\phi$, $y = \rho\sin\theta\sin\phi$, and $z = \rho\cos\phi$.

Definition 5.2 The *spherical coordinates*, (ρ, θ, ϕ), of a point (x, y, z) satisfy

$$x = \rho \sin \phi \cos \theta, \ y = \rho \sin \phi \sin \theta, \ \text{and} \ z = \rho \cos \phi,$$

where $\rho \geq 0$, $0 \leq \theta \leq 2\pi$, and $0 \leq \phi \leq \pi$. ◆

The spherical coordinates of a point $\mathbf{x}$ can be expressed in terms of its Cartesian coordinates. We already know that $\rho = \sqrt{x^2 + y^2 + z^2}$ and $\phi = \cos^{-1}(\frac{z}{\rho})$, so that $\phi = \cos^{-1}(\frac{z}{\sqrt{x^2+y^2+z^2}})$. From Figure 5.26, we have that $\tan \theta = y/x$, so that $\theta = \tan^{-1}(y/x)$ as in cylindrical coordinates. For example, if $\mathbf{x} = (1, \sqrt{3}, -1)$, $\rho = \sqrt{1^2 + \sqrt{3}^2 + (-1)^2} = \sqrt{5}$, $\tan \theta = \frac{\sqrt{3}}{1}$, and $\cos \phi = \frac{-1}{\sqrt{5}}$. Since $\mathbf{x}$ lies below the first quadrant of the xy-plane, we choose $\theta = \pi/3$ and $\phi \approx 2.03444$.

As we did for cylindrical coordinates, let us investigate subsets of space that are described by holding one or two spherical coordinates fixed. Note that we are measuring longitude and latitude in radians and that 0 latitude corresponds to the "north pole" and not the "equator."

Example 5.18 **Subsets Described in Spherical Coordinates**

A. In each of the following cases, let k be a constant.

a. The set $\{(\rho, \theta, \phi) : \rho = k \geq 0 \}$ consists of all points k units from the origin, so that it is the sphere of radius k centered at the origin.

b. As was the case for cylindrical coordinates, the set $\{(\rho, \theta, \phi) : \theta = k\}$ is the half-plane that contains the z-axis and forms an angle k with the half-plane $y = 0$ and $x \geq 0$.

c. The set of points $\{(\rho, \theta, \phi) : \phi = k \}$ consists of points with the property that the angle between the positive z-axis and the line through the point and the origin measures k radians. If $k = \pi/2$, this set of points is the xy-coordinate plane. If $k = 0$, this is the positive z-axis, and if $k = \pi$, this is the negative z-axis. If $0 < k < \pi/2$, this is a cone, lying above the xy-plane with vertex at the origin and central axis the z-axis. Finally, if $\pi/2 < k < \pi$, this is a cone, lying below the xy-plane with vertex at the origin and central axis the negative z-axis. Notice that since $\tan \phi = r/z = \sqrt{x^2 + y^2}/z$, in Cartesian coordinates these cones are the graph of $f(x, y) = (\cot k)\sqrt{x^2 + y^2}$.

B. We can describe the circles of latitude and longitude on a sphere by fixing two spherical coordinates, the radius ρ and one of the angles θ or ϕ. For example, the set of points of the form $\{(\rho, \theta, \phi) : \rho = k_1 \text{ and } \phi = k_2 \}$ is a circle of latitude k_2 radians from the positive z-axis on a sphere of radius k_1. (See Figure 5.27(a).) The set of points of the form $\{(\rho, \theta, \phi) : \rho = k_1 \text{ and } \theta = k_2 \}$ is half of a circle of longitude k_2 radians from the

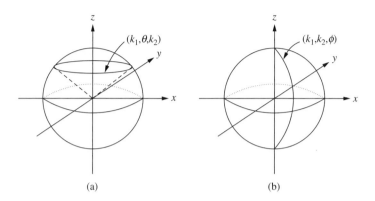

Figure 5.27   (a) The circle of latitude consisting of points of the form (k_1, θ, k_2) in spherical coordinates. (b) The line of longitude consisting of points of the form (k_1, k_2, ϕ) in spherical coordinates. (See Example 5.18B.)

positive x-axis on a sphere of radius k_1. (See Figure 5.27(b).) Finally, fixing the angular coordinates, $\{(\rho, \theta, \phi) : \theta = k_1 \text{ and } \phi = k_2 \}$, defines a ray from the origin.

In order to construct Riemann sums in spherical coordinates, we will use **spherical boxes** to partition regions with spherical coordinate descriptions. Of course, we will need a symbolic expression for the volume of spherical box.

Example 5.19   **Spherical Boxes**

A. **Description of a Spherical Box.** A *spherical box* is a region in space described in spherical coordinates by inequalities $0 \leq \rho_1 \leq \rho \leq \rho_2$, $\theta_1 \leq \theta \leq \theta_2$, and $\phi_1 \leq \phi \leq \phi_2$. Intuitively, we should think of a spherical box as the region of space lying between concentric spheres of radius ρ_1 and ρ_2 centered at the origin, between longitudes θ_1 and θ_2, and between latitudes ϕ_1 and ϕ_2. (See Figure 5.28(a).)

B. **Volume of a Spherical Box.** Unfortunately, there is no simple formula for the volume of an arbitrary spherical box. However, we can approximate the volume a spherical box by the product of its length, width, and height when the dimensions are small. This will be sufficient for our purposes in the next section, since the approximation is more accurate as the dimensions of the cell approach 0, which is the case when we take a limit of Riemann sums. We will use the notation from part A.

   The length of the box can be approximated by $\Delta\rho = \rho_2 - \rho_1$. The width of the box is a portion of the circumference of a circle of radius $r = \rho \sin \phi$. This measurement is

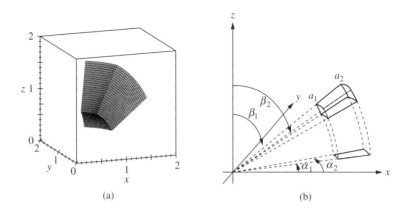

Figure 5.28   (a) A spherical box defined by $1 \leq \rho \leq 2$, $\pi/9 \leq \theta \leq \pi/3$, and $\pi/6 \leq \phi \leq \pi/3$. (See Example 5.19A.) (b) Approximating the volume of a spherical box. (See Example 5.19B.)

approximately $\overline{\rho} \sin(\overline{\phi}) \Delta \theta$, where $\overline{\rho} = \frac{\rho_1 + \rho_2}{2}$ and $\overline{\phi} = \frac{\phi_1 + \phi_2}{2}$, the averages of the ρ and ϕ bounds, and $\Delta \theta = \theta_2 - \theta_1$. The height of the box is a portion of the circumference of a circle of radius ρ. This can be approximated by $\overline{\rho} \Delta \phi$, where $\Delta \phi = \phi_2 - \phi_1$. Thus the **volume of the spherical box** is **approximated** by $\overline{\rho}^2 \sin(\overline{\phi}) \Delta \rho \Delta \theta \Delta \phi$. (See Figure 5.28(b).)

We have seen a variety of physical systems that display spherical symmetry. Here we reconsider the model for airborne diffusion.

Example 5.20   **Diffusion in Space**

A. If an airborne substance diffuses or spreads by random molecular motion from a point source in space in the absence of wind, we would expect that the concentration of the substance at an instant in time and a point in space depends on the distance of the point from the source. If we choose the coordinate system so the source is located at the origin, then the concentration at $\mathbf{x}$ is given by

$$c(\mathbf{x}, t) = \frac{M}{(4\pi\delta t)^{3/2}} e^{-\rho^2/(4\delta t)},$$

where M is the amount of the substance that is released from the source at time $t = 0$, δ is the diffusion constant, and ρ is the distance from $\mathbf{x}$ to the origin.

Because of the dependence of c on ρ, we can see that at time $t = t_0$, the level sets of S correspond to fixed values of ρ. That is, the sphere with radius ρ_0 is the level set of c for the concentration $\frac{M}{(4\pi\delta t_0)^{3/2}} e^{-\rho_0^2/(4\delta t_0)}$.

B. If the source of the substance is located on level ground, then all the particles will diffuse in the direction of positive z. So if M is the number of particles released at time $t = 0$, the concentration at time t is given by

$$c(\mathbf{x}, t) = \frac{2M}{(4\pi\delta t)^{3/2}} e^{-\rho^2/(4\delta t)}$$

for points above the xy-plane. Points that lie below the xy-plane are not part of the domain of S, even though the formula for c makes sense everywhere. In this case, at time $t = t_0$, the hemisphere of radius ρ_0 with $0 \leq \phi \leq \pi/2$ is the level set for c for the concentration $\frac{2M}{(4\pi\delta t_0)^{3/2}} e^{-\rho_0^2/(4\delta t_0)}$.

Riemann Sums in Spherical Coordinates

Our development of Riemann sums in spherical coordinates will closely follow the construction in cylindrical coordinates. Here the region $\mathcal{R}$ will be described by inequalities in spherical coordinates. In particular, we will focus on regions that are spherical boxes, so that the bounds of the inequalities will be constants. While it is not necessary to make this restriction, these regions are sufficient for our purposes.

Let f be a continuous function defined on a spherical box $\mathcal{R}$ that is given by

$$\mathcal{R} = \{(\rho, \theta, \phi) \, : a \leq \rho \leq b, \, \alpha_1 \leq \theta \leq \alpha_2, \text{ and } \beta_1 \leq \phi \leq \beta_2 \}.$$

We will assume that f is expressed in spherical coordinates, so that $f = f(\rho, \theta, \phi)$. If f is expressed in Cartesian coordinates, the spherical coordinate expression for f is obtained from the Cartesian expression using $f^*(\rho, \theta, \phi) = f(\rho \cos\theta \sin\phi, \rho \sin\theta \sin\phi, \rho \cos\phi)$.

Previously, we have described partitions of three-dimensional regions as consisting of columns of cells parallel to a coordinate axis. We have also chosen particular sampling schemes so that the sample points within a column of the partition lie along a line parallel to the same axis. In spherical coordinates, we will replace the idea of being "parallel to an axis" with the idea of sampling "along rays from the origin." Let us make this precise.

We will partition $\mathcal{R}$ into $L \times M \times N$ spherical boxes by choosing $L + 1$ values ρ_i with

$$a = \rho_0 < \rho_1 < \rho_2 < \ldots < \rho_{L-1} < \rho_L = b,$$

$M + 1$ values θ_j with

$$\alpha_1 = \theta_0 < \theta_1 < \theta_2 < \ldots < \theta_{M-1} < \theta_M = \alpha_2,$$

and $N + 1$ values ϕ_k with

$$\beta_1 = \phi_0 < \phi_1 < \phi_2 < \ldots < \phi_{N-1} < \phi_N = \beta_2.$$

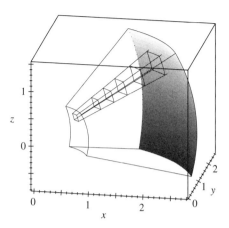

Figure 5.29   One column of a spherical partition of a spherical box. The column lies along a ray emanating from the origin. (See Riemann Sums in Spherical Coordinates in Section 5.6.)

The ijk^{th} cell of the partition $\mathcal{R}_{ijk}$ is the spherical box defined by

$$\mathcal{R}_{ijk} = \{(\rho, \theta, \phi) : \rho_{i-1} \leq \rho \leq \rho_i,\ \theta_{j-1} \leq \theta \leq \theta_j,\ \text{and}\ \phi_{k-1} \leq \phi \leq \phi_k\ \}.$$

The cells $\mathcal{R}_{ijk}$ for $i = 1, \ldots, L$ form a "column" of cells extending out from the origin in the radial direction from $\rho = a$ to $\rho = b$. (See Figure 5.29.) Using Example 5.19, we have that

$$\text{Vol}(\mathcal{R}_{ijk}) \approx \overline{\rho}_i^2 \sin(\overline{\phi}_k)\Delta\rho_i \Delta\theta_j \Delta\phi_k,$$

where $\overline{\rho}_i = \frac{1}{2}(\rho_{i-1} + \rho_i)$, $\overline{\phi}_k = \frac{1}{2}(\phi_{k-1} + \phi_k)$, $\Delta\rho_i = \rho_i - \rho_{i-1}$, $\Delta\theta_j = \theta_j - \theta_{j-1}$, and $\Delta\phi_k = \phi_k - \phi_{k-1}$. We define the ***mesh*** of the partition P to be the maximum of all the values $\Delta\rho_i$, $\Delta\theta_j$, and $\Delta\phi_k$.

In order to approximate the accumulation of f in the cell $\mathcal{R}_{ijk}$, we will sample f at a point $P_{ijk} = (\rho_{ijk}^*, \theta_{ijk}^*, \phi_{ijk}^*)$. As above, we will make a convenient choice of sample points. Here we will choose our sample points so that within a "radial" column of cells, we will sample f along a ray emanating from the origin. We will do this by choosing $\rho_{ijk}^* = \overline{\rho}_i = \rho_i^*$, $\theta_{ijk}^* = \theta_j^*$, with $\theta_{j-1} \leq \theta_j^* \leq \theta_j$, and $\phi_{ijk}^* = \overline{\phi}_k = \phi_k^*$. Note that we have chosen $\rho_i^* = \overline{\rho}_i$ and $\phi_k^* = \overline{\phi}_k$ because of the above formula for the volume of a cell. The sampled value for f in $\mathcal{R}_{ijk}$ is $f(\rho_i^*, \theta_j^*, \phi_k^*)$. We will approximate the accumulation of f in $\mathcal{R}_{ijk}$ by the scaled value

$$f(P_{ijk})\text{Vol}(\mathcal{R}_{ijk}) \approx f(\rho_i^*, \theta_j^*, \phi_k^*)(\rho_i^*)^2 \sin(\phi_k^*)\Delta\rho_i \Delta\theta_j \Delta\phi_k.$$

Note that the scaled value we are using is itself an approximation because we are not using an exact value for the volume of a cell. This is not an issue because we are in fact only constructing an approximation to the total accumulation, and, as the mesh of the partition approaches zero, our approximation to the volume of the cell approaches the true volume. The **Riemann sum** approximation to the total accumulation of f on $\mathcal{R}$ is the triply indexed sum

$$R(f, P) = \sum_{i,j,k=1}^{L,M,N} f(\rho_i^*, \theta_j^*, \phi_k^*)(\rho_i^*)^2 \sin(\phi_k^*)\Delta\rho_i\Delta\theta_j\Delta\phi_k.$$

We obtain an exact value for the total accumulation by taking the limit of Riemann sums $R(f, P)$ as mesh$(P) \to 0$. If this limit exists and is independent of the choice partition and sampling points, we define it to be the **spherical coordinate integral** of f over $\mathcal{R}$. We denote this limit by

$$\int\int\int_{\mathcal{R}} f \, dV_{\rho,\theta,\phi}.$$

The value of this spherical coordinate integral of f over $\mathcal{R}$ is equal to the value of the Cartesian coordinate integral of f over $\mathcal{R}$. We state this result without proof.

Theorem 5.11 Let $f = f(x, y, z)$ be a continuous function on the region $\mathcal{R}$ described in spherical coordinates by

$$\mathcal{R} = \{(\rho, \theta, \phi) : a \leq \rho \leq b, \ \alpha_1 \leq \theta \leq \alpha_2, \text{ and } \beta_1 \leq \phi \leq \beta_2 \}.$$

Let f^* be the spherical coordinate expression for f, that is, $f^*(\rho, \theta, \phi) = f(\rho \cos\theta \sin\phi, \rho\sin\theta\sin\phi, \rho\cos\phi)$. Then

$$\int\int\int_{\mathcal{R}} f^* \, dV_{\rho,\theta,\phi} = \int\int\int_{\mathcal{R}} f \, dV_{x,y,z}. \quad \blacklozenge$$

In order to express the spherical coordinate integral of f over $\mathcal{R}$ as an iterated integral, we must return to the Riemann sum $R(f, P)$. Comparing this to the form of the Cartesian coordinate Riemann sums that we used in Section 5.4, we see that it takes the form of a Riemann sum for the function

$$f(\rho, \theta, \phi)\rho^2 \sin\phi$$

on the region $[a, b] \times [\alpha_1, \alpha_2] \times [\beta_1, \beta_2]$ in a three-dimension Euclidean space with coordinates ρ, θ, and ϕ. Thus we are in a position to apply Fubini's theorem for functions of three variables. In this case, it says that the limit of spherical Riemann sums is equal to an iterated integral of $f(\rho, \theta, \phi) \cdot \rho^2 \sin \phi$. We state this as a theorem.

Theorem 5.12 Fubini's Theorem in Spherical Coordinates. Let f be a continuous function on the region

$$\mathcal{R} = \{(\rho, \theta, \phi) : a \le \rho \le b,\ \alpha_1 \le \theta \le \alpha_2,\ \text{and}\ \beta_1 \le \phi \le \beta_2\ \}.$$

Then the spherical coordinate integral for f on $\mathcal{R}$ is equal to each of the six possible triple iterated integrals of the function $f(\rho, \theta, \phi) \rho^2 \sin \phi$. For example,

$$\int \int \int_{\mathcal{R}} f\, dV_{\rho,\theta,\phi} = \int_a^b \int_{\alpha_1}^{\alpha_2} \int_{\beta_1}^{\beta_2} f(\rho, \theta, \phi) \rho^2 \sin \phi\, d\phi\, d\theta\, d\rho. \ \blacklozenge$$

In the following example, we will use Fubini's theorem to symbolically evaluate the total accumulation of a function f over a spherical box $\mathcal{R}$.

Example 5.21

A Spherical Integral. Let us use spherical coordinates to evaluate the total accumulation of the function $f(x, y, z) = x^2 + y^2$ over the region $\mathcal{R}$ consisting of the portion of the ball of radius r_0 centered at the origin and lying in the first octant. In spherical coordinates,

$$f(\rho, \theta, \phi) = (\rho \sin \phi \cos \theta)^2 + (\rho \sin \phi \sin \theta)^2 = \rho^2 \sin^2 \phi,$$

and $\mathcal{R} = \{(\rho, \theta, \phi) : 0 \le \rho \le r_0, 0 \le \theta \le \pi/2,\ \text{and}\ 0 \le \phi \le \pi/2\ \}$. The total accumulation of f over $\mathcal{R}$ is

$$\int \int \int_{\mathcal{R}} f\, dV_{\rho,\theta,\phi} = \int_0^{r_0} \int_0^{\pi/2} \int_0^{\pi/2} (\rho^2 \sin^2 \phi) \rho^2 \sin \phi\, d\phi\, d\theta\, d\rho$$

$$= \int_0^{r_0} \int_0^{\pi/2} \int_0^{\pi/2} \sin^3 \phi\, d\phi\, d\theta\, \rho^4 d\rho$$

$$= \int_0^{r_0} \int_0^{\pi/2} \frac{2}{3} d\theta\, \rho^4 d\rho$$

$$= \int_0^{r_0} \frac{\pi}{3} \rho^4 d\rho = \frac{\pi}{15} r_0^5.$$

In this last example, we use spherical coordinates to model the position of an electron orbiting an atomic nucleus. Rather than speak of the orbit of an electron, we speak of the *orbital* of an electron. This refers to a region of space that may be occupied by an electron in motion about a nucleus.

Example 5.22

Quantum Probability. In the quantum theory of the atom, the position of an electron orbiting about the nucleus is not given directly, but is expressed in terms of the probability of finding an electron in a region of space. This probability is the accumulation over the region of the square of what is known as the **wave function** for the orbital. Because of the symmetry of electron orbitals, the wave function Ψ is usually expressed in spherical coordinates with the nucleus of the atom located at the origin. The probability, or likelihood, of finding an electron in a region $\mathcal{R}$ of space is

$$\int\int\int_{\mathcal{R}} \Psi^2 \, dV_{\rho,\theta,\phi}.$$

This integral is always nonnegative. In addition, Ψ is scaled so that the integral of Ψ^2 over all of $\mathbb{R}^3$ is equal to 1. This integral is interpreted as follows: the closer its value is to 1 for a given region $\mathcal{R}$, the more likely it is that an electron can be found in $\mathcal{R}$; and the closer the value is to 0, the less likely it is that an electron can be found in $\mathcal{R}$.

For simple atoms it is possible to given an explicit formula for the wave functions Ψ. For example, the wave function for the orbital closest to the nucleus of the hydrogen atom, the $1s$ orbital, is

$$\Psi(\rho, \theta, \phi) = \frac{2}{(4\pi)^{1/2} a^{3/2}} e^{-\rho/a},$$

where $a \approx 52.9 \times 10^{-12}$ m, which is known as the Bohr radius.[4] More generally, a can be expressed in terms of the mass and charge of an electron. Since the $1s$ orbital is spherically symmetric, Ψ depends on ρ and not θ and ϕ. Let us compute the probability of finding an electron in the shell $\mathcal{R}$ given by $\rho_1 \leq \rho \leq \rho_2$.

$$\int\int\int_{\mathcal{R}} \Psi^2 \, dV_{\rho,\theta,\phi} = \int_{\rho_1}^{\rho_2} \int_0^\pi \int_0^{2\pi} \frac{4}{4\pi a^3} e^{-2\rho/a} \rho^2 \sin\phi \, d\theta \, d\phi \, d\rho$$

$$= \frac{4}{a^3} \int_{\rho_1}^{\rho_2} \rho^2 e^{-2\rho/a} \, d\rho$$

$$= I(\rho_2) - I(\rho_1),$$

where $I(\rho) = -(2\frac{\rho^2}{a^2} + 2\frac{\rho}{a} + 1)e^{-2\rho/a}$. The last integral was evaluated using integration by parts. In Exercise 7, we will investigate this result.

---

[4]The Danish atomic physicist Niels Bohr (1885–1962) made fundamental contributions to our understanding of quantum mechanics. He used this value for the radius of the orbit of an electron in circular orbit about a proton in his model of the hydrogen atom.

Summary

In this section, we developed *integration in spherical coordinates*. We constructed a *Riemann sum* approximation to the total accumulation of a function f over a "box" in the new coordinate system.

We followed our usual procedure, *partitioning* the region $\mathcal{R}$, *sampling* f in each cell of the partition, *scaling* the sampled value by the volume of a cell, and *summing* the scaled values. Applying the work of Section 5.4, we evaluated the limit of Riemann sums to obtain a *spherical integral* equal to the triple integral in rectangular coordinates of f over $\mathcal{R}$. We then applied Fubini's theorem to obtain a *triple iterated integral*.

If $\mathcal{R} = \{(\rho, \theta, \phi) : a \leq \rho \leq b,\ \alpha_1 \leq \theta \leq \alpha_2,\ \beta_1 \leq \phi \leq \beta_2\}$, we have, for example,

$$\iiint_{\mathcal{R}} f \, dV_{\rho,\theta,\phi} = \int_a^b \int_{\alpha_1}^{\alpha_2} \int_{\beta_1}^{\beta_2} f(\rho, \theta, \phi)\rho^2 \sin \phi \, d\phi \, d\theta \, d\rho.$$

The triple integral is equal to each of six iterated integrals corresponding to the six orderings of the three variables.

Section 5.6 Exercises

1. **Coordinates of Points**

    (a) Find the spherical coordinates of the points $(1, -1, 2)$, $(2, 1, -2)$, $(\sqrt{3}, 1, 1)$, $(1, 1, \sqrt{2})$, and $(0, 0, 2)$.

    (b) Find the Cartesian coordinates of the points with spherical coordinates $(1, \pi/2, \pi)$, $(\sqrt{2}, \pi/4, \pi/4)$, $(2, \pi, \pi/3)$, $(\sqrt{3}, 0, \pi/6)$, and $(1, \pi/3, \pi/2)$.

2. **Sets Given in Spherical Coordinates.** Describe and sketch each of the following sets given in spherical coordinates.

    (a) $\{(\rho, \theta, \phi) : \rho = 3,\ 0 < \theta < \pi/2\}$.
    (b) $\{(\rho, \theta, \phi) : \theta = \pi/2\}$.
    (c) $\{(\rho, \theta, \phi) : \theta = \pi/4,\ \phi = \pi/2\}$.
    (d) $\{(\rho, \theta, \phi) : \rho \geq 2\}$.
    (e) $\{(\rho, \theta, \phi) : 0 \leq \rho \leq 1,\ \pi/4 < \phi < 3\pi/4\}$.

3. **Regions in Spherical Coordinates.** Describe each of the following using spherical coordinates. All of the spheres are centered at the origin.

    (a) The upper hemisphere of a sphere of radius k, k a positive constant.
    (b) The cone $z = k\sqrt{x^2 + y^2}$, k a positive constant.
    (c) The region bounded by the sphere of radius 2 in the first octant.

(d) The region lying between the sphere of radius 1 and the sphere of radius 3 and above the xy-plane.

(e) The half-space bounded by the xy-plane and containing the positive z-axis.

(f) The region bounded below by the cone $z = 2\sqrt{x^2 + y^2}$ and above by the sphere of radius 5.

4. **Solar Mass.** A recent spherically symmetric model for physical structure of the sun gives the accompanying values (see Figure 5.30) for the density of the sun as a function of the distance from the center of the sun. The radial distance is written as a fraction of the solar radius, so that $\rho = R/R_\odot$, where $R_\odot = 6.96 \times 10^{10}$ cm is the solar radius and R is the actual distance to the sun's center. The density δ is given in grams per cubic centimeter. Describe how you would use this data to produce a Riemann sum to approximate the total mass of the sun.

5. **Total Heat.** For each of the following temperature distributions $T(x, y, z)$ in a homogeneous medium and containers $\mathcal{R}$, sketch the container and find the total heat in the container by computing $\int \int \int_\mathcal{R} \delta C T \, dV_{\rho,\theta,\phi}$, where δ and C are constants representing the density and the specific heat capacity of the medium, respectively.

(a) $T(x, y, z) = \ln(x^2 + y^2 + z^2)$ and the container $\mathcal{R}$ occupies the region between a sphere of radius 1 and a sphere of radius 2 centered at the origin.

ρ	δ	ρ	δ	ρ	δ
0	148	0.08533	97	0.2876	14.2
0.0039	148	0.1147	76.4	0.3176	10.1
0.0083	147	0.1346	64.5	0.3344	8.43
0.012	146	0.1551	54	0.3737	5.32
0.0158	145	0.1719	46.4	0.3975	4.06
0.0197	144	0.1881	39.9	0.4597	2.03
0.0237	142	0.2047	34	0.5038	1.27
0.0277	140	0.2212	28.8	0.6559	0.28
0.0317	137	0.2381	24.2	0.8015	0.08
0.04	132	0.2555	20.1	0.8573	0.04
0.0484	126	0.2628	18.6	0.9093	0.02
0.0708	108	0.2739	16.5	1	0

Figure 5.30   Computed values for solar density δ as a function of the radial distance from the center of the sun. The values for δ are given in units of grams per cubic centimeter. (See Exercise 4.)

(b) $T(x, y, z) = x$ and the container $\mathcal{R}$ occupies the region between the right hemisphere of radius 1 and the right hemisphere of radius 3 centered at the origin.

(c) $T(x, y, z) = z$ and the container $\mathcal{R}$ occupies the region bounded below by the cone $z = \sqrt{x^2 + y^2}$ and above by a sphere of radius 2 centered at the origin.

(d) $T(x, y, z) = xyz$ and the container $\mathcal{R}$ occupies the region in the first octant outside the cone $\sqrt{3}z = \sqrt{x^2 + y^2}$ and inside a sphere of radius 1 centered at the origin.

6. **Spherical Boxes.** Let $\mathcal{R}$ be the spherical box given by

$$\{(\rho, \theta, \phi) : a_1 \le \rho \le a_2, \ \alpha_1 \le \theta \le \alpha_2, \text{ and } \beta_1 \le \phi \le \beta_2 \}.$$

In Example 5.19, we argued that when the dimensions of the box are small, the volume can be approximated by the quantity $V_0 = \bar{\rho}^2 \sin(\bar{\phi}) \Delta \rho \Delta \theta \Delta \phi$, where $\Delta \rho = \rho_2 - \rho_1$, $\Delta \theta = \theta_2 - \theta_1$, $\Delta \phi = \phi_2 - \phi_1$, $\bar{\rho} = \frac{\rho_1 + \rho_2}{2}$, and $\bar{\phi} = \frac{\phi_1 + \phi_2}{2}$. In this exercise, we will explore this approximation.

(a) Use a spherical coordinates integral to compute the volume of this box. (Your answer will be in terms of ρ_1, ρ_2, θ_1, θ_2, ϕ_1, and ϕ_2.) Call this quantity V_1.

(b) To say that V_0 is a good approximation for V_1 when the box is small means that $\lim_{\Delta \rho, \Delta \theta, \Delta \phi \to 0} \frac{V_1}{V_0} = 1$. In order to evaluate this limit, it will be helpful to make the following substitutions in the formulas for V_0 and V_1: $\rho_2 = \rho_1 + \Delta \rho$, $\phi_2 = \phi_1 + \Delta \phi$, $\bar{\rho} = \rho_1 + \frac{1}{2} \Delta \rho$, and $\bar{\phi} = \phi_1 + \frac{1}{2} \Delta \phi$. In addition, it will be helpful to use the second order approximations to the sine and cosine functions. That is, $\sin(x) \approx x$ and $\cos(x) \approx 1 - \frac{x^2}{2}$. Using these substitutions and approximations, show that $\lim_{\Delta \rho, \Delta \theta, \Delta \phi \to 0} \frac{V_1}{V_0} = 1$.

7. **The Hydrogen 1s Orbital.** In Example 5.22 we introduced the wave function for the $1s$ orbital of hydrogen:

$$\Psi(\rho, \theta, \phi) = \frac{2}{(4\pi)^{1/2} a^{3/2}} e^{-\rho/a},$$

where $a = 52.9 \times 10^{-12}$ m. We expressed the spherical integral of Ψ in terms of the function I,

$$I(\rho) = -(2\frac{\rho^2}{a^2} + 2\frac{\rho}{a} + 1) e^{-2\rho/a}.$$

(a) Verify the integration in Example 5.22: $\int_{\rho_1}^{\rho_2} \int_0^{2\pi} \int_0^{\pi} \Psi(\rho, \theta, \phi) \rho^2 \sin \phi \, d\phi \, d\theta \, d\rho = I(\rho_2) - I(\rho_1)$.

(b) If $\rho_2 > \rho_1 \ge 0$, is the difference $I(\rho_2) - I(\rho_1)$ always positive, always negative, or sometimes positive and sometimes negative? Explain your answer.

(c) Verify the claim in Example 5.22 that $\int \int \int_{\mathbb{R}^3} \Psi^2 dV \rho, \theta, \phi = 1$. (*Hint:* This is an improper integral.)

(d) Because of (b) and (c), we know that $0 \le \rho_1 \le \rho_2$, $0 \le I(\rho_2) - I(\rho_1) \le 1$. The probability of finding an electron between the radii ρ_1 and ρ_2 is the difference $I(\rho_2) - I(\rho_1)$. One way to investigate this is to consider intervals of fixed widths, that is, of the form $\rho - \epsilon$ to $\rho + \epsilon$, and plot the difference $I(\rho + \epsilon) - I(\rho - \epsilon)$ as a function of ρ for fixed ϵ. Plot this difference for several values of ϵ, say $\epsilon = a, a/2, a/3, \ldots$. How do the plots change as the denominator is increased? Based on your plots, between which radii are you most likely to find an electron? Explain. (*Hint:* Plot on a horizontal scale from 0 to $20a$.)

8. **The Hydrogen 2s Orbital.** Here we consider the wave function for the second, or $2s$, orbital of hydrogen. In spherical coordinates, it is given by

$$\Psi(\rho, \theta, \phi) = \frac{2 - \frac{\rho}{a}}{2^{5/2}\sqrt{\pi}a^{3/2}} e^{-\rho/(2a)},$$

where $a \approx 52.9 \times 10^{-12}$ m. As in Example 5.22, the integral of Ψ^2 over a region gives the probability of finding an electron in the region. In particular, the probability of finding an electron in the spherical shell lying between the radii ρ_1 and ρ_2 is:

$$\int_{\rho_1}^{\rho_2} \int_0^{2\pi} \int_0^{\pi} \Psi(\rho, \theta, \phi)^2 \, \rho^2 \sin(\phi) \, d\phi \, d\theta \, d\rho = I(\rho_2) - I(\rho_1),$$

where $I(\rho) = -\left(\frac{1}{8}\left(\frac{\rho}{a}\right)^4 + \frac{1}{2}\left(\frac{\rho}{a}\right)^2 + \frac{\rho}{a} + 1\right)e^{-\rho/a}$.

(a) Verify the claim that the spherical integral of Ψ^2 can be expressed in terms of I.
(b) If $\rho_2 > \rho_1 \ge 0$, is the difference $I(\rho_2) - I(\rho_1)$ always positive, always negative, or sometimes positive and sometimes negative? Explain your answer.
(c) Verify that $\int\int\int_{\mathbb{R}^3} \Psi^2 dV_{\rho,\theta,\phi} = 1$. (*Hint:* This is an improper integral.)
(d) As in Exercise 7(d), we can investigate the probability of finding an electron in a spherical shell of fixed width as we change the radius of the shell. Plot the difference $I(\rho + \epsilon) - I(\rho - \epsilon)$ as a function of ρ for fixed ϵ. Plot this difference for several values of ϵ, say $\epsilon = a, a/2, a/3, \ldots$. How do the plots change as the denominator is increased? Based on your plots, between which radii are you most likely to find an electron? Explain. (*Hint:* Plot on a horizontal scale from 0 to $20a$.)
(e) Compare your answers to Exercise 7(d) and part (d) above.

■ 5.7 End of Chapter Exercises

---

1. **Polar Regions.** Give a symbolic description for each of the following regions in the plane using polar coordinates and set notation.

   (a) The third and fourth quadrants not including the x-axis.

(b) A semicircle of radius 3 centered at the origin in the first and fourth quadrants.
(c) A ray beginning at the origin and passing through the point $(1,1)$.
(d) $\{(x,y) : y \geq x\}$.
(e) $\{(x,y) : x^2 + 2y^2 = 1\}$.

2. **Regions in Cylindrical Coordinates.** Sketch each of the following regions and express them in terms of cylindrical coordinates. All of the cylinders have central axis along the z-axis.

(a) The paraboloid $z = x^2 + y^2 + k$, k a constant.
(b) The cone $z = k\sqrt{x^2 + y^2}$, k a positive constant.
(c) The half-space bounded by the yz-plane and containing the negative x-axis.
(d) The region bounded by the half-cylinder of radius 1 lying over the second and third quadrants in the xy-plane, the yz-plane, the xy-plane, and the plane $z = 2$.
(e) The region between the half-planes $y = x$ with $y \geq 0$, and $y = -x$ with $y \geq 0$, the xy-plane, the plane $z = 1$, and the cylinders of radius 1 and radius 3.
(f) The region bounded by the sphere of radius 1 centered at the origin.
(g) The region bounded below by the cone $z = 2\sqrt{x^2 + y^2}$ and above by the sphere of radius 5.

3. **Regions in Spherical Coordinates.** Describe each of the following using spherical coordinates. All of the spheres are centered at the origin.

(a) The half-space bounded by the xy-plane and containing the positive z-axis.
(b) The half-space bounded by the xz-plane and containing the negative y-axis.
(c) The region lying between the hemisphere of radius 1 and the hemisphere of radius 3 with y nonpositive.
(d) The region bounded below by the cone $z = 2\sqrt{x^2 + y^2}$ and above by the sphere of radius 5.
(e) The half-plane $y = kx$, $y \geq 0$, k a constant.
(f) The cylinder of radius k with central axis the z-axis.

4. **Riemann Sum Calculations.** For each of the following functions f, regions $\mathcal{R}$, and values for M and N, (i) construct a partition $P_{M,N}$ of $\mathcal{R}$; (ii) sketch the domain, the partition, and a collection of sample points; and (iii) evaluate the corresponding Riemann sum approximation to the total accumulation of f on $\mathcal{R}$.

(a) $f(x,y) = \cos(x - y)$, $\mathcal{R} = \{(x,y) : 0 \leq x \leq 2 \text{ and } -x \leq y \leq x^2\}$, and $M = 3$, $N = 3$.
(b) $f(x,y) = y2^x$, $\mathcal{R} = \{(x,y) : -2 \leq y \leq 0 \text{ and } y \leq x \leq -y\}$, and $M = 5$, $N = 2$.

5. **Double Integral Calculations.** Evaluate the following integrals using Cartesian or polar coordinates.

(a) $\int \int_{\mathcal{R}} (2x + y)^3 \, dA_{x,y}$, where $\mathcal{R} = [1,2] \times [1,2]$.
(b) $\int \int_{\mathcal{R}} \sin(x + y) \, dA_{x,y}$, where $\mathcal{R} = [0, \pi/2] \times [0, \pi/2]$.

(c) $\int\int_{\mathcal{R}} 1 + xy\, dA_{x,y}$, where $\mathcal{R} = \{(x,y) : -1 \le x \le 1, 0 \le y \le \sqrt{1-x^2}\}$.

(d) $\int\int_{\mathcal{R}} \sqrt{y^2+1}\, dA_{x,y}$, where $\mathcal{R} = \{(x,y) : 0 \le y \le 2, 0 \le x \le y\}$.

(e) $\int\int_{\mathcal{R}} 9 - x^2 - y^2\, dA_{x,y}$, where $\mathcal{R} = \{(x,y) : 0 \le y \le 3, 0 \le x \le \sqrt{9-y^2}\}$.

6. **Volumes.** Compute the volume of the following regions in space.

   (a) The region in the first octant lying above the quarter-disk of radius 1 and below the plane $z = 1 - y$.

   (b) $\{(x,y,z) : -2 \le x \le 2, -1 \le y \le 3, \text{ and } y - 2 \le z \le x^2 + y^2\}$.

   (c) $\{(x,y,z) : 0 \le x \le 1, xe^z \le y \le 4, \text{ and } 0 \le z \le 1\}$.

7. **Triple Integral Calculations.** For each of the following, sketch the region $\mathcal{R}$ of integration and set up and evaluate an iterated integral to compute the desired quantity using Cartesian, cylindrical, or spherical coordinates. Briefly explain how you selected an appropriate coordinate system.

   (a) $f(x,y,z) = e^{x+y-z}$ and $\mathcal{R} = [0,1] \times [0,1] \times [0,1]$.

   (b) $f(x,y,z) = x\cos(xy)/z$ and $\mathcal{R} = [0, \pi/2] \times [0,1] \times [1,2]$.

   (c) $f(x,y,z) = xy$ and $\mathcal{R} = \{(x,y,z) : z^2 - 1 \le x \le 1 - z^2, zx \le y \le zx + 1, \text{ and } -1 \le z \le 1\}$.

   (d) $f(x,y,z) = xy$, $\mathcal{R}$ is the region in the first octant inside the cylinder of radius 1 centered on the z-axis and under the sphere of radius 2 centered at the origin.

   (e) Compute the volume of the region inside the sphere of radius 3 centered at the origin and outside the cylinder of radius 1 centered on the z-axis.

   (f) The temperature of a homogeneous liquid is given by $T(x,y,z) = x^2 + y^2$. Compute the average temperature of the liquid in a container $\mathcal{R}$ that occupies the region bounded below by the cone $z^2 = x^2 + y^2$ and above by a sphere of radius $2\sqrt{2}$.

   (g) Compute the total mass of a spherical shell with inner radius 1 and outer radius 3 that has density $\delta(x,y,z) = (x^2 + y^2 + z^2)^{-1}$.

   (h) Compute the volume of the region inside the cylinder of radius 1 centered on the z-axis, bounded below by the paraboloid $z = 1 + x^2 + y^2$ and above by the plane $z = 4 - x - y$.

8. **Average Value at a Point in $\mathbb{R}^2$.** Suppose that $f : \mathcal{D} \subset \mathbb{R}^2 \to \mathbb{R}$ is continuous and that $\mathcal{R} \subset \mathcal{D}$. The average value of f on $\mathcal{R}$ is the quotient

$$\frac{\int\int_{\mathcal{R}} f\, dA_{x,y}}{\text{Area}(\mathcal{R})}.$$

   (a) Suppose that the dimensions of $\mathcal{R}$ are small and that $(x_0, y_0) \in \mathcal{R}$. Give an intuitive argument for the statement "the average value of f on $\mathcal{R}$ is approximately equal to $f(x_0, y_0)$."

(b) Based on the statement of part (a), give an intuitive argument justifying the following limit:

$$\lim_{\text{Area}(\mathcal{R}) \to 0} \frac{\int\int_{\mathcal{R}} f \, dA_{x,y}}{\text{Area}(\mathcal{R})} = f(x_0, y_0).$$

This result will be useful in later chapters.

9. **Average Value at a Point in $\mathbb{R}^3$.** Suppose that $f : \mathcal{D} \subset \mathbb{R}^3 \to \mathbb{R}$ is continuous and that $\mathcal{R} \subset \mathcal{D}$. The average value of f on $\mathcal{R}$ is the quotient

$$\frac{\int\int\int_{\mathcal{R}} f \, dV_{x,y,z}}{\text{vol}(\mathcal{R})}.$$

(a) Suppose that the dimensions of $\mathcal{R}$ are small and that $(x_0, y_0, z_0) \in \mathcal{R}$. Give an intuitive argument for the statement "the average value of f on $\mathcal{R}$ is approximately equal to $f(x_0, y_0, z_0)$."

(b) Based on the statement in (a), give an intuitive argument justifying the following limit:

$$\lim_{\text{vol}(\mathcal{R}) \to 0} \frac{\int\int\int_{\mathcal{R}} f \, dV_{x,y,z}}{\text{vol}(\mathcal{R})} = f(x_0, y_0, z_0).$$

This result will be useful in later chapters.

10. **Integral Properties.** Let f and g be continuous functions on an open set containing the region $\mathcal{R} = [a, b] \times [c, d]$.

(a) Let α be a scalar. Use the definition of the definite integral to justify the following equality.

$$\int\int_{\mathcal{R}} \alpha f \, dA_{x,y} = \alpha \int\int_{\mathcal{R}} f \, dA_{x,y}.$$

(*Hint:* Can a scalar be factored out of a sum or limit?)

(b) Use the definition of the definite integral to justify the following equality.

$$\int\int_{\mathcal{R}} (f + g) \, dA_{x,y} = \int\int_{\mathcal{R}} f \, dA_{x,y} + \int\int_{\mathcal{R}} g \, dA_{x,y}.$$

(*Hint:* Begin by writing out the definition of the integral for the left-hand side of this equality.)

11. **Improper Integrals.** It is possible to use the concepts that have been developed in this chapter to define the total accumulation of a function on an unbounded subset of the

plane. For example, let $\mathcal{R} = [a, b] \times [c, \infty)$, a vertical strip that extends to ∞ parallel to the positive y-axis. We define the total accumulation of f on $\mathcal{R}$ as a limit of the total accumulation of f on $[a, b] \times [c, d]$ as d approaches ∞. That is,

$$\int\int_{\mathcal{R}} f \, dA_{x,y} = \lim_{d \to \infty} \int\int_{[a,b] \times [c,d]} f \, dA_{x,y}.$$

The integral of f over $\mathcal{R}$ is called an **improper** integral. If the above limit exists, we say that the improper integral is **convergent**. If the limit does not exist, we say that the improper integral is **divergent**. Thus the total accumulation of f on an unbounded region $\mathcal{R}$ is finite if the corresponding improper integral of f over $\mathcal{R}$ is convergent. Which of the following functions f have finite total accumulation on the given regions $\mathcal{R}$?

(a) $f(x, y) = x/(1 + y^2)$ and $\mathcal{R} = [0, 2] \times [1, \infty)$.
(b) $f(x, y) = xe^{-y}$ and $\mathcal{R} = [0, 1] \times [0, \infty)$.
(c) $f(x, y) = x/(1 + y^2)$ and $\mathcal{R} = [0, 2] \times (-\infty, \infty)$.
(d) $f(x, y) = e^{-x-y}$ and $\mathcal{R} = [0, \infty) \times [0, \infty)$.

12. **An Improper Integral.** Let $f(x, y) = e^{-x^2 - y^2}$.

(a) Find the total accumulation of f on a disk of radius R centered at the origin.
(b) Use your answer to (a) to find the total accumulation of f on the entire plane.
(c) Since $f(x, y) = e^{-x^2 - y^2} = e^{-x^2} e^{-y^2}$, we can apply Exercise 7 of Section 5.2. Use this result and your answer to (b) to evaluate $\int_{-\infty}^{\infty} e^{-x^2} dx$.

13. **Upper and Lower Riemann Sums.** A general problem that arises when constructing an approximation is the determination of the error in the approximation. It is possible to obtain a bound on the error in a Riemann sum approximation using **upper** and **lower** Riemann sums. Let f be a continuous function on the rectangle $\mathcal{R} = [a, b] \times [c, d]$, and let $P_{M,N}$ be a regular partition of $\mathcal{R}$, that is, the cells all have dimensions $\Delta x \times \Delta y$, where $\Delta x = (b - a)/M$ and $\Delta y = (d - c)/N$. (Regular partitions will simplify the arguments below.) An **upper** Riemann sum for f on $\mathcal{R}$ is a Riemann sum for which the sample points (x_{ij}^*, y_{ij}^*) are chosen so that the value $f(x_{ij}^*, y_{ij}^*)$ is a **maximum** value for f on the cell R_{ij}. We denote this sum by $U(f, P_{M,N})$. Similarly, we define a **lower** Riemann sum to be a Riemann sum for which the sample points (x_{ij}^*, y_{ij}^*) are chosen so that the value $f(x_{ij}^*, y_{ij}^*)$ is a **minimum** value for f on the cell R_{ij}. We denote this sum by $L(f, P_{M,N})$.

(a) Show that if $R(f, P_{M,N})$ is any Riemann sum for f on $\mathcal{R}$ for the partition $P_{M,N}$, then

$$L(f, P_{M,N}) \leq R(f, P_{M,N}) \leq U(f, P_{M,N}).$$

(*Hint:* Let $(x_{ij}^{max}, y_{ij}^{max})$ be the maximum of f on R_{ij} and $(x_{ij}^{min}, y_{ij}^{min})$ be the minimum of f on R_{ij}. Use these points to demonstrate the inequality on R_{ij} first.)

(b) Let $P_{2M,2N}$ denote the regular partition of $\mathcal{R}$ with $2M \times 2N$ cells. The ij^{th} cell $\mathcal{R}_{ij}$ of $P_{M,N}$ is made up of four cells of $P_{2M,2N}$. Show that the term corresponding to $\mathcal{R}_{ij}$ in the upper sum $U(f, P_{M,N})$ is greater than or equal to the sum of the four terms of $U(f, P_{2M,2N})$ corresponding to $\mathcal{R}_{ij}$. (*Hint:* Think about the maximum values of f on $\mathcal{R}_{ij}$ and the four smaller cells.) Use this fact to conclude that $U(f, P_{2M,2N}) \leq U(f, P_{M,N})$. (Similarly, it can be shown that $L(f, P_{2M,2N}) \geq L(f, P_{M,N})$.)

(c) Construct an argument to show that $\lim_{k \to \infty} U(f, P_{2^k M, 2^k N}) = \int \int_{\mathcal{R}} f \, dA_{x,y}$. (Similarly, it can be shown that $\lim_{k \to \infty} L(f, P_{2^k M, 2^k N}) \leq \int \int_{\mathcal{R}} f \, dA_{x,y}$.)

(d) Use (c) to show that $L(f, P_{M,N}) \leq \int \int_{\mathcal{R}} f \, dA_{x,y} \leq U(f, P_{M,N})$.

(e) Show that if $R(f, P_{M,N})$ is any Riemann sum for f on $\mathcal{R}$ for the partition $P_{M,N}$, then

$$\left| \int \int_{\mathcal{R}} f \, dA_{x,y} - R(f, P_{M,N}) \right| \leq |U(f, P_{M,N}) - L(f, P_{M,N})|.$$

This gives a bound on the error of approximating the integral by a Riemann sum in terms of upper and lower sums.

14. **Additivity of Integrals over Regions.** Let f be a continuous function on the region $\mathcal{R} = [a, b] \times [c, d]$. Suppose that $a < x_* < b$. Let $\mathcal{R}_1 = [a, x_*] \times [c, d]$ and $\mathcal{R}_2 = [x_*, b] \times [c, d]$. Use the definition of the integral to justify the following equality:

$$\int \int_{\mathcal{R}} f \, dA_{x,y} = \int \int_{\mathcal{R}_1} f \, dA_{x,y} + \int \int_{\mathcal{R}_2} f \, dA_{x,y}.$$

(*Hint:* How do partitions of $\mathcal{R}_1$ and $\mathcal{R}_2$ give rise to a partition of $\mathcal{R}$?)

15. **Integrals with Parameters I.** In this exercise, we explore a concept introduced in the discussion of Fubini's theorem. If the integrand f depends on the variables of integration *and* other variables, it is still possible to make sense of the integral of f over a region in the xy-plane. For example, if f is a function of x, y, and w, and for each w, the function $\tilde{f}$ defined by $\tilde{f}(x, y) = f(x, y, w)$ is an integrable function of x and y on the region $\mathcal{R}$, then

$$F(w) = \int \int_{\mathcal{R}} f(x, y, w) \, dA_{x,y}$$

is a function of w. When evaluating integrals of this form, w is treated as a constant. We will use this idea in subsequent exercises. Here are several examples of integrals of this type. For each of the following functions f and regions $\mathcal{R}$, express the function F defined by integration of f over $\mathcal{R}$ as a function of the specified variables.

(a) Let $f(x, y, w) = x^2 - wy^2$, $\mathcal{R} = [-1, 1] \times [-1, 1]$. Compute $F(w)$.

(b) Let $f(x, y, w) = \sin(x - w) \cos(y - w)$, $\mathcal{R} = [0, \pi] \times [0, \pi]$. Compute $F(w)$.

(c) Let $f(x, y, s, t) = (x - s)(y - t)$, $\mathcal{R} = \{(x, y) : -1 \leq x \leq 1 \text{ and } x - 1 \leq y \leq x + 1\}$. Compute $F(s, t)$.

(d) Let $f(x, y, s, t, u) = (x - s)^2 + (y - t)^2 + u^2$, $\mathcal{R}$ is the triangle with vertices $(0, 0)$, $(1, 0)$, and $(0, 1)$. Compute $F(s, t, u)$.

16. **Characteristic Functions.** Suppose that the set S is a union of nonoverlapping regions of the type described in Fubini's theorem. We can define the **characteristic function** of S, which we denote by ch_S, by

$$\mathrm{ch}_S(x, y) = \begin{cases} 1 & \text{if } (x, y) \in S \\ 0 & \text{otherwise.} \end{cases}$$

That is, the characteristic function takes the value 1 when (x, y) is an element of S and the value 0 when (x, y) is not an element of S.

(a) Construct an argument to show that for any rectangle $\mathcal{R} = [a, b] \times [c, d]$ that contains S,

$$\int\int_{\mathcal{R}} \mathrm{ch}_S(x, y) \, dA_{x,y} = \mathrm{Area}(S).$$

(*Hint:* Split the integral into a sum of an integral over S and an integral over the complement of S in $\mathcal{R}$.)

(b) Each of the following functions h is the characteristic function of a set related to the sets S, S_1, or S_2. Describe the set in terms of S, S_1, or S_2. Note that (x_0, y_0) is a fixed point in the plane.

(i) $h(x, y) = \mathrm{ch}_{S_1} \mathrm{ch}_{S_2}$.

(ii) $h(x, y) = \mathrm{ch}_S(x - x_0, y - y_0)$.

(iii) $h(x, y) = \mathrm{ch}_S(-x, -y)$.

(iv) $h(x, y) = \mathrm{ch}_S(x_0 - x, y_0 - y)$.

17. **An Image Processing Filter.** Here we want to apply the ideas of Exercises 15 and 16 to explore a concept from the mathematics of image processing. Let $\mathcal{S} = [-\epsilon, \epsilon] \times [-\epsilon, \epsilon]$ be a square centered at the origin with sides of length 2ϵ, and let $g = \frac{1}{4\epsilon^2} \mathrm{ch}_S$. Further, let f be a continuous function.

(a) Evaluate

$$\int\int_{\mathcal{R}} g \, dA_{x,y},$$

where $\mathcal{R}$ is a region that contains the square S.

(b) What quantity related to f is represented by the integral

$$\int\int_{\mathcal{R}} g(x,y)f(x,y)\,dA_{x,y}?$$

Explain.

(c) Let (x_0, y_0) be a fixed point. What quantity related to f is represented by the integral

$$\int\int_{\mathcal{R}} g(x_0 - x, y_0 - y)f(x,y)\,dA_{x,y}?$$

Explain.

(d) Suppose that f is a numerical representation of the grayscale levels in a black-and-white photograph. For each point (x_0, y_0) in the photograph, replace the existing shading with the shading represented by the value of the integral in part (c). How will this alter the original photograph? Does your answer depend on the size of the square S? Explain.

The transformation of the photographic image represented by the construction of part (d) is an example of filtering an image. Transformations like this can be implemented by a computer.

18. **Integrals with Parameters II.** In Exercise 15, we introduced the idea of the integral of a function that depends on additional variables, beyond the variables of integration. Here we investigate the concept further. Suppose that f is a function of $\mathbf{x}$ and t and that f is an integrable function on $\mathcal{R}$ for each value of t *and* that f is a differentiable function of $\mathbf{x}$ and t. Define a function $F = F(t)$ by integrating f over $\mathcal{R}$ holding t fixed,

$$F(t) = \int\int\int_{\mathcal{R}} f(\mathbf{x}, t)\,dV_{x,y,z}.$$

Then it can be shown that F is a differentiable function of t and

$$\frac{dF}{dt}(t) = \int\int\int_{\mathcal{R}} \frac{\partial f}{\partial t}(\mathbf{x}, t)\,dV_{x,y,z}.$$

Note: This is *not* related to the fundamental theorem of calculus.

If $\mathcal{R}$ represents a container and f represents the concentration of a substance inside the container, then $F(t)$ represents the total amount of substance in the container. Note that in this case $f(\mathbf{x}, t) \geq 0$. Refer to this interpretation of f and F in answering the following questions.

(a) What does $\frac{dF}{dt}(t)$ represent?

(b) If $\frac{\partial f}{\partial t}(\mathbf{x}, t_0) = 0$ for all $\mathbf{x}$, what can we say about $F(t_0)$? Explain.

(c) If $\frac{\partial f}{\partial t}(\mathbf{x}, t_0) < 0$ for all $\mathbf{x}$, what does this say about F at t_0? Explain.

(d) Let $f(\mathbf{x}, t) = c(\mathbf{x}, t)$, the function of Example 5.13 of Section 5.4, and let $\mathcal{R}$ be a region of space containing the origin. Describe the behavior of the corresponding function F on a half-ball centered at the origin. Explain your answer in terms of $\frac{\partial f}{\partial t}$.

(e) Suppose that $\mathcal{R}$ is a hemispherical shell,

$$\mathcal{R} = \{(x, y, z) : r_1 \le x^2 + y^2 + z^2 \le r_2 \}.$$

Describe the behavior of the function F for f on $\mathcal{R}$. Explain your answer in terms of $\frac{\partial f}{\partial t}$.

19. **Partial Derivatives in Polar Coordinates.** If $f = f(x, y)$, we have used the notation f^* to denote the polar coordinate expression of f. That is, $f^*(r, \theta) = f(r \cos \theta, r \sin \theta)$. The partial derivative $\frac{\partial f^*}{\partial r}(r, \theta)$ is the instantaneous rate of change of f^* along a radial line passing through (r, θ). The partial derivative $\frac{\partial f^*}{\partial \theta}(r, \theta)$ is the instantaneous rate of change of f^* along a circle centered at the origin that passes through (r, θ). Because of the relationship between f^* and f, these derivatives are related to the partial derivatives of f.

Using the fact that r and θ are functions of x and y, we can apply the chain rule to express the partial derivatives of f in terms of the partial derivatives of f^*.

$$\frac{\partial f}{\partial x} = \frac{\partial f^*}{\partial r} \frac{\partial r}{\partial x} + \frac{\partial f^*}{\partial \theta} \frac{\partial \theta}{\partial x} \quad \text{and} \quad \frac{\partial f}{\partial y} = \frac{\partial f^*}{\partial r} \frac{\partial r}{\partial y} + \frac{\partial f^*}{\partial \theta} \frac{\partial \theta}{\partial y}.$$

Let us begin with these formulas.

(a) Use the relationship $r^2 = x^2 + y^2$ to show that $\frac{\partial r}{\partial x} = \cos \theta$ and $\frac{\partial r}{\partial y} = \sin \theta$.

(b) By differentiating the relationship $y = r \sin \theta$ with respect to x, show that $\frac{\partial \theta}{\partial x} = -\frac{\sin \theta}{r}$ and, similarly, by differentiating the relationship $x = r \sin \theta$, show that $\frac{\partial \theta}{\partial y} = \frac{\cos \theta}{r}$. (*Hint:* Remember, r and θ are functions of x and y.)

(c) The formulas $\frac{\partial \theta}{\partial x} = -\frac{\sin \theta}{r}$ and $\frac{\partial \theta}{\partial y} = \frac{\cos \theta}{r}$ tell us that the rates of change of θ as a function of x and y decrease as r increases. Explain why this makes sense geometrically.

(d) By substituting the results of parts (a) and (b) into the chain rule expressions above, we can express $\frac{\partial f}{\partial x}$ and $\frac{\partial f}{\partial y}$ in terms of r and θ. Use these formulas to express $\frac{\partial f}{\partial x}$ and $\frac{\partial f}{\partial y}$ in terms of r and θ for $f(x, y) = e^{-x^2 - y^2}$.

Integration on Curves

$\mathbf{I}$n this chapter, we will continue with the discussion of integration. We will be concerned with the total accumulation of a function defined on a curve in the plane or in space. In particular, we will investigate the integral of a vector field along a curve.

In Section 6.1, we will introduce the idea of the total accumulation of a function on a curve, which will lead to the concept of the path integral of a function along a curve. Following the program we developed in Chapter 5, we will first approximate the total accumulation by a Riemann sum and then define the integral to be a limit of Riemann sums. Once again, we will partition the region of integration, which is now a curve, sample the values of the function in each cell of the partition, scale the sampled values, and sum them to produce our Riemann sum. We express the partition and the Riemann sum in terms of a parametrization of the curve. An important question will be to understand how the integral depends on the choice of parametrization.

In Section 6.2, we will introduce a new concept, the line integral of a vector field along a curve, which is the total accumulation of the component of the vector field in the direction of the curve. This will be motivated by a discussion of the concept of work in physics, and we will apply this to a number of examples. In particular, we will explore the relationship between work and energy and define the concepts of kinetic and potential energy.

In Section 6.3, we will focus on the line integral of a vector field around a closed curve. This will lead to a discussion of Green's theorem, a surprising result that connects the line integral of a vector field around a closed curve to a double integral of derivatives of the vector field over the region enclosed by the curve. We will apply these results in a discussion of the flux of a vector field across a curve, where we revisit diffusion in the plane. The section ends with a proof of Green's theorem.

A Collaborative Exercise—A Charged Wire

In Chapter 5, we used Riemann sums to approximate the total accumulation of a function defined over a region in the plane or in space. The process we used to create the Riemann sum was to *partition* the region, *sample* the function in each cell of the partition, *scale* our sampled function value by multiplying it by the size (area or volume) of the cell, and then *sum* these sampled values over all cells in our partition of the region. We saw that we could improve our approximation by *refining* our partition, that is, shrinking the size of the cells and using more sampled values.

Now we will turn our attention to the problem of approximating the total accumulation of a function defined on a path in the plane or in space.

We begin with an example from physics. Let us consider the charge on an insulated copper wire in the presence of a stationary (not changing in time) electric field. Since the wire is insulated, the electric charge cannot leave the wire. However, since copper is a conductor, the charge will move freely in the wire in response to the electric field until it reaches an equilibrium distribution. The charge will generally be unevenly distributed in the wire and can be represented by a *charge density* function ρ that is defined on the wire. The total charge of the wire is the total accumulation of the charge density function over the length of the wire. In this discussion, we will try to develop a method to approximate the total charge of the wire.

1. Assume that a charged wire in space is in the presence of a stationary electric field. Further, assume we can measure the equilibrium charge density ρ at any point (x, y, z) on the wire and that the charge is measured in charge per unit length.

   a. Following the process we used in Chapter 5 (and outlined above), describe how you would set up a Riemann sum to approximate the total charge in the wire.
   b. What would you use for the *scaling factor*?
   c. Write out your Riemann sum in terms of ρ and points (x, y, z) on the wire.

2. Assume that the wire is represented by the image of a parametrization $\alpha : [a, b] \subset \mathbb{R} \to \mathbb{R}^3$. Thus points on the wire are of the form $\alpha(t) = (x(t), y(t), z(t))$ for $a \leq t \leq b$. Here we want to express the Riemann sum from 1c in terms of t.

   a. How would you represent your partition of the wire into N cells in terms of the parameter t?
   b. How would you represent your sampling scheme in terms of the parameter t?
   c. What would the charge density be at the sampled points?
   d. Can you represent the scaling factor for the ith cell in terms of t? If not, can you use the parametrization to approximate the scaling factor?
   e. Using your answers to a through d, express your Riemann sum from 1c in terms of t.

3. Based on our previous work with Riemann sums, what would you expect to do to calculate an exact value for the total charge on the wire?

■ 6.1 Path Integrals

In this section, we will turn our attention to the problem of determining the total accumulation of a function defined on a curve in the plane or in space. Before we begin, let us consider several examples of physical phenomena that can be modeled by functions on curves.

Example 6.1 **Total Accumulation on a Curve**

A. Total Charge. As in the collaborative exercise, consider an electric charge in an insulated wire. In the presence of an electric field, the electric charge in the wire will be unevenly distributed in the wire. This can be represented by a *charge density* function ρ that is defined on the wire. The total charge of the wire is the total accumulation of the charge density function over the length of the wire.

B. Total Mass. Suppose we have a function that represents a density distribution along a curve, for example, a wire coated by ice of varying thickness. The total accumulation of the density function along the curve is the *total mass*.

C. Arc Length. If a function takes a constant value 1 everywhere on a curve, the total accumulation of the function is the *arc length* of the curve.

D. Potential Energy. Consider a system that consists of a uniformly charged wire and a charged particle located at a point **x** not on the wire. The potential energy function for the system can be expressed as the total accumulation of the potential energy function for the particle in space and a charged particle on the wire, where the accumulation is taken over the particles in the wire.

As we did in Chapter 5, we will begin by constructing a Riemann sum to approximate the total accumulation of a function defined on a curve. In particular, we will express the Riemann sum in terms of the coordinate of a parametrization of the curve. That is, we will partition the domain of the parametrization, which will then give rise to a partition of the curve that is the image of the parametrization. Then we will sample the function on the curve, scale the sampled values, and sum the scaled values. In this case, since the domain is a subset of $\mathbb{R}$, the scaling factor will be the length of a cell rather than the area of a cell. The result will be a Riemann sum for a function of one variable. Thus, we will be able to use the fundamental theorem of calculus to evaluate the integral. Initially, we will work with curves in the plane; the generalization to curves in space is immediate.

Let us begin with a function f defined on a curve $\mathcal{C}$ in the plane. We will assume that we have been given or have constructed a parametrization $\alpha : [a, b] \to \mathbb{R}^2$ of $\mathcal{C}$. If $\alpha(t) = (x(t), y(t))$,

$$\mathcal{C} = \{(x(t), y(t)) : a \le t \le b\} \subset \mathbb{R}^2.$$

We want to construct a Riemann sum to approximate the total accumulation of f on $\mathcal{C}$, the image of α. In order to partition $\mathcal{C}$, we start with a partition P of $[a, b]$, the domain of α, into N subintervals. We choose $N + 1$ points t_i, $i = 0, \dots, N$, so that

$$a = t_0 < t_1 < t_2 < \dots < t_{N-1} < t_N = b.$$

The ith subinterval of the partition, $[t_{i-1}, t_i]$, has length $\Delta t_i = t_i - t_{i-1}$. The image of the partition P of $[a, b]$ gives rise to a partition of $\mathcal{C}$ into N cells. The ith cell of this partition is the image of the ith subinterval of P,

$$\{(x(t), y(t)) : t_{i-1} \le t \le t_i\}.$$

(See Figure 6.1.) In order to sample f in the ith cell of $\mathcal{C}$, choose a point t_i^* in the ith subinterval of the partition P of $[a, b]$, that is, $t_{i-1} \le t_i^* \le t_i$. We sample f at the point $\alpha(t_i^*)$, so that the sampled value of f in the ith cell is $f(\alpha(t_i^*))$.

Following the procedure we developed for constructing Riemann sums in Chapter 5, we scale the sampled value $f(\alpha(t_i^*))$ by the length Δs_i of the ith cell of $\mathcal{C}$. Thus the Riemann sum for the partition P is

$$R(f, \alpha, P) = \sum_{i=1}^{N} f(\alpha(t_i^*)) \Delta s_i.$$

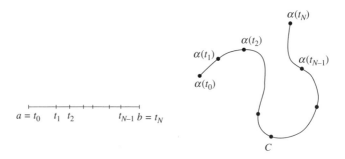

Figure 6.1 A partition of the interval $[a, b]$ and the corresponding partition of the curve $\mathcal{C}$ that is the image of α.

In order to obtain an exact value for the total accumulation of f on α, we evaluate the limit of $R(f, \alpha, P)$ as the mesh of P approaches 0, where the mesh of P, mesh(P), is the largest length Δs_i in the partition. Once again, if this limit exists and is independent of the choices of partitions and sampling schemes, we define the limit to be an integral of f. Since this construction depends on the choice of parametrization α, we will say that the integral is over α rather than $\mathcal{C}$. Note that the analogous construction applies to functions f of three variables and parametrizations α of curves in space. If we assume f is continuous, and α is continuously differentiable, that is, α is differentiable and α' is continuous, then one can prove this limit exists and is independent of choices. We summarize this in the following proposition:

Proposition 6.1 Let $\alpha : [a, b] \rightarrow \mathbb{R}^2$ or $\mathbb{R}^3$ be a continuously differentiable function, and let f be a continuous function defined on the image of α. Then

$$\lim_{\text{mesh}(P) \to 0} R(f, \alpha, P)$$

exists and is independent of the choices of partitions and sampling schemes. We call this limit the **path integral** of f on α, and we denote it by $\int_\alpha f \, ds$. Thus,

$$\int_\alpha f \, ds = \lim_{\text{mesh}(P) \to 0} R(f, \alpha, P). \; \blacklozenge$$

Before we develop a method for evaluating path integrals, let us consider an example of a calculation of a Riemann sum and a path integral for the potential energy of a charged particle and a uniformly charged wire. (See Example 6.1D.)

Example 6.2

Potential Energy of a Uniformly Charged Wire. Consider a uniformly charged wire of charge density C in the shape of a circle of radius 1 and a particle of charge q that is located at a point not on the wire. If the wire is fixed in space, potential energy depends only on the position of the charged particle. Intuitively, the potential energy of the system is obtained by summing the potential energy of pairs of points consisting of the point in space and a point in the wire. The potential energy of two charged particles with charges q_1 and q_2 is

$$\frac{q_1 q_2}{4\pi\epsilon_0} \frac{1}{r},$$

where r is the distance between the two particles and ϵ_0 is a constant.

To simplify matters, we will identify the wire with the circle parametrized by $\alpha(t) = (\cos t, \sin t, 0)$, $0 \le t \le 2\pi$, and will locate the particle of charge q at $(0, 1, 1)$. We will partition the domain $[0, 2\pi]$ into N subintervals, and we will let Δt_i be the length of the ith subinterval and t_i^* be a point in the ith subinterval of this partition. The partition of

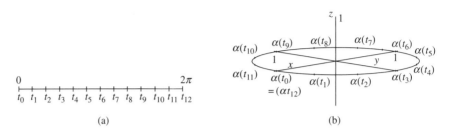

Figure 6.2 (a) A partition of the domain $[0, 2\pi]$ of Example 6.2 into $N = 12$ cells. (b) The image under α of the partition of $[0, 2\pi]$ is a partition of the unit circle into $N = 12$ cells.

the wire is the image of this partition under α. (See Figure 6.2.) For this parametrization, the length of the ith cell in the partition of the wire is also Δt_i. (Why?)

Since the length of a cell is small, the charge in the cell acts like a point charge of magnitude $C\Delta t_i$. Thus the potential energy due to the ith cell is approximately

$$\frac{qC\Delta t_i}{4\pi\epsilon_0}\frac{1}{r} = \frac{qC\Delta t_i}{4\pi\epsilon_0}\frac{1}{\|(0,1,1) - \alpha(t_i^*)\|}$$

$$= \frac{qC}{4\pi\epsilon_0}\frac{\Delta t_i}{\sqrt{\cos(t_i^*)^2 + (1 - \sin(t_i^*))^2 + 1}}.$$

The potential energy can be approximated by the Riemann sum

$$\frac{Cq}{4\pi\epsilon_0}\sum_{i=1}^{N}\frac{\Delta t_i}{\sqrt{\cos(t_i^*)^2 + (1 - \sin(t_i^*))^2 + 1}}.$$

We must now evaluate the limit of this Riemann sum as the mesh of the partition approaches 0, that is, as Δt_i approaches 0. The limit is equal to

$$\frac{Cq}{4\pi\epsilon_0}\int_0^{2\pi}\frac{1}{\sqrt{\cos^2 t + (1 - \sin t)^2 + 1}}\,dt = \frac{Cq}{4\pi\epsilon_0}\int_0^{2\pi}\frac{1}{\sqrt{3 - 2\sin t}}\,dt.$$

This defines the potential energy U of the system when the charged particle is located at $(0, 1, 1)$.

The Path Integral

Returning to the general case, let us assume that f is continuous and that α is continuously differentiable on $[a, b]$. In order to evaluate the path integral of f on α we must equate it with a definite integral in one variable.

We begin by approximating the Riemann sum

$$R(f, \alpha, P) = \sum_{i=1}^{N} f(\alpha(t_i^*)) \Delta s_i$$

with a Riemann sum for a function of t on $[a, b]$. Since we cannot, in general, compute the length Δs_i of a cell explicitly, we will approximate it by the length of the straight line segment between the endpoints of the cell,

$$\Delta s_i \approx \|\alpha(t_i) - \alpha(t_{i-1})\|.$$

However, this is not yet in the form that we want. Since α is differentiable, we can approximate $\alpha'(t)$ by the difference quotient $\frac{1}{\Delta t}(\alpha(t + \Delta t) - \alpha(t))$ for small values of Δt. It follows that

$$\alpha(t + \Delta t) - \alpha(t) \approx \alpha'(t) \Delta t.$$

Since $t_i = t_{i-1} + \Delta t_i$ where Δt_i is positive, we have

$$\begin{aligned}
\|\alpha(t_i) - \alpha(t_{i-1})\| &= \|\alpha(t_{i-1} + \Delta t_i) - \alpha(t_{i-1})\| \\
&\approx \|\alpha'(t_{i-1}) \Delta t_i\| \\
&= \|\alpha'(t_{i-1})\| \Delta t_i.
\end{aligned}$$

Returning to our Riemann sum, if we choose $t_i^* = t_{i-1}$, then $\Delta s_i \approx \|\alpha'(t_i^*)\| \Delta t_i$ and

$$R(f, \alpha, P) \approx \sum_{i=1}^{N} f(\alpha(t_i^*)) \|\alpha'(t_i^*)\| \Delta t_i.$$

This sum is a Riemann sum for the function of t given by $f(\alpha(t)) \|\alpha'(t)\|$. Since f, α, and α' are continuous, the limit of Riemann sums of this form exists and is equal to

$$\int_a^b f(\alpha(t)) \|\alpha'(t)\| \, dt.$$

It is possible to make this intuitive argument rigorous, though we will not do so here. We summarize this discussion in the following theorem. Notice that the criteria for the existence of the integral of f on α depend on both f and α.

Theorem 6.1 Let $\alpha : [a, b] \to \mathbb{R}^2$ be a continuously differentiable function, and let f be a continuous function defined on the image of α. Then

$$\int_\alpha f \, ds = \int_a^b f(\alpha(t)) \|\alpha'(t)\| \, dt. \quad \blacklozenge$$

If α is a parametrization of a curve in space, the substance of the argument remains the same and the Riemann sums and limit take the same form. We will use the same notation for the integral of a function on a space curve.

We use this result in the following example.

Example 6.3   **Path Integrals**

A. Total Mass on a Curve. Let $\alpha(t) = (t, t^3)$, $0 \le t \le 2$, be a parametrization of the curve $y = x^3$ between $(0, 0)$ and $(2, 8)$, and let $\delta(x, y) = y$ be a function that represents a density distribution on the curve. The total mass along α is the total accumulation of δ. Thus the total mass is

$$\int_\alpha \delta \, ds = \int_0^2 \delta(\alpha(t)) \|\alpha'(t)\| \, dt$$

$$= \int_0^2 t^3 \sqrt{1 + 9t^4} \, dt$$

$$= \frac{1}{54} \left. (1 + 9t^4)^{3/2} \right|_0^2$$

$$= \frac{(145)^{3/2} - 1}{54}.$$

B. Total Accumulation on a Helix. Let $f(x, y, z) = x^2 + y^2 + z^2$, and let $\alpha(t) = (\cos t, \sin t, t)$, $0 \le t \le 2\pi$, a parametrization of a spiral helix of radius 1 and height 2π. The total accumulation of f on α is

$$\int_\alpha f \, ds = \int_0^{2\pi} f(\alpha(t)) \|\alpha'(t)\| \, dt$$

$$= \int_0^{2\pi} f(\cos t, \sin t, t) \sqrt{2} \, dt$$

$$= \int_0^{2\pi} \left(1 + t^2 \right) \sqrt{2} \, dt$$

$$= \sqrt{2} \left[t + t^3/3 \right]_0^{2\pi}$$

$$= \sqrt{2} \left(2\pi + \frac{8}{3}\pi^3 \right).$$

If the function f is the constant function equal to 1, $f(x, y) = 1$ for all x and y, then the Riemann sum for f over α is

$$R(f, \alpha, P) = \sum_{i=1}^{N} \Delta s_i \approx \sum_{i=1}^{N} \|\alpha(t_i) - \alpha(t_{i-1})\|.$$

Each term is an approximation of the length of an individual cell of the partition of the image of α. The limit of these Riemann sums is defined to be the arc length of the parametrization α. Thus we have the following definition.

Definition 6.1 Let $\alpha : [a, b] \to \mathbb{R}^2$ or $\mathbb{R}^3$ be a continuously differentiable function. We define the **arc length** of α to be the path integral $\int_\alpha 1 \, ds = \int_a^b \|\alpha'(t)\| \, dt$. ◆

If $\alpha : [a, b] \to \mathbb{R}^2$ or $\mathbb{R}^3$ is continuously differentiable and f is a continuous function defined on the image of α, then we define the **average value** of f on α to be the total accumulation of f on α divided by the arc length α. In the following example, we use this to compute the average value of the function $f(x, y) = x$ on a parametrization α.

Example 6.4 **Arc Length and Average Value.** Let $\alpha(t) = (t, t^2)$, $0 \le t \le 2$, which parametrizes an arc $\mathcal{C}$ of the parabola given by $y = x^2$, and let $f(x, y) = x$. To find the average value of f on $\mathcal{C}$, we first compute the arc length of α.

$$\int_\alpha ds = \int_0^2 \|\alpha'(t)\| \, dt$$

$$= \int_0^2 \|(1, 2t)\| \, dt$$

$$= \int_0^2 \sqrt{1 + 4t^2} \, dt$$

$$= \left[\frac{t}{2}\sqrt{1 + 4t^2} + \frac{1}{4}\ln(2t + \sqrt{1 + 4t^2}) \right]_0^2$$

$$= \left[\sqrt{17} + \frac{1}{4}\ln(4 + \sqrt{17}) \right].$$

The total accumulation of $f(x, y) = x$ over α is

$$\int_\alpha f \, ds = \int_0^2 t\sqrt{1 + 4t^2} \, dt$$

$$= \left[\frac{2}{3 \cdot 8}(1 + 4t^2)^{3/2} \right]_0^2$$

$$= \frac{1}{12}(17^{3/2} - 1).$$

Thus the average value of f on α is the quotient of these two values, which is approximately 1.239.

Independence of Parametrization

Since our calculations depend on a parametrization of a curve, it would appear that a different choice of parametrization might yield a different value for the integral. Fortunately, this is not the case, since, as we will now demonstrate, the value of a path integral does not depend on the choice of the parametrization of the curve.

First, we must consider the relationship between different parametrizations of the same curve. Let us begin with a differentiable parametrization $\alpha = \alpha(u)$ of a curve C, $c \leq u \leq d$, that is one-to-one. This ensures that α traces C just once. If $u = h(t)$ is a differentiable function of one variable, $h : [a, b] \rightarrow [c, d]$, with $h(a) = c$, $h(b) = d$, and $h'(t) > 0$ on $[a, b]$, then the composition $\beta(t) = (\alpha \circ h)(t) = \alpha(h(t))$ is also a differentiable parametrization of the curve C that traces C just once. We claim that the total accumulation of a function f over α and over β are the same. That is, we have the following proposition, whose proof is left as an exercise. (See Exercise 8.)

Proposition 6.2   Under the above hypotheses on the parametrizations β and α of C,

$$\int_{\beta} f \ ds = \int_{\alpha} f \ ds. \ \blacklozenge$$

Notice that since $\beta'(t) = \alpha'(h(t))h'(t)$ and $h'(t) > 0$, β' is a positive multiple of α', so that β and α parametrize C in the same direction. If instead h had satisfied $h(a) = d$, $h(b) = c$, and $h'(t) < 0$, β' would be a negative multiple of α', and β and α would parametrize C in opposite directions. The above proposition also holds in this case.

Finally, if α and β are one-to-one continuously differentiable parametrizations of the same curve C, then it can be shown that there is a differentiable function h such that $\beta = \alpha \circ h$. Combining this fact with the proposition, it follows that the total accumulation of a function on the image of a parametrization depends on the image curve C and not on the choice of parametrization, as long as the parametrization is one-to-one. In particular, if C is a curve and α is a one-to-one parametrization of C, then we can define the arc length of the curve C to be the arc length of α. Thus we have the following definitions of the path integral of a function on a curve and the arc length of a curve.

Definition 6.2   Let C be a curve in the plane or in space, and let f be a continuous function whose domain contains C.

We define the **path integral** of f on $\mathcal{C}$ to be the path integral of f on α, where α is **any** one-to-one continuously differentiable parametrization of $\mathcal{C}$. We denote the path integral of f on $\mathcal{C}$ by $\int_{\mathcal{C}} f\, ds$.

We define the **arc length** of $\mathcal{C}$ to be the arc length of α, where α is **any** one-to-one continuously differentiable parametrization of $\mathcal{C}$. Thus the arc length of $\mathcal{C}$ is $\int_{\mathcal{C}} ds = \int_{\alpha} ds$. ◆

In the following example of total accumulation on a curve, we are given $\mathcal{C}$ and must find a convenient parametrization in order to compute the path integral.

Example 6.5

Total Accumulation on a Curve. Let us compute the total accumulation of $f(x,y) = xy$ on the portion of the curve $x^2 + 4y^2 = 4$ in the first quadrant. First we must parametrize the curve. Let us use the parametrization $\alpha(t) = (2\cos t, \sin t)$, $0 \le t \le \pi/2$. Then the total accumulation of f on $\mathcal{C}$ can be computed as follows.

$$
\begin{aligned}
\int_{\mathcal{C}} f\, ds &= \int_0^{\pi/2} f(\alpha(t)) \| \alpha'(t) \| \, dt \\
&= \int_0^{\pi/2} 2\cos t \sin t \sqrt{4\sin^2 t + \cos^2 t}\, dt \\
&= \int_0^{\pi/2} 2\cos t \sin t \sqrt{3\sin^2 t + 1}\, dt \\
&= \left[(2/9)\left(3\sin^2 t + 1\right)^{3/2} \right]_0^{\pi/2} \\
&= 14/9.
\end{aligned}
$$

Summary

In this section, we introduced the concept of the **path integral** of a function, which extends integration to functions defined on curves in the plane or in space.

We began by constructing a Riemann sum for the total accumulation of a function f on the curve $\mathcal{C}$. We first **partitioned the domain of a parametrization** $\alpha : [a, b] \to \mathbb{R}^2$ of $\mathcal{C}$. This gave rise to a partition of $\mathcal{C}$, which we used to construct a Riemann sum for the total accumulation of f. By taking the limit of Riemann sums constructed in this manner, we obtained an exact value for the total accumulation of f on α. This value is called the **path integral** of f over α, which we denoted by $\int_{\alpha} f\, ds$.

The integral exists when f and α are continuously differentiable. Further,

$$\int_{\alpha} f \, ds = \int_a^b f(\alpha(t)) \|\alpha'(t)\| \, dt.$$

Using this form, it is possible to show that the total accumulation of f over a parametrization of $\mathcal{C}$ does not depend on the choice of the parametrization α. Thus $\int_{\mathcal{C}} f \, ds = \int_{\alpha} f \, ds$ for any one-to-one parametrization α of $\mathcal{C}$.

Section 6.1 Exercises

1. **A Riemann Sum Calculation.** Figure 6.3 is a contour plot of a trail up Stinson Mountain in New Hampshire from the parking lot, located at approximately 1500 ft above sea level, to the peak, which is approximately 2850 ft above sea level. The contours are at 40-ft intervals.

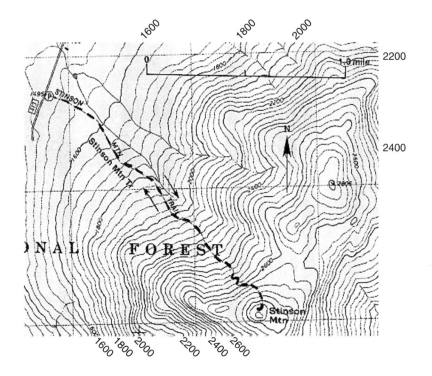

Figure 6.3   The contour plot of Stinson Mountain, New Hampshire, for Exercise 1. Map courtesy of the U.S. Geological Survey.

(a) Explain how you would use the contour map to construct a Riemann sum to approximate the length of this trail. (*Hint:* In order to do this, we must partition the path and then approximate Δs_i, the length of each cell in the partition. Keep in mind that the contour plot is a two-dimensional representation of the mountain, and thus the path drawn on the contour map is a two-dimensional representation of a path in space.)

(b) Use a Riemann sum with $N = 4$ cells to approximate the length of this trail. Carefully describe the endpoints of the cells in your partition of the trail, give the approximating value of Δs_i for each cell, and compute the sum that will approximate the length.

(c) We know we can improve the approximation by increasing the number of cells in the partition. Use a Riemann sum with $N = 8$ cells to approximate the length of this trail. Carefully describe the endpoints of the cells in your partition of the trail, give the approximating value of Δs_i for each cell, and compute the sum that will approximate the length.

(d) Why should your approximation in (c) be better than your approximation in (b)?

2. Arc Length. Compute the arc length of the following parametrizations.

(a) $\alpha(t) = (t, t^{3/2})$ for $1 \leq t \leq 4$.

(b) $\alpha(t) = (\cos(3t), \sin(3t), t)$ for $0 \leq t \leq \pi$.

(c) $\alpha(t) = (3t^2, 2t^3, \frac{3}{4}t^4)$ for $0 \leq t \leq 2$.

3. Path Integrals over $\mathcal{C}$. For each of the following functions f and curves $\mathcal{C}$, (i) construct a parametrization α of $\mathcal{C}$ and (ii) set up and evaluate an integral to compute the total accumulation of f on $\mathcal{C}$.

(a) $\mathcal{C}$ is the line segment from $(1, -1, 2)$ to $(3, 0, 1)$ and $f(x, y, z) = e^{x+y+z}$.

(b) $\mathcal{C}$ is the circle of radius 2 in the xy-plane centered at the origin and $f(x, y) = x^2 y$.

4. Total Mass. For each of the following parametrizations α and density functions δ, compute the total mass, $\int_\alpha \delta \, ds$.

(a) $\alpha(t) = (t, \ln(t))$ for $1 \leq t \leq 3$ with density function $\delta(x, y) = x^2$.

(b) $\alpha(t) = (t, t^2, t^2)$ for $0 \leq t \leq 2$ with density function $\delta(x, y, z) = x$.

5. Average Value. Compute the average value of $f(x, y, z) = x^2 + y^2 + z^2$ over each of the following parametrizations.

(a) $\alpha(t) = (\cos(3t), \sin(3t), t)$ for $0 \leq t \leq \pi$. (See Exercise 2(b).)

(b) $\alpha(t) = (3t^2, 2t^3, \frac{3}{4}t^4)$ for $0 \leq t \leq 2$. (See Exercise 2(c).)

6. Piecewise Defined Curves. Let $\mathcal{C}$ be a curve in the plane or space, where $\mathcal{C} = \mathcal{C}_1 \cup \mathcal{C}_2 \cup \ldots \cup \mathcal{C}_n$ with $\mathcal{C}_i$ parametrized by $\alpha_i = \alpha_i(t)$, $a_i \leq t \leq b_i$. Assume that each α_i is continuously differentiable. Then we define the integral of f over $\mathcal{C}$ by

$$\int_{\mathcal{C}} f \, ds = \int_{\mathcal{C}_1} f \, ds + \int_{\mathcal{C}_2} f \, ds + \cdots + \int_{\mathcal{C}_n} f \, ds.$$

That is, to compute the total accumulation of f, identify the pieces C_i of the curve C, parametrize each one separately, evaluate the integral of f over each piece, and then sum the results to obtain the total accumulation of f over C.

(a) Let $f(x, y) = xy$. Compute the total accumulation of f over the curve that is the boundary of the square with vertices at $(0, 0)$, $(2, 0)$, $(2, 2)$, and $(0, 2)$.

(b) Let $f(x, y) = x^2 + y^2$. Compute the total accumulation of f over the curve that is the boundary of the upper half of the disk of radius 2 centered at the origin.

(c) Let $f(x, y, z) = (x + y)e^z$. Compute the total accumulation of f over the curve that is the boundary of the triangle with vertices $(1, 0, 0)$, $(0, 0, 1)$, and $(0, 1, 0)$.

7. **Properties of Path Integrals.** Let $\alpha : [a, b] \to \mathbb{R}^2$ or $\mathbb{R}^3$ be a continuously differentiable function and let f and g be continuous functions defined on the image of α. Use the definition of the path integral to justify the following equations.

(a) If c is a constant, $\int_\alpha cf \, ds = c \int_\alpha f \, ds$.

(b) $\int_\alpha (f + g) \, ds = \int_\alpha f \, ds + \int_\alpha g \, ds$.

8. **Independence of Parametrization I.** Let $\alpha(u)$ be a differentiable parametrization of a curve C, $c \le u \le d$, that is one-to-one and has nonzero derivative on (c, d). Suppose $h(t)$ is a differentiable function of one variable, $h : [a, b] \to [c, d]$.

(a) Assume $h(a) = c$, $h(b) = d$, and $h'(t) > 0$ on $[a, b]$. Let $\beta(t) = \alpha(h(t))$; then β and α parametrize C in the same direction. Show that

$$\int_\beta f \, ds = \int_\alpha f \, ds.$$

(*Hint:* Evaluate $\int_\alpha f \, ds$ using the substitution $u = h(t)$.)

(b) Assume $h(a) = d$, $h(b) = c$, and $h'(t) < 0$ on $[a, b]$. Let $\beta(t) = \alpha(h(t))$; then β and α parametrize C in opposite directions. Show that

$$\int_\beta f \, ds = \int_\alpha f \, ds.$$

9. **Independence of Parametrization II.** In the previous exercise, we proved that the path integral of f on a curve C is independent of the choice of parametrization of C as long as the parametrizations are one-to-one. Explain why it is necessary that the parametrizations be one-to-one.

10. **Potential Energy.** In Example 6.2, suppose that the particle of charge q is located at the point $(0, 0, z)$ on the z-axis. Show that the potential energy is given by $\frac{Cq}{4\pi\epsilon_0} \int_0^{2\pi} \frac{1}{\sqrt{1+z^2}} \, dt$.

11. **Potential Energy of a Charged Segment.** Consider a system that consists of a charged straight wire of length 2 and a charged particle not on the wire. Assume that the wire is modeled by the line segment $\{(x, 0, 0) : -1 \le x \le 1\}$ and has constant charge density C and that the particle of charge q is located at $\mathbf{x} = (x, y, z)$.

(a) Set up a Riemann sum to approximate the potential energy function $U(\mathbf{x})$ of this system.
(b) If you consider the limit of these Riemann sums as the mesh of the partition approaches 0, what is the resulting integral?
(c) Evaluate the integral from part (b) when $(x, y, z) = (0, 0, 1)$.
(d) Evaluate the integral from part (b) when $(x, y, z) = (x_0, y_0, z_0)$.

■ 6.2 Line Integrals

In this section, we want to explore an important use of path integration in physics, the definition and calculation of the work done by a force on an object. Later, we will relate work to kinetic and potential energy. This definition will also give rise to the more general notion of a line integral of a vector field over the image of a parametrization. In order to simplify our presentation, we will state our results for the plane, but keep in mind that they also apply in space.

We begin by recalling the definition of the work done by a constant force on a particle moving in a straight line in the plane. (See Section 1.3.)

Definition 6.3 Let $\mathbf{F}$ be a constant force field; that is, $\mathbf{F}(x, y) = \mathbf{F}_0$ for all (x, y). If $\mathbf{F}$ moves a particle a distance d_0 in the direction $\mathbf{u}$, where $\|\mathbf{u}\| = 1$, then the ***work done by*** $\mathbf{F}$ ***on the particle*** is the product of the component of $\mathbf{F}$ in the direction of motion with the distance that the particle moves, that is,

$$W = (\mathbf{F} \cdot \mathbf{u})d_0. \quad \blacklozenge$$

Now let us suppose that an object moves through a continuous nonconstant force field

$$\mathbf{F}(x, y) = (u(x, y), v(x, y))$$

from a point P to a point Q. Suppose also that the motion of the object is described by a parametrization $\alpha(t) = (x(t), y(t))$, $a \leq t \leq b$, with $\alpha(a) = P$, $\alpha(b) = Q$, and $\alpha'(t) \neq \mathbf{0}$. Since $\mathbf{F}$ is not constant and α may not be linear, the above definition does not apply to this situation. However, since α is differentiable and $\mathbf{F}$ is continuous, over small segments of the image, the motion is approximately linear and the force is approximately constant. Thus we can use the above definition to approximate the work done by the force on small segments of the image. Summing these approximations over the image, we will arrive at an approximation to the total work done by $\mathbf{F}$ as the particle moves according to α.

We can carry out such an approximation by following the procedure we established in Section 6.1. We partition the interval $[a, b]$ into N subintervals $[t_{i-1}, t_i]$ of length

$\Delta t_i = t_i - t_{i-1}$ and select a point t_i^* in the ith subinterval. We will approximate the force acting on the object as it moves from $\alpha(t_{i-1})$ to $\alpha(t_i)$ by $\mathbf{F}(\alpha(t_i^*))$, and we will approximate the direction of the motion of the particle by the unit vector

$$\mathbf{T}(\alpha(t_i^*)) = \alpha'(t_i^*)/\|\alpha'(t_i^*)\|.$$

(See Figure 6.4.) Note that since $\alpha'(t) \neq 0$, this expression is well defined. Thus the component of the force in the direction of motion on the ith cell is approximated by $\mathbf{F}(\alpha(t_i^*)) \cdot \mathbf{T}(\alpha(t_i^*))$.

The work done by $\mathbf{F}$ in moving the particle along the ith cell of the partition is approximated by the product of $\mathbf{F}(\alpha(t_i^*)) \cdot \mathbf{T}(\alpha(t_i^*))$ with the distance Δs_i that the particle moves in going from $\alpha(t_{i-1})$ to $\alpha(t_i)$. Thus the approximation of the work for a particular cell is

$$(\mathbf{F}(\alpha(t_i^*)) \cdot \mathbf{T}(\alpha(t_i^*))) \, \Delta s_i.$$

The sum of these values gives an approximation to the total work done by the force as the particle moves according to α. It is a Riemann sum for the total accumulation of the function $\mathbf{F} \cdot \mathbf{T}$ over α:

$$R(\mathbf{F} \cdot \mathbf{T}, \alpha, P) = \sum_{i=1}^{N} (\mathbf{F}(\alpha(t_i^*)) \cdot \mathbf{T}(\alpha(t_i^*))) \, \Delta s_i.$$

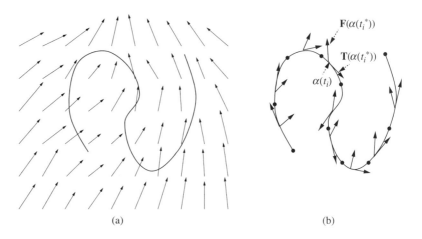

(a)   (b)

Figure 6.4   (a) A curve in a vector field. (b) The vectors $\mathbf{F}(\alpha(t_i^*))$ and $\mathbf{T}(\alpha(t_i^*))$ at the points $\alpha(t_i^*)$ on the image of α.

Since $\mathbf{F}$, α, and $\mathbf{T}$ are continuous, $\mathbf{F}(\alpha(\mathbf{t})) \cdot \mathbf{T}(\alpha(t))$ is also continuous. Consequently, the $\lim_{\mathrm{mesh}(P) \to 0} R(\mathbf{F} \cdot \mathbf{T}, \alpha, P)$ exists and is equal to the path integral of the function $\mathbf{F} \cdot \mathbf{T}$ on α, $\int_\alpha \mathbf{F} \cdot \mathbf{T} \, ds$. The path integral is equal to

$$\int_\alpha \mathbf{F} \cdot \mathbf{T} \, ds = \int_a^b (\mathbf{F}(\alpha(t)) \cdot \mathbf{T}(\alpha(t))) \, \|\alpha'(t)\| \, dt.$$

This discussion leads to the following general definition of work:

Definition 6.4 If $\mathbf{F} = \mathbf{F}(x, y)$ is a force field, then the **work** done by $\mathbf{F}$ on a particle moving according to α, W_α, is the path integral over α of the component of $\mathbf{F}$ in the direction of α. That is,

$$W_\alpha = \int_\alpha \mathbf{F} \cdot \mathbf{T} \, ds,$$

where $\mathbf{T}$ is the unit tangent vector to α.   ◆

More generally, suppose $\mathbf{F}$ is a vector field defined on a domain in $\mathbb{R}^2$ that contains the image of the parametrization α.

Definition 6.5 The **line integral** of $\mathbf{F}$ over the image of α is $\int_\alpha \mathbf{F} \cdot \mathbf{T} \, ds$, where $\mathbf{T}$ is the unit tangent vector to α.   ◆

Thus the line integral of $\mathbf{F}$ over α is defined to be the path integral of $\mathbf{F} \cdot \mathbf{T}$ over α. This integral can then be evaluated as a one-variable integral in terms of the parameter t. Since $\mathbf{T}(\alpha(t)) = \alpha'(t)/\|\alpha'(t)\|$, we can simplify the resulting definite integral:

$$\int_\alpha \mathbf{F} \cdot \mathbf{T} \, ds = \int_a^b \mathbf{F}(\alpha(t)) \cdot \mathbf{T}(\alpha(t)) \|\alpha'(t)\| \, dt$$

$$= \int_a^b \mathbf{F}(\alpha(t)) \cdot \frac{\alpha'(t)}{\|\alpha'(t)\|} \|\alpha'(t)\| \, dt.$$

Canceling the $\|\alpha'\|$ terms, we have

$$\int_\alpha \mathbf{F} \cdot \mathbf{T} = \int_a^b \mathbf{F}(\alpha(t)) \cdot \alpha'(t) \, dt,$$

which is considerably easier to evaluate than the original path integral formulation. The following example illustrates the use of this form for a line integral.

Example 6.6   **A Line Integral.** Let $\mathbf{F}(x, y) = (-y, x)$, and let $\alpha(t) = (t^2, t^3)$, $0 \le t \le 1$. The line integral of $\mathbf{F}$ over α is given by

$$
\int_\alpha \mathbf{F} \cdot \mathbf{T} \ ds = \int_0^1 \mathbf{F}(\alpha(t)) \cdot \alpha'(t) \ dt
$$

$$
= \int_0^1 (-t^3, t^2) \cdot (2t, 3t^2) \ dt
$$

$$
= \int_0^1 (-2t^4 + 3t^4) \ dt
$$

$$
= \int_0^1 t^4 \ dt
$$

$$
= \left[\frac{1}{5} t^5 \right]_0^1 = 1/5.
$$

Now let us consider an example of work involving a charged particle moving through an electric field. The force exerted by an electric field $\mathbf{E}$ on a particle of charge q is $q\mathbf{E}$. If the particle moves through $\mathbf{E}$ according to α, the work done by the force on the particle is

$$
W_\alpha = \int_\alpha q\mathbf{E} \cdot \mathbf{T} \ ds = \int_a^b q\mathbf{E}(\alpha(t)) \cdot \alpha'(t) \ dt.
$$

Here we consider the particular case of an electric field generated by a charged particle.

Example 6.7   **Work Done by an Electric Field.** A particle of charge q_0 located at the origin gives rise to an electric field

$$
\mathbf{E}(\mathbf{x}) = \frac{q_0}{4\pi\epsilon_0} \frac{\mathbf{x}}{\|\mathbf{x}\|^3},
$$

where ϵ_0 is a constant. Thus the force exerted by the field on a particle of charge q located at $\mathbf{x}$ is

$$
\mathbf{F}(\mathbf{x}) = q\mathbf{E}(\mathbf{x}) = \frac{qq_0}{4\pi\epsilon_0} \frac{\mathbf{x}}{\|\mathbf{x}\|^3}.
$$

If a particle of charge q moves through $\mathbf{E}$ according to $\alpha : [a, b] \to \mathbb{R}^2$, then the work done by $\mathbf{F}$ on the particle in motion is given by

$$W_\alpha = \int_a^b q\mathbf{E}(\alpha(t)) \cdot \alpha'(t) \, dt = \int_a^b \left(\frac{qq_0}{4\pi\epsilon_0} \frac{\alpha(t)}{\|\alpha(t)\|^3} \right) \cdot \alpha'(t) \, dt.$$

For a particle located at $\alpha(t)$, $\|\alpha(t)\| = \sqrt{\alpha(t) \cdot \alpha(t)}$. We can rewrite the integrand in terms of the dot product and use the fact that $(\alpha(t) \cdot \alpha(t))' = 2\alpha(t) \cdot \alpha'(t)$ to evaluate the integral:

$$\frac{qq_0}{4\pi\epsilon_0} \int_a^b \frac{\alpha(t) \cdot \alpha'(t)}{(\alpha(t) \cdot \alpha(t))^{3/2}} \, dt = \frac{qq_0}{4\pi\epsilon_0} \frac{-1}{(\alpha(t) \cdot \alpha(t))^{1/2}} \Big|_a^b$$

$$= \frac{qq_0}{4\pi\epsilon_0} \frac{-1}{\|\alpha(t)\|} \Big|_a^b$$

$$= \frac{-qq_0}{4\pi\epsilon_0} \left(\frac{1}{\|\alpha(b)\|} - \frac{1}{\|\alpha(a)\|} \right).$$

Notice that this quantity depends on the endpoints of α and not on any of the intermediate positions of the moving particle. This is an important property of electric fields that we will return to later in this section. For now, let us note that the quantity $-W_\alpha/q$ is called the **electric potential difference** along α. It depends on the electric field and not on the charge of the moving particle.

The Orientation of a Curve

In Section 6.1, we stated that if α and β are parametrizations of the same curve $\mathcal{C}$, then the path integral of a function f over α is equal to the path integral of f over β. That is, the total accumulation of a function on a curve does not depend on which parametrization we use to parametrize the curve, as long as the parametrization traces the curve just once. There is a corresponding result for line integrals, but it requires the added assumption that α and β trace $\mathcal{C}$ in the same direction. Intuitively, this makes sense because the amount of work done moving a particle through a force field should depend on whether the particle moves "with the force" or "against the force." Let us see how this works.

Suppose that $\alpha : [a, b] \to \mathbb{R}^2$, with $\alpha = \alpha(u)$, and $\beta : [c, d] \to \mathbb{R}^2$, with $\beta = \beta(t)$, are continuously differentiable parametrizations of the same curve $\mathcal{C}$ that are related by composition with a differentiable function h of one variable. That is, $\beta = \alpha \circ h$, where $u = h(t)$ so that $\beta(t) = \alpha(h(t)) = \alpha(u)$. Thus $\beta'(t) = \alpha'(h(t))h'(t)$. Since we want α and β to trace $\mathcal{C}$ in the same direction, we will also require that $h : [a, b] \to [c, d]$ with $h(a) = c$

and $h(b) = d$. With these hypotheses, let us compute the line integral of a vector field $\mathbf{F}$ along α using the substitution $u = h(t)$:

$$
\int_\alpha \mathbf{F} \cdot \mathbf{T}\, ds = \int_a^b \mathbf{F}(\alpha(u)) \cdot \alpha'(u)\, du
$$

$$
= \int_c^d \mathbf{F}(\alpha(h(t))) \cdot \alpha'(h(t)) h'(t)\, dt
$$

$$
= \int_c^d \mathbf{F}(\beta(t)) \cdot \beta'(t)\, dt
$$

$$
= \int_\beta \mathbf{F} \cdot \mathbf{T}\, ds.
$$

It follows that the line integral of $\mathbf{F}$ over a curve $\mathcal{C}$ does not depend on the parametrization, as long as the parametrizations trace the curve in the same direction.

Now let us see what happens if the parametrizations trace $\mathcal{C}$ in opposite directions. Thus we assume that $h(a) = d$ and $h(b) = c$. Repeating the above calculation, we have

$$
\int_\alpha \mathbf{F} \cdot \mathbf{T}\, ds = \int_a^b \mathbf{F}(\alpha(u)) \cdot \alpha'(u)\, du
$$

$$
= \int_d^c \mathbf{F}(\alpha(h(t))) \cdot \alpha'(h(t)) h'(t)\, dt
$$

$$
= \int_d^c \mathbf{F}(\beta(t)) \cdot \beta'(t)\, dt.
$$

Since $d > c$, the next step is to reverse the order of integration, which changes the sign of the integral.

$$
\int_d^c \mathbf{F}(\beta(t)) \cdot \beta'(t)\, dt = -\int_c^d \mathbf{F}(\beta(t)) \cdot \beta'(t)\, dt
$$

$$
= -\int_\beta \mathbf{F} \cdot \mathbf{T}\, ds.
$$

It follows that if α and β trace a curve $\mathcal{C}$ in opposite directions, the line integrals of $\mathbf{F}$ over α and β have the same absolute value but opposite signs.

Before we formally state these results, we would like to introduce the concept of orientation.

Definition 6.6 Let α and β be parametrizations of a curve $\mathcal{C}$.

1. If α and β trace $\mathcal{C}$ in the same direction, we say that α and β parametrize $\mathcal{C}$ with the ***same orientation***. If α and β trace $\mathcal{C}$ in opposite directions, we say that α and β parametrize $\mathcal{C}$ with ***opposite orientations***.

2. We say that $\mathcal{C}$ is **oriented** if we have chosen a direction for a parametrization of $\mathcal{C}$. Thus a given curve has two possible orientations. We say that a parametrization **agrees with the orientation of the curve** if its direction agrees with the direction we have chosen for the curve. If $\mathcal{C}$ is oriented, we will denote the **opposite** or **reverse orientation** of $\mathcal{C}$ by $-\mathcal{C}$. ◆

Using this new terminology, we summarize our earlier calculations.

Proposition 6.3 Let α and β be parametrizations of $\mathcal{C}$. If α and β parametrize C with the same orientation, then

$$\int_\beta \mathbf{F} \cdot \mathbf{T} \, ds = \int_\alpha \mathbf{F} \cdot \mathbf{T} \, ds.$$

If α and β parametrize C with opposite orientations, then

$$\int_\beta \mathbf{F} \cdot \mathbf{T} \, ds = -\int_\alpha \mathbf{F} \cdot \mathbf{T} \, ds. \quad ◆$$

Since the line integral of $\mathbf{F}$ over a parametrization of a curve $\mathcal{C}$ depends only on the orientation of the parametrization, we can define the line integral of a vector field over a curve independent of the choice of parametrization as long as we specify the orientation of the curve.

Definition 6.7 Let $\mathcal{C}$ be an oriented curve in the domain of the vector field $\mathbf{F}$. Then we define the **line integral** of $\mathbf{F}$ over $\mathcal{C}$ to be the line integral of $\mathbf{F}$ over α, where α is any parametrization of C that agrees with the orientation of C. We denote the line integral of $\mathbf{F}$ over the oriented curve C by

$$\int_C \mathbf{F} \cdot \mathbf{T} \, ds.$$

Thus, we have

$$\int_C \mathbf{F} \cdot \mathbf{T} \, ds = \int_\alpha \mathbf{F} \cdot \mathbf{T} \, ds. \quad ◆$$

Notice that combining the notation of Definitions 6.6 and 6.7, we have

$$\int_{-C} \mathbf{F} \cdot \mathbf{T} \, ds = -\int_C \mathbf{F} \cdot \mathbf{T} \, ds.$$

Returning to the concept of work, this result implies that if $\mathbf{F}$ is a force field, the work done on a particle moving along a curve $\mathcal{C}$ in a particular direction is independent of the way in which the particle moves along the curve. The work done on a particle moving along $\mathcal{C}$ in the opposite direction is the opposite of this value.

Independence of Path

In Example 6.7, we carried out the calculation of the work done by an electric field on a charged particle. We showed that for the electric field $\mathbf{E}$ due to a particle of charge q_0 located at the origin, the work done by the field on a particle of charge q moving along α is given by

$$W_\alpha = \frac{-qq_0}{4\pi\epsilon_0}\left(\frac{1}{\|\alpha(b)\|} - \frac{1}{\|\alpha(a)\|}\right).$$

We made the point that this expression depends only on the endpoints of α, not on any of the intermediate points. To clarify what is going on, let us define a function f by

$$f(\mathbf{x}) = -\frac{qq_0}{4\pi\epsilon_0\|\mathbf{x}\|},$$

so that

$$W_\alpha = f(\alpha(b)) - f(\alpha(a)).$$

A calculation shows that the gradient of f is the force field $q\mathbf{E}$, that is, $\nabla f = q\mathbf{E}$. (See Exercise 3.) Thus, in this case, we have expressed the integral of a gradient vector field of a function as the difference between the values of the function at the endpoints. This leads us to consider the general question of computing the line integral of a gradient vector field along a curve.

Let f be a differentiable function, let $\mathbf{F}$ be its gradient vector field, and let $\alpha : [a, b] \to \mathbb{R}^2$ be a differentiable parametrization of a curve $\mathcal{C}$ contained in the domain of f. We can use the chain rule in the form $\frac{d}{dt}f(\alpha(t)) = \nabla f(\alpha(t)) \cdot \alpha'(t)$ and the fundamental theorem of calculus to evaluate the line integral of $\mathbf{F} = \nabla f$ over α:

$$\int_\alpha \mathbf{F} \cdot \mathbf{T} \, ds = \int_a^b \mathbf{F}(\alpha(t)) \cdot \alpha'(t) \, dt$$
$$= \int_a^b \nabla f(\alpha(t)) \cdot \alpha'(t) \, dt$$
$$= \int_a^b \frac{d}{dt}\left(f(\alpha(t))\right) dt$$
$$= f(\alpha(t))|_{t=a}^{t=b}$$
$$= f(\alpha(b)) - f(\alpha(a)).$$

We see that the line integral of a gradient vector field depends only on the endpoints of the parametrization. That is, if $\alpha : [a, b] \to \mathbb{R}^2$ and $\beta : [c, d] \to \mathbb{R}^2$ are parametrizations with the same initial point and the same endpoint, $\alpha(a) = \beta(c)$ and $\alpha(b) = \beta(d)$, then

$$\int_\alpha \nabla f \cdot \mathbf{T} \, ds = f(\alpha(b)) - f(\alpha(a))$$
$$= f(\beta(d)) - f(\beta(c))$$
$$= \int_\beta \nabla f \cdot \mathbf{T} \, ds.$$

Alternatively, if α is a parametrization of any path from P to Q,

$$\int_\alpha \nabla f \cdot \mathbf{T} \, ds = f(Q) - f(P).$$

In particular, the work done in moving an object through a force field ∇f from one point to another is independent of the choice of path for the motion of the object, as we saw for the electric force field of Example 6.7.

This fact about gradient vector fields leads us to make the following definition.

Definition 6.8 A vector field $\mathbf{F} = \mathbf{F}(\mathbf{x})$ is said to have the **path independence** property if $\int_\alpha \mathbf{F} \cdot \mathbf{T} \, ds$ depends only on the endpoints of α; that is, if $\alpha : [a, b] \to \mathbb{R}^2$ and $\beta : [c, d] \to \mathbb{R}^2$ satisfy $\alpha(a) = \beta(c)$ and $\alpha(b) = \beta(d)$, then

$$\int_\alpha \mathbf{F} \cdot \mathbf{T} \, ds = \int_\beta \mathbf{F} \cdot \mathbf{T} \, ds. \quad \blacklozenge$$

Since every gradient vector field has the path independence property, it is natural to ask if this property is sufficient to imply that a vector field is a gradient vector field. That is, if $\mathbf{F}(x, y) = (u(x, y), v(x, y))$ has the path independence property, is there a differentiable function f so that $\nabla f = \mathbf{F}$? In fact, there is such a function, and the path independence property is exactly the condition that is required to construct the function. The construction of f is given in the proof of the following theorem.

Theorem 6.2 Let $\mathbf{F}(\mathbf{x}, \mathbf{y}) = (\mathbf{u}(\mathbf{x}, \mathbf{y}), \mathbf{v}(\mathbf{x}, \mathbf{y}))$ be a continuous vector field that satisfies the path independence property, and let $\mathbf{x}_0$ be a fixed point in the domain of $\mathbf{F}$. Let f be the function defined by

$$f(\mathbf{x}) = \int_\alpha \mathbf{F} \cdot \mathbf{T} \, ds,$$

where α is a parametrization of a curve with initial point $\mathbf{x}_0$ and final point $\mathbf{x}$. Then f is a function whose gradient is $\mathbf{F}$, $\nabla f = \mathbf{F}$. $\quad \blacklozenge$

Proof: In order to construct f, we must choose a fixed point $\mathbf{x}_0 = (x_0, y_0)$. Given any other point $\mathbf{x} = (x, y)$, we define f by

$$f(\mathbf{x}) = \int_\alpha \mathbf{F} \cdot \mathbf{T} \, ds,$$

where $\alpha : [a, b] \to \mathbb{R}^2$ is a parametrization that satisfies $\alpha(a) = \mathbf{x}_0$ and $\alpha(b) = \mathbf{x}$. Since $\mathbf{F}$ has the path independence property, $f(\mathbf{x})$ is uniquely defined independent of the choice of α. It remains to show that $\nabla f = \mathbf{F}$, that is,

$$(\frac{\partial f}{\partial x}(x, y), \frac{\partial f}{\partial y}(x, y)) = (u(x, y), v(x, y)).$$

To verify this, we will use the limit definition of the partial derivative and make use of the path independence property.

First, let us show that $\frac{\partial f}{\partial x}(x, y) = u(x, y)$. Using the definition of f,

$$\frac{\partial f}{\partial x}(x, y) = \lim_{h \to 0} \frac{f(x + h, y) - f(x, y)}{h}$$

$$= \lim_{h \to 0} \frac{\int_{\alpha_h} \mathbf{F} \cdot \mathbf{T} \, ds - \int_{\alpha_0} \mathbf{F} \cdot \mathbf{T} \, ds}{h}.$$

Here α_0 denotes a parametrization with initial point $\mathbf{x}_0$ and final point $\mathbf{x}$, and α_h denotes a parametrization with initial point $\mathbf{x}_0$ and final point $(x + h, y)$. Since $\mathbf{F}$ has the path independence property, we can choose α_h in such a way as to simplify the calculation of the partial derivative. In particular, we will choose α_h so that its image is the union of the image of α_0 and the horizontal line segment from $\mathbf{x} = (x, y)$ to $(x + h, y)$. (See Figure 6.5.) We will parametrize the segment of the image from $\mathbf{x}_0$ to $\mathbf{x}$ by α_0 and the

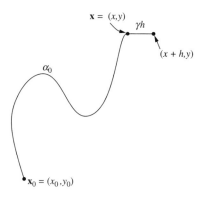

Figure 6.5   The image of the curve α_h consists of the union of the image of α_0 and the image of γ_h, which is the line segment from $\mathbf{x} = (x, y)$ to $(x + h, y)$.

horizontal segment by γ_h, where $\gamma_h(t) = (x + t, y)$, with $0 \leq t \leq h$ if h is positive and $h \leq t \leq 0$ if h is negative. Thus

$$\int_{\alpha_h} \mathbf{F} \cdot \mathbf{T} \, ds = \int_{\alpha_0} \mathbf{F} \cdot \mathbf{T} \, ds + \int_{\gamma_h} \mathbf{F} \cdot \mathbf{T} \, ds,$$

and

$$\int_{\alpha_h} \mathbf{F} \cdot \mathbf{T} \, ds - \int_{\alpha_0} \mathbf{F} \cdot \mathbf{T} \, ds = \int_{\gamma_h} \mathbf{F} \cdot \mathbf{T} \, ds.$$

Using this, we are able to simplify our limit,

$$\frac{\partial f}{\partial x}(x, y) = \lim_{h \to 0} \frac{\int_{\gamma_h} \mathbf{F} \cdot \mathbf{T} \, ds}{h}.$$

Since $\gamma_h'(t) = (1, 0)$, the unit tangent vector of γ_h is also equal to $(1, 0)$, and

$$\mathbf{F} \cdot \mathbf{T} = (u(x, y), v(x, y)) \cdot (1, 0) = u(x, y).$$

Therefore, the limit is equal to

$$\lim_{h \to 0} \frac{\int_0^h u(x + t, y) \, dt}{h}.$$

Notice that the quotient inside the limit is equal to the average value of u on the line segment from (x, y) to $(x + h, y)$. Since u is a continuous function, the limit of its average value on the line segment as the length of the segment approaches 0 is equal to the value of the function at the endpoint, that is,

$$\lim_{h \to 0} \frac{\int_0^h u(x + t, y) \, dt}{h} = u(x, y).$$

Thus we have shown that for f defined as above,

$$\frac{\partial f}{\partial x}(x, y) = u(x, y).$$

In order to compute the partial derivative of f with respect to y, we must replace γ_h in this argument by a curve that parametrizes the vertical line segment from $Q = (x, y)$ to $(x, y + h)$. Then a similar argument shows that

$$\frac{\partial f}{\partial y}(x, y) = v(x, y).$$

We conclude that every vector field satisfying the path independence property is a gradient vector field. ∎

The function f of the theorem depends on the choice of $\mathbf{x}_0$ as well as the vector field $\mathbf{F}$. Suppose that $\tilde{\mathbf{x}}_0 \neq \mathbf{x}_0$ and that we use $\tilde{\mathbf{x}}_0$ as in the theorem to define a function $\tilde{f}$. Then $\tilde{f}$ also satisfies $\nabla \tilde{f} = \mathbf{F}$. In Exercise 4 we show that f and $\tilde{f}$ differ by a constant, that is, there is a constant C so that $\tilde{f}(\mathbf{x}) = f(\mathbf{x}) + C$ for all $\mathbf{x}$. Since this construction yields $f(\mathbf{x}_0) = 0$, the different choices of the basepoint are equivalent to fixing a point where f has the value 0. We will use this fact in the discussion that follows.

Application: Work, Energy, and Conservative Forces

We are now in a position to explore the relationship between work and energy, which plays a fundamental role in physics. Let us suppose that $\mathbf{F}$ is a force field and that a particle of mass m moves through the field according to α. Here we assume that no other forces act on the object, so that the motion of the object is due solely to this force. From Newton's second law, we know that the force acting on the particle at an instant in time is equal to the product of the mass of the particle and its acceleration, $\mathbf{F} = m\alpha''$. Let us substitute this into the integral expression for work.

$$\begin{aligned}
W_\alpha &= \int_\alpha \mathbf{F} \cdot \mathbf{T} \, ds \\
&= \int_a^b m\alpha''(t) \cdot \alpha'(t) \, dt \\
&= \left[\frac{m}{2} \alpha'(t) \cdot \alpha'(t) \right]_a^b \\
&= \frac{m}{2} \alpha'(b) \cdot \alpha'(b) - \frac{m}{2} \alpha'(a) \cdot \alpha'(a).
\end{aligned}$$

Note that we used the fundamental theorem of calculus and Exercise 3(c) of Section 2.2 to evaluate the integral.

The quantity $K(t) = \frac{m}{2}\alpha'(t) \cdot \alpha'(t)$ is called the **kinetic energy** of the particle. In physics, the velocity α' of an object is denoted by $\mathbf{v}$, so that $K = \frac{m}{2}\mathbf{v} \cdot \mathbf{v}$. This calculation says that the amount of work done by the force in moving the object is equal to the difference between its kinetic energy at the initial point of its motion and at the final point of its motion, $W = K(b) - K(a)$. This is known as the **work-energy** theorem for a particle. (See Exercise 11.)

Certain forces have the property that if an object moves subject to the force through a closed loop, then the kinetic energy of the object upon return to its initial position is the same as it was initially. For example, neglecting air resistance, gravity has this property. If an object is launched vertically from the surface of the earth with velocity v_0, it will return to earth with velocity $-v_0$. Consequently, the work done by gravity on the object is zero, $W = \frac{m}{2}v_0^2 - \frac{m}{2}(-v_0)^2 = 0$. Forces with this property are called **conservative** forces. In addition to gravity, the electric force of Example 6.7 is a conservative force. If a force is conservative, it follows that it has the path independence property. (See Exercise 8.) Thus from the theorem, there is a function f with the property that $\nabla f = \mathbf{F}$.

The function $U = -f$, which satisfies $-\nabla U = \mathbf{F}$, is called the **potential energy** of the system. Along any path α from Q_1 to Q_2, $W_\alpha = -(U(Q_2) - U(Q_1))$. Since W depends on Q_1 and Q_2, and not on α, we will use the notation W_{Q_1,Q_2} to denote the work done by a conservative field in moving a particle from Q_1 to Q_2. Because the basepoint P that is used to define U will satisfy $U(P) = 0$, the choice of P is determined by the choice of a convenient location for the 0 value of the potential energy. After choosing a basepoint, $U(Q) = -W_{P,Q}$.

Since for a conservative system work is equal to the change in kinetic energy *and* the change in potential energy, we have for any Q_1 and Q_2,

$$K(Q_2) - K(Q_1) = -(U(Q_2) - U(Q_1)).$$

This implies that

$$K(Q_2) + U(Q_2) = K(Q_1) + U(Q_1).$$

This holds for any Q_1 and Q_2, so it follows that the sum $K(Q) + U(Q)$ must be a constant. This constant is called the **total energy** of the system. If a conservative force acts on a particle in motion, changing its kinetic energy, since the total energy is constant, there is a corresponding change in the potential energy that compensates for the change in the kinetic energy. (In Section 3.5 we assumed that $E = K + U$ is constant and then proved that $\nabla U = -\mathbf{F}$.)

An interesting application of this discussion is to determine the escape velocity for the earth, which we do in the following example.

Example 6.8 | **Escape Velocity of the Earth**

A. Assume that the center of the earth is located at the origin of a three-dimensional coordinate system and that an object of mass m is located at $\mathbf{x}$. Then the gravitational force acting on the object is

$$\mathbf{F} = -GM_e m \frac{\mathbf{x}}{\|\mathbf{x}\|^3},$$

where G is the gravitational constant and M_e is the mass of the earth. A calculation similar to the one in Example 6.7 shows that the work done by gravity in moving the object from $\mathbf{x}_1$ to $\mathbf{x}_2$, which we denote by $W_{\mathbf{x}_1,\mathbf{x}_2}$, is

$$W_{\mathbf{x}_1,\mathbf{x}_2} = GM_e m \left(\frac{1}{\|\mathbf{x}_2\|} - \frac{1}{\|\mathbf{x}_1\|} \right),$$

so that $\mathbf{F}$ has the path independence property, which confirms that gravity is a conservative force. As above, the **gravitational potential energy** U is defined by $U(\mathbf{x}) = -W_{\mathbf{x}_0,\mathbf{x}}$, where $\mathbf{x}_0$ is a basepoint chosen so that $U(\mathbf{x}_0) = 0$. If our focus is on the behavior of gravity near the surface of the earth, it is convenient to choose $\mathbf{x}_0$ to be a point on the ground. However, if we are concerned with a larger distance scale, it is convenient to have the potential energy at ∞ be equal to 0. To do this, we define the potential energy to be the limit of $-W_{\mathbf{x}_0,\mathbf{x}}$ as $\|\mathbf{x}_0\| \to \infty$. Thus

$$U(\mathbf{x}) = \lim_{\|\mathbf{x}_0\| \to \infty} -W_{\mathbf{x}_0,\mathbf{x}} = -\frac{GM_e m}{\|\mathbf{x}\|}.$$

Note that U satisfies $-\nabla U = \mathbf{F}$. Intuitively, the potential energy of the object at $\mathbf{x}$ is the negative of the amount of work required to bring the object from ∞ to $\mathbf{x}$.

B. The potential energy of an object on the surface of the earth is $-GM_e m/R_e$, where R_e is the radius of the earth. The amount of work required to move the object to infinity is the negative of the amount of work required to bring the object from infinity, so it is also equal $-GM_e m/R_e$. Neglecting air resistance, we can use the work-energy theorem to find the initial velocity necessary to propel the object from the surface of the earth to a point at infinity with zero velocity. Since the final velocity is zero, the initial velocity $\mathbf{v}_0$ must satisfy $-GM_e m/R_e = -m\mathbf{v}_0 \cdot \mathbf{v}_0/2$, so that the speed $\|\mathbf{v}_0\|$ must satisfy $\|\mathbf{v}_0\| = \sqrt{2GM_e/R_e} \approx 7 \text{ mi/s}$. (Note that $G = 6.67 \times 10^{-11} \text{ N-m/kg}^2$, $M_e \approx 6.0 \times 10^{24} \text{ kg}$, and $R_e \approx 6.37 \times 10^6 \text{ m}$.) This speed is known as the **escape velocity** of the earth.

Summary   In this section, we introduced the *line integral* of a vector field $\mathbf{F}$ over a curve in the plane or space. The construction was motivated by the effort to extend the concept of *work* in physics to the work done on a particle by a continuous nonconstant vector field.

If α is a parametrization of the curve and $\mathbf{T}$ is the unit tangent vector to α, the *line integral* of $\mathbf{F}$ along α is the path integral of $\mathbf{F} \cdot \mathbf{T}$ along α. Thus,

$$\int_{\alpha} \mathbf{F} \cdot \mathbf{T} \, ds = \int_{a}^{b} \left(\mathbf{F}(\alpha(t)) \cdot \mathbf{T}(\alpha(t)) \right) \|\alpha'(t)\| \, dt.$$

The integral on the right-hand side can be simplified to obtain the formula

$$\int_{\alpha} \mathbf{F} \cdot \mathbf{T} \, ds = \int_{a}^{b} \mathbf{F}(\alpha(t)) \cdot \alpha'(t) \, dt.$$

If α is the path of a particle moving through the force field $\mathbf{F}$, this integral defines the *work done by* $\mathbf{F}$ on the particle as it moves along α.

We defined the *orientation* of a path and showed that the line integral of $\mathbf{F}$ over two parametrizations of the same curve is the same if the parametrizations trace the curve with the same orientation.

Building on these ideas, we showed that if $\mathbf{F}$ is a gradient field, $\mathbf{F} = \nabla f$, for a continuously differentiable function f, then the line integral of $\mathbf{F}$ along a curve depends only on the values of f at the endpoints of the curve. We called this property *path independence*, and we proved the converse. *Every vector field with the path independence property is a gradient vector field.* We used these constructions to develop several ideas from physics, including the *work-energy theorem*, *conservative forces*, *potential energy*, and *total energy*.

Section 6.2 Exercises

1. **Line Integrals.** For each of the following force fields $\mathbf{F}$, compute the line integral of $\mathbf{F}$ over the parametrization α.

   (a) $\mathbf{F}(x, y) = (-y, x)$ and $\alpha(t) = (e^t, e^{-t})$, $0 \le t \le 1$.
   (b) $\mathbf{F}(x, y) = (x, y)$ and $\alpha(t) = (\cos^3 t, \sin^3 t)$, $0 \le t \le 2\pi$.
   (c) $\mathbf{F}(x, y, z) = (x + y, y, y)$ and $\alpha(t) = (t, t, t^2)$, $0 \le t \le 3$.

2. **Line Integrals over Curves.** For each of the following vector fields **F**, evaluate the line integral of **F** along the given curve.

   (a) $\mathbf{F}(x, y) = (x, y)$ along the circle of radius 2 centered at the origin traversed clockwise.
   (b) $\mathbf{F}(x, y) = (y^2, x^2)$ along the parabola $y = x^2$ from $(0, 0)$ to $(2, 4)$.
   (c) $\mathbf{F}(x, y) = (x^2, xy)$ along the perimeter of the unit square from $(0, 0)$ to $(1, 0)$ to $(1, 1)$ to $(0, 1)$ and back to $(0, 0)$.

3. **A Gradient Calculation.** Show that the gradient of $f(\mathbf{x}) = -\frac{qq_0}{4\pi\epsilon_0} \frac{1}{\|\mathbf{x}\|}$ is equal to the force field $\mathbf{F} = q\mathbf{E}$, where $\mathbf{E}$ is the electric field of Example 6.7, $\mathbf{E}(\mathbf{x}) = \frac{q_0}{4\pi\epsilon_0} \frac{\mathbf{x}}{\|\mathbf{x}\|^3}$.

4. **Path Independence.** Suppose that the vector field **F** has the path independence property. Let $\mathbf{x}_0$ be a fixed point and define the function f by

$$f(\mathbf{x}) = \int_\alpha \mathbf{F} \cdot \mathbf{T} \, ds,$$

   where α is a parametrization of a curve with initial point $\mathbf{x}_0$ and final point $\mathbf{x}$.

   (a) Show that if $\mathcal{C}$ is any curve in the domain of **F** from a point $\mathbf{x}_1$ to a point $\mathbf{x}_2$, then

$$\int_{\mathcal{C}} \mathbf{F} \cdot \mathbf{T} \, ds = f(\mathbf{x}_2) - f(\mathbf{x}_1).$$

   (*Hint:* Since **F** has the path independence property, the integral on the left-hand side can be evaluated along a path C that goes from $\mathbf{x}_1$ to $\mathbf{x}_0$ to $\mathbf{x}_2$.)
   (b) Suppose that $\tilde{\mathbf{x}}_0 \neq \mathbf{x}_0$ is another point in the domain of **F** and that $\tilde{f}$ is defined by

$$\tilde{f}(\mathbf{x}) = \int_\alpha \mathbf{F} \cdot \mathbf{T} \, ds,$$

   where α is a parametrization of a curve with initial point $\tilde{\mathbf{x}}_0$ and final point $\mathbf{x}$. Show that f and $\tilde{f}$ differ by

$$\int_{\mathcal{C}} \mathbf{F} \cdot \mathbf{T} \, ds$$

   where $\mathcal{C}$ is a curve from $\mathbf{x}_0$ to $\tilde{\mathbf{x}}_0$. (*Hint:* To define $f(\mathbf{x})$, use a path from $\mathbf{x}_0$ to $\tilde{\mathbf{x}}_0$ to $\mathbf{x}$.)
   (c) Conclude that f and $\tilde{f}$ differ by a constant.

5. **Mixed Partials and Path Independence.** Let **F** be a continuous and differentiable vector field on a domain $\mathcal{D}$ that satisfies the path independence property.

   (a) If $\mathbf{F}(x, y) = (F_1(x, y), F_2(x, y))$, show that $\frac{\partial F_1}{\partial y} = \frac{\partial F_2}{\partial x}$.

(b) If $\mathbf{F}(x, y, z) = (F_1(x, y, z), F_2(x, y, z), F_3(x, y, z))$, show that $\frac{\partial F_1}{\partial y} = \frac{\partial F_2}{\partial x}$, $\frac{\partial F_1}{\partial z} = \frac{\partial F_3}{\partial x}$, and $\frac{\partial F_2}{\partial z} = \frac{\partial F_3}{\partial y}$.

As a consequence of this exercise, if $\mathbf{F}$ is a vector field in the plane and $\frac{\partial F_1}{\partial y} \neq \frac{\partial F_2}{\partial x}$, then $\mathbf{F}$ does not satisfy the path independence property. Similarly, if $\mathbf{F}$ is a vector field in space and $\frac{\partial F_1}{\partial y} \neq \frac{\partial F_2}{\partial x}$, $\frac{\partial F_1}{\partial z} \neq \frac{\partial F_3}{\partial x}$, or $\frac{\partial F_2}{\partial z} \neq \frac{\partial F_3}{\partial y}$, then $\mathbf{F}$ does not satisfy the path independence property.

6. **Potential Functions.** A function f with the property that $\nabla f = \mathbf{F}$ is called a *potential function* for $\mathbf{F}(x, y) = (u(x, y), v(x, y))$. Since this means that $\frac{\partial f}{\partial x}(x, y) = u(x, y)$ and $\frac{\partial f}{\partial y}(x, y) = v(x, y)$, an ad hoc method for finding f is to evaluate and compare the indefinite integrals $\int u(x, y)\, dx$ and $\int v(x, y)\, dy$. However, since the partial derivative with respect to x of a term involving only y is zero, for example, $\frac{\partial \sin y}{\partial x} = 0$, the constant of integration for $\int u(x, y)\, dx$ may depend on y. Similarly, the constant of integration for $\int v(x, y)\, dy$ may depend on x. Call these C_y and C_x, respectively. By comparing the antiderivatives, we can find all the terms of f. For example, if $F(x, y) = (1 + y, x + y^2)$, then $\int (1 + y)\, dx = x + xy + C_y$ and $\int (x + y^2)\, dy = xy + y^3/3 + C_x$. Equating these two expressions, we see that the term xy appears in both expressions and that $C_y = y^3/3$ and $C_x = x$. We conclude that $f(x, y) = x + xy + y^3/3$ is a potential function for $\mathbf{F}$.

For each of the following vector fields, use Exercise 5 to determine if $\mathbf{F}$ is not path independent. For each vector field with equal mixed partial derivatives, use this technique to find a potential function for $\mathbf{F}$.

(a) $\mathbf{F}(x, y) = (-y, x)$, $\mathcal{D} = \mathbb{R}^2$.
(b) $\mathbf{F}(x, y) = (3x^2 + y, e^y + x)$, $\mathcal{D} = \mathbb{R}^2$.
(c) $\mathbf{F}(x, y, z) = (y^2, 2xy, z)$, $\mathcal{D} = \mathbb{R}^3$.
(d) $\mathbf{F}(x, y, z) = (z^2, y^2 z, 2xz)$, $\mathcal{D} = \mathbb{R}^3$.

7. **Path Independent Vector Fields.** Each of the following vector fields $\mathbf{F}$ satisfies the path independence property. Evaluate $\int_C \mathbf{F} \cdot \mathbf{T}\, ds$ by finding a potential function f with $\nabla f = \mathbf{F}$ and using f to evaluate the integral.

(a) $\mathbf{F}(x, y) = (x, y)$ and C is the upper half of a circle of radius 2 centered at the origin traversed counterclockwise.
(b) $\mathbf{F}(x, y) = (\cos x \cos y, -\sin x \sin y)$ and C is the path along $y = x^3$ from $(-1, 1)$ to $(1, 1)$.
(c) $\mathbf{F}(x, y, z) = (x + yz, y + xz, z + xy)$ and C is the path along the helix parametrized by $\alpha(t) = (\cos t, \sin t, t)$ from $(1, 0, 0)$ to $(0, 1, \pi/2)$.

8. **Integrals around Closed Curves.** We know that every vector field with the path independence property is a conservative vector field. Here we show the converse: that every conservative vector field has the path independence property. Suppose that $\mathbf{F}$ is a conservative vector field, so that $\int_C \mathbf{F} \cdot \mathbf{T}\, ds = 0$ for every closed curve C. Let C_1 and C_2 be oriented curves from the P to Q in the domain of $\mathbf{F}$.

(a) Let $-C_2$ denote C_2 with the opposite orientation. Then $C = C_1 \cup (-C_2)$ is a closed curve with initial point and endpoint equal to P. Show that $\int_{C_1} \mathbf{F} \cdot \mathbf{T}\, ds = -\int_{-C_2} \mathbf{F} \cdot \mathbf{T}\, ds$.

(b) Use part (a) to show that $\int_{\mathcal{C}_1} \mathbf{F} \cdot \mathbf{T} \, ds = \int_{\mathcal{C}_2} \mathbf{F} \cdot \mathbf{T} \, ds$. Since this argument holds for any P, Q, $\mathcal{C}_1$, and $\mathcal{C}_2$, this shows that $\mathbf{F}$ has the path independence property.

9. **Sum of Conservative Forces.** If forces $\mathbf{F}_1$ and $\mathbf{F}_2$ act on an object, then the total force $\mathbf{F}$ acting on the object is $\mathbf{F} = \mathbf{F}_1 + \mathbf{F}_2$. Show that if $\mathbf{F}_1$ and $\mathbf{F}_2$ are conservative forces, then $\mathbf{F}$ is a conservative force.

10. **Friction.** Assume that the force of friction acts in a direction opposite to the direction of motion with magnitude proportional to the speed. Thus if the path of a particle is parametrized by $\alpha = \alpha(t)$, the force due to friction at $\alpha(t)$ is $\mathbf{F}(\alpha(t)) = -k\alpha'(t)$ where k is a constant. Compute $\int_\alpha \mathbf{F} \cdot \mathbf{T} \, ds$ and show that $\mathbf{F}$ is not a conservative force.

11. **Work-Energy Theorem.** In this exercise, we want to investigate the consequences of the work-energy theorem.

(a) Suppose that a force acts on a particle and the motion of the particle is described by a parametrization α with constant speed, that is, $\|\alpha'(t)\|$ is constant. How much work was done by the field in moving the particle?

(b) Suppose that a force acts on a particle so that the initial speed of the particle is equal to the final speed of the particle. How much work was done by the field in moving the particle?

12. **Work and Conservative Fields.** If a force does a quantity of work W on a particle, it is said that the particle does work on whatever produced the force in the amount $-W$. It follows that in a conservative field, when a particle moves through a closed loop, the **particle does no work** on the field and that the particle's ability to do work is **conserved**. This interpretation is the source of the name "conservative" field. Use the work-energy theorem to explain the following statements:

(a) The kinetic energy of a particle decreases by an amount equal to the amount of work that a particle does.

(b) The kinetic energy of a body in motion is equal to the work it can do in being brought to rest.

13. **Inverse Square Fields.** An *inverse square force field* $\mathbf{F}$ takes the form

$$\mathbf{F}(\mathbf{x}) = \frac{K}{r^2} \mathbf{u},$$

where $r = \|\mathbf{x}\|$ and $\mathbf{u} = \mathbf{x}/\|\mathbf{x}\|$. For example, the electric force field generated by a point charge and a gravitational field generated by an object are inverse square force fields.

(a) Show that $\mathbf{F}$ is conservative (for all $\mathbf{x} \neq \mathbf{0}$) by finding a potential function f with $\nabla f = \mathbf{F}$.

(b) Show that the work done in moving a particle from $\mathbf{x}$ to a point infinitely far from the origin is K/r.

14. **Gravity.** Since gravity is a conservative force, the total energy of an object subject only to gravity is constant. This applies, for example, to a satellite orbiting the earth above

over the region $\mathcal{R}$ and the double integral of $\frac{\partial u}{\partial y}$ over $\mathcal{R}$. We will use the two descriptions of $\mathcal{R}$ to express each of these double integrals as an iterated integral.

$$\int\int_{\mathcal{R}} \left(\frac{\partial v}{\partial x} - \frac{\partial u}{\partial y} \right) dA_{x,y} = \int\int_{\mathcal{R}} \frac{\partial v}{\partial x} dA_{x,y} - \int\int_{\mathcal{R}} \frac{\partial u}{\partial y} dA_{x,y}$$

$$= \int_c^d \int_a^{h(y)} \frac{\partial v}{\partial x}(x, y) \, dx \, dy - \int_a^b \int_c^{g(x)} \frac{\partial u}{\partial y}(x, y) \, dy \, dx$$

$$= \int_c^d (v(h(y), y) - v(a, y)) \, dy - \int_a^b (u(x, g(x)) - u(x, c)) \, dx.$$

The boundary of the region $\mathcal{R}$ consists of three pieces: $\mathcal{C}_1$, the line segment $x = a$, $c \le y \le d$; $\mathcal{C}_2$, the line segment $y = c$, $a \le x \le b$; and $\mathcal{C}_3$, the portion of the curve $x = h(y)$, $c \le y \le d$. These pieces can be parametrized by $\alpha_1(y) = (a, y)$, $c \le y \le d$, $\alpha_2(x) = (x, c)$, $a \le x \le b$, and $\alpha_3(y) = (h(y), y)$, $c \le y \le d$, respectively. The boundary $\mathcal{C}$ is oriented counterclockwise by $-\alpha_1$, α_2, and α_3, where $-\alpha_1$ means that we parametrize $\mathcal{C}_1$ from (a, d) to (a, c). (See Figure 6.9.) Then,

$$\int_{\mathcal{C}} \mathbf{F} \cdot \mathbf{T} \, ds = \int_{-\alpha_1} \mathbf{F} \cdot \mathbf{T} \, ds + \int_{\alpha_2} \mathbf{F} \cdot \mathbf{T} \, ds + \int_{\alpha_3} \mathbf{F} \cdot \mathbf{T} \, ds$$

$$= \int_d^c (u(a, y), v(a, y)) \cdot (0, 1) \, dy + \int_a^b (u(x, c), v(x, c)) \cdot (1, 0) \, dx$$

$$+ \int_c^d (u(h(y), y), v(h(y), y)) \cdot (h'(y), 1) \, dy$$

$$= \int_d^c v(a, y) \, dy + \int_a^b u(x, c) \, dx + \int_c^d u(h(y), y) h'(y) \, dy + \int_c^d v(h(y), y) \, dy.$$

If we substitute $x = h(y)$ into the third integral above and use the fact that for all points on $\mathcal{C}_3$, $y = g(x)$, we have

$$\int_c^d u(h(y), y) h'(y) \, dy = \int_b^a u(x, g(x)) \, dx = -\int_a^b u(x, g(x)) \, dx.$$

Combining this with our first calculation, we have

$$\int_{\mathcal{C}} \mathbf{F} \cdot \mathbf{T} \, ds = \int_c^d (v(h(y), y) - v(a, y)) \, dy + \int_a^b (u(x, c) - u(x, g(x))) \, dx$$

$$= \int\int_{\mathcal{R}} \frac{\partial v}{\partial x} - \frac{\partial u}{\partial y} dA_{x,y},$$

which proves Green's theorem for three-sided regions of the specified type.

Thus

$$\int_{\mathcal{C}_1} \mathbf{F} \cdot \mathbf{T}\, ds + \int_{\mathcal{C}_2} \mathbf{F} \cdot \mathbf{T}\, ds = \int_{\alpha} \mathbf{F} \cdot \mathbf{T}\, ds + \int_{\beta} \mathbf{F} \cdot \mathbf{T}\, ds = \int_{\mathcal{C}} \mathbf{F} \cdot \mathbf{T}\, ds,$$

which verifies the claim when $\mathcal{R}$ is the union of two regions. If $\mathcal{R} = \mathcal{R}_1 \cup \mathcal{R}_2 \cup \cdots \cup \mathcal{R}_m$, apply the preceding result to $\mathcal{R}_1 \cup \mathcal{R}_2$, then $\mathcal{R}_1 \cup \mathcal{R}_2 \cup \mathcal{R}_3$, and so on. Since we have shown that the interior of every simple closed curve can be partitioned into regions for which Green's theorem holds, this completes the proof of Green's theorem. ∎

The proof of Green's theorem for more general regions follows from Green's theorem by partitioning a region $\mathcal{R}$ into a union of regions bounded by simple closed curves. Since Green's theorem applies to a region bounded by a simple closed curve, it applies to each of the regions making up $\mathcal{R}$. Then, by applying the conclusions of Step 3 of the proof of Green's theorem, we can show that Green's theorem applies to $\mathcal{R}$.

Summary

In this section, we stated and proved *Green's theorem*, which relates the line integral of a vector field $\mathbf{F}$ around a closed curve $\mathcal{C}$ to the double integral of partial derivatives of $\mathbf{F}$ over $\mathcal{R}$, the interior of $\mathcal{C}$.

If $\mathbf{F}(x,y) = (u(x,y), v(x,y))$ and $\mathcal{C}$ is *positively oriented*, Green's theorem states

$$\int_{\mathcal{C}} \mathbf{F} \cdot \mathbf{T}\, ds = \int\int_{\mathcal{R}} \left(\frac{\partial v}{\partial x} - \frac{\partial u}{\partial y} \right) dA_{x,y}.$$

By positive orientation, we mean that when $\mathcal{C}$ is traced in a *counterclockwise* manner, $\mathcal{R}$ is to the *left* of $\mathcal{C}$. For the theorem to hold, it is sufficient that $\mathcal{C}$ be *piecewise differentiable*. The theorem can also be applied to a region $\mathcal{R}$ whose boundary is a union of simple closed curves.

We presented two physical applications of Green's theorem involving fluids. First, we showed how Green's theorem could be used to relate the *total flux* of a fluid through a closed curve to the integral of the *infinitesimal flux* over the interior of the curve. Second, we showed how this result could be applied to a *diffusion process* that is governed by the *diffusion equation*. Here the total flux is equal to the double integral of the rate change of the concentration over the interior.

Section 6.3 Exercises

1. **Positive Orientation.** For each of the following regions with boundary, give a parametrization of the boundary of the region that positively orients the boundary.

   (a) The unit square, $[0,1] \times [0,1]$.

| Example 7.1 | **Parametrizations of Surfaces** |

A. A Paraboloid. Here we demonstrate a surface $\mathcal{S}$ that is a portion of the graph of a function of two variables. Let $f : \mathcal{D} \subset \mathbb{R}^2 \rightarrow \mathbb{R}$ and consider

$$\mathcal{S} = \{(x, y, z) : z = f(x, y), (x, y) \in \mathcal{D}\}.$$

The simplest parametrization of $\mathcal{S}$ is given by $\mathbf{\Phi}(x, y) = (x, y, f(x, y))$. The domain of $\mathbf{\Phi}$ is the domain of f, and x and y parametrize the coordinate curves on $\mathcal{S}$. For example, suppose $\mathcal{S}$ is the portion of the paraboloid given by $z = x^2 + y^2$, $0 \le z \le 1$. Then $\mathcal{S}$ is the portion of the graph of $f(x, y) = x^2 + y^2$ with (x, y) in the disk of radius 1 centered at the origin in the xy-plane. Thus $\mathcal{S}$ can be parametrized by

$$\mathbf{\Phi}(x, y) = (x, y, x^2 + y^2).$$

(See Figure 7.1(a).) The domain of $\mathbf{\Phi}$ is the disk $\mathcal{D} = \{(x, y) : x^2 + y^2 \le 1 \}$. In this case, the coordinate curves of $\mathcal{S}$, which are parametrized by $\alpha(x) = \mathbf{\Phi}(x, y_0)$ and $\beta(y) = \mathbf{\Phi}(x_0, y)$, are the vertical slices of the paraboloid parallel to the xz-plane and the yz-plane, respectively.

B. Cylinder. Suppose that $\mathcal{S}$ is a cylinder of radius 2 and height 5 with base in the xy-plane and centered on the z-axis. That is,

$$\mathcal{S} = \{(x, y, z) : x^2 + y^2 = 4, \ 0 \le z \le 5 \}.$$

(See Figure 7.1(b).) In this case, we can use the representation of the cylinder in cylindrical coordinates to construct a parametrization. The cylindrical coordinates (r, θ, z) of

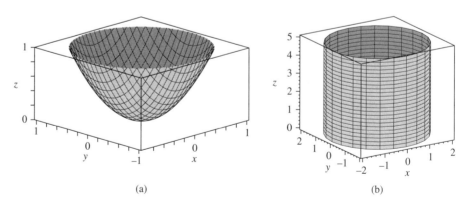

(a)   (b)

Figure 7.1   (a) The paraboloid $z = x^2 + y^2$ of Example 7.1A. (b) The cylinder of $\mathcal{S} = \{(x, y, z) : x^2 + y^2 = 4, \ 0 \le z \le 5 \}$ of Example 7.1B. Curves of the coordinate grid are shown on each surface.

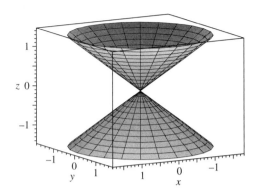

Figure 7.4   A double cone with vertex at the origin. (See Example 7.3.)

Example 7.3   **A Double Cone.** Consider the parametrization

$$\mathbf{\Phi}(\rho, \theta) = (\rho\cos(\theta)\sin(\tfrac{\pi}{3}), \rho\sin(\theta)\sin(\tfrac{\pi}{3}), \rho\cos(\tfrac{\pi}{3})),$$

where $-2 \leq \rho \leq 2$, $0 \leq \theta \leq 2\pi$. The image of $\mathbf{\Phi}$ is shown in Figure 7.4. It consists of two cones touching at their respective vertices. The parametrization maps all points of the form $(0, \theta)$, $0 \leq \theta \leq 2\pi$, to the origin. Based on the discussion preceding the example, we would say that $\mathbf{\Phi}$ fails to be one-to-one on this segment. However, more could be said. The tangent vectors to the coordinate curves when $\rho = 0$ are $\mathbf{\Phi}_\rho(0, \theta) = (\cos(\theta)\sin(\tfrac{\pi}{3}), \sin(\theta)\sin(\tfrac{\pi}{3}), \cos(\tfrac{\pi}{3}))$ and $\mathbf{\Phi}_\theta(0, \theta) = (0, 0, 0)$. Calculating further, the normal vector to the surface is $\mathbf{n} = \mathbf{\Phi}_\rho(0, \theta) \times \mathbf{\Phi}_\theta(0, \theta) = (0, 0, 0)$. Consequently, it is impossible to identify a normal direction to the surface at the vertex.

The vanishing of the normal vector is an issue in constructing integrals in Section 7.3, and we will want to rule it out from consideration. Thus we make the following definition.

Definition 7.2   A parametrization $\mathbf{\Phi} = \mathbf{\Phi}(s, t)$ of a surface $\mathcal{S}$, $\mathbf{\Phi} : \mathcal{D} \subset \mathbb{R}^2 \rightarrow \mathbb{R}^3$, is called a **regular** parametrization of $\mathcal{S}$ if it is continuously differentiable and $\mathbf{n} = \mathbf{\Phi}_s \times \mathbf{\Phi}_t$ is nonzero on the interior of $\mathcal{D}$.   ◆

In later sections, we will require parametrizations to be regular.

Summary

In this section, we introduced the concept of a parametrization of a surface. A *parametrization* is a function $\mathbf{\Phi} : \mathcal{D} \subset \mathbb{R}^2 \to \mathbb{R}^3$, defined by an expression $\mathbf{\Phi}(s,t) = (x(s,t), y(s,t), z(s,t))$ for $(s,t) \in \mathcal{D}$, where the coordinate functions are differentiable functions of two variables. The *coordinate curves* of the parametrization are obtained by holding one or the other of the independent variables s and t constant, $s = s_0$ or $t = t_0$. Using the tangent vectors to the coordinate curves, $\mathbf{\Phi}_s(s_0, t_0)$ and $\mathbf{\Phi}_t(s_0, t_0)$, we can construct a *parametrization of the tangent plane* to the surface at the point $\mathbf{\Phi}(s_0, t_0)$. The cross product of these tangent vectors is a *normal vector* to the surface at $\mathbf{\Phi}(s_0, t_0)$, $\mathbf{n} = \mathbf{\Phi}_s(s_0, t_0) \times \mathbf{\Phi}_t(s_0, t_0)$. Using $\mathbf{n}$, we can construct a *coordinate equation* for the tangent plane at a point. We defined a *regular parametrization* to be one that is continuously differentiable with nonzero normal vectors. Generally, we will be interested in regular parametrizations that are one-to-one on the interior of their domains.

Section 7.1 Exercises

1. **Parametrizing Planes.** In the collaborative exercise, we observed that if a plane $\mathcal{P}$ contains the noncollinear points P, Q, and R, it can be parametrized by

$$\mathbf{\Phi}(s,t) = \mathbf{p} + s\mathbf{u} + t\mathbf{v},$$

where $\mathbf{p} = P$, $\mathbf{u} = Q - P$, and $\mathbf{v} = R - P$.

(a) What are the tangent vectors to the coordinate curves, $\mathbf{\Phi}_s(s_0, t_0)$ and $\mathbf{\Phi}_t(s_0, t_0)$?
(b) Does your answer to (a) depend on the point (s_0, t_0)? Does this make sense in light of the coordinate curves?
(c) Does the normal vector $\mathbf{n}$ depend on the point (s_0, t_0)? Does this make sense in light of the surface?

2. **Parametrizing Polygons.** Use the formula given in Exercise 1 to construct parametrizations of the following polygons in space.

(a) The rectangle with vertices located at $(1,1,2)$, $(4,1,2)$, $(1,3,2)$, and $(4,3,2)$.
(b) The triangle with vertices located at $(4,1,2)$, $(1,3,2)$, and $(4,3,2)$.
(c) The parallelogram with vertices located at $(3,1,-1)$, $(5,4,-1)$, $(3,2,1)$, and $(8,6,0)$.
(d) The triangle with vertices $(1,1,0)$, $(1,0,1)$, and $(0,1,1)$.

3. **Parametrizing Surfaces.** For each of the following surfaces $\mathcal{S}$, find a parametrization of the surface. In each case, describe the domain of the parametrization.

(a) $\mathcal{S}$ is a half-cylinder of radius 3 and height 2 with base in the xy-plane, central axis on the z-axis, and positive y coordinate.
(b) $\mathcal{S}$ is the portion of a sphere of radius 2 centered at the origin contained in the first octant.
(c) $\mathcal{S}$ is the portion of the paraboloid $z = 9 - x^2 - y^2$ lying above the xy-plane.

(d) $\mathcal{S}$ is the plane containing the vectors $(1, 2, 1)$ and $(-1, 0, 3)$ and the point $(0, 2, -1)$.

(e) $\mathcal{S}$ is the portion of the sphere of radius 5 centered at the origin lying above the plane $z = 3$.

4. **Surfaces and Coordinate Curves.** For each of the following functions $\boldsymbol{\Phi} : \mathcal{D} \subset \mathbb{R}^2 \to \mathbb{R}^3$, describe and sketch the surface $\mathcal{S}$ that is the image of $\boldsymbol{\Phi}$. Describe the coordinate curves $\boldsymbol{\Phi}(s, t_0)$ and $\boldsymbol{\Phi}(s_0, t)$ and sketch these curves on $\mathcal{S}$.

   (a) $\boldsymbol{\Phi} : [0, 1] \times [0, 1] \to \mathbb{R}^3$ with $\boldsymbol{\Phi}(s, t) = (1 + s + 2t, s - t, 2 - s + t)$.
   (b) $\boldsymbol{\Phi} : [0, \pi/2] \times [0, 3] \to \mathbb{R}^3$ with $\boldsymbol{\Phi}(s, t) = (2 \cos s, 2 \sin s, t)$.
   (c) $\boldsymbol{\Phi} : [0, 2] \times [0, 3] \to \mathbb{R}^3$ with $\boldsymbol{\Phi}(s, t) = (s \cos(\pi/4), s \sin(\pi/4), t)$.
   (d) $\boldsymbol{\Phi} : [-2, 2] \times [0, 1] \to \mathbb{R}^3$ with $\boldsymbol{\Phi}(s, t) = (s, t, 4 - s^2)$.
   (e) $\boldsymbol{\Phi} : [0, 3] \times [0, 2\pi] \to \mathbb{R}^3$ with $\boldsymbol{\Phi}(s, t) = (s \cos t \sin(\pi/6), s \sin t \sin(\pi/6), s \cos(\pi/6))$.

5. **Tangent Planes.** For each of the parametrizations in Exercise 4, compute $\boldsymbol{\Phi}_s(s, t)$, $\boldsymbol{\Phi}_t(s, t)$, and $\mathbf{n} = \boldsymbol{\Phi}_s(s, t) \times \boldsymbol{\Phi}_t(s, t)$, and find a coordinate expression for the tangent plane at the following points, respectively.

   (a) $(s_0, t_0) = (\frac{1}{2}, \frac{1}{2})$.
   (b) $(s_0, t_0) = (\frac{\pi}{4}, 1)$.
   (c) $(s_0, t_0) = (1, 2)$.

   (d) $(s_0, t_0) = (0, \frac{1}{2})$.
   (e) $(s_0, t_0) = (2, \pi)$.

6. **One-to-One and Regular I.** For each of the parametrizations in Exercise 4, determine if the parametrization is (i) one-to-one and/or (ii) regular. (See Exercise 5.)

7. **Surfaces of Revolution I.** Surfaces of revolution can be parametrized using cylindrical coordinates. If the axis of revolution is the z-axis, we can use the following parametrization:

$$\boldsymbol{\Phi}(\theta, z) = (\cos(\theta)f(z), \sin(\theta)f(z), z),$$

where f is a non-negative differentiable function of z, so that the radius r at height z is $f(z)$. The domain of $\boldsymbol{\Phi}$ consists of (θ, z) with $0 \leq \theta \leq 2\pi$ and z in the domain of f. For each of the following functions $r = f(z)$, sketch the corresponding surface of revolution for the given interval for z. Show the coordinate curves on the surface.

   (a) $f(z) = 2$, $-3 \leq z \leq 3$.
   (b) $f(z) = \cos z$, $-\pi/2 \leq z \leq \pi/2$.

   (c) $f(z) = \sqrt{4 - z^2}$, $-2 \leq z \leq 2$.
   (d) $f(z) = \sqrt{1 + z^2}$, $-1 \leq z \leq 1$.

8. **Surfaces of Revolution II.** Consider a surface of revolution

$$\boldsymbol{\Phi}(\theta, z) = (\cos(\theta)f(z), \sin(\theta)f(z), z),$$

where f is a non-negative differentiable function of z with $0 \leq \theta \leq 2\pi$ and z in the domain of f.

   (a) Find the tangent vectors $\boldsymbol{\Phi}_\theta(\theta_0, z_0)$ and $\boldsymbol{\Phi}_z(\theta_0, z_0)$ to the coordinate curves of $\boldsymbol{\Phi}$ in terms of f and its derivative.

12. **Surfaces as Boundaries.** Each of the following surfaces $\mathcal{S}$ is the boundary of a region in space. For each surface, describe $\mathcal{S}$ as the union of surfaces $\mathcal{S}_i$, where any two surfaces intersect at most along their edges. Then give a parametrization of each surface $\mathcal{S}_i$.

    (a) $\mathcal{S}$ is the boundary of the cylindrical region of radius 1 and height 3 with base on the xy-plane and central axis on the z-axis.

    (b) $\mathcal{S}$ is the boundary of the box centered at the origin with length and width 2 and height 4 and faces parallel to the coordinate planes.

    (c) $\mathcal{S}$ is the boundary of the solid hemisphere of radius 2 centered at the origin lying in the lower half-space.

    (d) $\mathcal{S}$ is the boundary of the tetrahedron formed by the coordinate planes and the plane $z = 4 - x - 2y$.

    (e) $\mathcal{S}$ is the boundary of the region in the upper half-space between the cylinder of radius 1 and height 2 centered on the z-axis with base on the xy-plane and the sphere of radius 1 centered at the origin.

13. **Spherical Surfaces.** In spherical coordinates, the radius ρ may be replaced by a function $\rho = \rho(\theta, \phi)$. This gives rise to a parametrization

$$\mathbf{\Phi}(\theta, \phi) = (\rho(\phi, \theta) \cos\theta \, \sin\phi, \rho(\phi, \theta) \sin\theta \, \sin\phi, \rho(\phi, \theta) \cos\phi),$$

where $0 \le \phi \le \pi$ and $0 \le \theta \le 2\pi$. (A sphere of radius ρ_0 centered at the origin is obtained by letting $\rho(\theta, \phi) = \rho_0$.) Consider the spherical surface parametrized by $\mathbf{\Phi}$ with $\rho(\theta, \phi) = \cos(2\phi)$. (See Figure 7.5.)

    (a) Is $\mathbf{\Phi}$ one-to-one on the interior of its domain? If so, explain why. If not, which points in the interior are mapped to the same point?

    (b) Is $\mathbf{\Phi}$ a regular parametrization? Justify your answer with a calculation.

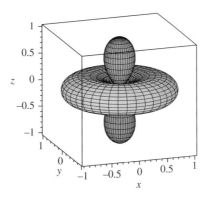

Figure 7.5 The spherical surface given by $\rho = \cos(2\phi)$ of Exercise 13.

■ 7.2 Surface Integrals

In this section, we will consider the problem of computing the total accumulation of a function defined on a surface. Let us begin by considering several examples of physical phenomena that can be modeled by functions on surfaces.

Example 7.4 **Total Accumulation on Surfaces**

  A. **Surface Area.** Suppose that the function f is constant and equal to 1 on $\mathcal{S}$, $f(x, y, z) = 1$ for $(x, y, z) \in \mathcal{S}$. Then the total accumulation of f on $\mathcal{S}$ is the **surface area** of $\mathcal{S}$.

  B. **Total Charge.** An electrical conductor is a substance in which electrical charge moves easily from one place to another. It can be shown both mathematically and experimentally that if the charge in a conductor is in electrostatic equilibrium, that is, the charges are at rest and the corresponding electric field is not changing, then the excess charges will necessarily lie on the surface of the conductor. Thus if $\mathcal{S}$ denotes the surface of the conductor, the **total charge** on the conductor is equal to the total accumulation of charge on the surface of the conductor.

  C. **Total Heat.** Suppose that an object consists of a thin shell of thickness h of a homogeneous material in the shape of a surface $\mathcal{S}$. If the temperature T at any point in the object is determined by its location on the shell and does not depend on its depth within the shell, then the **total heat** contained in the object is equal to the total accumulation of the function $\delta c T \cdot h$ on $\mathcal{S}$, where δ is the density of the material and c is the specific heat of the material.

  D. **Total Acoustic Power.** Suppose that a source emits a sound at a given frequency equally in all directions, which has the effect of changing the pressure to $p_a(x, y, z)$. Since the sound is the same in all directions, a level set $\mathcal{S}$ of p_a is a sphere. The intensity of the sound is defined to be the function $I = p_a^2/(\delta c)$, where δ is the density of the medium and c is the speed of sound in the medium. The **total acoustic power** of the acoustic source is the total accumulation of I on $\mathcal{S}$.

Riemann Sums

Let us assume that $\mathcal{S}$ is a surface that is the image of a regular parametrization $\boldsymbol{\Phi}$, so that the partial derivatives of $\boldsymbol{\Phi}$ exist and are continuous. Further, we will assume that

exists and is independent of the choices of partitions and sampling schemes. We call this limit the **surface integral** of f on $\mathcal{S}$ and we denote it by

$$\int\int_{\mathcal{S}} f \, dS. \quad \blacklozenge$$

Notice that although the limit appears to depend on the choice of parametrization $\boldsymbol{\Phi}$, it can be shown that if this limit exists, it is independent of the choice of parametrization. Thus we are justified in defining the limit to be an integral over $\mathcal{S}$.

In order to evaluate the limit, we must express the Riemann sum in a form that we recognize as a Riemann sum for a function of the variables s and t. Thus we must express the area ΔS_{ij} of the ij^{th} cell on $\mathcal{S}$ as a function of s and t. As was the case for parametric curves, it is not possible to compute this area exactly. We will instead approximate this area by the area of the parallelogram spanned by the scaled tangent vectors to the coordinate curves at the point $\boldsymbol{\Phi}(s_{i-1}, t_{j-1})$. These are $\boldsymbol{\Phi}_s(s_{i-1}, t_{j-1})\Delta s$ and $\boldsymbol{\Phi}_t(s_{i-1}, t_{j-1})\Delta t$, where Δs and Δt are the lengths of the sides of the cells of the partition of the domain of $\mathcal{S}$. (See Figure 7.8.) The area of this parallelogram is the length of the cross product of these vectors,

$$\|\boldsymbol{\Phi}_s(s_{i-1}, t_{j-1}) \times \boldsymbol{\Phi}_t(s_{i-1}, t_{j-1})\|\Delta s \Delta t.$$

Note that $\boldsymbol{\Phi}_s \times \boldsymbol{\Phi}_t$ is the normal vector to the surface given by the parametrization.

If we choose $(s_i^*, t_j^*) = (s_{i-1}, t_{j-1})$, we have

$$R(f, \boldsymbol{\Phi}, P) \approx \sum_{i,j=1}^{M,N} f(\boldsymbol{\Phi}(s_i^*, t_j^*))\|\boldsymbol{\Phi}_s(s_i^*, t_j^*) \times \boldsymbol{\Phi}_t(s_i^*, t_j^*)\|\Delta s \Delta t.$$

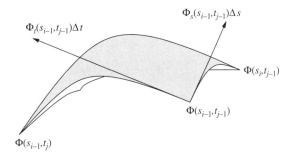

Figure 7.8 The ij^{th} cell of the partition of $\mathcal{S}$ and the tangent vectors $\boldsymbol{\Phi}_s(s_{i-1}, t_{j-1})\Delta s$ and $\boldsymbol{\Phi}_t(s_{i-1}, t_{j-1})\Delta t$ to the coordinate curves of $\boldsymbol{\Phi}$ through $\boldsymbol{\Phi}(s_{i-1}, t_{j-1})$.

for $\phi \in [0, \pi/2]$. Thus the total accumulation of f over $\mathcal{S}$ is given by

$$\int\int_{\mathcal{S}} f\, dS = \int_0^{2\pi} \int_0^{\pi/2} f(\mathbf{\Phi}(\theta, \phi)) \|\mathbf{\Phi}_\theta(\theta, \phi) \times \mathbf{\Phi}_\phi(\theta, \phi)\|\, d\phi\, d\theta$$

$$= \int_0^{2\pi} \int_0^{\pi/2} 2\cos\phi\, 4\sin\phi\, d\phi\, d\theta$$

$$= 4 \int_0^{2\pi} \left[\sin^2 \phi\right]_0^{\pi/2} d\theta$$

$$= 4 \int_0^{2\pi} 1\, d\theta$$

$$= 8\pi.$$

We can also apply the techniques of this section to compute the total accumulation of a function on more complicated surfaces, which cannot be represented as the image of a single parametrization $\mathbf{\Phi}$. If $\mathcal{S}$ is a union of surfaces, $S = S_1 \cup S_2 \cup \ldots \cup S_p$, where any two of the surfaces intersect at most along their edges, then the total accumulation of f on $\mathcal{S}$ is equal to the sum of the accumulations of f on each of the surfaces. That is,

$$\int\int_{\mathcal{S}} f\, dS = \int\int_{\mathcal{S}_1} f\, dS + \int\int_{\mathcal{S}_2} f\, dS + \cdots + \int\int_{\mathcal{S}_p} f\, dS.$$

In the following example, we will compute the average value of a function over the surface $\mathcal{S} = \mathcal{S}_1 \cup \mathcal{S}_2$, where $\mathcal{S}_1$ is a cone with vertex at the origin and opening upward with angle $\pi/4$ and slant height 1, and $\mathcal{S}_2$ is the portion of the sphere of radius 1 centered at the origin that "caps" the opening of the cone. (See Figure 7.9.) The average value of f is the total accumulation of f over $\mathcal{S}$ divided by the surface area of $\mathcal{S}$.

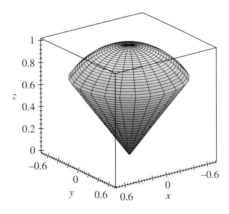

Figure 7.9   The surface $S = S_1 \cup S_2$ of Example 7.6 consisting of a cone with a spherical "cap."

the earth's atmosphere. Thus if α parametrizes the orbit of a satellite, the kinetic energy and the potential energy of the satellite satisfy

$$K(\alpha(t)) + U(\alpha(t)) = \frac{m}{2}\mathbf{v}(t) \cdot \mathbf{v}(t) - \frac{GM_e m}{\|\alpha(t)\|} = E,$$

where E is constant. Assuming that the orbit is a closed curve, where in the orbit will the speed of the satellite be greatest and where will it be least? Explain your answer.

■ 6.3 Integration over Closed Curves

In this section, we will continue to study line integrals focusing on the integral of a vector field over a simple closed curve in the plane. The primary result will be Green's[1] theorem, which relates the line integral of a vector field over a simple closed curve to a double integral over the region enclosed by the curve. Green's theorem applies to vector fields $\mathbf{F}$ whose coordinate functions are continuously differentiable. The generalization of Green's theorem to closed curves in space, Stokes' theorem, will be considered in the next chapter. We will introduce and discuss Green's theorem first, reserving the proof for the end of the section.

Intuitively, a curve is closed if it begins and ends at the same point. We say the curve is a simple closed curve if it does not intersect itself at any other points. More precisely, we have the following definition.

Definition 6.9 A curve is called **closed** if it has a parametrization $\alpha : [a, b] \rightarrow \mathbb{R}^2$ with $\alpha(a) = \alpha(b)$. A closed curve is called a **simple closed curve** if for any distinct points $t_1, t_2 \in [a, b)$, $\alpha(t_1) \neq \alpha(t_2)$.   ◆

A simple closed curve $\mathcal{C}$ in $\mathbb{R}^2$ divides the plane into two regions, the **interior** of $\mathcal{C}$ and the **exterior** of $\mathcal{C}$. The interior of $\mathcal{C}$ is the region contained inside $\mathcal{C}$, and the exterior of $\mathcal{C}$ is the region outside $\mathcal{C}$. Each of these regions is an open set in the plane. The curve $\mathcal{C}$ is the boundary of these regions. (See Section 3.3 for the definition of open sets and boundary.) We will consider curves that can be expressed as the finite union of curves can be parametrized by differentiable functions. We call these **piecewise differentiable** curves.

In Section 6.2, we showed that a continuous vector field $\mathbf{F}$ has the path independence property, or is conservative, on its domain, if and only if it is a gradient vector field. If a vector field has the path independence property, it follows that its line integral around a

---

[1]G. Green (1793–1841) was a self-taught British mathematician and physicist known for creating the first mathematical theory of electricity and magnetism.

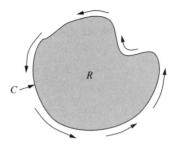

Figure 6.6   A region $\mathcal{R}$ with boundary a simple closed curve $\mathcal{C}$ oriented positively.

closed curve is zero. Conversely, in Exercise 8 of Section 6.2, we showed that if the line integral of a vector field around every closed curve is zero, then the vector field has the path independence property. This required that the interior of $\mathcal{C}$ was contained in the domain of $\mathbf{F}$. Combining these two results, a vector field $\mathbf{F}$ is a gradient vector field if and only if the line integral of $\mathbf{F}$ over every closed curve whose interior is contained in the domain of $\mathbf{F}$ is zero.

Green's theorem extends this result in that it makes a claim about the line integral of any continuously differentiable vector field $\mathbf{F}$ around a simple closed curve. It expresses the line integral of $\mathbf{F}$ around a simple closed curve $\mathcal{C}$ as a double integral of the partial derivatives of the coordinate functions of $\mathbf{F}$ over the interior of $\mathcal{C}$. To state Green's theorem, we must orient $\mathcal{C}$. The conventional orientation of $\mathcal{C}$ orients it so that the interior remains on the left as the curve is traced in the direction of the orientation. For simple closed curves in the plane, this orients the boundary curve in a *counterclockwise* direction around the interior $\mathcal{R}$. We will call this the *positive orientation* of the boundary. (See Figure 6.6.) We can now state Green's theorem. The proof appears at the end of the section.

Theorem 6.3   Green's Theorem. Let $\mathcal{R}$ be a region in the plane whose boundary $\mathcal{C}$ is a piecewise differentiable, simple closed curve that is positively oriented with respect to $\mathcal{R}$. Let $\mathbf{F}(x,y) = (u(x,y), v(x,y))$ be a vector field that is defined and continuously differentiable on an open set containing $\mathcal{C}$ and $\mathcal{R}$. Then

$$\int_{\mathcal{C}} \mathbf{F} \cdot \mathbf{T} \, ds = \int\int_{\mathcal{R}} \left(\frac{\partial v}{\partial x} - \frac{\partial u}{\partial y} \right) dA_{x,y}. \; \blacklozenge$$

In the special case when $\mathbf{F} = \nabla f$ is a gradient vector field, $u = \frac{\partial f}{\partial x}$ and $v = \frac{\partial f}{\partial y}$. Because the mixed partial derivatives of f are equal, it follows immediately that the integrand of the double integral in the theorem is zero.

$$\frac{\partial v}{\partial x} - \frac{\partial u}{\partial y} = \frac{\partial}{\partial x}\frac{\partial f}{\partial y} - \frac{\partial}{\partial y}\frac{\partial f}{\partial x} = 0.$$

Thus Green's theorem gives us an alternate way to understand the fact that the line integral of a gradient vector field around a closed curve is zero.

In addition, there is an interpretation of Green's theorem when $\mathbf{F}$ is the velocity field of a fluid flow. In this case, the line integral of $\mathbf{F}$ around a closed curve $\mathcal{C}$ gives the total accumulation of the component of the flow in the direction of $\mathcal{C}$. This quantity is called the **circulation** of the fluid flow around $\mathcal{C}$. If the velocity field $\mathbf{F}$ is continuous and differentiable on $\mathcal{R}$ and $\frac{\partial v}{\partial x} - \frac{\partial u}{\partial y} = 0$ on $\mathcal{R}$, then the circulation around any closed curve $\mathcal{C} \subset \mathcal{R}$ is 0. In this context, we say the fluid flow is **irrotational**. To check if a flow is irrotational, we need only check that $\frac{\partial v}{\partial x} - \frac{\partial u}{\partial y} = 0$.

Since Green's theorem equates a line integral with a double integral, it can be used in two ways as a calculational tool. For example, if we are interested in the line integral of a vector field around a closed curve, it might be the case that it is simpler to evaluate the corresponding double integral over the interior of the closed curve rather than the line integral. Conversely, we might prefer to evaluate the line integral over $\mathcal{C}$ in place of the double integral over the interior of $\mathcal{C}$. The following example illustrates both types of calculations.

Example 6.9	**Green's Theorem Calculations**

A. Let $\mathcal{R}$ be the region in the first quadrant bounded by the x-axis, the y-axis, the line $y = x + 2$, and the line $y = 2x - 4$. Let $\mathbf{F}(x, y) = (xy, x^2 - y^2)$. (See Figure 6.7.) The boundary $\mathcal{C}$ of this region consists of four straight line segments. In order to compute the line integral of $\mathbf{F}$ around $\mathcal{C}$ in a counterclockwise direction, we would parametrize each of these line segments with the appropriate orientation and then compute the corresponding line integrals. Using Green's theorem, we can replace this calculation by the calculation of the double integral over the region $\mathcal{R}$ of

$$\frac{\partial v}{\partial x} - \frac{\partial u}{\partial y} = 2x - x = x.$$

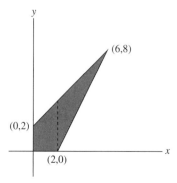

Figure 6.7 The region in Example 6.9A.

Thus we have

$$\int_{\mathcal{C}} \mathbf{F} \cdot \mathbf{T} \, ds = \int\int_{\mathcal{R}} x \, dA_{x,y}$$

$$= \int_0^2 \int_0^{x+2} x \, dy \, dx + \int_2^6 \int_{2x-4}^{x+2} x \, dy \, dx$$

$$= 20/3 + 80/3 = 100/3.$$

B. If $\mathcal{C}$ is a simple closed curve enclosing a region $\mathcal{R}$, and $\mathbf{F}(x,y) = \frac{1}{2}(-y, x)$, then using Green's theorem, we have

$$\int_{\mathcal{C}} \mathbf{F} \cdot \mathbf{T} \, ds = \int\int_{\mathcal{R}} 1 \, dA_{x,y} = \text{Area}(\mathcal{R}).$$

Thus we can use the line integral, $\int_{\mathcal{C}} \mathbf{F} \cdot \mathbf{T} \, ds$, to compute the area of the region enclosed by $\mathcal{C}$. For example, suppose $\mathcal{R}$ is the interior of the ellipse $\frac{x^2}{a^2} + \frac{y^2}{b^2} = 1$. The ellipse can be parametrized by $\alpha(t) = (a \cos t, b \sin t)$, $t \in [0, 2\pi]$. The area enclosed by the ellipse can be computed by evaluating the iterated integral

$$\int_{-a}^{a} \int_{-\frac{b}{a}\sqrt{a^2-x^2}}^{\frac{b}{a}\sqrt{a^2-x^2}} 1 \, dA.$$

Using Green's theorem with $\mathbf{F} = \frac{1}{2}(-y, x)$, we know the area enclosed by the ellipse is

$$\int_{\mathcal{C}} \mathbf{F} \cdot \mathbf{T} \, ds = \int_0^{2\pi} \frac{1}{2}(-b \sin t, a \cos t) \cdot (-a \sin t, b \cos t) \, dt$$

$$= \frac{1}{2} \int_0^{2\pi} ab \, dt = \pi ab.$$

Green's theorem can be extended to regions whose boundary consists of a collection of simple closed curves. A region of this type is obtained by removing the interiors of a collection of simple closed curves $\mathcal{C}_2, \mathcal{C}_3, ..., \mathcal{C}_n$ from the interior of a simple closed curve $\mathcal{C}_1$. It is important that these curves be oriented correctly. Each component of the boundary of $\mathcal{R}$ must be oriented so that when it is traversed in the direction of the orientation, the region stays on the left. Thus the curve $\mathcal{C}_1$ must be traced counterclockwise and the curves $\mathcal{C}_2, \mathcal{C}_3, \ldots, \mathcal{C}_n$ must be traced clockwise. (See Figure 6.8(a).)

We are now in a position to state the generalization of Green's theorem and to sketch a proof of the result.

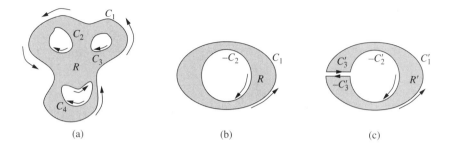

Figure 6.8 (a) A region that is the interior of a simple closed curve with a number of holes. (b) and (c) The region between the ellipse $x^2 + 2y^2 = 4$ and the circle $x^2 + y^2 = 1$ used in the proof of Green's theorem for more general regions. (See Theorem 6.4 and Example 6.10.)

Theorem 6.4 **Green's Theorem for More General Regions.** Let $\mathcal{R}$ be a region in $\mathbb{R}^2$ whose boundary $\mathcal{C}$ consists of the union of a finite number of piecewise differentiable, simple closed curves that are positively oriented with respect to $\mathcal{R}$. Let $\mathbf{F}(x, y) = (u(x, y), v(x, y))$ be a vector field that is defined and continuously differentiable on an open set containing $\mathcal{C}$ and $\mathcal{R}$. Then

$$\int_{\mathcal{C}} \mathbf{F} \cdot \mathbf{T}\, ds = \int \int_{\mathcal{R}} \left(\frac{\partial v}{\partial x} - \frac{\partial u}{\partial y} \right)\, dA_{x,y}. \quad \blacklozenge$$

Proof Sketch: We consider a special case to demonstrate the general situation. Suppose $\mathcal{C}_1$ is the ellipse $x^2 + 2y^2 = 4$ oriented counterclockwise and $\mathcal{C}_2$ is the circle $x^2 + y^2 = 1$, also oriented counterclockwise. Let $\mathcal{R}$ be the region between them. (See Figure 6.8(b).) We want to replace $\mathcal{R}$ by a region $\mathcal{R}'$ arbitrarily close to $\mathcal{R}$ whose boundary is a simple closed curve. To do so, we insert a curve $\mathcal{C}_3$ connecting $\mathcal{C}_1$ to $\mathcal{C}_2$ as shown in Figure 6.8(c). Parametrize $\mathcal{C}_3$ from $\mathcal{C}_1$ to $\mathcal{C}_2$, so that $-\mathcal{C}_3$ is traced from $\mathcal{C}_2$ to $\mathcal{C}_1$. Separating $\mathcal{C}_3$ and $-\mathcal{C}_3$ slightly, we obtain a region $\mathcal{R}'$, whose boundary $\mathcal{C}_1' \cup \mathcal{C}_3' \cup -\mathcal{C}_2' \cup -\mathcal{C}_3'$ is a simple closed curve oriented counterclockwise, or with the region on the left. (The prime notation indicates the modification that results from separating $\mathcal{C}_3$ and $-\mathcal{C}_3$.) Thus we may apply Green's theorem to $\mathcal{R}'$ and its boundary:

$$\int \int_{\mathcal{R}'} \left(\frac{\partial v}{\partial x} - \frac{\partial u}{\partial y} \right)\, dA_{x,y} = \int_{\mathcal{C}_1'} \mathbf{F} \cdot \mathbf{T}\, ds + \int_{\mathcal{C}_3'} \mathbf{F} \cdot \mathbf{T}\, ds + \int_{-\mathcal{C}_2'} \mathbf{F} \cdot \mathbf{T}\, ds + \int_{-\mathcal{C}_3'} \mathbf{F} \cdot \mathbf{T}\, ds.$$

Intuitively, if we allow the separation between $\mathcal{C}_3'$ and $-\mathcal{C}_3'$ to approach 0, the integrals over $\mathcal{C}_3'$ and $-\mathcal{C}_3'$ are opposites of each other, and we may remove the primes from the remaining terms. Separating these steps, we have

$$\int \int_{\mathcal{R}'} \left(\frac{\partial v}{\partial x} - \frac{\partial u}{\partial y} \right)\, dA_{x,y} = \int_{\mathcal{C}_1'} \mathbf{F} \cdot \mathbf{T}\, ds + \int_{\mathcal{C}_3'} \mathbf{F} \cdot \mathbf{T}\, ds - \int_{\mathcal{C}_2'} \mathbf{F} \cdot \mathbf{T}\, ds - \int_{\mathcal{C}_3'} \mathbf{F} \cdot \mathbf{T}\, ds,$$

and then

$$\int\int_{\mathcal{R}} \left(\frac{\partial v}{\partial x} - \frac{\partial u}{\partial y} \right) dA_{x,y} = \int_{\mathcal{C}_1} \mathbf{F} \cdot \mathbf{T} \, ds - \int_{\mathcal{C}_2} \mathbf{F} \cdot \mathbf{T} \, ds.$$

This is the generalized version of Green's theorem for $\mathcal{R}$.

To extend this approach to regions of the type shown in Figure 6.8(a), we can proceed inductively, increasing the number of boundary components, or we can introduce sufficiently many segments to produce a region whose boundary is a simple closed curve as we did above. ∎

The extension of Green's theorem can be used to simplify the evaluation of line integrals around closed curves, because it allows us to replace more complicated line integrals by less complicated line integrals. We demonstrate this for the circle and ellipse used in the proof sketch.

Example 6.10   **Green's Theorem on a Punctured Region.** Let $\mathbf{F}(x,y) = (\frac{-y}{x^2+y^2}, \frac{x}{x^2+y^2})$. The vector field $\mathbf{F}$ is continuous, differentiable, and satisfies $\frac{\partial v}{\partial x} - \frac{\partial u}{\partial y} = 0$ at every point except the origin. Thus if $\mathcal{C}$ is a simple closed curve that does not enclose the origin, so that $\mathbf{F}$ is defined and differentiable everywhere on the interior of $\mathcal{C}$, then Green's theorem tells us that $\int_{\mathcal{C}} \mathbf{F} \cdot \mathbf{T} \, ds = 0$. However, it does not apply directly to curves that circle the origin. Nevertheless, here we see how it can be of use.

A. Suppose $\mathcal{C}_2$ is the circle $x^2 + y^2 = 1$ oriented counterclockwise. We cannot apply Green's theorem to evaluate the line integral of $\mathbf{F}$ around $\mathcal{C}_2$, but we can evaluate the line integral directly. If we parametrize the unit circle by $\alpha(t) = (\cos t, \sin t)$, $t \in [0, 2\pi]$, we have

$$\int_{\mathcal{C}_2} \mathbf{F} \cdot \mathbf{T} \, ds = \int_0^{2\pi} \mathbf{F}(\alpha(t)) \cdot \alpha'(t) \, dt$$

$$= \int_0^{2\pi} (-\sin t, \cos t) \cdot (-\sin t, \cos t) \, dt = 2\pi.$$

B. Now suppose $\mathcal{C}_1$ is the ellipse $x^2 + 2y^2 = 4$ oriented counterclockwise. Again, we cannot apply Green's theorem directly to evaluate the line integral of $\mathbf{F}$ around $\mathcal{C}_1$, because the origin is also contained in the interior of $\mathcal{C}_1$. If we parametrize $\mathcal{C}_1$ and try to evaluate the line integral directly, we find that the integral is difficult to evaluate symbolically. Instead we can use the generalized version of Green's theorem to evaluate the integral. Let $\mathcal{R}$ be the region between $\mathcal{C}_1$ and $\mathcal{C}_2$. (See Figure 6.8(b).) In this case, $\mathbf{F}$ is defined on $\mathcal{R}$ and $\frac{\partial v}{\partial x} - \frac{\partial u}{\partial y} = 0$ on $\mathcal{R}$. Thus, applying the generalized form of Green's theorem, we have

$$0 = \int\int_{\mathcal{R}} \left(\frac{\partial v}{\partial x} - \frac{\partial u}{\partial y} \right) dA_{x,y} = \int_{\mathcal{C}_1} \mathbf{F} \cdot \mathbf{T} \, ds - \int_{\mathcal{C}_2} \mathbf{F} \cdot \mathbf{T} \, ds.$$

It follows that

$$\int_{\mathcal{C}_1} \mathbf{F} \cdot \mathbf{T} \, ds = \int_{\mathcal{C}_2} \mathbf{F} \cdot \mathbf{T} \, ds = 2\pi.$$

We can see from the structure of the argument that we could replace $\mathcal{C}_1$ by any simple closed curve that circles the origin once and is oriented counterclockwise.

The Flux of a Vector Field

There is an important application of Green's theorem to fluid flows in the plane when the velocity field $\mathbf{F}$ of the flow does not vary in time. If $\mathcal{C}$ is a simple closed curve in the plane, we are interested in the amount of fluid that crosses $\mathcal{C}$ in a unit of time. This quantity is called the **total flux** of $\mathbf{F}$ across $\mathcal{C}$. This is equal to the total accumulation of the component of $\mathbf{F}$ in a direction normal to $\mathcal{C}$. Initially, we will express the total flux as a path integral, and then we will see how to rewrite it as a line integral.

First, let us establish our notation. Let $\alpha(t) = (x(t), y(t))$, $t \in [a, b]$, be a continuously differentiable parametrization of $\mathcal{C}$ with a nonzero derivative, so that $\|\alpha'(t)\| \neq 0$. The unit tangent vector to α is given by

$$\mathbf{T}(t) = \frac{\alpha'(t)}{\|\alpha'(t)\|} = \frac{1}{\|\alpha'(t)\|}(x'(t), y'(t)).$$

The unit vector $\mathbf{N}$ given by

$$\mathbf{N}(t) = \frac{1}{\|\alpha'(t)\|}(y'(t), -x'(t))$$

is orthogonal to $\mathbf{T}$. It can be shown that $\mathbf{N}$ always points to the same side of $\mathcal{C}$. Thus we can use $\mathbf{N}$ to specify a direction from one side of $\mathcal{C}$ to the other everywhere on $\mathcal{C}$ in a consistent manner. (See Exercise 11.) At each point of $\mathcal{C}$, the quantity $\mathbf{F} \cdot \mathbf{N}$ is the component of the velocity of the flow in a direction normal to or across $\mathcal{C}$. The **total flux** of $\mathbf{F}$ across $\mathcal{C}$ in the direction $\mathbf{N}$ is defined to be the path integral

$$\int_{\mathcal{C}} \mathbf{F} \cdot \mathbf{N} \, ds.$$

Keep in mind that although this is the integral of the dot product of vectors, it is *not* a line integral because $\mathbf{T}$ is not one of the vectors. However, after a brief calculation, we will be able to express the total flux as a line integral.

Let $\mathbf{F}(x, y) = (u(x, y), v(x, y))$. Then the total flux of $\mathbf{F}$ across $\mathcal{C}$ in the direction $\mathbf{N}$ is

$$
\begin{aligned}
\int_{\mathcal{C}} \mathbf{F} \cdot \mathbf{N} \, ds &= \int_a^b (u(\alpha(t)), v(\alpha(t))) \cdot \frac{1}{||\alpha'(t)||} (y'(t), -x'(t)) ||\alpha'(t)|| \, dt \\
&= \int_a^b -v(\alpha(t)) x'(t) + u(\alpha(t)) y'(t) \, dt \\
&= \int_a^b (-v(\alpha(t)), u(\alpha(t))) \cdot \alpha'(t) \, dt.
\end{aligned}
$$

This last integral is the line integral of the vector field $(-v, u)$ along $\mathcal{C}$. Thus

$$
\int_{\mathcal{C}} \mathbf{F} \cdot \mathbf{N} \, ds = \int_{\mathcal{C}} (-v, u) \cdot \mathbf{T} \, ds.
$$

If $\mathcal{C}$ is a simple closed curve with interior $\mathcal{R}$, we can use Green's theorem to express the total flux across $\mathcal{C}$ as a double integral. If α parametrizes $\mathcal{C}$ in a counterclockwise manner, then the vector $\mathbf{N}$ given above points out of $\mathcal{R}$, so that the total flux of $\mathbf{F}$ across $\mathcal{C}$ in the direction $\mathbf{N}$ measures the amount of fluid flowing out of $\mathcal{R}$ in a unit of time. Applying Green's theorem, we have

$$
\int_{\mathcal{C}} \mathbf{F} \cdot \mathbf{N} \, ds = \int_{\mathcal{C}} (-v, u) \cdot \mathbf{T} \, ds = \int\int_{\mathcal{R}} \left(\frac{\partial u}{\partial x} + \frac{\partial v}{\partial y} \right) dA_{x,y}.
$$

This result is a two-dimensional version of the divergence theorem, which we will encounter in Section 7.4. Since the double integral represents the total flux of $\mathbf{F}$ out of $\mathcal{R}$ and the integrand is defined at every point of $\mathcal{R}$, it makes sense on an intuitive level to interpret the integrand

$$
\frac{\partial u}{\partial x} + \frac{\partial v}{\partial y}
$$

as a measure of the flux of the flow at a point. Thus we will call this quantity the *pointwise* or *infinitesimal flux* of $\mathbf{F}$. Following through on this idea, if the integrand or infinitesimal flux is always positive, then the total flux is also positive. We will give a precise definition of infinitesimal flux in Section 7.3.

Application: Diffusion

This result can also be applied to a time-dependent diffusion process to measure the flux of the solute out of a region. The velocity of the fluid is assumed to be zero so that the movement of the solute is due solely to diffusion. This is an important consideration because diffusion is a microscopic process that is due to the random motion of molecules, whereas a fluid flow is a macroscopic process, so that the effects of fluid velocity are more significant than those of diffusion. The movement of the solute is represented by the flux vector. We will assume that the diffusion process satisfies ***Fick's*[2] *law***, that is, the flux vector is a multiple of the gradient vector of the concentration of the solute. The following example computes the flux of a source centered at the origin.

Example 6.11 **Total Flux of a Diffusion Process.** If an amount M of solute is released instantaneously from the origin in the plane at time $t = 0$, the concentration of the solute at time t at (x, y) is given by

$$c(x, y, t) = \frac{M}{4\pi\delta t}e^{-(x^2+y^2)/(4\delta t)},$$

where the constant δ is the diffusivity of the solute in the solvent. According to Fick's law, the flux vector $\mathbf{J}$ of the diffusion process is given by

$$\mathbf{J}(x, y, t) = -\delta\nabla c(x, y, t)$$
$$= -\delta(\frac{\partial c}{\partial x}(x, y, t), \frac{\partial c}{\partial y}(x, y, t)).$$

Also, a calculation shows that c satisfies the equation

$$\frac{\partial c}{\partial t}(x, y, t) = \delta\left(\frac{\partial^2 c}{\partial x^2}(x, y, t) + \frac{\partial^2 c}{\partial y^2}(x, y, t)\right).$$

This equation is called the ***diffusion equation***. The expression on the right-hand side is usually written $\Delta c(x, y, t)$ where $\Delta = \frac{\partial^2}{\partial x^2} + \frac{\partial^2}{\partial y^2}$. This combination of second derivatives of c is called the ***Laplacian*[3]** derivative of c.

Let us compute the flux of the concentration through a circle $\mathcal{C}$ of radius 1 centered at the origin in the outward direction. Here we assume that $\mathcal{C}$ is positively oriented relative to its interior $\mathcal{R}$, the unit disk centered at the origin. Thus, we want to compute

$$\int_{\mathcal{C}}\mathbf{J}\cdot\mathbf{N}\,ds = \int_{\mathcal{C}}-\delta(\nabla c\cdot\mathbf{N})\,ds = \int_{\mathcal{C}}-\delta(-\frac{\partial c}{\partial y}, \frac{\partial c}{\partial x})\cdot\mathbf{T}\,ds.$$

Applying Green's theorem to the right-most integral, we see that

$$\int_{\mathcal{C}}\mathbf{J}\cdot\mathbf{N}\,ds = \int\int_{\mathcal{R}}-\delta\left(\frac{\partial^2 c}{\partial x^2} + \frac{\partial^2 c}{\partial y^2}\right)dA_{x,y}$$
$$= -\int\int_{\mathcal{R}}\frac{\partial c}{\partial t}(x, y, t)\,dA_{x,y},$$

---

[2] A. E. Fick (1829–1901) was a German physiologist interested in diffusion of gas through a membrane in physiology and physics.

[3] Pierre-Simon Laplace, marquis de Laplace (1749–1827), was a French mathematician and astronomer known for his groundbreaking work in analysis and applied mathematics.

where we have used the fact that c satisfies the diffusion equation to make the last substitution. Thus the total flux of the diffusion process is equal to the negative of the double integral of the time derivative of the concentration.

For the particular c we started with,

$$\frac{\partial c}{\partial t}(x, y, t) = \frac{M}{4\pi\delta}\left(\frac{-1}{t^2} + \frac{x^2 + y^2}{4\delta t^3}\right)e^{-(x^2+y^2)/(4\delta t)}.$$

Substituting this expression into the double integral for the total flux, a calculation shows that the flux at time t is equal to

$$\frac{M}{4\delta t^2}e^{-1/(4\delta t)}.$$

We will explore this function in Exercise 13.

In the course of this example, we have shown that if $c = c(x, y, t)$ satisfies the diffusion equation, then the flux of c across $\mathcal{C}$ at time t is the negative total accumulation of $\frac{\partial c}{\partial t}$ on the interior of $\mathcal{C}$,

$$\int_{\mathcal{C}} \mathbf{J} \cdot \mathbf{N}\, ds = -\int\int_{\mathcal{R}} \frac{\partial c}{\partial t}(x, y, t)\, dA_{x,y}.$$

Notice the importance of the minus sign. Assuming that there is no addition or removal of substance within $\mathcal{R}$, if the integral of $\frac{\partial c}{\partial t}$ is positive, then the total amount of substance in $\mathcal{R}$ is increasing, so that there must be a net influx of substance into $\mathcal{R}$. It follows that the total flux of $\mathbf{J}$ *out* of $\mathcal{R}$ must be ***negative***. Conversely, if the integral of $\frac{\partial c}{\partial t}$ is negative, then the total amount of substance in $\mathcal{R}$ is decreasing, so that there must be a net efflux of substance out of $\mathcal{R}$ and the total flux of $\mathbf{J}$ out of $\mathcal{R}$ must be positive.

If $\mathbf{F}$ represents the velocity field of a flow or the flux field of a diffusion process, then the total flux across $\mathcal{C}$ represents the total amount of substance that leaves $\mathcal{R}$ in one unit of time. If no substance is added or lost at a point inside $\mathcal{R}$, the total flux is also the total change in the amount of substance in $\mathcal{R}$ per unit time. In this case, we might interpret the integrand of the double integral as the change in the amount of substance per unit area per unit time, so that its total accumulation is the total change in the amount of substance in $\mathcal{R}$ per unit time. We will give a more careful justification of this interpretation in Section 7.3, when we consider the generalization of this result to vector fields in space.

The Proof of Green's Theorem

Let us begin by recalling the hypothesis of Green's theorem: $\mathcal{R}$ is a region in the plane whose boundary $\mathcal{C}$ is a piecewise differentiable, simple closed curve that is positively oriented with respect to $\mathcal{R}$, and $\mathbf{F}(x, y) = (u(x, y), v(x, y))$ is a vector field that is defined and differentiable on an open set containing $\mathcal{C}$ and $\mathcal{R}$. We want to show that

$$\int_{\mathcal{C}} \mathbf{F} \cdot \mathbf{T}\, ds = \int\int_{\mathcal{R}} \left(\frac{\partial v}{\partial x} - \frac{\partial u}{\partial y} \right)\, dA.$$

The proof of Green's theorem involves a calculation of both sides of this expression to show they are equal. We will carry out this calculation in three steps: First we will show that the result holds for a particularly simple type of region that is bounded by the graphs of functions of x or functions of y, then we will indicate how to partition a general region into subregions of this type, and lastly we will show that Green's theorem holds on the general region if it holds on the subregions.

Step 1

There are two special cases for $\mathcal{R}$ that we want to consider. The simpler of the two cases, when $\mathcal{R}$ is a rectangle, is considered in Exercise 4. We will consider the more involved case here. Assume that $\mathcal{R}$ is a region that is bounded by a horizontal line, a vertical line, and a curve that can be written both in the form $y = g(x)$ and $x = h(y)$. (See Figure 6.9.)

The region $\mathcal{R}$ can be described in the form $a \leq x \leq b$ and $c \leq y \leq g(x)$ or in the form $c \leq y \leq d$ and $a \leq x \leq h(y)$. Using the additivity of the double integral, the double integral of $\frac{\partial v}{\partial x} - \frac{\partial u}{\partial y}$ over $\mathcal{R}$ can be expressed as the difference of the double integral of $\frac{\partial v}{\partial x}$

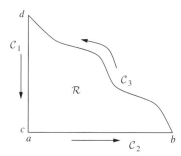

Figure 6.9   Region bounded by two lines and a curve that can be written both in the form $y = g(x)$ and $x = h(y)$.

over the region $\mathcal{R}$ and the double integral of $\frac{\partial u}{\partial y}$ over $\mathcal{R}$. We will use the two descriptions of $\mathcal{R}$ to express each of these double integrals as an iterated integral.

$$
\iint_{\mathcal{R}} \left(\frac{\partial v}{\partial x} - \frac{\partial u}{\partial y} \right) dA_{x,y} = \iint_{\mathcal{R}} \frac{\partial v}{\partial x} dA_{x,y} - \iint_{\mathcal{R}} \frac{\partial u}{\partial y} dA_{x,y}
$$

$$
= \int_c^d \int_a^{h(y)} \frac{\partial v}{\partial x}(x,y)\, dx\, dy - \int_a^b \int_c^{g(x)} \frac{\partial u}{\partial y}(x,y)\, dy\, dx
$$

$$
= \int_c^d (v(h(y),y) - v(a,y))\, dy - \int_a^b (u(x,g(x)) - u(x,c))\, dx.
$$

The boundary of the region $\mathcal{R}$ consists of three pieces: $\mathcal{C}_1$, the line segment $x = a$, $c \le y \le d$; $\mathcal{C}_2$, the line segment $y = c$, $a \le x \le b$; and $\mathcal{C}_3$, the portion of the curve $x = h(y)$, $c \le y \le d$. These pieces can be parametrized by $\alpha_1(y) = (a,y)$, $c \le y \le d$, $\alpha_2(x) = (x,c)$, $a \le x \le b$, and $\alpha_3(y) = (h(y),y)$, $c \le y \le d$, respectively. The boundary $\mathcal{C}$ is oriented counterclockwise by $-\alpha_1$, α_2, and α_3, where $-\alpha_1$ means that we parametrize $\mathcal{C}_1$ from (a,d) to (a,c). (See Figure 6.9.) Then,

$$
\int_{\mathcal{C}} \mathbf{F} \cdot \mathbf{T}\, ds = \int_{-\alpha_1} \mathbf{F} \cdot \mathbf{T}\, ds + \int_{\alpha_2} \mathbf{F} \cdot \mathbf{T}\, ds + \int_{\alpha_3} \mathbf{F} \cdot \mathbf{T}\, ds
$$

$$
= \int_d^c (u(a,y), v(a,y)) \cdot (0,1)\, dy + \int_a^b (u(x,c), v(x,c)) \cdot (1,0)\, dx
$$

$$
+ \int_c^d (u(h(y),y), v(h(y),y)) \cdot (h'(y),1)\, dy
$$

$$
= \int_d^c v(a,y)\, dy + \int_a^b u(x,c)\, dx + \int_c^d u(h(y),y)h'(y)\, dy + \int_c^d v(h(y),y)\, dy.
$$

If we substitute $x = h(y)$ into the third integral above and use the fact that for all points on $\mathcal{C}_3$, $y = g(x)$, we have

$$
\int_c^d u(h(y),y)h'(y)\, dy = \int_b^a u(x,g(x))\, dx = -\int_a^b u(x,g(x))\, dx.
$$

Combining this with our first calculation, we have

$$
\int_{\mathcal{C}} \mathbf{F} \cdot \mathbf{T}\, ds = \int_c^d (v(h(y),y) - v(a,y))\, dy + \int_a^b (u(x,c) - u(x,g(x)))\, dx
$$

$$
= \iint_{\mathcal{R}} \frac{\partial v}{\partial x} - \frac{\partial u}{\partial y}\, dA_{x,y},
$$

which proves Green's theorem for three-sided regions of the specified type.

Step 2

Now let us suppose that $\mathcal{R}$ is the interior of a simple closed curve $\mathcal{C}$. We would like to partition $\mathcal{R}$ into rectangles and region of the form given in Step 1. Figure 6.10 indicates how to partition such a region. Intuitively, we fill sufficiently much of the interior of $\mathcal{R}$ with nonoverlapping rectangles, so that we can fill the remaining portion of $\mathcal{R}$ with three-sided regions of the type specified in Step 1, where the boundary of any three-sided region is made up of a vertical and a horizontal line segment and a portion of $\mathcal{C}$. By using sufficiently many rectangles and choosing sufficiently small three-sided regions, it is possible to guarantee that the third side of each triangular region can be written both as the graph of a function of x and as the graph of a function of y. From Step 1 and Exercise 4, we have that Green's theorem holds for each of these subregions.

Step 3

Now let us assume that the interior $\mathcal{R}$ of a simple closed curve $\mathcal{C}$ has been partitioned into regions $\mathcal{R}_1, \ldots, \mathcal{R}_m$ and that Green's theorem holds on each of these regions. Since $\mathcal{R} = \mathcal{R}_1 \cup \mathcal{R}_2 \cup \ldots \cup \mathcal{R}_m$, the double integral of a function over $\mathcal{R}$ is the sum of the double integrals of the function over each $\mathcal{R}_i$. Thus we can calculate the right-hand side of the expression in Green's theorem by calculating

$$\int\int_{\mathcal{R}_i} \left(\frac{\partial v}{\partial x} - \frac{\partial u}{\partial y} \right) \, dA_{x,y}$$

for each i and summing the results.

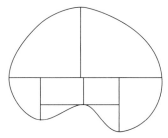

Figure 6.10 A region partitioned into regions for which Green's theorem can be verified directly.

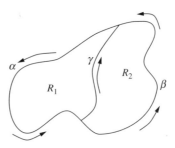

Figure 6.11   A region $\mathcal{R} = \mathcal{R}_1 \cup \mathcal{R}_2$ with positively oriented boundaries. The boundary $\mathcal{C}_1$ of $\mathcal{R}_1$ is parametrized by α and γ. The boundary $\mathcal{C}_2$ of $\mathcal{R}_2$ is parametrized by β and $-\gamma$.

After we have partitioned $\mathcal{R}$, let $\mathcal{C}_i$ be the boundary of $\mathcal{R}_i$ oriented counterclockwise. In order to compute the left-hand side of the expression in Green's theorem, we must relate the line integral of $\mathbf{F}$ over $\mathcal{C}$ to the line integrals of $\mathbf{F}$ over the $\mathcal{C}_i$. We claim that

$$\int_{\mathcal{C}} \mathbf{F} \cdot \mathbf{T} \, ds = \int_{\mathcal{C}_1} \mathbf{F} \cdot \mathbf{T} \, ds + \int_{\mathcal{C}_2} \mathbf{F} \cdot \mathbf{T} \, ds + \cdots + \int_{\mathcal{C}_m} \mathbf{F} \cdot \mathbf{T} \, ds.$$

We will demonstrate this for the case where $\mathcal{R}$ is partitioned into two regions. (See Figure 6.11.) If $\mathcal{R} = \mathcal{R}_1 \cup \mathcal{R}_2$ with boundaries $\mathcal{C}_1$ and $\mathcal{C}_2$, then we want to show that

$$\int_{\mathcal{C}} \mathbf{F} \cdot \mathbf{T} \, ds = \int_{\mathcal{C}_1} \mathbf{F} \cdot \mathbf{T} \, ds + \int_{\mathcal{C}_2} \mathbf{F} \cdot \mathbf{T} \, ds.$$

Let $\tilde{\mathcal{C}}$ denote the common portion of $\mathcal{C}_1$ and $\mathcal{C}_2$. Let γ be a parametrization of $\tilde{\mathcal{C}}$ oriented so that it agrees with the orientation of $\mathcal{C}_1$ and let $-\gamma$ denote the parametrization in the opposite direction so that $-\gamma$ agrees with the orientation of $\mathcal{C}_2$. Let α parametrize the portion of $\mathcal{C}_1$ not including $\tilde{\mathcal{C}}$ and let β parametrize the portion of $\mathcal{C}_2$ not including $\tilde{\mathcal{C}}$. We will assume that α and β have the orientation inherited from $\mathcal{C}$. Notice that these orientations agree with the counterclockwise orientations of $\mathcal{C}_1$ and $\mathcal{C}_2$.

Then we have

$$\int_{\mathcal{C}_1} \mathbf{F} \cdot \mathbf{T} \, ds + \int_{\mathcal{C}_2} \mathbf{F} \cdot \mathbf{T} \, ds = \int_{\alpha} \mathbf{F} \cdot \mathbf{T} \, ds$$
$$+ \int_{\gamma} \mathbf{F} \cdot \mathbf{T} \, ds + \int_{-\gamma} \mathbf{F} \cdot \mathbf{T} \, ds + \int_{\beta} \mathbf{F} \cdot \mathbf{T} \, ds.$$

From Section 6.2, we know that

$$\int_{-\gamma} \mathbf{F} \cdot \mathbf{T} \, ds = -\int_{\gamma} \mathbf{F} \cdot \mathbf{T} \, ds.$$

Thus

$$\int_{\mathcal{C}_1} \mathbf{F} \cdot \mathbf{T} \, ds + \int_{\mathcal{C}_2} \mathbf{F} \cdot \mathbf{T} \, ds = \int_{\alpha} \mathbf{F} \cdot \mathbf{T} \, ds + \int_{\beta} \mathbf{F} \cdot \mathbf{T} \, ds = \int_{\mathcal{C}} \mathbf{F} \cdot \mathbf{T} \, ds,$$

which verifies the claim when $\mathcal{R}$ is the union of two regions. If $\mathcal{R} = \mathcal{R}_1 \cup \mathcal{R}_2 \cup \cdots \cup \mathcal{R}_m$, apply the preceding result to $\mathcal{R}_1 \cup \mathcal{R}_2$, then $\mathcal{R}_1 \cup \mathcal{R}_2 \cup \mathcal{R}_3$, and so on. Since we have shown that the interior of every simple closed curve can be partitioned into regions for which Green's theorem holds, this completes the proof of Green's theorem. ∎

The proof of Green's theorem for more general regions follows from Green's theorem by partitioning a region $\mathcal{R}$ into a union of regions bounded by simple closed curves. Since Green's theorem applies to a region bounded by a simple closed curve, it applies to each of the regions making up $\mathcal{R}$. Then, by applying the conclusions of Step 3 of the proof of Green's theorem, we can show that Green's theorem applies to $\mathcal{R}$.

Summary

In this section, we stated and proved ***Green's theorem***, which relates the line integral of a vector field $\mathbf{F}$ around a closed curve $\mathcal{C}$ to the double integral of partial derivatives of $\mathbf{F}$ over $\mathcal{R}$, the interior of $\mathcal{C}$.

If $\mathbf{F}(x, y) = (u(x, y), v(x, y))$ and $\mathcal{C}$ is ***positively oriented***, Green's theorem states

$$\int_{\mathcal{C}} \mathbf{F} \cdot \mathbf{T} \, ds = \int \int_{\mathcal{R}} \left(\frac{\partial v}{\partial x} - \frac{\partial u}{\partial y} \right) dA_{x,y}.$$

By positive orientation, we mean that when $\mathcal{C}$ is traced in a ***counterclockwise*** manner, $\mathcal{R}$ is to the ***left*** of $\mathcal{C}$. For the theorem to hold, it is sufficient that $\mathcal{C}$ be ***piecewise differentiable***. The theorem can also be applied to a region $\mathcal{R}$ whose boundary is a union of simple closed curves.

We presented two physical applications of Green's theorem involving fluids. First, we showed how Green's theorem could be used to relate the ***total flux*** of a fluid through a closed curve to the integral of the ***infinitesimal flux*** over the interior of the curve. Second, we showed how this result could be applied to a ***diffusion process*** that is governed by the ***diffusion equation***. Here the total flux is equal to the double integral of the rate change of the concentration over the interior.

Section 6.3 Exercises

1. **Positive Orientation.** For each of the following regions with boundary, give a parametrization of the boundary of the region that positively orients the boundary.

   (a) The unit square, $[0, 1] \times [0, 1]$.

(b) The region bounded by the ellipse $x^2/4 + y^2 = 1$.

(c) The region in the upper half-plane bounded by $y = |x|$ and $x^2 + y^2 = 4$.

(d) The annulus $1 \leq x^2 + y^2 \leq 4$.

2. **Green's Theorem Calculations I.** Using Green's theorem, set up and evaluate a double integral that is equal to $\int_{\mathcal{C}} \mathbf{F} \cdot \mathbf{T} \, ds$, for each of the following vector fields $\mathbf{F}$ and curves $\mathcal{C}$.

(a) $\mathbf{F}(x, y) = (x^2, xy)$ and $\mathcal{C}$ is the triangle with vertices $(0,0)$, $(2,0)$, and $(2,3)$.

(b) $\mathbf{F}(x, y) = (-y^3, x^3)$ and $\mathcal{C}$ is the boundary of the region in the upper half-plane bounded by $y = |x|$ and $x^2 + y^2 = 4$.

(c) $\mathbf{F}(x, y) = (x + y, \cos x)$ and $\mathcal{C}$ is the square with vertices $(0,0)$, $(4,0)$, $(4,4)$, and $(0,4)$.

3. **Green's Theorem Calculations II.** Using Green's theorem, set up and evaluate a line integral to compute the area of each of the following regions. (See Example 6.9B. Be sure to parametrize the boundary in a counterclockwise manner.)

(a) The region enclosed by the cycloid $\alpha(t) = (2t - 2\sin t, 2 - 2\cos t)$ for $0 \leq t \leq 2\pi$ and the x-axis between $x = 0$ and $x = 4\pi$.

(b) The region that is the intersection of the interiors of the circles $x^2 + y^2 = 1$ and $x^2 + (y - 1)^2 = 1$.

4. **Proof of Green's Theorem.** Complete the proof of Green's theorem by verifying Green's theorem for a rectangle $\mathcal{R} = [a, b] \times [c, d]$. Parametrize the boundary $\mathcal{C}$ of $\mathcal{R}$ counterclockwise and show that the line integral of $\mathbf{F}(x, y) = (u(x, y), v(x, y))$ around $\mathcal{C}$ is equal to the integral of $\frac{\partial v}{\partial x} - \frac{\partial u}{\partial y}$ over $\mathcal{R}$.

5. **Green's Theorem and Area.** In Example 6.9B, we showed that if $\mathbf{F}(x, y) = \frac{1}{2}(-y, x)$, then by applying Green's theorem

$$\int_{\mathcal{C}} \mathbf{F} \cdot \mathbf{T} \, ds = \text{Area}(\mathcal{R}),$$

where $\mathcal{R}$ is the interior of the simple closed curve $\mathcal{C}$.

(a) Find another vector field $\mathbf{G}$ with the property that $\int_{\mathcal{C}} \mathbf{G} \cdot \mathbf{T} \, ds = \text{Area}(\mathcal{R})$.

(b) What is the line integral of $\mathbf{F} - \mathbf{G}$ around $\mathcal{C}$?

(c) Based on (b), what can we say about a vector field $\mathbf{G}$ that satisfies $\int_{\mathcal{C}} \mathbf{G} \cdot \mathbf{T} \, ds = \text{Area}(\mathcal{R})$?

6. **Verifying Green's Theorem.** Verify Green's theorem for each of the following regions $\mathcal{R}$ and vector fields $\mathbf{F}$.

(a) $\mathcal{R} = \{(x, y) : 1 \leq x^2 + y^2 \leq 4\}$ and the vector field $\mathbf{F}(x, y) = (xy^2, -yx^2)$.

(b) $\mathcal{R} = \{(x, y) : 0 \leq x \leq 1, 1 - x \leq y \leq 2 - x\}$ and the vector field $\mathbf{F}(x, y) = (x + y, -yx)$.

7. **Generalized Green's Theorem.** Consider the region $a^2 \leq x^2 + y^2 \leq b^2$. Prove the general case of Green's theorem for this region by partitioning the annulus into regions for which the first version of Green's theorem applies.

8. **Total Flux.** For each of the following vector fields $\mathbf{F}$ and regions with boundary, compute the total flux of the vector field across the boundary of the region.

   (a) The region is a unit disk centered at the origin and $\mathbf{F}(x, y) = (e^y + x^3, y^3)$.
   (b) The region is a unit square centered at the origin and $\mathbf{F}(x, y) = (x^2 - y^3, x^3 + y^2)$.
   (c) The region is the annulus $1 \leq x^2 + y^2 \leq 4$ and $\mathbf{F}(x, y) = (yx^2, xy^2)$.

9. **The Plane Minus the Origin.** (See Example 6.10.) Let $\mathbf{F}(x, y) = \left(\frac{-y}{x^2+y^2}, \frac{x}{x^2+y^2} \right)$.

   (a) Let $\mathcal{C}$ be an oriented simple closed curve that encloses the origin. Show that the line integral of $\mathbf{F}$ around $\mathcal{C}$ is 2π if $\mathcal{C}$ is oriented counterclockwise and -2π if $\mathcal{C}$ is oriented clockwise.
   (b) Suppose that $\mathcal{C}$ is a closed curve that wraps twice around the origin in a counterclockwise manner. Show that the line integral of $\mathbf{F}$ around $\mathcal{C}$ is 4π.

10. **A Punctured Region.** Let $\mathbf{F} = (u, v)$ be a continuously differentiable vector field that is defined everywhere except at a single point (x_0, y_0) in $\mathbb{R}^2$. Suppose also that $\frac{\partial v}{\partial x}(x, y) - \frac{\partial u}{\partial y}(x, y) = 0$ for all $(x, y) \neq (x_0, y_0)$.

   (a) Show that if $\mathcal{C}$ is any simple closed curve that does not enclose the point (x_0, y_0), then $\int_{\mathcal{C}} \mathbf{F} \cdot \mathbf{T} \, ds = 0$.
   (b) Show that if $\mathcal{C}_1$ and $\mathcal{C}_2$ are simple closed curves oriented counterclockwise, which enclose the point (x_0, y_0), then $\int_{\mathcal{C}_1} \mathbf{F} \cdot \mathbf{T} \, ds = \int_{\mathcal{C}_2} \mathbf{F} \cdot \mathbf{T} \, ds$.

11. **Normal Field to a Curve.** Let α be a continuously differentiable parametrization of a curve from P to Q, $\alpha(t) = (x(t), y(t))$ with $\|\alpha(t)\| \neq 0$. (Recall that α is continuously differentiable if the coordinate functions of α are differentiable and have continuous derivatives.) Let the vector field $\mathbf{N} = \mathbf{N}(t)$ at $\alpha(t)$ be defined by

$$\mathbf{N}(t) = \frac{1}{\|\alpha'(t)\|} (y'(t), -x'(t)).$$

   (a) Show that $\mathbf{N}$ is normal to the image of α.
   (b) Construct an intuitive argument to justify the claim that $\mathbf{N}$ always points to the same side of the image of α.
   (c) If $\mathcal{C}$ is a simple closed curve in the plane parametrized in a counterclockwise direction by $\alpha(t) = (x(t), y(t))$, show that $\mathbf{N}$ points out of the interior of $\mathcal{C}$.

12. **Irrotational Flows.** Let $\mathbf{F}$ be a differentiable vector field that represents the velocity field of a fluid flow. Discuss the relationship between the property of path independence for $\mathbf{F}$ and the property of $\mathbf{F}$ being irrotational.

13. **Total Diffusive Flux.** In Example 6.11, we outlined the calculation of the total flux through the unit circle of the diffusion process whose concentration is given by

$$c(x, y, t) = \frac{M}{4\pi\delta t} e^{-(x^2+y^2)/(4\delta t)}.$$

Here we fill in the details of that calculation and explore the resulting expression for the total flux.

(a) Verify the calculation of $\frac{\partial c}{\partial t}$.
(b) Use polar coordinates to show that

$$\int\int_{\mathcal{R}} -\frac{\partial c}{\partial t}(x, y, t)\, dA_{x,y} = \frac{M}{4\delta t^2} e^{-1/(4\delta t)},$$

where $\mathcal{R}$ is the disk of radius 1 centered at the origin.
(c) Setting $M = 1$ and $\delta = 1$, describe the behavior of the total flux $\frac{M}{4\delta t^2} e^{-1/(4\delta t)}$ as a function of time t. What does this tell us about the diffusion process as a function of time? Explain.

■ 6.4 End of Chapter Exercises

1. **Arc Length and Average Value.** Let $\alpha(t) = (e^t \cos t, e^t \sin t, 2)$ with $0 \le t \le 1$.

(a) Compute the arc length of α.
(b) Compute the average value of $f(x, y, z) = 2x^2 + 2y^2 + 2z^2$ along α.

2. **Total Mass.** For each of the following parametrizations α and density functions δ, compute the total mass along the parametrization.

(a) $\alpha(t) = (\cos(2t), \sin(2t), t^2)$ for $0 \le t \le 2\pi$ with density function $\delta(x, y, z) = \sqrt{z}$.
(b) $\alpha(t) = (t \cos t, t \sin t, t)$ for $0 \le t \le \pi/2$ with density function $\delta(x, y, z) = z$.

3. **Path Integrals over $\mathcal{C}$.** For each of the following functions f and curves $\mathcal{C}$, (i) construct a parametrization α of $\mathcal{C}$ and (ii) set up and evaluate an integral to compute the total accumulation of f on $\mathcal{C}$.

(a) $\mathcal{C}$ is the semicircle of radius $1/2$ with $x \ge 0$ in the xy-plane centered at the origin and $f(x, y) = xy^3$.
(b) $\mathcal{C}$ is the portion of the parabola $y = x^2 - 1$ from $(1, 0)$ to $(2, 3)$ and $f(x, y) = xy$.

4. **Potential Energy of a Charged Ellipse.** Consider a system that consists of a charged wire in the shape of an ellipse with axes of length $2a$ and $2b$ and a charged particle not on

the wire. Assume that the wire is modeled by the ellipse $\frac{x^2}{a^2} + \frac{y^2}{b^2} = 1$ in the xy-plane and has constant charge density C and that the particle of charge q is located at $\mathbf{x} = (x, y, z)$.

(a) Set up a Riemann sum to approximate the potential energy function $U = U(\mathbf{x})$ of this system.

(b) If you consider the limit of these Riemann sums as the mesh of the partition approaches zero, what is the resulting integral?

5. **Total Accumulation by Computer.** Use a computer algebra system to compute the following:

(a) The length of the ellipse parametrized by $\alpha(t) = (\cos t, 2\sin t)$.

(b) The length of the curve $y = x^3$ from $(0, 0)$ to $(2, 8)$.

(c) The average value of $f(x, y, z) = z$ on the curve parametrized by $\alpha(t) = (\cos t, \sin t, t^2)$, $0 \le t \le 2\pi$.

6. **Line Integrals.** For each of the following vector fields $\mathbf{F}$, compute the line integral of $\mathbf{F}$ over the parametrization α.

(a) $\mathbf{F}(x, y, z) = (x, y, z)$ and $\alpha(t) = (\cos t, \sin t, t)$, $0 \le t \le 1$.

(b) $\mathbf{F}(x, y, z) = (y, z, x)$ and $\alpha(t) = (t, t^2, t^3)$, $0 \le t \le 1$.

7. **Line Integrals over Curves.** For each of the following vector fields $\mathbf{F}$, evaluate the line integral of $\mathbf{F}$ along the given curve. (*Hint:* Green's theorem might be helpful.)

(a) $\mathbf{F}(x, y, z) = (e^x, e^{-y}, e^z)$ along the straight line from $(-1, 1, 1)$ to $(1, 0, 1)$.

(b) $\mathbf{F}(x, y, z) = (yz, xz, xy)$ along the perimeter of the triangle from the vertex $(1, 0, 0)$ to $(0, 0, 1)$ to $(0, 1, 0)$ and back to $(1, 0, 0)$.

8. **Potential Functions.** Determine whether or not the following vector fields satisfy the path independence property on the given domain by finding a potential function f with $\mathbf{F} = \nabla f$ or by using Exercise 5 of Section 6.2.

(a) $\mathbf{F}(x, y) = (2xy, x^2)$, $\mathcal{D} = \mathbb{R}^2$.

(b) $\mathbf{F}(x, y) = \left(\frac{x}{\sqrt{x^2+y^2}}, \frac{y}{\sqrt{x^2+y^2}} \right)$, $\mathcal{D}$ is the annulus $1 \le x^2 + y^2 \le 4$.

(c) $\mathbf{F}(x, y, z) = (3x^2 - yz, 3y^2 - xz, -xy)$, $\mathcal{D} = \mathbb{R}^3$.

9. **Positive Orientation.** For each of the following regions with boundary, give a parametrization of the boundary of the region that positively orients the boundary.

(a) The triangular region with vertices $(0, 0)$, $(1, 1)$, and $(-1, 1)$.

(b) The unit disk centered at $(0, 1)$.

(c) The region outside the circle of radius $1/2$ centered at the origin and inside the square $[-1, 1] \times [-1, 1]$.

10. **Green's Theorem Calculations I.** Using Green's theorem, set up and evaluate a double integral to compute the line integral of **F** around the given curves.

    (a) $\mathbf{F}(x,y) = (-y, yx)$ and $\mathcal{C}$ is the quadrilateral with vertices $(0,0)$, $(4,0)$, $(4,2)$, and $(2,4)$.
    (b) $\mathbf{F}(x,y) = (x - y, x + y)$ and $\mathcal{C}$ is the curve consisting of the interval from $(-1,0)$ to $(1,0)$ on the x-axis and the graph of $y = 1 - x^2$ from -1 to 1.

11. **Green's Theorem Calculations II.** Using Green's theorem, set up and evaluate a line integral to compute the area of each of the following regions. (See Example 6.9B.)

    (a) The region enclosed by the hypocycloid $x^{2/3} + y^{2/3} = 1$.
    (b) The region inside the right half-plane between the circle of radius 2 centered at the origin and the ellipse $y^2/4 + x^2 = 1$.

12. **Total Flux.** For each of the following vector fields **F** and regions with boundary, compute the total flux of the vector field across the boundary of the region.

    (a) The region is a triangle with vertices $(0,0)$, $(3,0)$, and $(0,3)$ and $\mathbf{F}(x,y) = (y - 1, x - 1)$.
    (b) The region is bounded by the ellipse $x^2 + 4y^2 = 4$ and $\mathbf{F}(x,y) = (\sin y - x, \cos x + y)$.
    (c) The region is bounded by the graphs of $y = 4 - x^2$ and the lower half of the semicircle of radius 2 centered at the origin and $\mathbf{F}(x,y) = (2x - y, x + 2y)$.

13. **Line Integral of Acceleration.** Let $\alpha : [a,b] \to \mathbb{R}^2$ be a twice differentiable parametrization of a curve in the plane. Show that

$$\int_a^b \alpha''(t) \cdot \alpha'(t)\, dt = \frac{1}{2}\alpha'(b) \cdot \alpha'(b) - \frac{1}{2}\alpha'(a) \cdot \alpha'(a).$$

14. **Electrostatic Equilibrium.** If a solid conductor in electrostatic equilibrium carries a net charge, then the charge resides on the surface of the conductor and the electrostatic field is zero inside the conductor. On the surface of the conductor, the field is perpendicular to the surface of the conductor.

    (a) Show that every point on the surface of the conductor has the same potential. (*Hint:* If the potential energy is not constant on the surface, what can you say about the direction of the electric field?)
    (b) Show that the potential energy is constant inside the conductor. (*Hint:* If not, show the field is nonzero inside the conductor.)

15. **Central Force Fields.** A *central force field* is a field whose magnitude depends only on the distance between two particles and that acts in a direction along the line joining

the particles. Thus if $\mathbf{F}$ is a central force field and one particle is located at the origin and the other at $\mathbf{x}$, then the force on the particle at $\mathbf{x}$ is of the form

$$\mathbf{F}(\mathbf{x}) = \frac{g(\|\mathbf{x}\|)}{\|\mathbf{x}\|}\mathbf{x}$$

for some $g : \mathbb{R} \to \mathbb{R}$, where we assume that g is a continuous function. Let α be a parametrization of a curve from a point P to a point Q. Show that $\mathbf{F}$ is conservative by showing that it is a gradient vector field. (*Hint:* See Exercise 10 of Section 3.6.)

16. **Diffusion in Equilibrium.** In this exercise, we explore the diffusive flux of a diffusion process in equilibrium that follows Fick's law. Suppose that a solute is being introduced into a solvent at a constant rate at the origin in the plane so that on a small circle of radius r_1 centered at the origin a concentration Q_1 is maintained and at a distance r_2 from the origin a concentration of Q_2 is maintained. Let $\mathcal{R}$ denote the region that lies between the circle $\mathcal{C}_1$ of radius r_1 and the circle $\mathcal{C}_2$ of radius r_2 that are centered at the origin.

   (a) The function $c = c(x, y)$ given by

   $$c(x, y) = Q_1 + \frac{Q_2 - Q_1}{\ln(r_2/r_1)} \ln(\|\mathbf{x}\|/r_1)$$

   models this equilibrium diffusion process. Show that $c(x, y) = Q_1$ if $\|\mathbf{x}\| = r_1$ and $c(x, y) = Q_2$ if $\|\mathbf{x}\| = r_2$.
   (b) Compute the flux vector $\mathbf{J} = -\delta \nabla c$ of the diffusion process.
   (c) Compute the total flux of $\mathbf{J}$ out of $\mathcal{R}$ by computing either the appropriate line integral or the appropriate double integral.
   (d) What does your answer to (c) tell you about the flux *into* $\mathcal{R}$ across $\mathcal{C}_1$ and *out of* $\mathcal{R}$ across $\mathcal{C}_2$? Does this make sense for an equilibrium diffusion process? Explain.

17. **Diffusion and Growth.** A population that grows exponentially and spreads in the plane according to Fick's Law can be modeled by population density function

   $$p(x, y, t) = \frac{M}{4\pi\delta t} e^{\alpha t - (x^2 + y^2)/(4\delta t)},$$

   where M is the initial population of the species, which is concentrated at the origin, δ is the diffusivity of the population, and α is the growth rate of the population.

   (a) What is the relationship between the function p and the function c of Exercise 13 of Section 6.3?
   (b) Compute the flux vector $\mathbf{J}(x, y, t) = -\delta \nabla p(x, y, t)$ of the diffusion process. (*Hint:* Why is part (a) relevant?)

(c) Compute the outward flux of $\mathbf{J}$ through a circle of radius 1 centered at the origin by computing either the appropriate line integral or double integral. (*Hint:* Think about part (a).)

(d) Setting $M = 1$, $\delta = 1$, and $\alpha = 1$, describe the behavior of the total flux as a function of time t. What does this tell us about the diffusion process as a function of time? Explain.

(e) How does your answer to (d) compare to your answer to part (c) of Exercise 13 of Section 6.3? In particular, explain the effect on the total flux of including the exponential growth of the population.

7

Integration on Surfaces

In this chapter, we will extend our work on the total accumulation of a function to functions defined on surfaces. This will require the concept of a parametrization. A parametrization will give us a way to construct a coordinate system on a surface, so that we can easily describe the location of points and cells on a surface. We can then use the coordinate system to set up and evaluate a Riemann sum for the total accumulation of a function on a surface.

In Section 7.1, we introduce parametric representations of surfaces. In Section 7.2, we use a parametrization of a surface to construct a Riemann sum for the total accumulation of a function on the surface. As in previous chapters, the limit of the Riemann sum will give rise to the integral of the function on the surface.

In Section 7.3, we introduce the concept of a flux integral of a vector field through a surface. This extends the material in Section 6.3 on the flux of a planar vector field through a closed curve. Our motivation here is the problem of computing the total amount of a fluid that flows through a surface. The flux integral is the total accumulation of the component of a vector field normal to the surface. We will also consider several other applications of flux integrals.

In Sections 7.4 and 7.5, we present two major theorems on the integration of vector fields: the divergence theorem and Stokes' theorem. The divergence of a vector field is a measure of the instantaneous flux of a vector field $\mathbf{F}$ at a point. The divergence theorem relates the integral of the divergence of $\mathbf{F}$ over a region in space to the flux integral of $\mathbf{F}$ through the closed surface bounding the region. Stokes' theorem relates the line integral of $\mathbf{F}$ over a closed curve in space to the flux integral of a related vector field, the curl of $\mathbf{F}$, through a surface bounded by the closed curve. Both theorems have a number of physical applications. In Section 7.6, we give their proofs.

A Collaborative Exercise—Using Spherical Coordinates

Although we have considered surfaces in space in several contexts already, most notably in Chapter 3 and 4 as the graphs of functions $z = f(x, y)$ of two variables and as level sets of functions $w = f(x, y, z)$ of three variables, we require a third method of representing surfaces in order to continue the discussion of integration. This is the concept of a parametrization of a surface, which made a brief appearance in Section 1.4, when we considered the vector or parametric description of a plane.

Recall that if a plane $\mathcal{P}$ contains the noncollinear points P, Q, and R, then $\mathcal{P}$ is the set of points of the form $\mathbf{p} + s\mathbf{u} + t\mathbf{v}$, where $\mathbf{p} = P$, and $\mathbf{u} = Q - P$ and $\mathbf{v} = R - P$ are the respective displacement vectors from P to Q and from P to R. Here we want to view the plane as the image of a function $\boldsymbol{\Phi}$ of two variables whose image is contained in $\mathbb{R}^3$. In this case, we write

$$\boldsymbol{\Phi}(s, t) = \mathbf{p} + s\mathbf{u} + t\mathbf{v},$$

where s and t can be any pair of real numbers. If necessary, we can expand the right side to obtain a coordinate representation of points in the image.

When considering a parametrization $\boldsymbol{\Phi}$, it helps to understand the coordinate curves of $\boldsymbol{\Phi}$. Fixing a value of t, say $t = t_0$, we obtain an s coordinate curve, $\alpha(s) = \boldsymbol{\Phi}(s, t_0)$. Fixing a value of s, say $s = s_0$, we obtain a t coordinate curve, $\beta(t) = \boldsymbol{\Phi}(s_0, t)$. The coordinate curves of the above parametrization of a plane consist of two families of parallel lines, one in the direction of $\mathbf{u}$ and the other in the direction of $\mathbf{v}$. If $\mathbf{u}$ and $\mathbf{v}$ are orthogonal, the lines of the two families meet at right angles.

This exercise will focus on another familiar coordinate formula: the expression for spherical coordinates. Recall that a point (x, y, z) can be represented by a spherical coordinate triple (ρ, ϕ, θ), where

$$x = \rho \sin \phi \, \cos \theta, \ y = \rho \sin \phi \, \sin \theta, \ \text{and} \ z = \rho \cos \phi$$

for $\rho \geq 0$, $0 \leq \theta \leq 2\pi$, and $0 \leq \phi \leq \pi$. By holding one of the three variables ρ, ϕ, or θ constant, we obtain a parametrization of a surface. The following questions explore each of the possibilities.

1. Fix $\rho = \rho_0 > 0$ and define a parametrization $\boldsymbol{\Phi}$ by

$$\boldsymbol{\Phi}(\phi, \theta) = (\rho_0 \sin \phi \, \cos \theta, \rho_0 \sin \phi \, \sin \theta, \rho_0 \cos \phi)$$

for $0 \leq \theta \leq 2\pi$ and $0 \leq \phi \leq \pi$. What is the image $\mathcal{S}$ of $\boldsymbol{\Phi}$? Why did we rule out $\rho_0 = 0$? Describe the ϕ and θ coordinate curves on $\mathcal{S}$. Do they meet in right angles? (*Hint:* Use basic geometric facts about $\mathcal{S}$ to answer the last question.)

2. Fix $\phi = \phi_0$, where $0 < \phi_0 < \pi$, and define a parametrization $\mathbf{\Phi}$ by

$$\mathbf{\Phi}(\rho, \theta) = (\rho \sin \phi_0 \, \cos \theta, \rho \sin \phi_0 \, \sin \theta, \rho \cos \phi_0)$$

for $\rho \geq 0$ and $0 \leq \theta \leq 2\pi$. What is the image $\mathcal{S}$ of $\mathbf{\Phi}$? Why did we rule out $\phi_0 = 0$ and $\phi_0 = \pi$? Does your answer depend on ϕ_0? Describe the ρ and θ coordinate curves on $\mathcal{S}$. Do they meet in right angles? (*Hint:* Use basic geometric facts about $\mathcal{S}$ to answer the last question.)

3. Fix $\theta = \theta_0$, where $0 \leq \theta_0 \leq 2\pi$, and define a parametrization $\mathbf{\Phi}$ by

$$\mathbf{\Phi}(\rho, \phi) = (\rho \sin \phi \, \cos \theta_0, \rho \sin \phi \, \sin \theta_0, \rho \cos \phi)$$

for $\phi \geq 0$ and $0 \leq \phi \leq \pi$. What is the image $\mathcal{S}$ of $\mathbf{\Phi}$? Does your answer depend on ϕ_0? Describe the ρ and ϕ coordinate curves on $\mathcal{S}$. Do they meet in right angles? (*Hint:* Use basic geometric facts about $\mathcal{S}$ to answer the last question.)

■ 7.1 Parametrization of Surfaces

We begin with the formal definition of a parametrization.

Definition 7.1 A function $\mathbf{\Phi} : \mathcal{D} \subset \mathbb{R}^2 \to \mathbb{R}^3$, $\mathbf{\Phi}(s, t) = (x(s,t), y(s,t), z(s,t))$, is a ***parametric representation*** or a ***parametrization*** of a surface $\mathcal{S} \subset \mathbb{R}^3$ if $\mathcal{S}$ is the image of $\mathbf{\Phi}$. That is,

$$\mathcal{S} = \{(x, y, z) : \ x = x(s,t), \ y = y(s,t), \ z = z(s,t), \ (s,t) \in \mathcal{D}\}.$$

The independent variables s and t are called the ***parameters*** of $\mathbf{\Phi}$, and the surface $\mathcal{S}$ is said to be ***parametrically defined.***

The curves $\alpha(s) = \mathbf{\Phi}(s, t_0)$ for a fixed t_0 and $\beta(t) = \mathbf{\Phi}(s_0, t)$ for a fixed s_0 are called s and t ***coordinate curves***, respectively. ◆

The coordinate functions of $\mathbf{\Phi}$ are real-valued functions defined on the domain $\mathcal{D}$ in $\mathbb{R}^3$. In most cases, the coordinate functions will be differentiable functions in the sense of Chapters 3 and 4 with continuous partial derivatives. Unless it is stated otherwise, we will assume that $\mathcal{D}$ is a closed bounded subset of $\mathbb{R}^2$. Together with the continuity of $\mathbf{\Phi}$, this implies that the image of $\mathbf{\Phi}$ is closed and bounded.

The coordinate curves on $\mathcal{S}$ are the images of the grid lines in $\mathcal{D}$ under $\mathbf{\Phi}$. For example, if we let $\alpha(s) = \mathbf{\Phi}(s, t_0)$ for a fixed t_0 and $\beta(t) = \mathbf{\Phi}(s_0, t)$ for a fixed s_0, the images of α and β are curves on $\mathcal{S}$ that pass through the point $\mathbf{\Phi}(s_0, t_0)$. These curves form a ***coordinate grid*** on the surface $\mathcal{S}$.

We begin with two familiar surfaces.

Example 7.1   **Parametrizations of Surfaces**

A. A Paraboloid. Here we demonstrate a surface $\mathcal{S}$ that is a portion of the graph of a function of two variables. Let $f : \mathcal{D} \subset \mathbb{R}^2 \to \mathbb{R}$ and consider

$$\mathcal{S} = \{(x, y, z) : z = f(x, y), (x, y) \in \mathcal{D}\}.$$

The simplest parametrization of $\mathcal{S}$ is given by $\Phi(x, y) = (x, y, f(x, y))$. The domain of Φ is the domain of f, and x and y parametrize the coordinate curves on $\mathcal{S}$. For example, suppose $\mathcal{S}$ is the portion of the paraboloid given by $z = x^2 + y^2$, $0 \le z \le 1$. Then $\mathcal{S}$ is the portion of the graph of $f(x, y) = x^2 + y^2$ with (x, y) in the disk of radius 1 centered at the origin in the xy-plane. Thus $\mathcal{S}$ can be parametrized by

$$\Phi(x, y) = (x, y, x^2 + y^2).$$

(See Figure 7.1(a).) The domain of Φ is the disk $\mathcal{D} = \{(x, y) : x^2 + y^2 \le 1\}$. In this case, the coordinate curves of $\mathcal{S}$, which are parametrized by $\alpha(x) = \Phi(x, y_0)$ and $\beta(y) = \Phi(x_0, y)$, are the vertical slices of the paraboloid parallel to the xz-plane and the yz-plane, respectively.

B. Cylinder. Suppose that $\mathcal{S}$ is a cylinder of radius 2 and height 5 with base in the xy-plane and centered on the z-axis. That is,

$$\mathcal{S} = \{(x, y, z) : x^2 + y^2 = 4,\ 0 \le z \le 5\}.$$

(See Figure 7.1(b).) In this case, we can use the representation of the cylinder in cylindrical coordinates to construct a parametrization. The cylindrical coordinates (r, θ, z) of

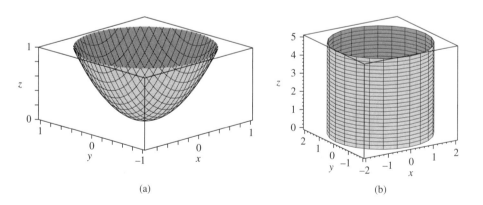

(a)   (b)

Figure 7.1   (a) The paraboloid $z = x^2 + y^2$ of Example 7.1A. (b) The cylinder of $\mathcal{S} = \{(x, y, z) : x^2 + y^2 = 4,\ 0 \le z \le 5\}$ of Example 7.1B. Curves of the coordinate grid are shown on each surface.

the points on $\mathcal{S}$ are $(2, \theta, z)$, where $0 \leq \theta \leq 2\pi$ and $0 \leq z \leq 5$. Using the relationship between Cartesian coordinates and cylindrical coordinates, $x = r \cos \theta$, $y = r \sin \theta$, $z = z$, we have

$$\mathcal{S} = \{(x, y, z) : x = 2 \cos \theta, \ y = 2 \sin \theta, \ 0 \leq \theta \leq 2\pi, \ 0 \leq z \leq 5 \}.$$

Thus the cylinder can be parametrized by

$$\mathbf{\Phi}(\theta, z) = (2 \cos \theta, 2 \sin \theta, z),$$

where $0 \leq \theta \leq 2\pi$ and $0 \leq z \leq 5$. The parameters of $\mathbf{\Phi}$ are the independent variables θ and z, and the domain of $\mathbf{\Phi}$ is the rectangle $[0, 2\pi] \times [0, 5]$.

The coordinate curve $\alpha(\theta) = \mathbf{\Phi}(\theta, z_0)$ is the circle on the cylinder at height $z = z_0$, which is a horizontal slice of the cylinder. The coordinate curve $\beta(z) = \mathbf{\Phi}(\theta_0, z)$ is a line on the cylinder parallel to the z-axis through the point $(2 \cos \theta_0, 2 \sin \theta_0, 0)$. Note that we can use the same parametrization $\mathbf{\Phi}$ to parametrize a subset of the cylinder of radius 2 by changing the domain of the parametrization. For example, to parametrize the portion of the cylinder in the first octant, we would restrict θ to $0 \leq \theta \leq \frac{\pi}{2}$.

A note of caution is in order when considering graphs of surfaces as in Example 7.1A. The variables x and y are being used in two ways: first, as domain variables, which are the independent variables of the parametrization, and second, as coordinates in the target, which are the dependent variables. We could resolve this ambiguity by rewriting the formula for $\mathbf{\Phi}$ using s and t, $\mathbf{\Phi}(s, t) = (s, t, s^2 + t^2)$ at the expense of introducing two new symbols. In this formulation, $x(s, t) = s$, $y(s, t) = t$, and $z(s, t) = s^2 + t^2$.

The tangent vectors to coordinate curves can be expressed in terms of the partial derivatives of the coordinate functions. At a point $\mathbf{\Phi}(s_0, t_0)$ on the surface $\mathcal{S}$ parametrized by $\mathbf{\Phi}$, the tangent vector to the s coordinate curve $\alpha(s) = \mathbf{\Phi}(s, t_0)$ is

$$\alpha'(s_0) = \left(\frac{\partial x}{\partial s}(s_0, t_0), \frac{\partial y}{\partial s}(s_0, t_0), \frac{\partial z}{\partial s}(s_0, t_0) \right).$$

We will denote this vector by $\mathbf{\Phi}_s(s_0, t_0)$. Similarly, the tangent vector to the t coordinate curve $\beta(t) = \mathbf{\Phi}(s_0, t)$ at $\mathbf{\Phi}(s_0, t_0)$ is

$$\beta'(t_0) = \left(\frac{\partial x}{\partial t}(s_0, t_0), \frac{\partial y}{\partial t}(s_0, t_0), \frac{\partial z}{\partial t}(s_0, t_0) \right).$$

We will denote this vector by $\mathbf{\Phi}_t(s_0, t_0)$.

The vectors $\mathbf{\Phi}_s(s_0, t_0)$ and $\mathbf{\Phi}_t(s_0, t_0)$ are tangent to $\mathcal{S}$ at $\mathbf{\Phi}(s_0, t_0)$. Thus if they are also linearly independent, they span the ***tangent plane*** to $\mathcal{S}$ at $\mathbf{\Phi}(s_0, t_0)$, and the tangent plane can be parametrized by

$$\Psi(s, t) = \mathbf{\Phi}(s_0, t_0) + s\mathbf{\Phi}_s(s_0, t_0) + t\mathbf{\Phi}_t(s_0, t_0).$$

Again, if the vectors $\mathbf{\Phi}_s(s_0, t_0)$ and $\mathbf{\Phi}_t(s_0, t_0)$ are linearly independent, their cross product $\mathbf{\Phi}_s(s_0, t_0) \times \mathbf{\Phi}_t(s_0, t_0)$ is nonzero and orthogonal to the tangent plane. That is, $\mathbf{\Phi}_s(s_0, t_0) \times \mathbf{\Phi}_t(s_0, t_0)$ is ***normal*** to $\mathcal{S}$ at $\mathbf{\Phi}(s_0, t_0)$. Thus, a coordinate equation for the tangent plane to $\mathcal{S}$ at $\mathbf{\Phi}(s_0, t_0)$ is

$$(\mathbf{\Phi}_s(s_0, t_0) \times \mathbf{\Phi}_t(s_0, t_0)) \cdot ((x, y, z) - \mathbf{\Phi}(s_0, t_0)) = 0.$$

We demonstrate this in the following example.

Example 7.2

A Hemisphere. Suppose $\mathcal{S}$ is the upper hemisphere of radius 3 centered at the origin,

$$\mathcal{S} = \{(x, y, z) : x^2 + y^2 + z^2 = 9, \; z \geq 0 \}.$$

(See Figure 7.2.) We can use spherical coordinates to construct a parametrization of $\mathcal{S}$. The spherical coordinates (ρ, θ, ϕ) of a point on $\mathcal{S}$ are $(3, \theta, \phi)$, where $0 \leq \theta \leq 2\pi$ and $0 \leq \phi \leq \frac{\pi}{2}$. The parametrization $\mathbf{\Phi}$ is given by

$$\mathbf{\Phi}(\theta, \phi) = (3 \cos \theta \, \sin \phi, 3 \sin \theta \, \sin \phi, 3 \cos \phi).$$

The domain of $\mathbf{\Phi}$ is the rectangle $[0, 2\pi] \times [0, \frac{\pi}{2}]$. The coordinate curves $\alpha(\theta) = \mathbf{\Phi}(\theta, \phi_0)$ and $\beta(\phi) = \mathbf{\Phi}(\theta_0, \phi)$ are, respectively, the lines of latitude and longitude on the sphere.

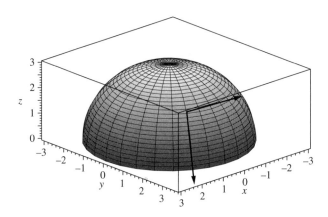

Figure 7.2   The hemisphere of Example 7.2. The tangent vectors to the coordinate curves at the point $\mathbf{\Phi}(\frac{\pi}{3}, \frac{\pi}{4}) = (\frac{3}{2^{3/2}}, \frac{3^{3/2}}{2^{3/2}}, \frac{3}{\sqrt{2}})$ are shown to scale.

A. The Tangent Plane. The tangent vectors to the coordinate curves passing through $\Phi(\theta_0, \phi_0)$ are

$$\alpha'(\theta_0) = \Phi_\theta(\theta_0, \phi_0) = (-3\sin\theta_0 \, \sin\phi_0, 3\cos\theta_0 \, \sin\phi_0, 0)$$

$$\beta'(\phi_0) = \Phi_\phi(\theta_0, \phi_0) = (3\cos\theta_0 \, \cos\phi_0, 3\sin\theta_0 \, \cos\phi_0, -3\sin\phi_0).$$

For example, at $\Phi(\frac{\pi}{3}, \frac{\pi}{4}) = (\frac{3}{2^{3/2}}, \frac{3^{3/2}}{2^{3/2}}, \frac{3}{\sqrt{2}})$, we have

$$\Phi_\theta\left(\frac{\pi}{3}, \frac{\pi}{4}\right) = \left(-\left(\frac{3}{2}\right)^{3/2}, \frac{3}{2^{3/2}}, 0\right) \text{ and } \Phi_\phi\left(\frac{\pi}{3}, \frac{\pi}{4}\right) = \left(\frac{3}{2^{3/2}}, \frac{3^{3/2}}{2^{3/2}}, -\frac{3}{\sqrt{2}}\right).$$

The tangent plane to the sphere at $\Phi(\frac{\pi}{3}, \frac{\pi}{4})$ can be parametrized by

$$\Psi(s, t) = \Phi\left(\frac{\pi}{3}, \frac{\pi}{4}\right) + s\Phi_\theta\left(\frac{\pi}{3}, \frac{\pi}{4}\right) + t\Phi_\phi\left(\frac{\pi}{3}, \frac{\pi}{4}\right)$$

$$= \left(\frac{3}{2^{3/2}}, \frac{3^{3/2}}{2^{3/2}}, \frac{3}{\sqrt{2}}\right) + s\left(-\left(\frac{3}{2}\right)^{3/2}, \frac{3}{2^{3/2}}, 0\right) + t\left(\frac{3}{2^{3/2}}, \frac{3^{3/2}}{2^{3/2}}, -\frac{3}{\sqrt{2}}\right).$$

B. The Normal Vector. The cross product of the tangent vectors to the coordinate curves at $\Phi(\frac{\pi}{3}, \frac{\pi}{4})$ is

$$\mathbf{n} = \Phi_\theta\left(\frac{\pi}{3}, \frac{\pi}{4}\right) \times \Phi_\phi\left(\frac{\pi}{3}, \frac{\pi}{4}\right)$$

$$= \left(-\left(\frac{3}{2}\right)^{3/2}, \frac{3}{2^{3/2}}, 0\right) \times \left(\frac{3}{2^{3/2}}, \frac{3^{3/2}}{2^{3/2}}, -\frac{3}{\sqrt{2}}\right)$$

$$= (-\frac{9}{4}, -\frac{9}{4}\sqrt{3}, -\frac{9}{2}).$$

Notice that the coordinates of $\mathbf{n}$ are negative. Since $\Phi(\frac{\pi}{3}, \frac{\pi}{4})$ lies in the first octant, this means that $\mathbf{n}$ is an *inward* pointing normal vector to the sphere. The coordinate equation for the tangent plane is

$$0 = \mathbf{n} \cdot \left((x, y, z) - \Phi(\frac{\pi}{3}, \frac{\pi}{4})\right)$$

$$= \left(-\frac{9}{4}, -\frac{9}{4}\sqrt{3}, -\frac{9}{2}\right) \cdot \left(x - (\frac{3}{2^{3/2}}, y - \frac{3^{3/2}}{2^{3/2}}, z - \frac{3}{\sqrt{2}})\right).$$

The vector $\mathbf{n}$ is shown in Figure 7.3 to scale along with the tangent vectors $\Phi_\theta\left(\frac{\pi}{3}, \frac{\pi}{4}\right)$ and $\Phi_\phi\left(\frac{\pi}{3}, \frac{\pi}{4}\right)$.

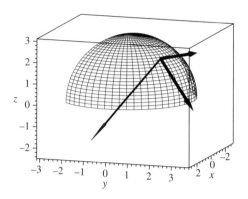

Figure 7.3   The hemisphere of Example 7.2. The inward pointing normal vector $\mathbf{n} = \mathbf{\Phi}_\theta\left(\frac{\pi}{3}, \frac{\pi}{4}\right) \times \mathbf{\Phi}_\phi\left(\frac{\pi}{3}, \frac{\pi}{4}\right)$ is shown to scale along with the tangent vectors to the coordinate curves at the point $\mathbf{\Phi}(\frac{\pi}{3}, \frac{\pi}{4})$. Notice the hemisphere is transparent and the view is through the hemisphere to the normal vector.

There are two rather subtle issues that must be kept in mind when considering parametrizations. The first is somewhat simpler and concerns whether the image of the parametrization intersects itself. The second concerns whether the image is even a surface according to our common understanding of a surface. We take these in order.

In Example 7.1B, we considered the parametrization of cylinder of height 5 and radius 2 centered on the z-axis given by

$$\mathbf{\Phi}(\theta, z) = (2\cos\theta, 2\sin\theta, z),$$

where $0 \leq \theta \leq 2\pi$ and $0 \leq z \leq 5$. Because sine and cosine are periodic with period 2π, we see that $\mathbf{\Phi}(0, z) = \mathbf{\Phi}(2\pi, z)$ for all z. That is, the parametrization covers the line segment $\{(2, 0, z) : 0 \leq z \leq 5\}$ twice. Other points in the image, when $\theta \neq 0, 2\pi$, occur only once. In precise terms, we say that $\mathbf{\Phi}$ is **one-to-one** for $\theta \neq 0, 2\pi$. This means that for $\theta \neq 0, 2\pi$, each point on the cylinder is mapped to by a single point in the domain. This fails when $\theta = 0$ and when $\theta = 2\pi$. While there are occasions when one might desire a parametrization that is one-to-one on the entire domain of the parametrization, this turns out not to be necessary in integration. Our domains of integration will be closed sets, and it will suffice to require that the parametrization is one-to-one on the interior of the closed set, that is, on the closed set with its boundary removed. This condition is satisfied by the parametrization of the cylinder in Example 7.1B. The parametrization is one-to-one on the interior of the domain, $0 < \theta < 2\pi, 0 < z < 5$.

To understand the second issue, we introduce a double cone.

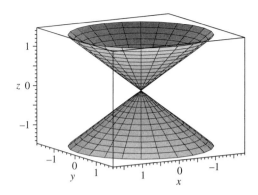

Figure 7.4   A double cone with vertex at the origin. (See Example 7.3.)

Example 7.3   **A Double Cone.** Consider the parametrization

$$\boldsymbol{\Phi}(\rho, \theta) = (\rho\cos(\theta)\sin(\frac{\pi}{3}), \rho\sin(\theta)\sin(\frac{\pi}{3}), \rho\cos(\frac{\pi}{3})),$$

where $-2 \le \rho \le 2$, $0 \le \theta \le 2\pi$. The image of $\boldsymbol{\Phi}$ is shown in Figure 7.4. It consists of two cones touching at their respective vertices. The parametrization maps all points of the form $(0, \theta)$, $0 \le \theta \le 2\pi$, to the origin. Based on the discussion preceding the example, we would say that $\boldsymbol{\Phi}$ fails to be one-to-one on this segment. However, more could be said. The tangent vectors to the coordinate curves when $\rho = 0$ are $\boldsymbol{\Phi}_\rho(0, \theta) = (\cos(\theta)\sin(\frac{\pi}{3}), \sin(\theta)\sin(\frac{\pi}{3}), \cos(\frac{\pi}{3}))$ and $\boldsymbol{\Phi}_\theta(0, \theta) = (0, 0, 0)$. Calculating further, the normal vector to the surface is $\mathbf{n} = \boldsymbol{\Phi}_\rho(0, \theta) \times \boldsymbol{\Phi}_\theta(0, \theta) = (0, 0, 0)$. Consequently, it is impossible to identify a normal direction to the surface at the vertex.

The vanishing of the normal vector is an issue in constructing integrals in Section 7.3, and we will want to rule it out from consideration. Thus we make the following definition.

Definition 7.2   A parametrization $\boldsymbol{\Phi} = \boldsymbol{\Phi}(s, t)$ of a surface $\mathcal{S}$, $\boldsymbol{\Phi} : \mathcal{D} \subset \mathbb{R}^2 \to \mathbb{R}^3$, is called a **_regular_** parametrization of $\mathcal{S}$ if it is continuously differentiable and $\mathbf{n} = \boldsymbol{\Phi}_s \times \boldsymbol{\Phi}_t$ is nonzero on the interior of $\mathcal{D}$.   ◆

In later sections, we will require parametrizations to be regular.

Summary

In this section, we introduced the concept of a parametrization of a surface. A ***parametrization*** is a function $\boldsymbol{\Phi} : \mathcal{D} \subset \mathbb{R}^2 \rightarrow \mathbb{R}^3$, defined by an expression $\boldsymbol{\Phi}(s,t) = (x(s,t), y(s,t), z(s,t))$ for $(s,t) \in \mathcal{D}$, where the coordinate functions are differentiable functions of two variables. The ***coordinate curves*** of the parametrization are obtained by holding one or the other of the independent variables s and t constant, $s = s_0$ or $t = t_0$. Using the tangent vectors to the coordinate curves, $\boldsymbol{\Phi}_s(s_0, t_0)$ and $\boldsymbol{\Phi}_t(s_0, t_0)$, we can construct a ***parametrization of the tangent plane*** to the surface at the point $\boldsymbol{\Phi}(s_0, t_0)$. The cross product of these tangent vectors is a ***normal vector*** to the surface at $\boldsymbol{\Phi}(s_0, t_0)$, $\mathbf{n} = \boldsymbol{\Phi}_s(s_0, t_0) \times \boldsymbol{\Phi}_t(s_0, t_0)$. Using $\mathbf{n}$, we can construct a ***coordinate equation*** for the tangent plane at a point. We defined a ***regular parametrization*** to be one that is continuously differentiable with nonzero normal vectors. Generally, we will be interested in regular parametrizations that are one-to-one on the interior of their domains.

Section 7.1 Exercises

1. **Parametrizing Planes.** In the collaborative exercise, we observed that if a plane $\mathcal{P}$ contains the noncollinear points P, Q, and R, it can be parametrized by

$$\boldsymbol{\Phi}(s,t) = \mathbf{p} + s\mathbf{u} + t\mathbf{v},$$

where $\mathbf{p} = P$, $\mathbf{u} = Q - P$, and $\mathbf{v} = R - P$.

   (a) What are the tangent vectors to the coordinate curves, $\boldsymbol{\Phi}_s(s_0, t_0)$ and $\boldsymbol{\Phi}_t(s_0, t_0)$?
   (b) Does your answer to (a) depend on the point (s_0, t_0)? Does this make sense in light of the coordinate curves?
   (c) Does the normal vector $\mathbf{n}$ depend on the point (s_0, t_0)? Does this make sense in light of the surface?

2. **Parametrizing Polygons.** Use the formula given in Exercise 1 to construct parametrizations of the following polygons in space.

   (a) The rectangle with vertices located at $(1, 1, 2)$, $(4, 1, 2)$, $(1, 3, 2)$, and $(4, 3, 2)$.
   (b) The triangle with vertices located at $(4, 1, 2)$, $(1, 3, 2)$, and $(4, 3, 2)$.
   (c) The parallelogram with vertices located at $(3, 1, -1)$, $(5, 4, -1)$, $(3, 2, 1)$, and $(8, 6, 0)$.
   (d) The triangle with vertices $(1, 1, 0)$, $(1, 0, 1)$, and $(0, 1, 1)$.

3. **Parametrizing Surfaces.** For each of the following surfaces $\mathcal{S}$, find a parametrization of the surface. In each case, describe the domain of the parametrization.

   (a) $\mathcal{S}$ is a half-cylinder of radius 3 and height 2 with base in the xy-plane, central axis on the z-axis, and positive y coordinate.
   (b) $\mathcal{S}$ is the portion of a sphere of radius 2 centered at the origin contained in the first octant.
   (c) $\mathcal{S}$ is the portion of the paraboloid $z = 9 - x^2 - y^2$ lying above the xy-plane.

(d) $\mathcal{S}$ is the plane containing the vectors $(1, 2, 1)$ and $(-1, 0, 3)$ and the point $(0, 2, -1)$.
(e) $\mathcal{S}$ is the portion of the sphere of radius 5 centered at the origin lying above the plane $z = 3$.

4. **Surfaces and Coordinate Curves.** For each of the following functions $\mathbf{\Phi} : \mathcal{D} \subset \mathbb{R}^2 \to \mathbb{R}^3$, describe and sketch the surface $\mathcal{S}$ that is the image of $\mathbf{\Phi}$. Describe the coordinate curves $\mathbf{\Phi}(s, t_0)$ and $\mathbf{\Phi}(s_0, t)$ and sketch these curves on $\mathcal{S}$.

   (a) $\mathbf{\Phi} : [0, 1] \times [0, 1] \to \mathbb{R}^3$ with $\mathbf{\Phi}(s, t) = (1 + s + 2t, s - t, 2 - s + t)$.
   (b) $\mathbf{\Phi} : [0, \pi/2] \times [0, 3] \to \mathbb{R}^3$ with $\mathbf{\Phi}(s, t) = (2 \cos s, 2 \sin s, t)$.
   (c) $\mathbf{\Phi} : [0, 2] \times [0, 3] \to \mathbb{R}^3$ with $\mathbf{\Phi}(s, t) = (s \cos(\pi/4), s \sin(\pi/4), t)$.
   (d) $\mathbf{\Phi} : [-2, 2] \times [0, 1] \to \mathbb{R}^3$ with $\mathbf{\Phi}(s, t) = (s, t, 4 - s^2)$.
   (e) $\mathbf{\Phi} : [0, 3] \times [0, 2\pi] \to \mathbb{R}^3$ with $\mathbf{\Phi}(s, t) = (s \cos t \sin(\pi/6), s \sin t \sin(\pi/6), s \cos(\pi/6))$.

5. **Tangent Planes.** For each of the parametrizations in Exercise 4, compute $\mathbf{\Phi}_s(s, t)$, $\mathbf{\Phi}_t(s, t)$, and $\mathbf{n} = \mathbf{\Phi}_s(s, t) \times \mathbf{\Phi}_t(s, t)$, and find a coordinate expression for the tangent plane at the following points, respectively.

   (a) $(s_0, t_0) = (\frac{1}{2}, \frac{1}{2})$.   (d) $(s_0, t_0) = (0, \frac{1}{2})$.
   (b) $(s_0, t_0) = (\frac{\pi}{4}, 1)$.   (e) $(s_0, t_0) = (2, \pi)$.
   (c) $(s_0, t_0) = (1, 2)$.

6. **One-to-One and Regular I.** For each of the parametrizations in Exercise 4, determine if the parametrization is (i) one-to-one and/or (ii) regular. (See Exercise 5.)

7. **Surfaces of Revolution I.** Surfaces of revolution can be parametrized using cylindrical coordinates. If the axis of revolution is the z-axis, we can use the following parametrization:

$$\mathbf{\Phi}(\theta, z) = (\cos(\theta)f(z), \sin(\theta)f(z), z),$$

   where f is a non-negative differentiable function of z, so that the radius r at height z is $f(z)$. The domain of $\mathbf{\Phi}$ consists of (θ, z) with $0 \le \theta \le 2\pi$ and z in the domain of f. For each of the following functions $r = f(z)$, sketch the corresponding surface of revolution for the given interval for z. Show the coordinate curves on the surface.

   (a) $f(z) = 2, -3 \le z \le 3$.   (c) $f(z) = \sqrt{4 - z^2}, -2 \le z \le 2$.
   (b) $f(z) = \cos z, -\pi/2 \le z \le \pi/2$.   (d) $f(z) = \sqrt{1 + z^2}, -1 \le z \le 1$.

8. **Surfaces of Revolution II.** Consider a surface of revolution

$$\mathbf{\Phi}(\theta, z) = (\cos(\theta)f(z), \sin(\theta)f(z), z),$$

   where f is a non-negative differentiable function of z with $0 \le \theta \le 2\pi$ and z in the domain of f.

   (a) Find the tangent vectors $\mathbf{\Phi}_\theta(\theta_0, z_0)$ and $\mathbf{\Phi}_z(\theta_0, z_0)$ to the coordinate curves of $\mathbf{\Phi}$ in terms of f and its derivative.

(b) Find the normal vector $\mathbf{n}$ to the surface in terms of f and its derivative.

(c) When will $\mathbf{n}$ be horizontal? (*Hint:* What do we know about a vector if it is horizontal?)

9. **Quadric Surfaces.** Quadric surfaces were introduced in Section 1.1. These are given implicitly by equations of the form

$$Ax^2 + By^2 + Cz^2 = R^2,$$

where A, B, C, and R are constants. Suppose that $\mathcal{S}$ is a quadric surface given by such an equation and two of A, B, and C are equal. It turns out that $\mathcal{S}$ is a surface of revolution in the sense of Exercise 7.

(a) Assuming two of A, B, and C are equal, which coordinate axis is the axis of rotation?

(b) Assuming two of A, B, and C are equal, find the radius r as a function f of the variable corresponding to the axis of rotation.

(c) For each of the following quadric surfaces $\mathcal{S}$, give a parametrization of S as a surface of revolution. (*Hint:* Is the surface a sphere, an ellipsoid, or a hyperboloid of one or two sheets?)

(i) $\mathcal{S}$ defined by $x^2 + y^2 - z^2 = 4$.   (iii) $\mathcal{S}$ defined by $9x^2 + y^2 + 9z^2 = 1$.

(ii) $\mathcal{S}$ defined by $\frac{1}{4}x^2 - y^2 + \frac{1}{4}z^2 = 1$.   (iv) $\mathcal{S}$ defined by $x^2 - y^2 - z^2 = 16$.

10. **One-to-One and Regular II.** For each of the parametrizations in Exercise 9(c), determine if the parametrization is (i) one-to-one and/or (ii) regular.

11. **Parametrizing Ellipsoids.** A quadric surface $\mathcal{S}$ given implicitly by an equation of the form

$$Ax^2 + By^2 + Cz^2 = R^2,$$

where A, B, and C are distinct constants (no two the same), is an ellipsoid without rotational symmetry. Consequently, it cannot be parametrized as a surface of revolution.

(a) Show that the following is a parametrization of $\mathcal{S}$:

$$\mathbf{\Phi}(\theta, \phi) = \left(\frac{R}{\sqrt{A}} \cos \theta \sin \phi, \ \frac{R}{\sqrt{B}} \sin \theta \sin \phi, \ \frac{R}{\sqrt{C}} \cos \phi \right),$$

where $0 \le \theta \le 2\pi$ and $0 \le \phi \le \pi$.

(b) Is $\mathbf{\Phi}$ one-to-one? If so, explain why; if not, where does it fail to be one-to-one?

(c) Is $\mathbf{\Phi}$ regular?

(d) Is the normal vector $\mathbf{n}$ inward or outward pointing?

12. **Surfaces as Boundaries.** Each of the following surfaces S is the boundary of a region in space. For each surface, describe S as the union of surfaces S_i, where any two surfaces intersect at most along their edges. Then give a parametrization of each surface S_i.

   (a) S is the boundary of the cylindrical region of radius 1 and height 3 with base on the xy-plane and central axis on the z-axis.
   (b) S is the boundary of the box centered at the origin with length and width 2 and height 4 and faces parallel to the coordinate planes.
   (c) S is the boundary of the solid hemisphere of radius 2 centered at the origin lying in the lower half-space.
   (d) S is the boundary of the tetrahedron formed by the coordinate planes and the plane $z = 4 - x - 2y$.
   (e) S is the boundary of the region in the upper half-space between the cylinder of radius 1 and height 2 centered on the z-axis with base on the xy-plane and the sphere of radius 1 centered at the origin.

13. **Spherical Surfaces.** In spherical coordinates, the radius ρ may be replaced by a function $\rho = \rho(\theta, \phi)$. This gives rise to a parametrization

$$\mathbf{\Phi}(\theta, \phi) = (\rho(\phi, \theta) \cos\theta \sin\phi, \rho(\phi, \theta) \sin\theta \sin\phi, \rho(\phi, \theta) \cos\phi),$$

where $0 \leq \phi \leq \pi$ and $0 \leq \theta \leq 2\pi$. (A sphere of radius ρ_0 centered at the origin is obtained by letting $\rho(\theta, \phi) = \rho_0$.) Consider the spherical surface parametrized by $\mathbf{\Phi}$ with $\rho(\theta, \phi) = \cos(2\phi)$. (See Figure 7.5.)

   (a) Is $\mathbf{\Phi}$ one-to-one on the interior of its domain? If so, explain why. If not, which points in the interior are mapped to the same point?
   (b) Is $\mathbf{\Phi}$ a regular parametrization? Justify your answer with a calculation.

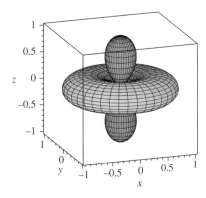

Figure 7.5 The spherical surface given by $\rho = \cos(2\phi)$ of Exercise 13.

14. **Tubular Surfaces.** An interesting collection of surfaces arises by creating a tube around or along a curve. These are generalizations of surfaces of revolution, which are tubes of varying radius around a line. Let $\gamma(t) = (x(t), y(t), z(t))$ be a curve in space and let $\mathbf{n}_1(t)$ and $\mathbf{n}_2(t)$ be unit vector fields normal to the curve and to each other along γ. Then if $r = r(t)$ is a differentiable function of t,

$$\boldsymbol{\Phi}(t, s) = \gamma(t) + r(t)\cos(s)\mathbf{n}_1(t) + r(t)\sin(s)\mathbf{n}_2(t)$$

parametrizes a tube around γ of radius $r(t)$, where $0 \leq t \leq 2\pi$ and $0 \leq s \leq 2\pi$. Here we explore two such surfaces. In each case, $\gamma(t) = (\cos(t), \sin(t), 0)$ is the unit circle in the xy-plane centered at the origin, $\mathbf{n}_1(t) = (0, 0, 1)$, and $\mathbf{n}_2(t) = (\cos(t), \sin(t), 0)$.

(a) Verify that $\mathbf{n}_1$ and $\mathbf{n}_2$ are unit vectors normal to each other and normal to the curve for all t.

(b) Let $r(t) = \frac{1}{2}$. The surface, known as a *torus*, is shown in Figure 7.6(a). For this r, is $\boldsymbol{\Phi}$ one-to-one on the interior of its domain and regular? Provide calculations to support your answer.

(c) Let $r(t) = \frac{1}{2}\sin(\frac{1}{2}t)$. The surface, known as a *pinched torus*, is shown in Figure 7.6(b). For this r, is $\boldsymbol{\Phi}$ one-to-one on the interior of its domain and regular? Provide calculations to support your answer.

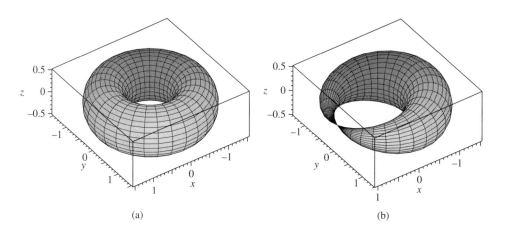

(a)   (b)

Figure 7.6   (a) The torus of Exercise 14(b). (b) The tubular surface of varying radius of Exercise 14(c).

■ 7.2 Surface Integrals

In this section, we will consider the problem of computing the total accumulation of a function defined on a surface. Let us begin by considering several examples of physical phenomena that can be modeled by functions on surfaces.

Example 7.4	**Total Accumulation on Surfaces**

A. **Surface Area.** Suppose that the function f is constant and equal to 1 on S, $f(x, y, z) = 1$ for $(x, y, z) \in S$. Then the total accumulation of f on S is the ***surface area*** of S.

B. **Total Charge.** An electrical conductor is a substance in which electrical charge moves easily from one place to another. It can be shown both mathematically and experimentally that if the charge in a conductor is in electrostatic equilibrium, that is, the charges are at rest and the corresponding electric field is not changing, then the excess charges will necessarily lie on the surface of the conductor. Thus if S denotes the surface of the conductor, the ***total charge*** on the conductor is equal to the total accumulation of charge on the surface of the conductor.

C. **Total Heat.** Suppose that an object consists of a thin shell of thickness h of a homogeneous material in the shape of a surface S. If the temperature T at any point in the object is determined by its location on the shell and does not depend on its depth within the shell, then the ***total heat*** contained in the object is equal to the total accumulation of the function $\delta c T \cdot h$ on S, where δ is the density of the material and c is the specific heat of the material.

D. **Total Acoustic Power.** Suppose that a source emits a sound at a given frequency equally in all directions, which has the effect of changing the pressure to $p_a(x, y, z)$. Since the sound is the same in all directions, a level set S of p_a is a sphere. The intensity of the sound is defined to be the function $I = p_a^2/(\delta c)$, where δ is the density of the medium and c is the speed of sound in the medium. The ***total acoustic power*** of the acoustic source is the total accumulation of I on S.

Riemann Sums

Let us assume that S is a surface that is the image of a regular parametrization $\mathbf{\Phi}$, so that the partial derivatives of $\mathbf{\Phi}$ exist and are continuous. Further, we will assume that

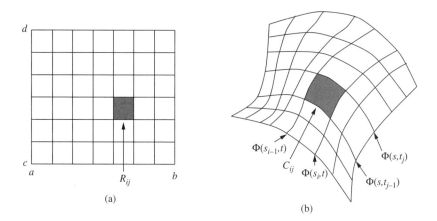

Figure 7.7   (a) A regular partition of the domain $\mathcal{D}$ of $\boldsymbol{\Phi}$. (b) The partition of the image of $\boldsymbol{\Phi}$ corresponding to the partition of $\mathcal{D}$. (See Riemann Sums in Section 7.2.)

f is defined and continuous on $\mathcal{S}$. In order to construct a Riemann sum for f on $\mathcal{S}$, we will proceed as we did for curves. We begin by partitioning $\mathcal{D}$, the domain of the parametrization $\boldsymbol{\Phi}$. Since $\mathcal{D}$ is a region in $\mathbb{R}^2$, we can partition it as we did in Section 5.2. The image of this partition under $\boldsymbol{\Phi}$ is a partition of $\mathcal{S}$. (See Figure 7.7.) We choose a point (s_{ij}^*, t_{ij}^*) in the ij^{th} cell of the partition of $\mathcal{D}$. Then $\boldsymbol{\Phi}(s_{ij}^*, t_{ij}^*)$ is in the ij^{th} cell of the partition of $\mathcal{S}$. The sampled value for f in the ij^{th} cell is $f(\boldsymbol{\Phi}(s_{ij}^*, t_{ij}^*))$. We must scale the sampled value by the area ΔS_{ij} of the ij^{th} cell. Then the total accumulation of f on $\mathcal{S}$ is approximated by the Riemann sum

$$R(f, \boldsymbol{\Phi}, P) = \sum_{i,j=1}^{M,N} f(\boldsymbol{\Phi}(s_{ij}^*, t_{ij}^*))\Delta S_{ij}.$$

To obtain an exact value for the total accumulation of f on $\mathcal{S}$, we must take the limit of this sum as the mesh of the partition approaches 0.

Theorem 7.1   Let $\boldsymbol{\Phi} : \mathcal{D} \to \mathbb{R}^3$ be a regular parametrization of the surface $\mathcal{S}$ and let f be defined and continuous on $\mathcal{S}$. Then

$$\lim_{\text{mesh}(P) \to 0} R(f, \boldsymbol{\Phi}, P)$$

exists and is independent of the choices of partitions and sampling schemes. We call this limit the **_surface integral_** of f on $\mathcal{S}$ and we denote it by

$$\int\int_{\mathcal{S}} f\, dS. \quad \blacklozenge$$

Notice that although the limit appears to depend on the choice of parametrization $\mathbf{\Phi}$, it can be shown that if this limit exists, it is independent of the choice of parametrization. Thus we are justified in defining the limit to be an integral over $\mathcal{S}$.

In order to evaluate the limit, we must express the Riemann sum in a form that we recognize as a Riemann sum for a function of the variables s and t. Thus we must express the area ΔS_{ij} of the ij^{th} cell on $\mathcal{S}$ as a function of s and t. As was the case for parametric curves, it is not possible to compute this area exactly. We will instead approximate this area by the area of the parallelogram spanned by the scaled tangent vectors to the coordinate curves at the point $\mathbf{\Phi}(s_{i-1}, t_{j-1})$. These are $\mathbf{\Phi}_s(s_{i-1}, t_{j-1})\Delta s$ and $\mathbf{\Phi}_t(s_{i-1}, t_{j-1})\Delta t$, where Δs and Δt are the lengths of the sides of the cells of the partition of the domain of $\mathcal{S}$. (See Figure 7.8.) The area of this parallelogram is the length of the cross product of these vectors,

$$\|\mathbf{\Phi}_s(s_{i-1}, t_{j-1}) \times \mathbf{\Phi}_t(s_{i-1}, t_{j-1})\|\Delta s \Delta t.$$

Note that $\mathbf{\Phi}_s \times \mathbf{\Phi}_t$ is the normal vector to the surface given by the parametrization.

If we choose $(s_i^*, t_j^*) = (s_{i-1}, t_{j-1})$, we have

$$R(f, \mathbf{\Phi}, P) \approx \sum_{i,j=1}^{M,N} f(\mathbf{\Phi}(s_i^*, t_j^*))\|\mathbf{\Phi}_s(s_i^*, t_j^*) \times \mathbf{\Phi}_t(s_i^*, t_j^*)\|\Delta s \Delta t.$$

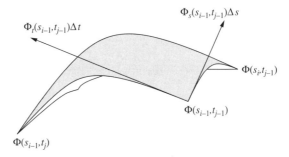

Figure 7.8   The ij^{th} cell of the partition of $\mathcal{S}$ and the tangent vectors $\mathbf{\Phi}_s(s_{i-1}, t_{j-1})\Delta s$ and $\mathbf{\Phi}_t(s_{i-1}, t_{j-1})\Delta t$ to the coordinate curves of $\mathbf{\Phi}$ through $\mathbf{\Phi}(s_{i-1}, t_{j-1})$.

This is a Riemann sum for the function $f(\mathbf{\Phi}(s,t))\|\mathbf{\Phi}_s(s,t) \times \mathbf{\Phi}_t(s,t)\|$ on $\mathcal{D}$. Since $\mathbf{\Phi}$ and f are continuous, the composition $f \circ \mathbf{\Phi}$ is also continuous. Since $\mathbf{\Phi}$ is continuously differentiable, $\mathbf{\Phi}_s$ and $\mathbf{\Phi}_t$ are continuous and the length of the cross product, $\|\mathbf{\Phi}_s(s,t) \times \mathbf{\Phi}_t(s,t)\|$, is also continuous. It follows that $f(\mathbf{\Phi}(s,t))\|\mathbf{\Phi}_s(s,t) \times \mathbf{\Phi}_t(s,t)\|$ is continuous. Thus the limit of this Riemann sum is equal to the integral

$$\int\int_{\mathcal{D}} f(\mathbf{\Phi}(s,t))\|\mathbf{\Phi}_s(s,t) \times \mathbf{\Phi}_t(s,t)\| dA_{s,t}.$$

As a consequence, we have the following theorem.

Theorem 7.2 If $\mathcal{S}$ is parametrized by the continuously differentiable function $\mathbf{\Phi} : \mathcal{D} \to \mathbb{R}^3$, and f is defined and continuous on $\mathcal{S}$, then

$$\int\int_{\mathcal{S}} f dS = \int\int_{\mathcal{D}} f(\mathbf{\Phi}(s,t))\|\mathbf{\Phi}_s(s,t) \times \mathbf{\Phi}_t(s,t)\| dA_{t,s}. \quad \blacklozenge$$

Since the double integral on the right-hand side in the theorem is an ordinary double integral over a domain in the plane, it is equal to an iterated integral by Fubini's theorem. (See Section 5.2.) In the following example, we use this result to compute the total accumulation of $f(x,y,z) = z$ over a hemisphere $\mathcal{S}$.

Example 7.5   **Total Accumulation on a Hemisphere.** Let us compute the total accumulation of $f(x,y,z) = z$ on the upper hemisphere of a sphere of radius 2 centered at the origin. First, we must parametrize the surface. Using Example 7.2, we see that the hemisphere $\mathcal{S}$ can be parametrized by

$$\mathbf{\Phi}(\theta, \phi) = (2\cos\theta \sin\phi, 2\sin\theta \sin\phi, 2\cos\phi),$$

where $0 \le \theta \le 2\pi$ and $0 \le \phi \le \pi/2$. Thus $f(\mathbf{\Phi}(\theta, \phi)) = 2\cos\phi$. In order to express the total accumulation as an iterated integral, we must compute $\|\mathbf{\Phi}_\theta(\theta, \phi) \times \mathbf{\Phi}_\phi(\theta, \phi)\|$. We have

$$\mathbf{\Phi}_\theta(\theta, \phi) = (-2\sin\theta \sin\phi, 2\cos\theta \sin\phi, 0),$$

$$\mathbf{\Phi}_\phi(\theta, \phi) = (2\cos\theta \cos\phi, 2\sin\theta \cos\phi, -2\sin\phi),$$

and

$$\|\mathbf{\Phi}_\theta(\theta, \phi) \times \mathbf{\Phi}_\phi(\theta, \phi)\| = \|(-4\cos\theta \sin^2\phi, -4\sin\theta \sin^2\phi, -4\sin\phi \cos\phi)\|$$

$$= |4\sin\phi|$$

$$= 4\sin\phi$$

for $\phi \in [0, \pi/2]$. Thus the total accumulation of f over $\mathcal{S}$ is given by

$$\int\int_{\mathcal{S}} f\, dS = \int_0^{2\pi} \int_0^{\pi/2} f(\boldsymbol{\Phi}(\theta, \phi)) \|\boldsymbol{\Phi}_\theta(\theta, \phi) \times \boldsymbol{\Phi}_\phi(\theta, \phi)\|\, d\phi\, d\theta$$

$$= \int_0^{2\pi} \int_0^{\pi/2} 2\cos\phi 4\sin\phi\, d\phi\, d\theta$$

$$= 4\int_0^{2\pi} \left[\sin^2\phi\right]_0^{\pi/2}\, d\theta$$

$$= 4\int_0^{2\pi} 1\, d\theta$$

$$= 8\pi.$$

We can also apply the techniques of this section to compute the total accumulation of a function on more complicated surfaces, which cannot be represented as the image of a single parametrization $\boldsymbol{\Phi}$. If $\mathcal{S}$ is a union of surfaces, $S = S_1 \cup S_2 \cup \ldots \cup S_p$, where any two of the surfaces intersect at most along their edges, then the total accumulation of f on $\mathcal{S}$ is equal to the sum of the accumulations of f on each of the surfaces. That is,

$$\int\int_{\mathcal{S}} f\, dS = \int\int_{S_1} f\, dS + \int\int_{S_2} f\, dS + \cdots + \int\int_{S_p} f\, dS.$$

In the following example, we will compute the average value of a function over the surface $S = S_1 \cup S_2$, where S_1 is a cone with vertex at the origin and opening upward with angle $\pi/4$ and slant height 1, and S_2 is the portion of the sphere of radius 1 centered at the origin that "caps" the opening of the cone. (See Figure 7.9.) The average value of f is the total accumulation of f over $\mathcal{S}$ divided by the surface area of $\mathcal{S}$.

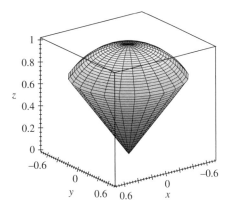

Figure 7.9 The surface $S = S_1 \cup S_2$ of Example 7.6 consisting of a cone with a spherical "cap."

Example 7.6

An Average Value Calculation. Let $f(x, y, z) = \sqrt{x^2 + y^2 + z^2}$, the distance from the point (x, y, z) to the origin. The average value of f over $\mathcal{S}$ will be the average distance of a point on $\mathcal{S}$ from the origin.

A. First, we must parametrize the surfaces $\mathcal{S}_1$ and $\mathcal{S}_2$. We will use a form of spherical coordinates to parametrize each one. The cone $\mathcal{S}_1$ is a portion of the cone that can be described in spherical coordinates by $\{(\rho, \theta, \phi) : \phi = \pi/4\}$. Thus to parametrize $\mathcal{S}_1$, we will use

$$\mathbf{\Phi}(\rho, \theta) = (\rho \cos \theta \, \sin(\pi/4), \rho \sin \theta \, \sin(\pi/4), \rho \cos(\pi/4)) = \frac{\sqrt{2}}{2}(\rho \cos \theta, \rho \sin \theta, \rho).$$

The domain of $\mathbf{\Phi}$ is $0 \le \rho \le 1$ and $0 \le \theta \le 2\pi$, since we are only considering the portion of the cone that lies inside the sphere of radius 1.

The cap $\mathcal{S}_2$ is a portion of a sphere of radius 1. This can be parametrized by

$$\mathbf{\Psi}(\theta, \phi) = (\cos \theta \, \sin \phi, \sin \theta \, \sin \phi, \cos \phi).$$

The appropriate domain of $\mathbf{\Psi}$ is $0 \le \theta \le 2\pi$ and $0 \le \phi \le \pi/4$, since we want to consider only the portion of the sphere lying inside the cone.

B. To calculate the surface area of $\mathcal{S}$, we must compute the integral of the constant function 1 over $\mathcal{S}_1$ and $\mathcal{S}_2$. Using the parametrization $\mathbf{\Phi}$ for $\mathcal{S}_1$, we have

$$\mathbf{\Phi}_\rho = \left(\frac{\sqrt{2}}{2} \cos \theta, \frac{\sqrt{2}}{2} \sin \theta, \frac{\sqrt{2}}{2} \right) \quad \text{and} \quad \mathbf{\Phi}_\theta = \left(-\frac{\sqrt{2}}{2} \rho \sin \theta, \frac{\sqrt{2}}{2} \rho \cos \theta, 0 \right).$$

It follows that $\|\mathbf{\Phi}_\rho \times \mathbf{\Phi}_\theta\| = \frac{\sqrt{2}}{2}\rho$, and the surface area of $\mathcal{S}_1$ is

$$\int\int_{\mathcal{S}_1} dS = \int_0^1 \int_0^{2\pi} \frac{\sqrt{2}}{2}\rho \, d\theta \, d\rho = \frac{\sqrt{2}}{2}\pi.$$

Using the parametrization $\mathbf{\Psi}$ for $\mathcal{S}_2$, we have

$$\mathbf{\Psi}_\theta = (-\cos \theta \, \sin \phi, \sin \theta \, \sin \phi, \cos \phi) \quad \text{and} \quad \mathbf{\Psi}_\phi = (\cos \theta \, \cos \phi, \sin \theta \, \cos \phi, -\sin \phi).$$

It follows that $\|\mathbf{\Psi}_\theta \times \mathbf{\Psi}_\phi\| = \sin(\phi)$, and the surface area of $\mathcal{S}_2$ is

$$\int\int_{\mathcal{S}_2} dS = \int_0^{2\pi} \int_0^{\pi/4} \sin \phi \, d\phi \, d\theta = \pi(2 - \sqrt{2}).$$

The surface area of $\mathcal{S}$ is equal to $\frac{\sqrt{2}}{2}\pi + \pi(2 - \sqrt{2}) \approx 4.06$.

C. To calculate the total accumulation of $f(x, y, z) = \sqrt{x^2 + y^2 + z^2}$ over $\mathcal{S}$, notice that $f(x, y, z) = \rho$ on $\mathcal{S}_1$ and $f(x, y, z) = 1$ on $\mathcal{S}_2$. Therefore, the total accumulation of f on $\mathcal{S}_1$ is

$$\int\int_{\mathcal{S}_1} f \, dS = \int\int_{\mathcal{S}_1} \rho \, dS$$

$$= \int_0^1 \int_0^{2\pi} \frac{\sqrt{2}}{2} \rho^2 \, d\theta d\rho$$

$$= \frac{\sqrt{2}}{3} \pi.$$

The total accumulation on $\mathcal{S}_2$ is

$$\int\int_{\mathcal{S}_2} f \, dS = \int\int_{\mathcal{S}_2} dS = \pi(2 - \sqrt{2})$$

from above. The total accumulation of f on $\mathcal{S}$ is equal to $\frac{\sqrt{2}}{3}\pi + \pi(2 - \sqrt{2}) \approx 3.32$. Thus the average distance of a point on $\mathcal{S}$ to the origin is approximately $3.32/4.06 \approx 0.82$.

Application: Acoustics

In acoustics, surface integrals are used in the computation of the acoustic absorption of surfaces. Here we indicate how this might be used in designing a physical space.

In analyzing the acoustic features of a room, one of the quantities acoustic designers measure is the time required for a sound to die out in the room. This is measured by the **reverberation time**, which is defined to be the amount of time required for the sound level to decrease by 60 decibels. The reverberation time depends on how often the sound encounters an absorbing surface and how much energy is lost in each encounter. The **Sabine equation**[1] for reverberation time is

$$\text{RT} = \frac{c \cdot \text{volume}}{\text{total absorption}},$$

where c is a constant depending on the speed of sound. (We will use 0.16 for the value of c.) The total absorption of a room will depend on the area and the building materials of the interior surfaces of the room. We define the absorption coefficient to be the fraction

---

[1] The American physicist W. C. Sabine (1868–1919) developed this formula empirically. He founded the field of architectural acoustics and was the acoustical architect for Symphony Hall, Boston. (See Wallace Clement Sabine, Wikipedia.)

Material	Coefficient	Material	Coefficient
Acoustical plaster	0.50	Acoustical tile	0.65
Carpeted floor	0.60	Concrete	0.01
Draperies	0.55	Paneling	0.20
Plaster	0.05	Vinyl floor on concrete	0.03
Wood floor	0.06	Adult person	0.45

Figure 7.10   Absorption coefficients for common building materials. These coefficients are measured for a sound frequency of 500 Hz. Values are for surfaces areas expressed in square meters. (See Application: Acoustics in Section 7.2.) (Data is from *Music, Speech, Audio* by W. Strong and G. Plitnik.)

of energy absorbed on each reflection of a sound wave. Thus the total absorption of a surface is the product of the absorption coefficient and the area of the surface. A table of absorption coefficients for several common surfaces is given in Figure 7.10. In the following example, we carry out a calculation of total absorption.

Example 7.7

Total Absorption. Suppose a lecture hall can be modeled by the region in the first octant bounded by the plane $x + y = 5$, the plane $z = 3 + 0.5x + 0.5y$, and the cylinder $x^2 + y^2 = 400$, where the units of measurement are meters. (See Figure 7.11.) Assume the back wall of the lecture hall is covered with wood paneling. We would like to compute the total absorption of the back wall. Since the absorption coefficient for paneling is 0.20, the total absorption of the wall is $0.20 \times$ surface area. The back wall can be parametrized by the function $\mathbf{\Phi}(\theta, z) = (20 \cos \theta, 20 \sin \theta, z)$ for $0 \leq \theta \leq \pi/2$ and $0 \leq z \leq 3 + 10 \cos \theta + 10 \sin \theta$. In this case, $\mathbf{\Phi}_\theta \times \mathbf{\Phi}_z = (20 \cos \theta, 20 \sin \theta, 0)$, so that $\|\mathbf{\Phi}_\theta \times \mathbf{\Phi}_z\| = 20$.

The total absorption of the back wall is

$$0.20 \int_0^{\pi/2} \int_0^{3+10\cos(\theta)+10\sin(\theta)} 20 \, dz \, d\theta = 4 \int_0^{\pi/2} 3 + 10 \cos(\theta) + 10 \sin(\theta) \, d\theta$$

$$= 6\pi + 80 \approx 98.85$$

Exercises 4 and 5 return to this example.

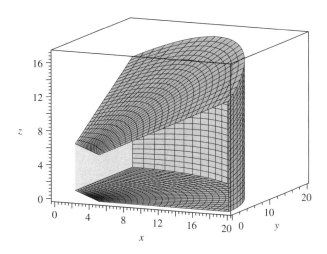

Figure 7.11 The model for the lecture hall in Example 7.7 with a side wall and the front wall removed.

Summary

In this section, we introduced **surface integration** in order to compute the total accumulation of a function on a surface. We focused on **parametrically defined** surfaces, which are the images of regular parametrizations $\mathbf{\Phi} : \mathcal{D} \subset \mathbb{R}^2 \to \mathbb{R}^3$. We then constructed a Riemann sum to approximate the total accumulation of a function f on $\mathcal{S}$. We used a partition of the domain $\mathcal{D}$ of $\mathbf{\Phi}$ to construct a partition of $\mathcal{S}$, which we used to construct the Riemann sum. As we have seen before, if f is continuous, then the limit of Riemann sums of this type exists and is independent of any of the choices we made in constructing the sum. We showed that integral of f over $\mathcal{S}$ is equal to an iterated integral over $\mathcal{D}$:

$$\int\int_{\mathcal{S}} f\,dS = \int\int_{\mathcal{D}} f(\mathbf{\Phi}(s,t)) \|\mathbf{\Phi}_s(s,t) \times \mathbf{\Phi}_t(s,t)\| dA_{s,t}.$$

If $\mathcal{D} = [a,b] \times [c,d]$, we have

$$\int\int_{\mathcal{S}} f\,dS = \int_a^b \int_c^d f(\mathbf{\Phi}(s,t)) \|\mathbf{\Phi}_s(s,t) \times \mathbf{\Phi}_t(s,t)\| \, dt\,ds.$$

We then applied this formula to calculate the **total accumulation** of a function, the **area of a surface**, the **average value** of a function on a surface, and the **total absorption** of a physical space in acoustics.

Section 7.2 Exercises

1. **Surface Area.** For each of the following parametrizations Φ and domains $\mathcal{D}$, describe the surface $\mathcal{S}$ that is the image of Φ and compute the surface area of Φ.

   (a) $\Phi(s,t) = (1 + s + t, 1 - s - 2t, 2s + 3t)$, $\mathcal{D} = \{(s,t) : 0 \le s \le 1 \text{ and } 0 \le t \le 1\}$.
   (b) $\Phi(r,\theta) = (r\cos\theta, r\sin\theta, 1)$, $\mathcal{D} = \{(r,\theta) : 0 \le r \le 2 \text{ and } 0 \le \theta \le \pi/2\}$.
   (c) $\Phi(\theta,y) = (3\cos\theta, y, 3\sin\theta)$, $\mathcal{D} = \{(\theta,y) : 0 \le \theta \le 3\pi/2 \text{ and } -2 \le y \le 2\}$.
   (d) $\Phi(\rho,\theta) = (\rho\cos\theta\sin(\pi/4), \rho\sin\theta\sin(\pi/4), \rho\cos(\pi/4))$, $\mathcal{D} = \{(\rho,\theta) : 0 \le \rho \le 3 \text{ and } 0 \le \theta \le 2\pi\}$.

2. **Total Accumulation.** For each of the following parametrizations Φ, domains $\mathcal{D}$, and functions f, describe the surface $\mathcal{S}$ that is the image of Φ and compute the total accumulation of f on $\mathcal{S}$.

   (a) $\Phi(s,t) = (s + t, 2 - s - t, s - t)$, $\mathcal{D} = \{(s,t) : 0 \le s \le 1 \text{ and } -1 \le t \le 1\}$, and $f(x,y,z) = x + y + z$.
   (b) $\Phi(\theta,z) = (2\cos\theta, 2\sin\theta, z)$, $\mathcal{D} = \{(\theta,z) : 0 \le \theta \le \pi \text{ and } -2 \le z \le 2\}$, and $f(x,y,z) = x^2 z$.
   (c) $\Phi(\rho,\phi) = (\rho\cos(\pi/4)\sin\phi, \rho\sin(\pi/4)\sin\phi, \rho\cos\phi)$, $\mathcal{D} = \{(\rho,\phi) : 0 \le \rho \le 2 \text{ and } 0 \le \phi \le \pi/4\}$, and $f(x,y,z) = y + z$.

3. **Surfaces of Revolution**. (See Exercise 8 of Section 7.1.) A surface of revolution $\mathcal{S}$ with the z-axis as the axis of rotation can be parametrized by

$$\Phi(\theta, z) = (\cos(\theta)f(z), \sin(\theta)f(z), z),$$

   where f is a non-negative differentiable function of z, with $0 \le \theta \le 2\pi$ and z in the domain of f. The radius r at height z is $f(z)$.

   (a) Set up the integral for the surface area of a surface of revolution given in the above form.
   (b) For each of the following functions $r = f(z)$, set up and evaluate the integral for the surface area of the corresponding surface of revolution.

   (i) $f(z) = 2$, $-3 \le z \le 3$.
   (ii) $f(z) = \cos(z)$, $-\pi/2 \le z \le \pi/2$.

   (iii) $f(z) = \sqrt{4 - z^2}$, $-2 \le z \le 2$.
   (iv) $f(z) = \sqrt{1 + z^2}$, $-1 \le z \le 1$.

   (c) For each function in (b), if possible, evaluate the integral to find the surface area directly. If not, use a computer algebra system to compute the integral and find the surface area.

4. **Total Absorption.** Suppose the lecture hall in Example 7.7 has acoustical tile on the ceiling, paneling on all of the walls, and carpeting on the floor. What is the total absorption of the hall?

5. **Reverberation Time.**

   (a) What is the reverberation time of the lecture hall in Exercise 4? Recall that reverberation time is $0.16 \cdot$ volume/total absorption.
   (b) An acceptable reverberation time for a classroom of this size is 1.0. Is the reverberation time of this room acceptable? If not, explain what modifications you could make to the room to make the reverberation time acceptable.

6. **Total Heat.** A sphere made of copper has radius 5 cm and thickness 0.01 cm. The specific heat of copper is 0.0923 cal/g$^\circ$C and the density of copper is 8.93 g/cm^3. If the temperature on the sphere is given in degrees centigrade by the formula $T(x, y, z) = 50 + x^2 - y^2 + z^2$, compute the total heat contained in the sphere. (See Example 7.4C.)

7. **Surface Area of a Torus.** Exercise 14(b) of Section 7.1 introduced the parametrization of a torus as a tube around a circle. This can be generalized as follows. Let $\gamma(t) = (R_0 \cos(t), R_0 \sin(t), 0)$ be a circle of radius $R_0 > 0$ in the plane. Then $\mathbf{n}_1(t) = (0, 0, 1)$ and $\mathbf{n}_2(t) = (\cos(t), \sin(t), 0)$ are vector fields along γ normal to γ. Define

$$\mathbf{\Phi}(t, s) = \gamma(t) + r_0 \cos(s)\mathbf{n}_1(t) + r_0 \sin(s)\mathbf{n}_2(t),$$

where r_0 is a constant, $0 < r_0 < R_0$, and $0 \le t \le 2\pi$ and $0 \le s \le 2\pi$. This generalizes the torus of Exercise 14(b). It creates a tube of radius r_0 about a circle of radius R_0.

   (a) Compute $\mathbf{\Phi}_t(t, s)$, $\mathbf{\Phi}_s(t, s)$, and $\mathbf{\Phi}_t(t, s) \times \mathbf{\Phi}_s(t, s)$. (Be careful to simplify your cross product as much as possible.)
   (b) Use your answer to (a) to compute the surface area of the image of $\mathbf{\Phi}$ in terms of R_0 and r_0.

■ 7.3 Flux Integrals

In Section 6.3, we investigated the flux or flow of a vector field across a curve in the plane. Here we would like to extend these ideas to vector fields in space and explore the flux of a vector field through a surface. In particular, we want to express the total flux of a vector field through a surface as the surface integral of the dot product of the vector field and a vector field normal to the surface. We will then see how to express surface integrals of this type as double integrals over the domain of a parametrization of the surface. Let us begin by briefly considering several examples of total flux.

| Example 7.8 | **Total Flux** |

A. Total Flux of a Fluid Flow. If a fluid is moving through a region of space, it is often of interest to determine the amount of fluid that moves through a surface in a unit of time.

For example, the rate of flow of water in a stream or river is the total flux of the velocity field **F** of the flow through a surface that spans the stream from stream bed to surface and from bank to bank.

B. Total Current. An electric current consists of charged particles in motion. The flow of particles can be represented by a vector field **J**, called the **current density**, with **J** $= nq$**F** where n is the number of particles per unit volume, q is the charge of an individual particle, and **F** is the velocity field of the flow. The total flux of **J** through a surface is the **total current** passing through the surface.

C. Diffusive Flux. If an airborne substance, for example a pollutant, spreads by diffusion through the atmosphere, the total amount of the substance that passes through a sphere centered at the source is the total flux of the flux field of the diffusion process. If the diffusion process obeys Fick's law, the substance flows from regions of high concentration to low concentration. Specifically, the flow is proportional to the negative of the gradient of the concentration. That is, the flux field **J** $= -\delta\nabla c$, where c is the concentration of the pollutant and δ is the **diffusivity** of the diffusion process. This model also applies to the flow of heat. In this case, the heat flux at a point is $-k\nabla T$, where T is the temperature and k is the **thermal conductivity** of the medium. The total flux of $-k\nabla T$ through a surface is a measure of the total heat that moves through the surface.

D. Total Acoustic Power. The phenomenon of sound is a consequence of variations in air pressure. Due to the physical properties of substances, and in particular air, sound waves propagate longitudinally. That is, as a sound wave moves through air, the local movement of the air is in the direction of the propagation of the sound. If the sound has constant frequency, we can represent the propagation of the sound by the **acoustic energy flux** **I**. The direction of **I** is the direction of the local movement of the air, and the magnitude of **I** is proportional to the square of the average change in pressure due to the sound. The total flux of **I** through a surface enclosing the source of the sound is the **total acoustic power** of the source of sound.

The concept of the flux of a vector field is most easily understood in the case where the vector field is the velocity field **F** of a moving fluid. Recall that at a point $(x, y, z) \in \mathbb{R}^3$, the vector **F**$(x, y, z)$ represents the velocity of a particle moving with the fluid when it passes through (x, y, z). Further, we are assuming that this does not depend on time, so that whenever a particle passes through (x, y, z), it has the same velocity. We begin with the flow of a fluid across a planar surface.

Let $\mathcal{S}$ be a plane, and let **N** be a unit normal vector to $\mathcal{S}$. Let us suppose that **F** $=$ **F**(x, y, z) is a continuous vector field that represents the velocity of a fluid flow. If **F** is a constant vector field, we can immediately compute the volume of fluid that passes through a given rectangle $\mathcal{R}$ in S in the direction of **N** in a unit of time. In one unit

19. **A Charged Shell.** A thin spherical shell (a hollow ball) of radius R_0 has total charge Q that is distributed uniformly over its surface.

    (a) What is the electric field outside the surface? Explain your answer.
    (b) What is the electric field inside the surface? Explain your answer.

20. **A Charge Combination.** A thin spherical shell (a hollow ball) of radius R_0 has total charge $2Q$ that is distributed uniformly over its surface. In addition, a point charge of charge $-Q$ is located at the center of the sphere.

    (a) What is the electric field outside the surface? Explain your answer.
    (b) What is the electric field inside the surface? Explain your answer.

21. **Infinite Charge Distribution.** Suppose that charge is distributed uniformly along the z-axis, so that the charge density is given by $\rho(0, 0, z) = q/\epsilon_0$ and $\rho(x, y, z) = 0$ off of the z-axis. Describe the direction of the corresponding electric field $\mathbf{E}$ at points not on the z-axis.

22. **The Differential Formulation of Ampère's Law.** In the text, we claimed that the fact that

$$\mu_0 \int \int_{\mathcal{S}} \mathbf{J} \cdot \mathbf{N} \, dS = \int \int_{\mathcal{S}} \operatorname{curl} \mathbf{B} \cdot \mathbf{N} \, dS$$

implies that $\mu_0 \mathbf{J} = \operatorname{curl} \mathbf{B}$. Here we complete this argument. Observe first that this equality is equivalent to

$$\int \int_{\mathcal{S}} (\mu_0 \mathbf{J} - \operatorname{curl} \mathbf{B}) \cdot \mathbf{N} \, dS = 0.$$

To show that this integral is zero, we will rely on a more general fact about integrals of vector fields. That is, we will need to show that if $\mathbf{F}$ satisfies $\int \int_{\mathcal{S}} \mathbf{F} \cdot \mathbf{N} \, dS = 0$ for every surface $\mathcal{S}$, then $\mathbf{F} = \mathbf{0}$. Here we work with the third coordinate of $\mathbf{F}$ and a disk $\mathcal{S}$ of radius r centered at the origin in the xy-plane.

    (a) First, consider S where S is a disk of radius r centered at the origin in the xy-plane. For this $\mathcal{S}$, show that $\mathbf{F} \cdot \mathbf{N} = \pm F_3$.
    (b) For this s, what is the average value of

$$\int \int_{\mathcal{S}} \mathbf{F} \cdot \mathbf{N} \, dS$$

    over $\mathcal{S}$ as $r \to 0$? Explain your answer.
    (c) Use parts (a) and (b) to show that the third coordinate of $\mathbf{F}$, F_3 is zero.
    (d) How would you show that the first two coordinates of $\mathbf{F}$ are zero?

Since $\int \int_{\mathcal{S}} (\mu_0 \mathbf{J} - \operatorname{curl} \mathbf{B}) \cdot \mathbf{N} \, ds = 0$ for all surfaces $\mathcal{S}$, this argument allows us to conclude $\mu_0 \mathbf{J} - \operatorname{curl} \mathbf{B} = 0$, hence $\mu_0 \mathbf{J} = \operatorname{curl} \mathbf{B}$.

23. **Flux Zero.** Show that if the flux of a vector field through every surface is equal to zero, the vector field has divergence equal to zero.

24. **Charges in Motion.** In physics, it is said that "a positive charge moving in one direction is equivalent in nearly all its external effects to a negative charge moving in the opposite direction." Given the fact that the physical properties of a current can be expressed in terms of the current density $\mathbf{J}$, justify this statement.

25. **Closed Surfaces.** Let $\mathcal{S}$ be a closed surface oriented by the outward pointing normal $\mathbf{N}$. Suppose that $\mathcal{C}$ is a simple closed curve lying in $\mathcal{S}$ that divides $\mathcal{S}$ into two surfaces, $\mathcal{S}_1$ and $\mathcal{S}_2$, so that $S = \mathcal{S}_1 \cup \mathcal{S}_2$ and $\mathcal{C}$ is the boundary of $\mathcal{S}_1$ and of $\mathcal{S}_2$. Use Stokes' theorem to show that for any continuously differentiable vector field $\mathbf{F}$,

$$\int\int_{\mathcal{S}_1} \text{curl } \mathbf{F} \cdot \mathbf{N} \, dS = -\int\int_{\mathcal{S}_2} \text{curl } \mathbf{F} \cdot \mathbf{N} \, dS.$$

Thus $\int\int_{\mathcal{S}} \text{curl } \mathbf{F} \cdot \mathbf{N} \, dS = 0$.

26. **Fluid in a Pipe.** In Exercise 8 of Section 7.3, we investigated the velocity field of a fluid in a pipe of radius r lying on the z-axis. It was given by $\mathbf{F}(x, y, z) = \left(0, 0, v_0(1 - \frac{r^2}{r_0^2})\right)$, where $r^2 = x^2 + y^2$.

   (a) Show that this flow is not irrotational; that is, it has nonzero curl.
   (b) Suppose an object was flowing with the fluid in the pipe and it was located at $(r_0/2, r_0/2, 0)$. Describe the rotation of the object at this location.
   (c) Is your answer to (b) expected based on the magnitude of $\mathbf{F}$? Explain.

Solutions to Selected Exercises

CHAPTER 1 SOLUTIONS

■ 1.1 The Cartesian Coordinate System

---

1. **Subsets of the Plane.** (a) $\{(x, y) : -1 \leq x \leq 1,\ 1 \leq y \leq 2\}$.
   (b) $\{(x, y) : 0 \leq x \leq 2,\ -\frac{1}{2}x + 1 \leq y \leq 1\}$.
   (c) $\{(x, y) : -1 \leq x \leq 1,\ -x - 1 \leq y \leq -\frac{1}{2}x + \frac{3}{2}\}$.
   (d) $\{(x, y) : 1 \leq x \leq 2,\ -\sqrt{1 - (x-1)^2} \leq y \leq \sqrt{1 - (x-1)^2}\}$, or $\{(x, y) : -1 \leq y \leq 1,\ 1 \leq x \leq 1 + \sqrt{1 - y^2}\}$.

5. **Subsets of $\mathbb{R}^3$.** (a) $\mathcal{S} = \{(x, y, z) : y < 0\}$.
   (c) $\mathcal{S} = \{(x, y, z) : z > 5\}$.
   (e) $\mathcal{S} = \{(x, y, z) : x > 2, y > 2\}$.

6. **A Cone.** (i) The horizontal slices of $\mathcal{S}$ are the circles parallel to the xy-plane centered on the positive z-axis given by the equations $x^2 + y^2 = z_0^2$ for $z > 0$. When $z_0 = 0$, the slice is the origin, $x = y = z = 0$.
   (ii) The vertical slices parallel to the xz-plane are the hyperbolas given by the equations $z^2 - x^2 = y_0^2$ for $y_0 \neq 0$. When $y_0 = 0$, the slice consists of the half-lines given by $y = 0$ and $z = x$ for $x \geq 0$ and $y = 0$ and $z = -x$ for $x \leq 0$.
   (iii) The vertical slices parallel to the yz-plane are the hyperbolas given by the equations $z^2 - y^2 = x_0^2$ for $x_0 \neq 0$. When $x_0 = 0$, the slice consists of the half-lines given by $x = 0$ and $z = y$ for $y \geq 0$ and $x = 0$ and $z = -y$ for $y \leq 0$.

7. **Positions of a Satellite.** $\mathcal{S} = \{(x, y, z) : (3959)^2 < x^2 + y^2 + z^2 \le (4159)^2\}$.

8. **A Hyperboloid of One Sheet.** (i) The horizontal slices of $\mathcal{S}$ for $z_0 > 1$ and for $z_0 < -1$ are the branches of the hyperbolas defined by $x^2 - y^2 = z_0^2 - 1$, which open out in the directions of positive and negative x. The horizontal slices of $\mathcal{S}$ for $-1 \le z_0 \le 1$ are the branches of the hyperbolas defined by $x^2 - y^2 = 1 - z_0^2$, which open out in the directions of positive and negative y. The horizontal slices for $z_0 = \pm 1$ consist of the pair of lines $y = x$ and $y = -x$ for $z_0 = 1$ and for $z_0 = -1$.
(ii) The vertical slices of $\mathcal{S}$ parallel to the xz-plane are analogous to the horizontal slices. For $y_0 > 1$ and for $y_0 < -1$, they are the branches of the hyperbolas defined by $x^2 - z^2 = y_0^2 - 1$, which open out in the directions of positive and negative x. The vertical slices of $\mathcal{S}$ for $-1 \le y_0 \le 1$ are the branches of the hyperbolas defined by $x^2 - z^2 = 1 - y_0^2$, which open out in the directions of positive and negative z. The vertical slices for $y_0 = \pm 1$ consist of the pair of lines $z = x$ and $z = -x$ for both $y_0 = 1$ and for $y_0 = -1$.
(iii) The vertical slices of $\mathcal{S}$ parallel to the yz-plane are the circles centered on the x-axis given by $y^2 + z^2 = 1 + x_0^2$.

9. **Slices of Quadric Surfaces.** (a) $\mathcal{S} \cap \mathcal{P} = \{(x, y, z) : y = 1, z = \pm 1\}$ is a pair of lines parallel to the x-axis passing through the points $(0, 1, 1)$ and $(0, 1, -1)$.
(b) $\mathcal{S} \cap \mathcal{P} = \{(x, y, z) : x = 1, y^2 - z^2 = 4\}$ is both branches of a hyperbola lying in the plane $x = 1$, and it opens out in the directions of positive and negative y.

12. **The Heart.** (a) Since the heart walls are modeled by cylinders centered on the z-axis, we use the equation $x^2 + y^2 = r^2$ for the inner and outer walls for $0 \le z \le h$: $\mathcal{S} = \{(x, y, z) : r_1^2 \le x^2 + y^2 \le r_2^2, 0 \le z \le h\}$.
(b) Since the volume of a cylinder of radius r and height h is $\pi r^2 h$, we must choose r_1 and h so that $\pi r_1^2 h = 40$. For example, letting $h = 5$ cm, we have $r_1 = \sqrt{8/\pi}$ cm. Since the volume of the region between the cylinders is $\pi r_2^2 h - \pi r_1^2 h = \pi h(r_2^2 - r_1^2)$, we can solve for r_2 to obtain $r_2 = \sqrt{28\pi}$ cm. These values are not unique.

■ 1.2 Vectors

1. **Distance.** The vector $\mathbf{v}$ can be represented in space as the vector with initial point (x_0, y_0, z_0) and endpoint $(x_0 + v_1, y_0 + v_2, z_0 + v_3)$. The distance between these points is: $\sqrt{((x_0 + v_1) - x_0)^2 + ((y_0 + v_2) - y_0)^2 + ((z_0 + v_3) - z_0)^2} = \sqrt{v_1^2 + v_2^2 + v_3^2}$.

4. **Vectors and Angles.** (a) Sketch the vectors in the same coordinate plane.
(i) $\mathbf{v} = (\cos(\pi/6), \sin(\pi/6))$, $\|\mathbf{v}\| = \sqrt{\cos^2(\pi/6) + \sin^2(\pi/6)} = 1$. (ii) $\mathbf{v} = (2\sin(\pi/6), \cos(\pi/6))$, $\|\mathbf{v}\| = \sqrt{4\sin^2(\pi/6) + 4\cos^2(\pi/6)} = 2$. (iii) $\mathbf{v} = (\cos(3\pi/4), \sin(3\pi/4))$, $\|\mathbf{v}\| = \sqrt{\cos^2(3\pi/4) + \sin^2(3\pi/4)} = 1$. (iv) $\mathbf{v} = (\frac{5}{2}\cos(-3\pi/4), \frac{5}{2}\sin(-3\pi/4))$, $\|\mathbf{v}\| = \sqrt{\frac{25}{4}\cos^2(-3\pi/4) + \frac{25}{4}\sin^2(-3\pi/4)} = \frac{5}{2}$.

(b) The vector $(R\cos\alpha, R\sin\alpha)$ is a vector of length $|R|$ that makes an angle α with the positive x-axis if $R > 0$ and $\alpha + \pi$ if $R < 0$.

6. **Vector Sums.** (a) Sketch all the vectors in the same coordinate plane.
(b) (i) $\mathbf{u} = (-2, 0)$, $\mathbf{v} = (1, \sqrt{3})$, and $\mathbf{w} = (1 - \sqrt{3})$, so $\mathbf{v} + \mathbf{w} = -\mathbf{u}$. (ii) $\mathbf{u} + \mathbf{v} = -\mathbf{w}$.
(iii) $\mathbf{u} + \mathbf{v} + \mathbf{w} = \mathbf{0}$. (iv) $(1, 0) = -1/2\mathbf{u}$.

7. **Properties of Vector Sums.** (a) $\mathbf{u} + \mathbf{v} = (u_1, u_2) + (v_1, v_2) = (u_1 + v_1, u_2 + v_2) = (v_1 + u_1, v_2 + u_2) = (v_1, v_2) + (u_1, u_2) = \mathbf{v} + \mathbf{u}$.
(b) $(\mathbf{u} + \mathbf{v}) + \mathbf{w} = (u_1 + v_1, u_2 + v_2) + (w_1, w_2) = ((u_1 + v_1) + w_1, (u_2 + v_2) + w_2) = (u_1 + (v_1 + w_1), u_2 + (v_2 + w_2)) = (u_1, u_2) + (v_1 + w_1, v_2 + w_2) = \mathbf{u} + (\mathbf{v} + \mathbf{w})$.

9. **Unit Vectors.** (a) If $\mathbf{v}$ is any nonzero vector, then $\frac{1}{\|\mathbf{v}\|}\mathbf{v}$ is a positive scalar multiple of $\mathbf{v}$, so it points in the same direction as $\mathbf{v}$, but it has length equal to $\frac{1}{\|\mathbf{v}\|}\|\mathbf{v}\| = 1$. Thus it is a unit vector that points in the same direction as $\mathbf{v}$.
(b) (i) $\frac{1}{\sqrt{5}}(-1, 2) = (\frac{-1}{\sqrt{5}}, \frac{2}{\sqrt{5}})$. (ii) $\frac{1}{3}(1, 2, 2) = (\frac{1}{3}, \frac{2}{3}, \frac{2}{3})$. (iii) $\frac{1}{R}(R\sin\theta, R\cos\theta) = (\sin\theta, \cos\theta)$.
(c) $(\cos\theta, \sin\theta)$.

10. **Forces on Charged Particles.** (a) The forces are $\frac{(5\times10^{-6})(4\times10^{-6})}{4\pi\epsilon_0 2^{3/2}}(1, 1)$ and $\frac{(-3\times10^{-6})(4\times10^{-6})}{4\pi\epsilon_0 1^{3/2}}(1, 0)$.
(b) The total force is the sum of the forces from part (a): $\frac{(5\times10^{-6})(4\times10^{-6})}{4\pi\epsilon_0 2^{3/2}}(1, 1) + \frac{(-3\times10^{-6})(4\times10^{-6})}{4\pi\epsilon_0 1^{3/2}}(1, 0) = \frac{10^{-12}}{4\pi\epsilon_0}(5\sqrt{2} - 12, 5\sqrt{2})$.

13. **Force in a Hydrogen Atom.** (a) The magnitude of the electrical force is (see Example 1.7A of Section 1.2): $|\frac{q_1 q_2}{4\pi\epsilon_0 r^2}| = |\frac{(1.6\times10^{-19})^2}{4\pi(8.9\times10^{-12})r^2}| = \frac{1.43\times10^{-28}}{r^2}$.
(b) The magnitude of the gravitational force is (see Exercise 11): $\frac{Gm_1 m_2}{r^2} = \frac{6.673\times10^{-11}(9\times10^{-31})(1.7\times10^{-27})}{r^2} = \frac{1.02\times10^{-67}}{r^2}$.
(c) The electrical force between the electron and the proton is approximately 1.4×10^{39} times greater than the gravitational force between them.

■ 1.3 Vector Products

1. **Dot Product Calculations.** (a) $\mathbf{u}\cdot\mathbf{v} = -1 + 2 = 1$; $\theta = \arccos\left(\frac{1}{\sqrt{5}\sqrt{2}}\right) = \arccos\left(\frac{1}{\sqrt{10}}\right) \approx 1.25$.
(c) $\mathbf{u}\cdot\mathbf{v} = 3 + 2 = 5$; $\theta = \arccos\left(\frac{5}{\sqrt{5}\sqrt{10}}\right) = \arccos\left(\frac{1}{\sqrt{2}}\right) = \frac{\pi}{4}$.
(e) $\mathbf{u}\cdot\mathbf{v} = \cos\alpha\sin\alpha + \sin\alpha\cos\alpha = \sin(2\alpha)$; $\theta = \arccos\left(\frac{\sin(2\alpha)}{(1)(1)}\right) = \frac{\pi}{2} - 2\alpha$.

3. **Properties of the Dot Product.** Let $\mathbf{u} = (u_1, u_2, u_3)$, $\mathbf{v} = (v_1, v_2, v_3)$, and $\mathbf{w} = (w_1, w_2, w_3)$. Each of these properties can be verified by expressing the equations in terms of these coordinates.

(a) $\mathbf{u} \cdot \mathbf{v} = (u_1, u_2, u_3) \cdot (v_1, v_2, v_3) = (u_1 v_1 + u_2 v_2 + u_3 v_3) = (v_1 u_1 + v_2 u_2 + v_3 u) = (v_1, v_2, v_3) \cdot (u_1, u_2, u_3) = \mathbf{v} \cdot \mathbf{u}$.

(b) (i) $\mathbf{u} \cdot (\mathbf{v} + \mathbf{w}) = (u_1, u_2, u_3) \cdot ((v_1, v_2, v_3) + (w_1, w_2, w_3)) = (u_1, u_2, u_3) \cdot (v_1 + w_1, v_2 + w_2, v_3 + w_3) = (u_1(v_1 + w_1) + u_2(v_2 + w_2) + u_3(v_3 + w_3)) = (u_1 v_1 + u_2 v_2 + u_3 v_3) + (u_1 w_1 + u_2 w_2 + u_3 w_3) = (\mathbf{u} \cdot \mathbf{v}) + (\mathbf{u} \cdot \mathbf{w})$; (ii) $c\mathbf{u} \cdot \mathbf{v} = (cu_1 v_1 + cu_2 v_2 + cu_3 v_3) = c(u_1 v_1 + u_2 v_2 + u_3 v_3) = c(\mathbf{u} \cdot \mathbf{v}) = (u_1 cv_1 + u_2 cv_2 + u_3 cv_3) = \mathbf{u} \cdot (c\mathbf{v})$.

5. **Triangle Inequality.** We will square both sides and use the result of Exercise 4 to show that $\|\mathbf{u} + \mathbf{v}\|^2 \leq (\|\mathbf{u}\| + \|\mathbf{v}\|)^2$: $\|\mathbf{u} + \mathbf{v}\|^2 = (\mathbf{u} + \mathbf{v}) \cdot (\mathbf{u} + \mathbf{v}) = \mathbf{u} \cdot \mathbf{u} + 2\mathbf{u} \cdot \mathbf{v} + \mathbf{v} \cdot \mathbf{v} \leq \|\mathbf{u}\|^2 + 2\|\mathbf{u}\|\,\|\mathbf{v}\| + \|\mathbf{v}\|^2 = (\|\mathbf{u}\| + \|\mathbf{v}\|)^2$.

7. **Cross Product Calculations.** (a) $\mathbf{u} \times \mathbf{v} = (2, -1, 4)$; $\|\mathbf{u} \times \mathbf{v}\| = \sqrt{21}$.

(b) $\mathbf{u} \times \mathbf{v} = (1, 5, -4)$; $\|\mathbf{u} \times \mathbf{v}\| = \sqrt{42}$.

(c) $\mathbf{u} \times \mathbf{v} = (-0.5, -3, 1)$; $\|\mathbf{u} \times \mathbf{v}\| = \sqrt{10.25}$.

(d) $\mathbf{u} \times \mathbf{v} = (0, 0, 0)$; $\|\mathbf{u} \times \mathbf{v}\| = 0$.

9. **Methane.** (a) The vectors representing the edges of the tetrahedron are $(1, 1, 0)$, $(1, 0, 1)$, $(0, 1, 1)$, $(0, -1, 1)$, $(-1, 0, 1)$, and $(-1, 1, 0)$. The length of each of these edges is $\sqrt{2}$.

(b) The angle between any two adjacent edges of the tetrahedron satisfies $\cos\theta = \frac{1}{\sqrt{2}\sqrt{2}} = \frac{1}{2}$; thus $\theta = \pi/3$.

(c) If the carbon atom is located at $(0.5, 0.5, 0.5)$, two vectors representing the bonds from the carbon atom to two of the hydrogen atoms located at $(0, 0, 0)$ and $(1, 1, 0)$ are $(-0.5, -0.5, -0.5)$ and $(0.5, 0.5, -0.5)$. The angle between these vectors satisfies $\cos\theta = \frac{-0.25}{\sqrt{0.75}\sqrt{0.75}} = -\frac{1}{3}$. Thus $\theta = \arccos(-1/3) \approx 1.91$.

11. **Orthogonality of $\mathbf{u} \times \mathbf{v}$.** We know that the vector $\mathbf{u} \times \mathbf{v}$ is orthogonal to $\mathbf{u}$ and $\mathbf{v}$. To show it is orthogonal to $s\mathbf{u} + t\mathbf{v}$ for any s and t, we must show that the dot product, $(\mathbf{u} \times \mathbf{v}) \cdot (s\mathbf{u} + t\mathbf{v})$, is zero: $(\mathbf{u} \times \mathbf{v}) \cdot (s\mathbf{u} + t\mathbf{v}) = (\mathbf{u} \times \mathbf{v}) \cdot (s\mathbf{u}) + (\mathbf{u} \times \mathbf{v}) \cdot (t\mathbf{v}) = s(\mathbf{u} \times \mathbf{v}) \cdot \mathbf{u} + t(\mathbf{u} \times \mathbf{v}) \cdot \mathbf{v} = (s)(0) + (t)(0) = 0$.

■ 1.4 Lines and Planes

1. **Descriptions of Lines.** (a) $\mathbf{x} = (1, 2) + t(-1, 1) = (1 - t, 2 + t)$, $t \in \mathbb{R}$.

(b) $\mathbf{x} = (-1, 3, 2) + t(0, -1, 1) = (-1, 3 - t, 2 + t)$, $t \in \mathbb{R}$.

(c) Two lines are parallel if their direction vectors are scalar multiples of one another. Thus $(3, 1, 2)$ is a direction vector for the line through P: $\mathbf{x} = (3, 2, 0) + t(3, 1, 2) = (3 + 3t, 2 + t, 2t)$, $t \in \mathbb{R}$.

3. **Direction Vectors of a Line.** The direction vector from $\mathbf{x}$ to $\mathbf{y}$ is $(y_1 - x_1, y_2 - x_2)$, and the direction vector from $\tilde{\mathbf{x}}$ to $\tilde{\mathbf{y}}$ is $(\tilde{y}_1 - \tilde{x}_1, \tilde{y}_2 - \tilde{x}_2)$. Since $m = \frac{y_2 - x_2}{y_1 - x_1}$ and $m = \frac{\tilde{y}_2 - \tilde{x}_2}{\tilde{y}_1 - \tilde{x}_1}$, the direction vector from $\tilde{\mathbf{x}}$ to $\tilde{\mathbf{y}}$ can be rewritten as follows: $(\tilde{y}_1 - \tilde{x}_1, \tilde{y}_2 - \tilde{x}_2) = (\tilde{y}_1 - \tilde{x}_1, m(\tilde{y}_1 - \tilde{x}_1)) = \frac{\tilde{y}_1 - \tilde{x}_1}{y_1 - x_1}(y_1 - x_1, m(y_1 - x_1)) = \frac{\tilde{y}_1 - \tilde{x}_1}{y_1 - x_1}(y_1 - x_1, y_2 - x_2)$. Thus the direction vector from $\tilde{\mathbf{x}}$ to $\tilde{\mathbf{y}}$ is the scalar product of $\frac{\tilde{y}_1 - \tilde{x}_1}{y_1 - x_1}$ and the direction vector from $\mathbf{x}$ to $\mathbf{y}$.

5. **Parallel Lines.** (a) From Exercise 4, if $v_1 \neq 0$, then points on L satisfy a point-slope equation with $m = v_2/v_1$. Similarly, points on $\tilde{L}$ satisfy a point-slope equation with $\tilde{m} = (cv_2)/(cv_1) = v_2/v_1 = m$, so that L and $\tilde{L}$ have the same slope. If $v_1 = 0$, then $cv_1 = 0$ and both lines are vertical.
(b) If both lines have the same slope and v_1 and $\tilde{v}_1 \neq 0$, then $\tilde{v}_1 = cv_1$, where $c = \tilde{v}_1/v_1$. Since $\tilde{v}_2/\tilde{v}_1 = m$ and $v_2/v_1 = m$, $\tilde{v}_2 = m\tilde{v}_1 = mcv_1 = cv_2$. It follows that $\tilde{\mathbf{v}} = c\mathbf{v}$. If both lines are vertical, then $v_1 = \tilde{v}_1 = 0$, $v_2 \neq 0$, and $\tilde{v}_2 \neq 0$. It follows that $\tilde{\mathbf{v}} = c\mathbf{v}$, where $c = \tilde{v}_2/v_2$.

7. **Intersecting Lines.** (a) If L_1 is the line through P with direction vector $\mathbf{v}$, then points on L_1 can be written in the form $\mathbf{p} + s\mathbf{v}$, where s is a real number. Similarly, if L_2 is the line through Q with direction vector $\mathbf{w}$, then points on L_2 can be written in the form $\mathbf{q} + t\mathbf{w}$, where t is a real number. The lines intersect if there are values for s and t that satisfy $\mathbf{p} + s\mathbf{v} = \mathbf{q} + t\mathbf{w}$. We can attempt to find such values by solving the corresponding coordinate equations for s and t.

8. **Angles Between Lines.** (a) If $\mathbf{v}_1$ is the direction vector of L_1 and $\mathbf{v}_2$ is the direction vector of L_2, then the angle θ between the direction vectors satisfies $\cos\theta = \frac{\mathbf{v}_1 \cdot \mathbf{v}_2}{\|\mathbf{v}_1\|\|\mathbf{v}_2\|}$. Compute this value and find its inverse cosine, choosing the value θ for the inverse cosine with $0 \le \theta \le \pi$.

9. **Orthogonal Lines II.** (a) $(-1, 2, 1) \cdot (2, 1, -1) = -1 \neq 0$, so the lines are not orthogonal.
(b) $(0, -2, -2) \cdot (2, 1, -1) = 0$, so the lines are orthogonal if they intersect. Since $(-1, 3 - 2s, 2 - 2s) = (2t, 2 + t, 1 - t)$ has no solution for s and t, the lines do not intersect and so are not orthogonal.
(c) $(-1, 2, 1) \cdot (1, 1, -1) = 0$, so the lines are orthogonal if they intersect. Solving $(2 - s, 1 + 2s, 3 + s) = (t, 5 + t, 5 - t)$ for s and t yields $s = 2$ and $t = 0$, so the lines intersect at $(0, 5, 5)$ and are orthogonal.

11. **Equations of Planes.** (a) $x + z = 3$.
(b) $-x + 2y + z = -2$.
(c) The normal vector to this plane is $\mathbf{n} = \mathbf{u} \times \mathbf{v} = (0, 1, 1)$, so the coordinate equation is $y + z = 3$.
(d) The plane contains the vectors $\mathbf{u} = \overrightarrow{PQ} = (-3, 3, 0)$ and $\mathbf{v} = \overrightarrow{PR} = (0, -1, -2)$, so that a normal vector is given by $\mathbf{n} = \mathbf{u} \times \mathbf{v} = (-6, -6, 3)$. The coordinate equation of the plane is $-6x - 6y + 3z = 0$.

13. **Distance of a Point from a Plane.** Use the formula from Example 1.20; the distance is $\left| \overrightarrow{QP} \cdot \frac{\mathbf{n}}{\|\mathbf{n}\|} \right|$, where $\mathbf{n}$ is the normal vector to the plane and P is a point on the plane.

   (a) $\mathbf{n} = (2, -1, 3)$, $P = (2, 0, 0)$, and the distance is $\frac{\sqrt{7}}{\sqrt{2}}$.

   (b) Q satisfies the equation for the plane, so it lies on the plane and the distance is zero.

   (c) $\mathbf{n} = (1, 1, -2)$, $P = (2, 0, 0)$, and the distance is $\frac{\sqrt{3}}{\sqrt{2}}$.

CHAPTER 2 SOLUTIONS

■ 2.1 Parametric Representations of Curves

---

1. **Line Segments.** (a) $\alpha(t) = (1, 0) + t(-1, 1) = (1 - t, t)$.

   (b) $\alpha(t) = (1, 0, 1) + \frac{1}{2}t(-1, 1, -2) = (1 - \frac{1}{2}t, \frac{1}{2}t, 1 - t)$.

3. **Arcs of Circles.** (a) α parametrizes the unit circle centered at the origin from $(\sqrt{2}/2, \sqrt{2}/2)$ to $(-\sqrt{2}/2, -\sqrt{2}/2)$ in a counterclockwise direction.

   (b) α parametrizes the top half of the unit circle centered at the origin from $(1, 0)$ to $(-1, 0)$ in a counterclockwise direction.

5. **Arcs of Ellipses.** (a) α parametrizes the portion of the ellipse $\frac{x^2}{4} + y^2 = 1$ from $(2, 0)$ to $(-\sqrt{2}, \frac{\sqrt{2}}{2})$ in a counterclockwise direction.

   (b) α parametrizes the portion of the ellipse $x^2 + 4y^2 = 1$ from $(-1, 0)$ to $(1, 0)$ in a counterclockwise direction.

7. **Circles and Ellipses.** (a) α parametrizes the portion of the unit circle centered at the origin in the first quadrant from $(0, 1)$ to $(1, 0)$ in a clockwise direction.

   (b) α parametrizes the portion of the unit circle centered at the origin from $(-1, 0)$ to $(0, -1)$ in a clockwise direction.

8. **Parametrizing Arcs of Circles.** (a) $\alpha(t) = (\cos(t + \pi/2), \sin(t + \pi/2))$.

   (b) $\alpha(t) = (\sin(\pi t - \pi/2), \cos(\pi t - \pi/2))$.

9. **Helices.** (a) The image of α is a portion of a helix centered on the z-axis lying over the unit circle in the xy-plane. The image traces the helix from the point $(1, 0, 0)$ to the point $(1, 0, 2\pi)$.

   (b) (i) The helix centered on the z-axis lying over the circle of radius 2 in the xy-plane traced from $(2, 0, 0)$ to $(2, 0, 2\pi)$.

11. **Graphs of Functions.** (a) Graph of $y = x$ for $x \geq 1$. (b) Graph of $y = x^2$ for $0 < x \leq 1$.

14. The Period of a Simple Pendulum. (a) The period of a pendulum is the length of time it takes for it to return to its starting position and velocity. If $\alpha(0) = (x_0, 0)$, $\alpha(t) = (x_0, 0)$ again if and only if $\sqrt{\frac{9.8}{l}}t + \pi/2 = 5\pi/2$. Thus the period is $t = 2\pi\sqrt{\frac{l}{9.8}}$.

(b) If we increase the length of the pendulum, we increase the amount of time it takes to return to its initial position and velocity; thus we "slow the pendulum down."

(c) The image of $\alpha(t)$ is an ellipse in the xv-plane with axes of length $2x_0$ and $2x_0\sqrt{9.8/l}$. Increasing the length of the pendulum decreases the length of the vertical axis of the image of α.

15. The Velocity of a Simple Pendulum. (a) Velocity and speed are zero at x_0 and $-x_0$. Velocity is negative when the pendulum is moving clockwise, positive when the pendulum is moving counterclockwise. The maximum speed occurs when the position is zero.

(b) The image of $\alpha(t)$ is an ellipse in the xv-plane with axes of length $2x_0$ and $2x_0\sqrt{9.8/l}$; thus changing x_0 changes both the horizontal and vertical axes of the image of α. Since the maximum and minimum positions of the pendulum are x_0 and $-x_0$, and the maximum and minimum velocities occur at the ends of the vertical axis of the image, changing x_0 changes both the maximum and minimum positions and the maximum and minimum velocities of the pendulum. The period of the pendulum is $2\pi\sqrt{\frac{l}{9.8}}$; thus changing x_0 does not change the period of the pendulum.

■ 2.2 The Derivative of a Parametrization

1. Velocity and Speed. (a) $\mathbf{v}(t) = (e^t, 3e^{3t})$, $\mathbf{v}(0) = (1, 3)$, $\|\mathbf{v}(0)\| = \sqrt{10}$.

(b) $\mathbf{v}(t) = (-2\sin t, 3\cos t)$, $\mathbf{v}(0) = (-2, 0)$, $\|\mathbf{v}(0)\| = 2$.

2. Tangent Lines. (a) $\mathbf{x}(t) = (1, 3) + t(-2, 0) = (1 - 2t, 3)$.

(b) $\mathbf{x}(t) = (e^2, e) + t(3e^2, 2e) = (e^2(1 + 3t), e(1 + 2t))$.

3. The Algebra of Differentiation. (a) $(c\alpha)'(t) = (cx(t), cy(t), cz(t))' = (cx'(t), cy'(t), cz'(t)) = c(\alpha'(t))$.

(b) $(\alpha + \beta)'(t) = (x_1(t) + x_2(t), y_1(t) + y_2(t), z_1(t) + z_2(t))' = ((x_1(t) + x_2(t))', (y_1(t) + y_2(t))', (z_1(t) + z_2(t))') = (x_1'(t) + x_2'(t), y_1'(t) + y_2'(t), z_1'(t) + z_2'(t)) = (\alpha'(t) + \beta'(t))$.

7. Constant Speed Motion. $(s(t))^2 = \mathbf{v}(t) \cdot \mathbf{v}(t) = k^2$. Differentiating, we have $2\mathbf{v}(t) \cdot \mathbf{a}(t) = 0$. Thus the angle between the velocity vector $\mathbf{v}(t)$ and the acceleration vector $\mathbf{a}(t)$ is $\pi/2$.

8. Projectile Motion. (a) If the initial position is $(0, 0)$ and the initial velocity is $(5, 10)$, the motion is parametrized by $\alpha(t) = (5t, -9.8t^2/2 + 10t)$. The object is on the ground when $-9.8t^2/2 + 10t = 0$, which occurs when $t = 0$ and $t = 10/4.9 \approx 2.04$. Thus the object is in the air for approximately 2.04 s. The maximum height occurs when the derivative of $-9.8t^2/2 + 10t$ is zero, thus $-9.8t + 10 = 0$, $t = 10/9.8 \approx 1.02$ s. The height at this time is $\frac{-9.8}{2}\left(\frac{10}{9.8}\right)^2 + 10\left(\frac{10}{9.8}\right) = 50/9.8 \approx 5.10$ m. The horizontal distance traveled is $5(10/4.9) = 50/4.9 \approx 10.4$ m.

(b) If the initial speed is 50 m/s, the initial velocity is $(50\cos\theta, 50\sin\theta)$, where θ is the angle at which the object is thrown. The motion is parametrized by $((50\cos\theta)t, -9.8t^2/2 + (50\sin\theta)t)$. The object reaches its maximum height when $-9.8t + 50\sin\theta = 0$, that is, when $t = (50\sin\theta)/9.8$. If the maximum height is 20, we must have $20 = \frac{-9.8}{2}\left(\frac{50\sin\theta}{9.8}\right)^2 + 50\sin\theta\left(\frac{50\sin\theta}{9.8}\right)$. Thus $20 = 1250\sin^2\theta/9.8$, $\sin\theta = 0.40$, and $\theta = 0.41$. The object is on the ground when $t = 100\sin\theta/9.8 \approx 4.04$ s. The horizontal distance traveled is $(50\cos\theta)4.04 \approx 185.5$ m.

■ 2.3 Modeling with Parametric Curves

1. **Forward-Velocity Kinematics.** (a) The x-coordinate of the position of the effector is increasing since $x'(t) = 2 + \cos(t) > 0$ for all t. The y-coordinate of effector increases for $0 < t < \pi$, then decreases for $\pi < t < 2\pi$, since $y'(t) = \sin(t)$. It reaches maximum height at $t = \pi$.
(c) The motion is similar to that of (a).

2. **Inverse-Velocity Kinematics I.** (a) Since v_0 is larger, g must be smaller. That is, if the horizontal motion of the slider is greater, the hinge must rotate less rapidly. The final position of O would be to the right of 1.3, since it would take longer for the effector to rotate up from height 0 to height 1.

3. **Inverse-Kinematics Velocity II.** (a) Notice the point $(\frac{1}{2}, \frac{1}{2})$ lies inside the circle of radius 1 centered at $(0, 1)$. Thus if the angle $\angle OPQ$ is to remain less than $\frac{\pi}{2}$, the slider must move to the left. Therefore, f' should be negative and g' should be positive.

6. **A Cartesian Robot.** (b) This is a right-handed coordinate system. To see this, translate the arrows labeled x, y, and z so they have a common base point.
(c) Parametrize the motion by $\alpha(t) = (z_1, x_1, y_1) + t(v_0/\|(z_1, x_2, y_2) - (z_1, x_1, y_1)\|)$ $((z_1, x_2, y_2) - (z_1, x_1, y_1))$. Then $\|\alpha'(t)\| = v_0$.

7. **A SCARA Robot I.** (a) (i) $R(0, \frac{\pi}{2}) = (1, 1)$. (b) $R(\theta, 2\theta) = (\cos\theta - \cos(-\theta), \sin\theta - \sin(-\theta)) = (\cos\theta - \cos\theta, \sin\theta + \sin\theta) = (0, 2\sin\theta)$, which lies on the y-axis.

9. **When Does an Epidemic End?** (a) Even though there are still susceptible individuals, the epidemic ends because the number of infected individuals is zero. As the susceptible population decreases, the rate at which susceptible individuals become infected is exceeded by the rate at which infected individuals recover, thus causing the infected population to decrease to zero.
(b) In general, not all susceptible people become infected because they do not all have contact with infected people.

(c) This implies that not everyone will be exposed to childhood diseases like chicken pox and measles during childhood. Since they are not infected, they will not build up an immunity to the disease. This is a problem because these diseases are usually much·more serious if contracted in adulthood instead of childhood.

10. **Initial Conditions.** (a) These curves model a disease that begins with a relatively small number of infected individuals and a large number of susceptible individuals. The epidemic increases as susceptible individuals come into contact with infected individuals. After the number of susceptible individuals drops below approximately 200, the epidemic begins to subside; that is, the number of infected individuals begins to decrease. The epidemic ends when the number of infected individuals reaches zero, even though there are still susceptible individuals in the population.
(b) The effect of changing the number of students initially infected is that we change the number of individuals who are infected at the peak of the epidemic and the number of susceptible individuals in the population at the end of the epidemic.
(c) For each of these curves, the epidemic begins to subside when $I'(t) = 0$. This occurs when $I(t) = 0$ or $rS(t) - a = 0$. Thus, in each case, the epidemic subsides when $S(t) = \frac{a}{r}$, that is, when the susceptible population is $\frac{4.4 \times 10^{-1}}{2.18 \times 10^{-3}} \approx 202$.

■ 2.4 Vector Fields

2. **Verifying Parametrizations of Flow Lines.** (a) (i) $\alpha'(t) = (3e^{3t}, 2e^{2t}) = \mathbf{F}(\alpha(t))$, $\alpha(0) = (1, 1)$. (ii) $\alpha'(t) = (-6e^{3t}, 8e^{2t}) = \mathbf{F}(\alpha(t))$, $\alpha(0) = (-2, 4)$.

4. **Flow Lines and Initial Points.** (a) $\alpha'(t) = (-Ae^{-t}, 2Be^{2t}) = \mathbf{F}(\alpha(t))$, $\alpha(0) = (A, B)$. (i) With initial point $(2, 1)$, $A = 2$, $B = 1$. (ii) With initial point $(-3, 3)$, $A = -3$, $B = 3$.
(b) $\alpha'(t) = (Ae^t, Be^t) = \mathbf{F}(\alpha(t))$, $\alpha(0) = (A - 2, B - 1)$. (i) With initial point $(1, 1)$, $A = 3$, $B = 2$. (ii) With initial point $(0, 1)$, $A = 2$, $B = 2$.

5. **Critical Points of Vector Fields.** (a) The critical point at $(\frac{1}{3}, \frac{1}{3})$ is a saddle.
(b) The critical point at $(0, 0)$ is a sink and the critical point at $(1, 1)$ is a saddle.
(c) The critical points are located at $(\frac{k\pi}{2}, 0)$ and $(0, \frac{k\pi}{2})$, where k is an odd integer. From the plot, we see that $(\pm\frac{\pi}{2}, 0)$ are saddle points, $(0, -\frac{\pi}{2})$ is a sink, and $(0, \frac{\pi}{2})$ is a source.

7. **Shifting and Scaling.** (a) $\beta'(t) = (-R\sin(t + k), R\cos(t + k)) = \mathbf{F}(\beta(t))$. Thus β is a flow line.
(b) $\beta'(t) = (-Rc\sin(ct), cR\cos(ct)) = c\mathbf{F}(\beta(t))$. If $c \neq 0, 1$ then β is not a flow line.

9. **The Vector Field $-\mathbf{F}$.** (a) The vectors of $\mathbf{F}$ and $\mathbf{G}$ have the same length but point in opposite directions.
(b) For any given point (x_0, y_0) in the plane, the flow line for $\mathbf{F}$ that passes through the point and the flow line for $-\mathbf{F}$ that passes through this point trace the same curve but in opposite directions.

(c) The only critical point of $\mathbf{F}$ is $(0,0)$, and it is a source. $(0,0)$ is also the only critical point of $-\mathbf{F}$, but in this case it is a sink.

■ 2.5 Modeling with Vector Fields

---

1. **Extreme Values of the Populations.** (a) The points where the prey population is maximized or minimized are the points where the vectors in the vector field are vertical, that is, when $x = 0$ or $y = a/b$. For the nonconstant flow lines, $x \neq 0$; thus the maximum and minimum prey populations occur when the predator population is a/b. Note that $ax - bxy$ is the expression for $x'(t)$; thus we are in fact computing the points where $x'(t) = 0$.

   (b) The points where the predator population is maximized or minimized are the points where the vectors in the vector field are horizontal, that is, when $y = 0$ or $x = c/d$. For the nonconstant flow lines, $y \neq 0$; thus the maximum and minimum predator populations occur when the prey population is c/d. Note that $-cy + dxy$ is the expression for $y'(t)$; thus we are in fact computing the points where $y'(t) = 0$.

   (c) We cannot find both coordinates of these points exactly. We know the y-coordinate, that is, the predator population, at the points where the prey population is maximized and minimized and the x-coordinate, that is, the prey population, at the points where the predator population is maximized and minimized. The actual maximum and minimum prey and predator populations at these points will depend on which flow line we are on, thus on the initial values of the populations.

   (d) The equilibrium point of the system is $(c/d, a/b)$. Thus the prey population at the equilibrium point is the same as the prey population at the point where the predator population is maximized and minimized. The predator population at the equilibrium point is the same as the predator population at the point where the prey population is maximized and minimized.

2. **Average Population Values.** (a) The predator population and the prey population vary more dramatically in the case when the prey population is initially 500 as opposed to the case when the prey population is initially 2000. It seems that if we start with an initial population near the equilibrium point of $(3000, 150)$, the populations remain near the equilibrium point, but if we start far away, the populations remain far away.

   (b) The average prey population is $x_{\text{avg}} = \frac{1}{dT} \int_0^T \left(\frac{y'(t)}{y(t)} + c \right) dt = \frac{1}{dT} \left(\ln(y(t)) + ct \right) \big|_0^T = \frac{1}{dT} \left(\ln(y(T)) - \ln(y(0)) + cT \right)$. Since the cycle begins and ends at the same point, $y(T) = y(0)$. Thus the integral evaluates to c/d.

   (c) The average predator population is $y_{\text{avg}} = \frac{1}{bT} \int_0^T \left(\frac{x'(t)}{x(t)} + a \right) dt = \frac{1}{bT} \left(\ln(x(t)) + at \right) \big|_0^T = \frac{1}{bT} \left(\ln(x(T)) - \ln(x(0)) + aT \right)$. Since the cycle begins and ends at the same point, $x(T) = x(0)$. Thus the integral evaluates to a/b.

(d) These values are consistent with the equilibrium point. If we begin at $(c/d, a/b)$, we will remain there for the entire time. Thus the average values in this case are c/d and a/b. If we begin at another point, the population values cycle around this point, keeping the average constant.

5. **The Tangential Component of Force.** (a) Differentiating the equation $\beta(t) \cdot \beta(t) = l^2$, we have $2\beta(t) \cdot \beta'(t) = 0$.
(b) $\beta(t) = (l\cos((x(t)/l) - \pi/2), l\sin((x(t)/l) - \pi/2))$ and $\beta'(t) = (-\sin((x(t)/l) - \pi/2)x'(t), \cos((x(t)/l) - \pi/2)x'(t))$.
$\beta''(t) = (\frac{-1}{l}\cos((x(t)/l) - \pi/2)(x'(t))^2 - \sin((x(t)/l) - \pi/2)x''(t), \frac{-1}{l}\sin((x(t)/l) - \pi/2)(x'(t))^2 + \cos((x(t)/l) - \pi/2)x''(t)) = -(\frac{x'(t)}{l})^2\beta(t) + \frac{x''(t)}{x'(t)}\beta'(t)$, if $x'(t) \neq 0$.

8. **Nonclosed Flow Lines.** Not all the flow lines of this vector field are closed curves. If the pendulum is given a sufficiently large initial velocity, the flow lines form curves that are sinusoidal. This corresponds to the pendulum swinging "over the top" in multiple revolutions.

CHAPTER 3 SOLUTIONS

■ 3.1 Representation and Graphical Analysis of Functions

1. **Worth Mountain.** (a) The portion of the hike up Sucker Brook Trail from the parking area to the intersection with Long Trail is gradual for the first 5/6 mi and then becomes steeper: For the first 1/2 mile, the elevation increases from 2020 ft to 2100 ft; for the next 1/3 mile, the elevation increases from 2100 ft to 2300 ft; and for the last 1/4 mile, the elevation increases from 2300 ft to 2680 ft.
(b) The trail begins at an elevation of 2680 ft and alternately climbs and descends three lesser peaks, passing through three saddles before reaching the summit of Worth Mountain. The trail ascends 1/10 mi to Sucker Brook Shelter at approximately 2700 ft; descends briefly to a saddle at 2660 ft; ascends to a local maximum at 2700 ft 1/10 mi after the first peak and approximately 80 ft below an unnamed peak; descends for 1/3 mi to a second saddle at 2460 ft before resuming the 4/10-mi climb to the third unnamed peak at 2828 ft; descends for 1/5 mi to a saddle at 2800 ft; and then climbs to the summit of Worth Mountain in the remaining 1/2 mi. This last climb is the steepest portion of the hike.

3. **A Perfect Gas.** (a) The vertical slices for $V = V_0$ satisfy $T = \frac{1}{8.3145}pV_0$, which is the equation of a line passing through $(p, T) = (0, 0)$ with slope $\frac{1}{8.3145}V_0$. Thus for fixed volume, temperature is a linear function of pressure.

(b) The vertical slices for $p = p_0$ satisfy $T = \frac{1}{8.3145} p_0 V$, which is the equation of a line passing through $(V, T) = (0, 0)$ with slope $\frac{1}{8.3145} p_0$. Thus for fixed pressure, temperature is a linear function of volume.

(c) The contour curve for $T = T_0$ satisfies $T_0 = \frac{1}{8.3145} p V$, which is the equation for a hyperbola. Since $p > 0$ and $V > 0$, the contour curve for $T = T_0$ is the branch of this hyperbola lying in the first quadrant. It satisfies $V = 8.3145 T_0 \frac{1}{p}$.

(d) The graph of g is the portion of a saddle lying over the first quadrant. The saddle point is located at the origin.

5. **A Density Plot.** (a) In each case, we might use six different shades of increasing brightness to represent the six values appearing, the data set, 0, 0.5, 1, 1.5, 2, and 3. In the 6×6 case, shade each pixel according to the value at the upper right vertex of the pixel. For example, the pixel $[0, 1] \times [0, 1]$ should be shaded for the value at $(1, 1)$, which is 1. For the 3×3 case, we might shade each pixel according to the data point in the middle of the pixel. For example, the pixel $[0, 2] \times [0, 2]$ should be shaded for the value at $(1, 1)$, which is 1. We might use similar systematic assignments of values in the other two cases. Notice, however, that these assignments are not unique; there are many other ways to assign values to pixels.

(b) The plots are not identical because of the differing number of pixels and the necessarily different schemes for assigning values.

8. **Extreme Points and Saddles.** (a) There is a local maximum located approximately at $(-0.6, 0)$ with value $f(-0.6, 0) = 0.384$; there is a saddle located approximately at $(0.6, 0)$ with value $f(0.6, 0) = -0.384$. The global maximum on the domain $[-1.5, 1.5] \times [-1, 1]$ occurs at $(1.5, 0)$ and is equal to $f(1.5, 0) = 1.875$; the global minimum occurs at $(-1.5, 0)$ and is equal to $f(-1.5, 0) = -1.875$.

(b) There are saddle points at approximately $(0, -1.6)$, $(-3.1, 1.6)$, and $(3.1, 1.6)$, which have function values of zero. The extreme points are at $(0, 1.6)$, $(-3.1, -1.6)$, and $(3.1, -1.6)$, with function values of 2, -2, and -2, respectively. Thus $(0, 1.6)$ is a relative maximum, and the other two points are relative minima.

(c) There is a saddle point at $(0, 0)$ with function value zero. The global maximum on the domain $[-2, 2] \times [-2, 2]$ occurs at $(-2, 2)$ and $(-2, -2)$ with function value 16; the global minimum occurs at $(2, 2)$ and $(2, -2)$ with function value -16. There is also a local maximum at $(2, 0)$ with value 8 and a local minimum at $(-2, 0)$ with value -8. The graph of this function is known as a *monkey saddle*. Why?

9. **Contour Plots.** (a) The contour curves are parallel lines satisfying equations of the form $y = \frac{1}{2} x - \frac{1}{2} z_0$, which is the equation of a line with slope $\frac{1}{2}$ and y-intercept $\frac{1}{2} z_0$.

(b) The contour curves are concentric ellipses centered at the origin with major axis lying along the x-axis and minor axis lying along the y-axis.

10. **Comparing Plots.** The graph shows more clearly the behavior of the function inside the three innermost contours on the contour plot. Not only is there a maximum located near $(1, 0)$, but there appears to be a local maximum located at $(0, 0)$ and a saddle point located approximately at $(0.25, 0)$.

12. The Method of Slices. (a) The vertical slice over the x-axis is $z = f(x, 0) = x^2$, which is a parabola that is concave up. The vertical slice over the y-axis is $z = f(0, y) = 2y^2$, which is also a concave up parabola. The level sets satisfy $x^2 + 2y^2 = c$ and are ellipses for $c > 0$ with major axis lying along the x-axis and minor axis lying along the y-axis. (b) The vertical slice over the x-axis is $z = -x^2$, which is a concave down parabola with maximum at $x = 0$. The vertical slice over the y-axis is $z = 3y$, which is a line passing through the origin with slope 3. The level sets satisfy $y = \frac{1}{3}x^2 + \frac{1}{3}c$, which is the equation for a concave up parabola with minimum value of $\frac{1}{3}c$ at $x = 0$.

14. Contour Surfaces. (a) The contour surfaces are planes parallel to the plane given by $x - 2y + z = 0$ and passing through the points $(0, 0, c)$. (b) The contour surfaces are the origin for $c = 0$ and concentric spheres of radius $\sqrt{c}$ for the other values of c.

15. Acoustic Monopole. (a) The collection of isobars remains the same, but the value attached to each isobar changes. For example, the sphere of radius 1 is an isobar for any A and the value is A. (b) Yes. Assuming the source produces the same intensity sound in all directions, say like a bell rather than the human voice, and the sound is directional, we would expect that the sound would radiate in the same manner in all directions and that as the distance from the source increases, the intensity of the sound would decrease.

■ 3.2 Directional and Partial Derivatives

1. Temperature in a Plate. For each point, estimate the direction in which the shading is changing most rapidly: $\frac{1}{\sqrt{2}}(-1, -1)$ for $(1, 1)$; $\frac{1}{\sqrt{2}}(1, -1)$ for $(1, -1)$; $\frac{1}{\sqrt{5}}(-1, 2)$ for $(-1, 1)$; and $\frac{1}{\sqrt{5}}(2, 1)$ for $(-1, -1)$.

2. Rates of Change by Isotherms. Using the same method as in Exercise 1, the temperature increases most rapidly in the direction $\frac{1}{\sqrt{5}}(-2, -1)$ at $(2, 2)$; $\frac{1}{\sqrt{17}}(1, 4)$ at $(2, -2)$; $\frac{1}{\sqrt{17}}(4, 1)$ at $(-2, 2)$; and $(-1, 0)$ at $(-2, -2)$.

6. Directional Derivative Calculations. For the given f and $\mathbf{u}$, compute the derivative with respect to t at $t = 0$ of the function $g(t) = f(\mathbf{x}_0 + t\mathbf{u})$.
(a) $g'(0) = x_0 u_2 + y_0 u_1$, $D_{(1/\sqrt{2}, 1/\sqrt{2})}f(\mathbf{x}_0) = \frac{x_0}{\sqrt{2}} + \frac{y_0}{\sqrt{2}}$, and $D_{(0,1)}f(\mathbf{x}_0) = x_0$.
(b) $g'(0) = \frac{x_0 u_2 - y_0 u_1}{x_0^2}$, $D_{(1/\sqrt{2}, 1/\sqrt{2})}f(\mathbf{x}_0) = \frac{x_0 - y_0}{\sqrt{2}x_0^2}$, and $D_{(0,1)}f(\mathbf{x}_0) = \frac{1}{x_0}$.

7. Directional Derivatives in Space. For the given f and $\mathbf{u}$, compute the derivative with respect to t at $t = 0$ of the function $g(t) = f(\mathbf{x}_0 + t\mathbf{u})$.
(a) $g'(0) = u_0 + 2y_0 z_0 u_2 + y_0^2 u_3$ and $D_{\mathbf{u}}f(x_0) = \frac{1}{\sqrt{2}}(1 - y_0^2)$.
(b) $g'(0) = y_0 z_0 u_1 + x_0 z_0 u_2 + x_0 y_0 u_3$ and $D_{\mathbf{u}}f(x_0) = \frac{x_0 z_0}{\sqrt{5}} + \frac{2x_0 y_0}{\sqrt{5}}$.

8. Directions of Increase and Decrease in $\mathbb{R}^2$. In each case, determine in which directions $D_{\mathbf{u}}f(\mathbf{x}_0)$ is positive, negative, or zero.
(a) $D_{\mathbf{u}}f(2,1) = u_1 + 2u_2 = (1,2) \cdot \mathbf{u}$. Thus f is increasing if the angle between $(1,2)$ and $\mathbf{u}$ is less than $\pi/2$, decreasing if this angle is between $\pi/2$ and π, and not changing in the directions of $\pm(2,-1)$.
(b) $D_{\mathbf{u}}f(1,-1) = -u_1 + u_2 = (-1,1) \cdot \mathbf{u}$. Thus f is increasing if the angle between $(-1,1)$ and $\mathbf{u}$ is less than $\pi/2$, decreasing if this angle is between $\pi/2$ and π, and not changing in the directions of $\pm(1,1)$.
(c) $D_{\mathbf{u}}f(1,0) = 6u_1$. Thus f is increasing if $u_1 > 0$, decreasing if $u_1 < 0$, and not changing if $u_1 = 0$.

10. Partial Derivatives. (a) $\frac{\partial f}{\partial x}(x,y) = 2xy$, $\frac{\partial f}{\partial y}(x,y) = x^2 + 3y^2$, $\frac{\partial^2 f}{\partial x^2}(x,y) = 2y$, $\frac{\partial^2 f}{\partial y \partial x}(x,y) = 2x$, and $\frac{\partial^2 f}{\partial y^2}(x,y) = 6y$.
(b) $\frac{\partial f}{\partial x}(x,y) = ye^{xy}$, $\frac{\partial f}{\partial y}(x,y) = xe^{xy}$, $\frac{\partial^2 f}{\partial x^2}(x,y) = y^2 e^{xy}$, $\frac{\partial^2 f}{\partial y \partial x}(x,y) = e^{xy}(1 + xy)$, and $\frac{\partial^2 f}{\partial y^2}(x,y) = x^2 e^{xy}$.

13. Properties of $D_{\mathbf{u}}$. (a) Using the additivity of limits, $D_{\mathbf{u}}(f + g)(\mathbf{x}_0) = \lim_{h \to 0} \frac{(f(\mathbf{x}_0 + h\mathbf{u}) + g(\mathbf{x}_0 + h\mathbf{u})) - (f(\mathbf{x}_0) + g(\mathbf{x}_0))}{h} = \lim_{h \to 0} \left(\frac{f(\mathbf{x}_0 + h\mathbf{u}) - f(\mathbf{x}_0)}{h} + \frac{g(\mathbf{x}_0 + h\mathbf{u}) - g(\mathbf{x}_0)}{h} \right) = \lim_{h \to 0} \frac{f(\mathbf{x}_0 + h\mathbf{u}) - f(\mathbf{x}_0)}{h} + \lim_{h \to 0} \frac{g(\mathbf{x}_0 + h\mathbf{u}) - g(\mathbf{x}_0)}{h} = D_{\mathbf{u}}(f)(\mathbf{x}_0) + D_{\mathbf{u}}(g)(\mathbf{x}_0)$.
(b) Using the linearity of limits, $D_{\mathbf{u}}(af)(\mathbf{x}_0) = \lim_{h \to 0} \frac{af(\mathbf{x}_0 + h\mathbf{u}) - af(\mathbf{x}_0)}{h} = \lim_{h \to 0} a\frac{f(\mathbf{x}_0 + h\mathbf{u}) - f(\mathbf{x}_0)}{h} = a \lim_{h \to 0} \frac{f(\mathbf{x}_0 + h\mathbf{u}) - f(\mathbf{x}_0)}{h} = aD_{\mathbf{u}}(f)(\mathbf{x}_0)$.

15. Using the Definition. (a) For all points $(x,y) \neq (0,0)$, $\frac{\partial f}{\partial x}(x,y) = \frac{(x^2 + y^2)y - xy(2x)}{(x^2 + y^2)^2} = \frac{y^3 - x^2 y}{(x^2 + y^2)^2}$, and $\frac{\partial f}{\partial y}(x,y) = \frac{(x^2 + y^2)x - xy(2y)}{(x^2 + y^2)^2} = \frac{x^3 - xy^2}{(x^2 + y^2)^2}$.
(b) To compute $\frac{\partial f}{\partial x}$ and $\frac{\partial f}{\partial y}$ at $(0,0)$, we use the definitions and the fact that $f(0,0) = 0$: $\frac{\partial f}{\partial x}(0,0) = \lim_{h \to 0} \frac{f(h,0) - f(0,0)}{h} = \lim_{h \to 0} \frac{0 - 0}{h} = 0$, and $\frac{\partial f}{\partial y}(0,0) = \lim_{h \to 0} \frac{f(0,h) - f(0,0)}{h} = \lim_{h \to 0} \frac{0 - 0}{h} = 0$.

■ 3.3 Limits and Continuous Functions

1. Constructing Open Balls. (a) The line $y = x$ divides the plane into two half-planes. The set $\mathcal{O}$ is the lower of these half-planes (which contains the positive x-axis).
(b) Any $r > 0$ that is less than the distance from P to the line $y = x$ will suffice. Since this distance is $\frac{3}{\sqrt{2}}$, choose, for example, $r = 1$.

3. **Open Sets in** $\mathbb{R}^2$. (a) This set is neither open nor closed. It is not open because it is does not contain an open ball centered at the point $(1, 0)$ or at any other point on the unit circle. It is not closed because it does not contain the origin.
(b) This set is closed. The boundary is the circle $x^2 + y^2 = 1$, which is contained in $\mathcal{M}$.

4. **The Boundary.** (a) The boundary consists of the vertical line $x = 2$.
(b) The boundary consists of the portion of the vertical line $x = 2$ with $y \leq 1$ and the portion of the horizontal line $y = 1$ with $x \geq 2$.

5. **Closed Sets.** The sets $\mathcal{S}$ in parts (a), (c), (d), and (f) are closed sets because they contain their boundaries. The sets $\mathcal{S}$ in parts (b) and (e) are not closed sets because they do not contain all of their boundaries.

7. **Open Sets in** $\mathbb{R}^3$. (a) Intuitively, $\mathcal{N}$ is open because it is defined by strict inequalities. More rigorously, for $(x_0, y_0, z_0) \in \mathcal{N}$, choose $r = \frac{1}{2} \min(x_0, y_0)$. Then the ball of radius r centered at (x_0, y_0, z_0) is contained in $\mathcal{N}$, and we conclude that $\mathcal{N}$ is open.
(b) Intuitively, $\mathcal{N}$ is neither open nor closed because it is defined by both a strict inequality, $-2 < z$, which means it is not closed, and a nonstrict inequality, $z \leq 1$, which means it is not open. That is, it is not closed because it does not contain the plane $z = -2$, which forms part of its boundary, and it is not open because it does not contain any open ball centered at points of the form $(x, y, 1)$.
(c) Intuitively, $\mathcal{N}$ is an open set because it is defined by a strict inequality, $x^2 + y^2 > 0$. More rigorously, for $(x_0, y_0, z_0) \in \mathcal{N}$, choose $r = \frac{1}{2} \sqrt{x_0^2 + y_0^2}$. Then the ball of radius r centered at (x_0, y_0, z_0) is contained in $\mathcal{N}$, and we conclude that $\mathcal{N}$ is open.

8. **The Boundary of a Subset of** $\mathbb{R}^3$. (a) The boundary is the horizontal plane $z = 1$, $\{(x, y, z) : z = 1\}$.
(b) The boundary is the horizontal plane $z = 1$, $\{(x, y, z) : z = 1\}$.

9. **A Limit That Does Not Exist.** (a) Since $h\frac{1}{\sqrt{2}} \cdot h\frac{1}{\sqrt{2}} > 0$, $f(h\mathbf{u}) = 1$ for $h \neq 0$. Therefore, $\lim_{h \to 0} f(h\mathbf{u}) = 1$.
(b) Since $h\frac{-1}{\sqrt{2}} \cdot h\frac{1}{\sqrt{2}} < 0$, $f(h\mathbf{u}) = 0$ for $h \neq 0$. Therefore, $\lim_{h \to 0} f(h\mathbf{u}) = 0$.
(c) If $\lim_{\mathbf{x} \to 0} f(\mathbf{x})$ exists, then we will obtain the limit no matter how $\mathbf{x}$ approaches $\mathbf{0}$. Parts (a) and (b) show that this is not the case; thus we conclude that the limit does not exist.

■ 3.4 Differentiable Functions

1. **Differentiability.** (a) To show that f is differentiable on $\mathcal{O}$, it would be sufficient to compute the partial derivatives of f and show they exist and are continuous everywhere on $\mathcal{O}$.

(b) The domain of f is all of $\mathbb{R}^2$. The partial derivatives of f are $\frac{\partial f}{\partial x}(x, y) = \cos(x + y)$ and $\frac{\partial f}{\partial y}(x, y) = \cos(x + y)$. These are also defined on all of $\mathbb{R}^2$. Since $x + y$ is a continuous function on $\mathbb{R}^2$ and cosine is a continuous function on $\mathbb{R}$, their composition, $\cos(x + y)$, is also continuous on $\mathbb{R}^2$. We conclude that f is differentiable on $\mathbb{R}^2$.

2. **Linear Approximations in $\mathbb{R}^2$.** (a) $f(1, -1) = 4$, $\frac{\partial f}{\partial x}(1, -1) = 7$, and $\frac{\partial f}{\partial y}(1, -1) = -1$, so that $l(x, y) = 7(x - 1) - 1(y + 1) + 4$.
   (b) $f(\pi/2, \pi/2) = 0$, $\frac{\partial f}{\partial x}(\pi/2, \pi/2) = -1$, and $\frac{\partial f}{\partial y}(\pi/2, \pi/2) = -1$, so that $l(x, y) = -(x - \pi/2) - (y - \pi/2)$.

3. **Linear Approximations in $\mathbb{R}^3$.** (a) $f(0, 1, -1) = 0$, $\frac{\partial f}{\partial x}(0, 1, -1) = 1$, $\frac{\partial f}{\partial y}(0, 1, -1) = 0$, and $\frac{\partial f}{\partial z}(0, 1, -1) = 0$, so that $l(x, y, z) = x$.
   (b) $f(0, 0, 0) = 1$, $\frac{\partial f}{\partial x}(0, 0, 0) = 0$, $\frac{\partial f}{\partial y}(0, 0, 0) = 0$, and $\frac{\partial f}{\partial z}(0, 0, 0) = 0$, so that $l(x, y, z) = 1$.

4. **Differentiable?** (a) The function f is a polynomial, so it is continuous at $\mathbf{x}_0$. Its partial derivatives, $\frac{\partial f}{\partial x}(x, y) = 2x + 3y$ and $\frac{\partial f}{\partial y}(x, y) = 3x + 4y$, are defined and continuous at $\mathbf{x}_0$, so that f is differentiable at $\mathbf{x}_0$.
   (b) The function f is the product of continuous functions (notice that $x^{1/3}$ is continuous at the origin), so it is continuous at $\mathbf{x}_0$. However, $\frac{\partial f}{\partial x}(x, y) = \frac{y^{1/3}}{3x^{2/3}}$ and $\frac{\partial f}{\partial y}(x, y) = \frac{x^{1/3}}{3y^{2/3}}$ do not exist at $(0, 0)$, so that f is not differentiable at $\mathbf{x}_0$.

5. **Tangent Planes.** (a) $z = 2(x - 1) - 4(y - 1) - 1$.   (b) $z = -(x - \pi/2) - y - 1$.

■ 3.5 The Chain Rule and the Gradient

1. **Chain Rule Calculations.** (a) $(f \circ \alpha)'(t) = 2t - 256t^3$.   (b) $(f \circ \alpha)'(t) = 0$.

2. **Gradient Calculations.** (a) $\nabla f(2, -1) = (9, -4)$.   (b) $\nabla f(1, 3) = (1/\sqrt{10}, 3/\sqrt{10})$.

3. **Directional Derivative Calculations.** (a) $D_{\mathbf{u}} f(1, 2) = 7/\sqrt{2}$.   (b) $D_{\mathbf{u}} f(1, 0) = \sqrt{2}/\sqrt{3}$.

4. **Rates of Change.** (a) Since $D_{\mathbf{u}} f(1, 2) = (4, 3) \cdot \mathbf{u} = 4u_1 + 3u_2$, $D_{\mathbf{u}} f(1, 2) = 0$ when $\mathbf{u} = \pm(\frac{3}{5}, -\frac{4}{5})$; $D_{\mathbf{u}} f(1, 2) > 0$, when the angle θ between $\mathbf{u}$ and $(4, 3)$ satisfies $0 \le \theta < \pi/2$; and $D_{\mathbf{u}} f(1, 2) < 0$, when the angle θ between $\mathbf{u}$ and $(4, 3)$ satisfies $\pi/2 < \theta \le \pi$.
   (b) Since $D_{\mathbf{u}} f(1, 0) = (0, 1) \cdot \mathbf{u} = u_2$, $D_{\mathbf{u}} f(1, 0) = 0$ when $\mathbf{u} = \pm(1, 0)$; $D_{\mathbf{u}} f(1, 0) > 0$ when the angle θ between $\mathbf{u}$ and $(0, 1)$ satisfies $0 \le \theta < \pi/2$; and $D_{\mathbf{u}} f(1, 0) < 0$ when the angle θ between $\mathbf{u}$ and $(0, 1)$ satisfies $\pi/2 < \theta \le \pi$.

5. **Horizontal Tangent Planes.** In order for the tangent plane to be horizontal, it must satisfy the equation $z = f(x_0, y_0)$; that is, $z = $ constant. Since $z = \frac{\partial f}{\partial x}(x_0, y_0)(x - x_0) + \frac{\partial f}{\partial y}(x_0, y_0)(y - y_0) + f(x_0, y_0)$, this occurs when $\frac{\partial f}{\partial x}(x_0, y_0) = \frac{\partial f}{\partial y}(x_0, y_0) = 0$. Thus $D_{\mathbf{u}} f(\mathbf{x}_0) = \nabla f(\mathbf{x}_0) \cdot \mathbf{u} = 0$ for any direction $\mathbf{u}$ when the tangent plane is horizontal.

6. **Extrema of Rates of Change.** The direction of maximum rate of increase is the direction of the gradient, and the direction of minimum rate of increase is the negative of this direction. These directions are:
   (a) $\nabla f(1, -1) = (4, -3)$ and $-\nabla f(1, -1) = -(4, -3)$.
   (b) $\nabla f(\pi/2, 0) = (-1, 1 + \pi/2)$ and $-\nabla f(\pi/2, 0) = -(-1, 1 + \pi/2)$.

7. **Tangent Planes to Level Surfaces.** Use the equation $\nabla f(\mathbf{x}_0) \cdot (\mathbf{x} - \mathbf{x}_0) = 0$ to obtain:
   (a) $2(x - 1) + 2(y - 1) - (z - 1) = 0$.
   (b) $2(x - 1) - 2(z - 1) = 0$.

9. **Potential Energy.** We can rewrite U as $U(\mathbf{x}) = \frac{q_1 q_2}{4\pi\epsilon_0}(x^2 + y^2 + z^2)^{-1/2}$. Differentiating,
$\nabla U(\mathbf{x}) = \frac{q_1 q_2}{4\pi\epsilon_0}\left(-x(x^2 + y^2 + z^2)^{-3/2}, -y(x^2 + y^2 + z^2)^{-3/2}, -z(x^2 + y^2 + z^2)^{-3/2}\right) = -\frac{q_1 q_2}{4\pi\epsilon_0(x^2+y^2+z^2)^{3/2}}(x, y, z) = -\frac{q_1 q_2}{4\pi\epsilon_0\|\mathbf{x}\|^{3/2}}\mathbf{x} = -\mathbf{F}$. Thus, $-\nabla U = \mathbf{F}$.

CHAPTER 4 SOLUTIONS

■ 4.1 Graphical Analysis of Critical Points

1. **Graphical Classification of Critical Points.** (a) There is an isolated global maximum at $(\frac{1}{\sqrt{2}}, 0)$ and an isolated global minimum at $(-\frac{1}{\sqrt{2}}, 0)$.
   (b) There is an isolated local minimum located at $(0, 1)$ and a saddle located at $(0, -1)$, which is also isolated.

2. **Isolated Critical Points.** (a) The origin is an isolated maximum. The level sets are concentric circles centered at the origin.
   (b) Note that $f(x, y) = (x - 2y)^2$. There is a line of nonisolated minima at $x = 2y$.

3. **Specified Critical Points.** (a) The contour curves should be concentric closed curves surrounding the origin.
   (b) The contour curves near $(1, 0)$ should be concentric closed curves surrounding the point. The contour curves near $(-1, 0)$ should take the characteristic shape of a family of hyperbolas, and the branches that open toward $(1, 0)$ should be closed curves that surround $(1, 0)$.

4. **Graphical Classification of Critical Points.** (a) The critical points are $(0, 1)$, $(0, -1)$, $(1, 0)$, and $(-1, 0)$. Using the plot, $(1, 0)$ is a maximum, $(-1, 0)$ is a minimum, and $(0, 1)$ and $(0, -1)$ are saddle points. There are no global extrema.
   (b) The critical points are $(0, 0)$, $(0, \sqrt{2})$, and $(0, -\sqrt{2})$. Using the plot, $(0, 0)$ is a local maximum, and the other two points are saddle points. There are no global extrema.

6. **f and ∇.** (a) If $\mathbf{x}_0$ is a local isolated maximum, then the level sets near $\mathbf{x}_0$ will be concentric closed curves with function values that increase as the curves close in on $\mathbf{x}_0$. Since ∇f is orthogonal to level sets away from critical points and points in the direction of maximum increase of f, ∇f will point toward the inside of each of these closed curves

at a point on the curve. That is, ∇f will point roughly toward $\mathbf{x}_0$, and we conclude that $\mathbf{x}_0$ is a sink of ∇f.

7. f and $-f$. (a) Minima of f are maxima of $-f$, because $f(\mathbf{x}_0) \leq f(\mathbf{x})$ for $\mathbf{x}$ near $\mathbf{x}_0$ is equivalent to $-f(\mathbf{x}_0) \geq -f(\mathbf{x})$. Similarly, maxima of f are minima of $-f$ and saddles of f are saddles of $-f$.

8. Isolated Minima. Since $\mathbf{x}_0$ is a minimum of f, there is an open ball $\mathcal{B}$ centered at $\mathbf{x}_0$, so that $f(\mathbf{x}_0) \leq f(\mathbf{x})$ for all $\mathbf{x} \in \mathcal{B}$. Since $\mathbf{x}_0$ is an isolated critical point, we can choose a possibly smaller open ball $\overline{\mathcal{B}}$ so that $\mathbf{x}_0$ is the only critical point of f inside $\overline{\mathcal{B}}$. If $\overline{\mathcal{B}}$ contains a point $\mathbf{x}$ with $f(\mathbf{x}) = f(\mathbf{x}_0)$, $\mathbf{x}$ is also a minimum for f in $\overline{\mathcal{B}}$. Since a minimum of a differentiable function is also a critical point, we have produced a critical point of f that is inside $\overline{\mathcal{B}}$ but is different from $\mathbf{x}_0$. However, this is impossible, since we assumed that $\mathbf{x}_0$ was an isolated critical point. Therefore, there is no such point $\mathbf{x}$ and $f(\mathbf{x}_0) < f(\mathbf{x})$ for all $\mathbf{x}$ in $\overline{\mathcal{B}}$. Note that a similar argument would prove the result for isolated maxima.

10. Critical Points in $\mathbb{R}^3$. (a) The only critical point is $(0, 0, 0)$, so it is isolated.
(b) The critical points are $(\frac{1}{\sqrt{3}}, 0, \frac{1}{\sqrt{3}})$, $(-\frac{1}{\sqrt{3}}, 0, \frac{1}{\sqrt{3}})$, $(\frac{1}{\sqrt{3}}, 0, -\frac{1}{\sqrt{3}})$, and $(-\frac{1}{\sqrt{3}}, 0, -\frac{1}{\sqrt{3}})$. For each critical point, we can find an open ball centered at the point that contains no other critical point ($r = \frac{1}{2}$ will do), so each is an isolated critical point.

■ 4.2 Algebraic Classification of Critical Points

1. A Quadratic Saddle. Assume that $A < 0$ and $AC - B^2 < 0$. Notice that $f(0, 0) = k$. After completing the square as in the text, $f(x, y) = k + \left(A\left(x + \frac{B}{A}y\right)^2 + \frac{1}{A}(AC - B^2)y^2\right)$. Since $A < 0$, the fact that $f(x, 0) = k + Ax^2$ means that there are points within any open ball centered at $(0, 0)$ with $f(x, 0) < f(0, 0)$ for $x \neq 0$. Similarly, since $A < 0$ and $AC - B^2 < 0$, the fact that $f(-\frac{B}{A}y, y) = k + \frac{1}{A}(AC - B^2)y^2$ means that there are points with any open ball centered at $(0, 0)$ with $f(-\frac{B}{A}y, y) > f(0, 0)$ for $y \neq 0$. Since any open ball centered at $(0, 0)$ contains points with function values less than $f(0, 0)$ and points with values greater than $f(0, 0)$, we conclude that $(0, 0)$ is a saddle point of f.

3. Taylor Polynomials. (a) $l(x, y) = 3 + 5x - 2y$; $q(x, y) = 3 + 5x - 2y + 3x^2 - xy$.
(b) $l(x, y) = 3 - 9(x + 1) + 4(y - 3)$; $q(x, y) = 3 - 9(x + 1) + 4(y - 3) + 6(x + 1)^2 - 9(x + 1)(y - 3) + (y - 3)^2$.

4. Second Derivative Test Calculations. (a) The only critical point is $(0, 0)$. At $(0, 0)$, $A = 6$, $B = -2$, $C = 2$, and $AC - B^2 = 8$, so that $AC - B^2 > 0$ and $A > 0$ and $(0, 0)$ is a minimum.
(b) The only critical point is $(-\frac{1}{4}, -\frac{5}{4})$. At $(-\frac{1}{4}, -\frac{5}{4})$, $A = 2$, $B = 2$, $C = -2$, and $AC - B^2 = -8$, so that $AC - B^2 < 0$ and $(-\frac{1}{4}, -\frac{5}{4})$ is a saddle point.

5. **Degenerate Critical Points.** (a) The only critical point is at $(0,0)$. Since $A = 0$, $B = 0$, and $C = -2$, $AC - B^2 = 0$, so that $(0,0)$ is a degenerate critical point. Evaluating f at points near $(0,0)$, we see that $f(x,0) = x^4 > 0 = f(0,0)$, while $f(0,y) = -y^2 < 0 = f(0,0)$. We conclude that $(0,0)$ is a saddle point.
(b) The only critical point is at $(0,0)$. Since $A = B = C = 0$, $AC - B^2 = 0$, so that $(0,0)$ is a degenerate critical point. Evaluating f at points near $(0,0)$, we see that $f(x,0) = x^3 > 0 = f(0,0)$ for $x > 0$ and $f(x,0) < 0$ for $x < 0$. We conclude that $(0,0)$ is a saddle point.

9. **Cubic Polynomials.** (a) $l(x,y) = c_1 x + c_2 y + d$ and $q(x,y) = d + c_1 x + c_2 y + \frac{1}{2}(2b_1 x^2 + 2b_2 xy + 2b_3 y^2)$.
(b) The linear approximation consists of the degree zero (constant) and degree one terms of f. The quadratic approximations consists of the degree zero, degree one, and degree two terms of f.

■ 4.3 Constrained Optimization in the Plane

1. **Finding Extrema.** (a) The level sets of f are the lines $x + y = c$. $\mathcal{S}$ is a rectangle with vertices $(1,2)$, $(-1,2)$, $(1,-2)$, and $(-1,-2)$. The largest value of f on $\mathcal{S}$ occurs when $x + y$ is largest, which is at the vertex $(1,2)$, where $f(1,2) = 3$. Similarly, the smallest value occurs at the vertex $(-1,-2)$, where $f(-1,-2) = -3$.
(b) The level sets of f are the circles $x^2 + y^2 = c^2$ of radius c centered at the origin. The extreme values of f on $\mathcal{S}$ are at the points of $\mathcal{S}$ that are farthest from or closest to the origin. Since $\mathcal{S}$ is a square with vertices $(0,0)$, $(0,-1)$, $(1,0)$, and $(1,-1)$, the maximum of f occurs at the vertex $(1,-1)$, where $f(1,-1) = 2$, and the minimum occurs at $(0,0)$, where $f(0,0) = 0$.

2. **Lagrange Multipliers.** (a) The Lagrange equations are $2x = \lambda 2x$, $-4y = \lambda 2y$, and $x^2 + y^2 = 1$. There are four possible extreme points: $(\pm 1, 0)$ with $f(\pm 1, 0) = 1$ and $(0, \pm 1)$ with $f(0, \pm 1) = -2$. Thus the maxima of f are at $(\pm 1, 0)$ and the minima are at $(0, \pm 1)$.
(b) The Lagrange equations are $2x = \lambda x$, $y = \lambda 2y$, and $\frac{x^2}{2} + y^2 = 1$. There are four possible extreme points: $(\pm \sqrt{2}, 0)$ with $f(\pm \sqrt{2}, 0) = 2$ and $(0, \pm 1)$ with $f(0, \pm 1) = 1/2$. Thus the maxima of f are at $(\pm \sqrt{2}, 0)$, and the minima are at $(0, \pm 1)$.

3. **Extrema on Closed Sets.** In each case, we first find the critical points of f on $\mathcal{S}$ and then locate the extrema of f on the boundary of $\mathcal{S}$.
(a) $\nabla f(x,y) = (2x - 2, 2y)$, so f has a single critical point located at $(1,0)$ with $f(1,0) = 0$. The Lagrange equations are $2x - 2 = \lambda 2x$, $2y = \lambda 2y$, and $x^2 + y^2 = 4$. There are two possible extreme points: $(\pm 2, 0)$ with $f(2,0) = 1$ and $f(-2,0) = 9$. Thus $(1,0)$ is a minimum, and $(-2,0)$ is a maximum.

(b) $\nabla f(x, y) = (y e^{xy}, x e^{xy})$, so f has a single critical point at $(0, 0)$ with $f(0, 0) = 1$. The Lagrange equations are $y e^{xy} = \lambda(x - 1)/2$, $x e^{xy} = \lambda 2y$, and $\frac{(x-1)^2}{4} + y^2 = 1$. There are four possible extrema: $p_1 \approx (2.186, 0.805)$ with $f(p_1) \approx 5.811$; $p_2 \approx (2.186, -0.805)$ with $f(p_2) \approx 0.172$; $p_3 \approx (-0.686, 0.538)$ with $f(p_3) \approx 0.692$; and $p_4 \approx (-0.686, -0.538)$ with $f(p_4) \approx 1.44$. Thus p_1 is a maximum, and p_2 is a minimum.

4. **Geometric Test for Extrema.** (a) From the plot, there are four extrema located approximately at $(0.2, 1.9)$, $(0.2, -1.9)$, $(1, -0.6)$, and $(-1, -0.6)$.

5. **Using the Gradient.** (a) From the plot, there are eight extrema located approximately at $(1.2, -1.2)$, $(1.4, -0.6)$, $(0.8, 0.4)$, $(-0.5, -1.3)$, $(-0.8, -.2)$, $(-1.0, 1.4)$, $(-1.5, 1.1)$, and $(0, -1)$.

7. **Extreme Points on Curves.** Not necessarily. The extrema of f on $\mathcal{B}$ will occur at points $\mathbf{x}_0 = \alpha(t_0)$ with $(f \circ \alpha)'(t_0) = 0$, but not all such points are extrema. For example, if $f(x, y) = x^3$ and $\mathcal{B}$ is the unit circle $x^2 + y^2 = 1$ and $\alpha(t) = (\cos t, \sin t)$, then $(0, 1) = \alpha(\frac{\pi}{2})$ satisfies $(f \circ \alpha)'(\frac{\pi}{2}) = 0$. However, it is neither a local minimum nor a local maximum of f on the circle, since for $x > 0$, $f(x, y) > f(0, 1)$ and for $x < 0$, $f(x, y) < f(0, 1)$.

9. **Hypotheses for the Maximum Principle I.** (a) The set $\mathcal{S}$ is not closed since it does not contain the origin, which is a boundary point of $\mathcal{S}$. To see this, notice that every open ball of radius $r > 0$ centered at $(0, 0)$ contains a point not in $\mathcal{S}$, the origin, and points in $\mathcal{S}$, namely, any point a positive distance less than r from the origin.
(b) The function f has a global maximum of 1 at the origin. Since $x^2 + y^2 > 0$ implies that $f(x, y) < 1$, the origin is the only point where f achieves the maximum. However, f takes values arbitrarily close to 1 on $\mathcal{S}$; by choosing ϵ small, $f(\epsilon, 0) = 1 - \epsilon^2$ can be made as close to 1 as we choose. This says that no smaller number than 1 can be a maximum for f on $\mathcal{S}$. Therefore f has no maximum on $\mathcal{S}$.

■ 4.4 Constrained Optimization in Space

1. **Extrema on a Constraint Surface.** (a) The Lagrange equations are $2x = 2\lambda x$, $-2y = 2\lambda y$, $4z = 2\lambda z$, and $x^2 + y^2 + z^2 = 1$. There are six solutions to these equations: $(0, 0, \pm 1)$ with $f(0, 0, \pm 1) = 2$; $(0, \pm 1, 0)$ with $f(0, \pm 1, 0) = -1$; and $(\pm 1, 0, 0)$ with $f(\pm 1, 0, 0) = 1$. Thus f has maxima at $(0, 0, \pm 1)$ and minima at $(0, \pm 1, 0)$.
(b) The Lagrange equations are $2x = 6\lambda x$, $-2y = 2\lambda y$, $-2z = 4\lambda z$, and $3x^2 + y^2 + 2z^2 = 1$. There are six solutions to these equations: $(0, 0, \pm \frac{1}{\sqrt{2}})$ with $f(0, 0, \pm \frac{1}{\sqrt{2}}) = -\frac{1}{2}$; $(0, \pm 1, 0)$ with $f(0, \pm 1, 0) = -1$; and $(\pm \frac{1}{\sqrt{3}}, 0, 0)$ with $f(\pm \frac{1}{\sqrt{3}}, 0, 0) = \frac{1}{3}$. Thus f has maxima at $(\pm \frac{1}{\sqrt{3}}, 0, 0)$ and minima at $(0, \pm 1, 0)$.

2. **Extrema on a Closed Set.** In each case, we first find the critical points of f on the interior of $\mathcal{S}$ and then find the extrema on the boundary of $\mathcal{S}$.

(a) $\nabla f(x, y, z) = (-2x, -2y, -2z/9)$, so that f has a single critical point at $(0,0,0)$, which is a maximum of f with $f(0,0,0) = 2$. The Lagrange equations are $-2x = 2\lambda x$, $-2y = \frac{1}{2}\lambda y$, $-\frac{2}{9}z = 2\lambda z$, and $x^2 + y^2/4 + z^2 = 1$. There are six solutions: $(\pm 1, 0, 0)$ with $f(\pm 1, 0, 0) = 1$; $(0, \pm 2, 0)$ with $f(0, \pm 2, 0) = -2$; and $(0, 0, \pm 1)$ with $f(0, 0, \pm 1) = \frac{17}{9}$. Thus f has a maximum at $(0,0,0)$ and minima at $(0, \pm 2, 0)$.

3. **Extrema on Closed Sets with Multiple Boundary Surfaces.** We first find the critical points of f on the interior of $\mathcal{S}$ and then find the extrema on each of the components of the boundary of $\mathcal{S}$.

(a) $\nabla f(x, y, z) = (-2x, -2y, 2z)$ so that f has a single critical point $(0,0,0)$ with $f(0,0,0) = 4$. The boundary of $\mathcal{S}$ consists of three surfaces: $x^2 + y^2 = 1$ with $0 \le z \le 1$; $z = 0$ with $x^2 + y^2 \le 1$, which contains the critical point at $(0,0,0)$; and $z = 1$ with $x^2 + y^2 \le 1$. The Lagrange equations on the first surface are $-2x = 2\lambda x$, $-2y = 2\lambda y$, $-2z = 0$, and $x^2 + y^2 = 1$ with $0 \le z \le 1$. These equations are solved by points of the form $(x, y, 0)$ with $x^2 + y^2 = 1$ for which $f(x, y, 0) = 3$. The Lagrange equations on the second surface are $-2x = 0$; $-2y = 0$; $2z = \lambda$; and $z = 0$. The only solution is $(0,0,0)$, which we identified above. The Lagrange equations on the third surface are $-2x = 0$, $-2y = 0$, $2z = \lambda$, and $z = 1$. The only solution is $(0,0,1)$ with $f(0,0,1) = 5$. We have already considered the curve of intersection of the first and second surfaces, $x^2 + y^2 = 1$ and $z = 0$. On the curve of intersection of the second and third surfaces, $x^2 + y^2 = 1$ and $z = 1$, so that $f(x, y, 1) = 4$. Thus f has a maximum at $(0,0,1)$ and minima on the curve of intersection of the first and second surfaces.

4. **Extrema on Curves.** (a) Note f has a critical point at $(1, 0, 1)$ with $f(1, 0, 1) = 0$. The Lagrange equations are $y = 2\lambda_1 x$, $x - z = 2\lambda_1 y$, $-y = \lambda_2$, $x^2 + y^2 = 1$, and $z = 1$. These are solved by the points $(-\frac{1}{2}, \pm\sqrt{3}/2, 1)$ with $f(-\frac{1}{2}, \sqrt{3}/2, 1) = -3\sqrt{3}/4$ and $f(-\frac{1}{2}, -\sqrt{3}/2, 1) = 3\sqrt{3}/4$. The first point is a minimum and the second is a maximum.

(b) The Lagrange equations are $2x + 2y = 2\lambda_1 x + \lambda_2$, $2x + 2y = -2\lambda_1 y$, $0 = -2\lambda_1 z$, $x^2 - y^2 - z^2 = 1$, and $x = 2$. These are solved by the points: $(2, \pm\sqrt{3}, 0)$ with $f(2, \sqrt{3}, 0) = (2 + \sqrt{3})^2 \approx 13.93$ and $f(2, -\sqrt{3}, 0) = (2 - \sqrt{3})^2 \approx 0.072$. The first point is a maximum, and the second point is a minimum.

5. **Potential Energy on a Curve.** (a) $\alpha(t) = (2\cos t, 2\sin t, 2\cos t + 2\sin t)$ for $0 \le t \le 2\pi$.

(b) α traces the curve counterclockwise when viewed from the positive z-axis. It passes through the points in the order $\mathbf{x}_6$, $\mathbf{x}_2$, $\mathbf{x}_1$, $\mathbf{x}_3$, $\mathbf{x}_5$, and $\mathbf{x}_4$.

(c) U begins at $t = 0$ with the value $-0.740/(4\pi\epsilon_0)$ at $\mathbf{x}_6$; decreases to $-0.743/(4\pi\epsilon_0)$ at $\mathbf{x}_2$; increases to $-0.356/(4\pi\epsilon_0)$ at $\mathbf{x}_1$; decreases to $-0.675/(4\pi\epsilon_0)$ at $\mathbf{x}_3$; increases to $-0.585/(4\pi\epsilon_0)$ at $\mathbf{x}_5$; decreases to $-0.758/(4\pi\epsilon_0)$ at $\mathbf{x}_4$; and increases to $-0.740/(4\pi\epsilon_0)$ at $t = 2\pi$ at $\mathbf{x}_6$.

6. **Coloring a Surface.** (a) The sphere should be brightest at $(1, 0, 0)$, should be darkest at $(-1, 0, 0)$, and should have a smooth gradation of shading between these two points in circular bands parallel to the yz-plane.

(b) The sphere should be brightest at $(\frac{1}{\sqrt{2}}, \frac{1}{\sqrt{2}}, 0)$, should be darkest at $(-\frac{1}{\sqrt{2}}, -\frac{1}{\sqrt{2}}, 0)$, and should have a smooth gradation of shading between these two points in circular bands parallel to the plane with normal vector $(1, 1, 0)$.

CHAPTER 5 SOLUTIONS

■ 5.1 Riemann Sums—An Intuitive Introduction

1. **Total Heat.** Given that we have the temperature at nine equally spaced points, use a 3×3 rectangular grid with each cell $\frac{4}{3}$ cm by $\frac{5}{3}$ cm with an area of $\frac{20}{9}$ cm^2. If (x_{ij}, y_{ij}) is chosen to be one of the corners, centers of a side, or center of the plate, the contribution of each cell to the Riemann sum will be $T(x_{ij}, y_{ij}) \cdot 1 \cdot 2.7 \cdot 0.215 \cdot \frac{20}{9}$. Then the Riemann sum for total heat is given by $\sum_{i,j}^{3,3} T(x_{ij}, y_{ij}) \cdot 1 \cdot 2.7 \cdot 0.215 \cdot \frac{20}{9} = 210.3$ cal. (Note that different partitions will yield different Reimann sum approximations.)

3. **Refining an Approximation.** Divide the plot in Figure 5.3 into 6 columns and 8 rows so that each cell is approximately 1.5 mi wide and 2 mi high. The following chart contains a sampling of the rainfall in each cell with the cell labeled (1,1) corresponding to the lower left corner of the plot:

8	12	9	6	4	3	2.5
7	12	10	6	4	3	2
6	12	10	8	5	3	2
5	10	11	8	6	4	2
4	10	12	10	6	3	1.5
3	8	12	12	5	2	1.5
2	8	11	8	5	2.5	2
1	7	9	6	4	2	1
	1	2	3	4	5	6

Summing the products of the sampled value for rainfall and the area of a cell yields a Riemann sum approximation of 909. Dividing by the total area of 144 yields an average rainfall of 6.31 in., which is less than the average of 6.92 in. computed in the text.

4. Upper and Lower Sums. The sampled values for the maximum rainfall and for the minimum rainfall in a cell are given in the following tables:

4	13.1	9	4
3	12.6	10.5	6
2	14	14.5	6
1	12	12	3
Max	1	2	3

4	8	4	2.5
3	8	4	2
2	7	3	1.5
1	5	2	1.5
Min	1	2	3

(a) Computing the approximation from maximum values yields an approximate average of 9.73 in.

(b) Computing the approximation from minimum values yields an approximate average of 4.04 in.

(c) The true average value is between 4.04 in. and 9.73 in. Note that both the value in the text and the value computed in Exercise 3 fall in this range.

5. Volume. Partition the region, which is a square, into 4 rows and 4 columns, so that each cell is 2.5 cm on a side. We approximate the height over each cell as follows:

4	1	2	2	1
3	2	5	5	2
2	2	5	5	2
1	1	2	2	1
	1	2	3	4

Since the area of each cell is 6.25 cm^2, the Riemann sum approximation for the volume is 225 cm^3.

7. Average Temperature III. Following the example on solar intensity, we partition the disk into two cells: a circle of radius 5 centered at the origin, which has area 25π, and an annulus of inner radius 5 and outer radius 10, which has area 75π. The average temperature on the griddle can be approximated by $\frac{1}{100\pi}(150\cdot25\pi+100\cdot75\pi) = 112.5°$ C.

9. Average Value I. $f(x,y) = \sin(x^2 + y^2)$ and $\mathcal{R} = [0,3] \times [0,3]$. We will partition this region into cells, each of which is a portion of an annulus. Thus Cell 1 is the quarter-circle of radius .75 centered at $(0,0)$, Cell 2 is the quarter-annulus with inner radius 0.75 and outer radius 1.5, Cell 3 is the quarter-annulus with inner radius 1.5 and outer radius 2.5, and Cell 4 is the quarter-annulus with inner radius 2.5 and outer radius 3.5. Cells 5 and 6 are smaller portions of an annulus. Cell 5 is the portion of the annulus with inner radius 3 and outer radius 3.5 between the angles of approximately 0.32 and 1.25. Cell

6 is the portion of the annulus with inner radius 3.5 and outer radius 4.5 between the angles of approximately 0.6 and 1. For each cell, we have

Cell	Area	sample point	temperature
1	$\frac{0.75^2\pi}{4}$	$(0.5, 0.5)$	$\sin(1/2)$
2	$\frac{(1.5^2-0.75^2)\pi}{4}$	$(1, 1)$	$\sin(2)$
3	$\frac{(2.5^2-1.5^2)\pi}{4}$	$(1.5, 1.5)$	$\sin(18/4)$
4	$\frac{(3^2-2.5^2)\pi}{4}$	$(2, 2)$	$\sin(4)$
5	$(3.5^2-3^2)0.93$	$(2.5, 2.5)$	$\sin(50/4)$
6	$(4.5^2-3.5^2)0.4$	$(3, 3)$	$\sin(18)$

The sum of the products of the sampled values and the areas of the cells is -5.89. The average value is $\frac{-5.89}{9} = -0.65$.

11. **Total Mass.** Partition the cube into 27 smaller cubes of side-length $l/3$. Assign the value 1 to the cells that lie along an edge, 2 to the cells at the center of a face, and 5 to the cell at the center of the cube. The Riemann sum for total mass is equal to $((1 \cdot 20) + (2 \cdot 6) + (5 \cdot 1))\frac{l^3}{27} = \frac{37}{27}l^3$.

■ 5.2 Integration of Functions of Two Variables

1. **Riemann Sum Calculations.** (a) Using cells with width $\frac{2}{3}$ and height $\frac{1}{4}$ and sampling f at the midpoint of each cell, we obtain the value 3 for the Riemann sum approximation. (b) Using 4 columns of width $\frac{1}{2}$, with the first and fourth columns consisting of cells of height $\frac{1}{4}f(-1/2, 0) = \frac{3}{16}$ and the second and third consisting of cells of height $\frac{1}{4}$, and sampling f at the midpoint of each cell, we obtain the value $\frac{73}{4096} \approx -0.018$ for the Riemann sum approximation.

2. **Integrals over Rectangles.** (a) $\int_{-1}^{1}\int_0^5 x^3y^2\, dy\, dx = 0$.   (b) $\int_0^1\int_1^2 (x-y)^3\, dy\, dx = -\frac{3}{2}$.

3. **Integrals over Nonrectangles.** (a) $\int_0^1\int_{x^4}^x (y-x)\, dy\, dx = -\frac{1}{18}$.
   (b) $\int_0^1\int_y^{\sqrt{y}} (y+x)\, dx\, dy = \frac{3}{20}$.

4. **Iterated Integrals.** (a) $\frac{23}{3}$.   (b) $\frac{e^2-1}{4}$.

7. **Product Functions.** (a) Since f and g are continuous, their product is a continuous function of two variables.

(b) $\int \int_{\mathcal{R}} f \, dA_{x,y} = \int_a^b \left(\int_c^d g(x)h(y) \, dy \right) dx$ by definition. Since $g(x)$ does not depend on y, it can be factored out of the inner integral to obtain $\int_a^b g(x) \left(\int_c^d h(y) \, dy \right) dx$. Since $\int_c^d h(y) \, dy$ does not depend on x, it can be factored out of the outer integral to obtain $\left(\int_a^b g(x) \, dx \right) \left(\int_c^d h(y) \, dy \right)$.

8. Volumes. (a) $\int_{-3}^3 \int_{-\sqrt{9-x^2}}^{\sqrt{9-x^2}} (9 - x^2 - y^2) \, dy \, dx = \frac{81\pi}{2} \approx 127.23$.
(b) Setting the two expressions for z equal to each other, we see that the paraboloids intersect when $x^2 + y^2 = 4$. So the volume is given by the following double integral over the disk $x^2 + y^2 \le 4$: $\int_{-2}^2 \int_{-\sqrt{4-x^2}}^{\sqrt{4-x^2}} (12 - 2x^2 - y^2) - (x^2 + 2y^2) \, dy \, dx = 24\pi$

9. Average Value. (a) Area $= \int_0^1 \int_0^x dy \, dx = \frac{1}{2}$; average value $= 2 \int_0^1 \int_0^x xe^y \, dy \, dx = 1$.
(b) Area $= \int_0^2 \int_{-y}^y dy dx = 4$; average value $= \frac{1}{4} \int_0^2 \int_{-y}^y (3 - x - 2y) \, dy dx = \frac{1}{3}$.

11. Center of Mass. (a) $M = \int_0^1 \int_{1-x}^1 c \, dy \, dx = \frac{c}{2}$; $\overline{x} = \frac{2}{c} \int_0^1 \int_{1-x}^1 xc \, dy \, dx = \frac{2}{3}$;
$\overline{y} = \frac{2}{c} \int_0^1 \int_{1-x}^1 yc \, dy \, dx = \frac{2}{3}$.

■ 5.3 Integration in Polar Coordinates

1. Arctan. For the interval $[0, 2\pi)$, use the branch passing through $(0,0)$ for (x,y) in the first quadrant; the branch passing through $(0, \pi)$ for (x,y) in the second and third quadrants; and the branch passing through $(0, 2\pi)$ for (x,y) in the fourth quadrant. For the interval $(-2\pi, 0]$, use the branch passing through $(0, -2\pi)$ for (x,y) in the first quadrant; the branch passing through $(0, -\pi)$ for (x,y) in the second and third quadrants; and the branch passing through $(0,0)$ for (x,y) in the fourth quadrant. For the interval $(-\pi, \pi]$, use the branch passing through $(0,0)$ for (x,y) in the first and fourth quadrants; the branch passing through $(0, -\pi)$ for (x,y) in the third quadrant; and the branch passing through $(0, \pi)$ for (x,y) in the second quadrant. These results can be read off a plot showing all the branches of the arc tangent function.

2. Polar Coordinates. (a) For $(1, 1)$: $(\sqrt{2}, \frac{\pi}{4} + 2k\pi)$, k any integer. For $(0, -3)$: $(3, \frac{3\pi}{2} + 2k\pi)$, k any integer. For $(-2, 1)$: $(\sqrt{5}, \pi + \arctan(-1/2) + 2k\pi) \approx (\sqrt{5}, 2.678 + 2k\pi)$, k any integer. For $(-1, 5)$: $(\sqrt{26}, \pi + \arctan(-5) + 2k\pi) \approx (\sqrt{26}, 1.768 + 2k\pi)$, k any integer.
(b) For $(1, \pi/4)$: $(\frac{1}{\sqrt{2}}, \frac{1}{\sqrt{2}})$. For $(3, -5\pi/3)$: $(\frac{3}{2}, \frac{3\sqrt{3}}{2})$. For $(0, \pi/9)$: $(0,0)$. For $(2, 1)$: $(2\cos(1), 2\sin(1)) \approx (1.081, 1.682)$.

3. Polar Regions I. (a) $\{(r, \theta) : r \ge 0, \, \pi/2 \le \theta \le \pi\}$.
(b) $\{(r, \theta) : r \ge 2\}$.
(c) $\{(r, \theta) : 2 \le r \le 4\}$.

4. **Polar Regions II.** (a) The region in the first quadrant outside the open disk of radius 2.
(b) The region inside the closed unit disk centered at the origin in the fourth, first, and second quadrants.
(c) The region in the first and fourth quadrants inside the closed disk of radius 5 centered at the origin and outside the open disk of radius 1 centered at the origin.

5. **Converting to Polar Coordinates.** (a) $f(r, \theta) = r^2$.   (b) $f(r, \theta) = r$.   (c) $f(r, \theta) = e^{-r^2}$.

6. **Total Accumulation.** (a) $\int_0^{2\pi} \int_0^1 (1 - r^2) r \, dr \, d\theta = \frac{\pi}{2}$.   (b) $\int_0^{\pi/2} \int_0^2 (r \cos \theta) r \, dr \, d\theta = \frac{8}{3}$.

7. **Volumes.** (a) $\int_0^{2\pi} \int_0^2 \sqrt{4 - r^2} r \, dr \, d\theta = \frac{16\pi}{3}$.   (b) $\int_0^{\pi/2} \int_0^3 4r \, dr \, d\theta = 9\pi$.

8. **Polar vs. Rectangular Integrals.** (a) In polar coordinates: $\int_0^\pi \int_0^2 (r \cos(\theta))^2 r \, dr \, d\theta = 2\pi$.
In rectangular coordinates: $\int_{-2}^2 \int_0^{\sqrt{4-x^2}} y^2 \, dy \, dx = 2\pi$.

■ 5.4 Integration of Functions of Three Variables

---

1. **Iterated Integrals on Boxes.** (a) $-\frac{27}{8}$.   (b) 0.

2. **Integrals from the Text.** (a) $\frac{8}{3} r_0^3$.   (b) $\frac{8}{3} r_0^3$.

3. **Iterated Integrals on More General Regions.** (a) $\frac{7}{12}$.   (b) 0.   (c) $-\frac{32}{105}$.

4. **Evaluating Triple Integrals.** (a) $\int_0^1 \int_{-1}^1 \int_0^2 (x^2 - yz) \, dz \, dy \, dx = \frac{4}{3}$.
(b) $\int_0^\pi \int_{-\pi/2}^{\pi/2} \int_{-1}^0 \sin(x) \cos(y) z \, dz \, dy \, dx = -2$.

5. **A Geometric Criterion.** (a) The criterion has two parts. First, each line parallel to one axis that intersects the region must intersect the region in a single line segment, and the endpoints of these line segments must vary continuously. Second, projecting the region to the coordinate plane orthogonal to the axis used in the first half yields a planar region with the property that each line in this plane parallel to one of the remaining axes intersects the planar region in a single line segment, and the endpoints of these line segments much vary continuously.
(b) Suppose, for example, that the z-axis and lines parallel to it satisfy the first condition. Then the tops of these line segments describe a surface that is the graph of a function $z = h_2(x, y)$, and the bottoms of these line segments describe a surface that is the graph of a function $z = h_1(x, y)$. From the first condition, we also know that h_1 and h_2 are continous and that $\mathcal{R}$ lies between the graphs of these functions. The second condition ensures that there is a region $\overline{\mathcal{R}}$ in the xy-plane so that every line parallel to the z-axis intersecting this region also intersects $\mathcal{R}$ and does so only in the segments given by the first condition. The second condition then ensures that $\overline{\mathcal{R}}$ lies between the graphs of two functions, for example, $x = g_1(y)$ and $x = g_2(x)$.

6. **Find an Iterated Integral.** (a) $\int_0^1 \int_0^{1-x} \int_0^{1-x-y} f(x,y,z)\,dz\,dy\,dx$.
   (b) $\int_0^1 \int_{z-1}^{1-z} \int_0^1 f(x,y,z)\,dy\,dx\,dz$.

7. **More Complicated Regions.** In each case, there are several possible ways to write the integral as an iterated integral.
   (a) Although the region satisfies the criterion of Exercise 5, the continous functions that we would use as the limits of integration cannot be described by a single formula, which we require to evaluate an iterated integral. The four faces of $\mathcal{R}$ are given by $x+y+z = -1$; $x-y+z = 1$; $-x+y+z = 1$; and $x+y-z = 1$, so that $\mathcal{R} = \{(x,y,z) : 0 \le x \le 1, -x \le y \le x,$ and $x+y-1 \le z \le -x+y+1\} \cup \{(x,y,z) : -1 \le x \le 0, x \le y \le -x,$ and $-x-y-1 \le z \le x-y+1\} \cup \{(x,y,z) : 0 \le y \le 1, -y \le x \le y,$ and $x+y-1 \le z \le x-y+1\} \cup \{(x,y,z) : -1 \le y \le 0, y \le x \le -y,$ and $-x-y-1 \le z \le -x+y+1\}$. The integral can be written as the sum of four iterated integrals integrating first with respect to z and then over four triangles in the xy-plane: $\int\int\int_\mathcal{R} f\,dV_{x,y,z} = \int_0^1 \int_{-x}^x \int_{x+y-1}^{-x+y+1} f(x,y,z)\,dz\,dy\,dx +$
   $\int_{-1}^0 \int_x^{-x} \int_{-x-y-1}^{x-y+1} f(x,y,z)\,dz\,dy\,dx + \int_0^1 \int_{-y}^y \int_{x+y-1}^{x-y+1} f(x,y,z)\,dz\,dx\,dy + \int_{-1}^0 \int_y^{-y} \int_{-x-y-1}^{-x+y+1} f(x,y,z)\,dz\,dx\,dy$.
   (b) As in (a), the limits of integration do not have a single expression as a function. The region can be written as the union of four regions as follows: $\mathcal{R} = \{(x,y,z) : 0 \le x \le 1, 0 \le y \le 1-x,$ and $x+y-1 \le z \le -x-y+1\} \cup \{(x,y,z) : 0 \le x \le 1, x-1 \le y \le 0,$ and $x-y-1 \le z \le -x+y+1\} \cup \{(x,y,z) : -1 \le x \le 0, 0 \le y \le 1+x,$ and $-x+y-1 \le z \le x-y+1\} \cup \{(x,y,z) : -1 \le x \le 0, -1-x \le y \le 0,$ and $-x-y-1 \le z \le x+y+1\}$. The integral can be written as the sum of four iterated integrals integrating first with respect to z and then over four triangles in the xy-plane: $\int\int\int_\mathcal{R} f\,dV_{x,y,z} = \int_0^1 \int_0^{1-x} \int_{x+y-1}^{-x-y+1} f(x,y,z)\,dz\,dy\,dx +$
   $\int_0^1 \int_{x-1}^0 \int_{x-y-1}^{-x+y+1} f(x,y,z)\,dz\,dy\,dx + \int_{-1}^0 \int_0^{1+x} \int_{-x+y-1}^{x-y+1} f(x,y,z)\,dz\,dy\,dx+$
   $\int_{-1}^0 \int_{-1-x}^0 \int_{-x-y-1}^{x+y+1} f(x,y,z)\,dz\,dy\,dx$.

8. **Center of Mass.** (a) $M = \int_{-1}^1 \int_{-\sqrt{1-x^2}}^{\sqrt{1-x^2}} \int_0^{1-x^2-y^2} z\,dz\,dy\,dx = \frac{\pi}{6}$; $(\overline{x}, \overline{y}, \overline{z}) = (0,0,\frac{1}{2})$.
   (b) It simplifies the calculation to use polar coordinates for the variables x and y: $M = \int_0^1 \int_0^{2\pi} \int_0^{2-2r} 2r\,dz\,d\theta\,dr = \frac{4}{3}\pi$; $(\overline{x}, \overline{y}, \overline{z}) = (0,0,\frac{1}{2})$.

■ 5.5 Integration in Cylindrical Coordinates

1. **Coordinates of Points.** (a) For $(1,-1,2)$: $(\sqrt{2}, -\frac{\pi}{4}, 2)$. For $(2,1,-2)$: $(\sqrt{5}, \tan^{-1}(\frac{1}{2}), -2)$.
   For $(\sqrt{3},1,1)$: $(2, \frac{\pi}{6}, 1)$. For $(1,1,\sqrt{2})$: $(\sqrt{2}, \frac{\pi}{4}, \sqrt{2})$. For $(0,0,2)$: $(0,0,2)$.
   (b) For $(2,\pi,2)$: $(-2,0,2)$. For $(1,\pi/4,-1)$: $(\frac{1}{\sqrt{2}}, \frac{1}{\sqrt{2}}, -1)$. For $(\sqrt{3}, \pi/6, -3)$: $(\frac{3}{2}, \frac{\sqrt{3}}{2}, -3)$.
   For $(3, 3\pi/4, 1)$: $(-\frac{3}{\sqrt{2}}, \frac{3}{\sqrt{2}}, 1)$. For $(1, -\pi/3, 2)$: $(\frac{1}{2}, -\frac{\sqrt{3}}{2}, 2)$.

2. **Surfaces in Cylindrical Coordinates.** (a) $r^2 + z^2 = k^2$. (b) $\theta = \tan^{-1} k$.

3. **Regions in Cylindrical Coordinates.** (a) $\{(r, \theta, z) : z \geq 0\}$.
   (b) $\{(r, \theta, z) : \pi \leq \theta \leq 2\pi\}$.
   (c) $\{(r, \theta, z) : 0 \leq r \leq 2, 0 \leq \theta \leq \pi/2 \text{ and } 1 \leq z \leq 4\}$.

4. **Sets Given in Cylindrical Coordinates.** (a) The set is the cone with central axis the positive z-axis, $z = \sqrt{x^2 + y^2}$.
   (b) The set is the half-line of intersection of the cone $z = \sqrt{x^2 + y^2}$ and the plane $y = x$.

7. **Cylindrical Integration.** (a) $\int_0^2 \int_0^\pi \int_0^3 (r \cos\theta \, r \sin\theta z) \, r \, dz \, d\theta \, dr = 0$.

8. **Set Up the Integral.** (a) $\int_0^3 \int_0^{2\pi} \int_{-\sqrt{9-r^2}}^{\sqrt{9-r^2}} \cos(r^2 \cos^2\theta) \, r \, dz \, d\theta \, dr$.
   (b) $\int_0^4 \int_0^{2\pi} \int_{-\sqrt{16-r^2}}^0 \sqrt{1 + (r \cos\theta)^3} r \sin\theta \, r \, dz \, d\theta \, dr$.

■ 5.6 Integration in Spherical Coordinates

1. **Coordinates of Points.** (a) For $(1, -1, 2)$: $(\sqrt{6}, -\frac{\pi}{4}, \sin^{-1}(\sqrt{\frac{2}{3}}))$. For $(2, 1, -2)$: $(3, \tan^{-1}(\frac{1}{2}), \pi - \sin^{-1}(\frac{\sqrt{5}}{3}))$. (Note that $\sin^{-1}$ must be adjusted to choose values from the correct branch.) For $(\sqrt{3}, 1, 1)$: $(\sqrt{5}, \frac{\pi}{6}, \sin^{-1}(\frac{2}{\sqrt{5}}))$. For $(1, 1, \sqrt{2})$: $(2, \frac{\pi}{4}, \frac{\pi}{4})$. For $(0, 0, 2)$: $(2, 0, 0)$.
   (b) For $(1, \frac{\pi}{2}, \pi)$: $(0, 0, -1)$. For $(\sqrt{2}, \frac{\pi}{4}, \frac{\pi}{4})$: $(\frac{1}{\sqrt{2}}, \frac{1}{\sqrt{2}}, 1)$. For $(2, \pi, \frac{\pi}{3})$: $(-\sqrt{3}, 0, 1)$. For $(\sqrt{3}, 0, \frac{\pi}{6})$: $(\frac{\sqrt{3}}{2}, 0, \frac{3}{2})$. For $(1, \frac{\pi}{3}, \frac{\pi}{2})$: $(\frac{1}{2}, \frac{\sqrt{3}}{2}, 0)$.

2. **Sets Given in Spherical Coordinates.** (a) The quarter-sphere of radius 3 centered at the origin with positive x- and y-coordinates.
   (b) The half-plane consisting of the portion of the yz-plane with $y \geq 0$.

3. **Regions in Spherical Coordinates.** (a) $\{(\rho, \theta, \phi) : \rho = k \text{ and } \phi \leq \frac{\pi}{2}\}$.
   (b) $\{(\rho, \theta, \phi) : \phi = \tan^{-1}(\frac{1}{k})\}$.
   (c) $\{(\rho, \theta, \phi) : 0 \leq \rho \leq 2, 0 \leq \theta \leq \frac{\pi}{2}, \text{ and } 0 \leq \phi \leq \frac{\pi}{2}\}$.

5. **Total Heat.** (a) $\delta C \int_1^2 \int_0^{2\pi} \int_0^\pi \ln(\rho^2) \, \rho^2 \sin(\phi) \, d\phi \, d\theta \, d\rho = \delta C \left(\frac{32\pi}{3} \ln(4) - \frac{56\pi}{9}\right) \approx 26.9 \delta C$.
   (b) $\delta C \int_1^3 \int_{-\frac{\pi}{2}}^{\frac{\pi}{2}} \int_0^\pi \rho \sin(\phi) \cos(\theta) \rho^2 \sin(\phi) \, d\phi \, d\theta \, d\rho = 20\pi \delta C$.

7. **The Hydrogen 1s Orbital.** (a) From Example 5.22, we must evaluate $\frac{4}{a^3} \int \rho^2 e^{-2\rho/a} \, d\rho$:
   $\frac{4}{a^3} \int \rho^2 e^{-2\rho/a} \, d\rho = \frac{4}{a^3} \left(\rho^2 (-\frac{a}{2}) e^{-2\rho/a} - 2(-\frac{a}{2}) \int \rho e^{-2\rho/a} d\rho\right) = \frac{4}{a^3} \left(\rho^2 (-\frac{a}{2}) e^{-2\rho/a} - 2(-\frac{a}{2})\right.$
   $\left.(-\frac{a}{2} \rho e^{-2\rho/a} - (-\frac{a}{2} \int e^{-2\rho/a} d\rho))\right) = -\left(2\frac{\rho^2}{a^2} e^{-2\rho/a} + 2\frac{\rho}{a} e^{-2\rho/a} - e^{-2\rho/a}\right) = I(\rho)$.

(b) The difference is always positive because I is an increasing function of ρ. This can be seen from the graph of I or by differentiating I to obtain $I'(\rho) = \frac{4}{a^3}\rho^2 e^{-2\rho/a}$, which is positive for $\rho > 0$.

(c) We need only verify that $\lim_{\rho \to \infty} I(\rho) - I(0) = 1$. Since $I(0) = -1$, we need only show that $\lim_{\rho \to \infty} I(\rho) = 0$. This follows from applying l'Hopital's rule twice.

(d) Plotting $I(\rho + \epsilon) - I(\rho - \epsilon)$ for $\epsilon = a/k$ for larger k, we see a single, well-defined maximum that occurs at approximately $\rho = a$, the Bohr radius. As the band narrows in width, the value at the maximum decreases, but it clearly remains the maximum. Thus an electron is most likely to be located near the Bohr radius.

CHAPTER 6 SOLUTIONS

■ 6.1 Path Integrals

1. **A Riemann Sum Calculation.** (a) To partition the trail, select $N + 1$ points where contours cross the trail with the first point occurring at the parking lot and the final point at the summit. Use the scale on the map to determine the horizontal distance between consecutive points on the list. Use the contours to determine the change in altitude between consecutive points. For each pair of consecutive points, use the preceding pair of values and the Pythagorean theorem to produce an estimate of the length of the trail between the points. Sum these values to obtain a Riemann sum approximation for the length of the trail.

(b) For example, in addition to the initial point at 1500 ft and the final point at 2850 ft, choose the points where the trail intersects the contours of height 1600, 2000, and 2400 ft. The vertical changes for the intervals are 100, 400, 400, and 450 ft, respectively. From the map, the horizontal changes for the intervals are approximately 1430, 3140, 1710, and 1140 ft, respectively. Then the Riemann sum approximation is: $\sqrt{100^2 + 1430^2} + \sqrt{400^2 + 3140^2} + \sqrt{400^2 + 1710^2} + \sqrt{450^2 + 1140^2} = 7586$ ft.

2. **Arc Length.** (a) $\int_1^4 \sqrt{1 + (9/4)t}\, dt = (80\sqrt{10} - 13\sqrt{13})/27 \approx 7.634$.

(b) $\int_0^\pi \sqrt{10}\, dt = \sqrt{10}\pi \approx 9.935$.

3. **Path Integrals over** $\mathcal{C}$. (a) (i) Let $\alpha(t) = (2t + 1, t - 1, -t + 2)$, $0 \le t \le 1$.

(ii) $\int_0^1 e^{(2t+1)+(t-1)+(-t+2)}\sqrt{6}\, dt = \frac{1}{2}e^4\sqrt{6} - \frac{1}{2}e^2\sqrt{6} \approx 57.82$.

(b) (i) $\alpha(t) = (2\cos t, 2\sin t)$, $0 \le t \le 2\pi$. (ii) $\int_0^{2\pi} 2\cos^2(t)2\sin(t)2\, dt = 0$.

5. **Average Value.** (a) $(\int_0^\pi (1+t^2)\sqrt{10}\, dt)/(\sqrt{10}\pi) = (\sqrt{10}\pi^3/3 + \sqrt{10}\pi)/(\sqrt{10}\pi) = \pi^2/3 + 1 \approx 4.29$.

6. Piecewise Defined Curves. (a) Parametrize $\mathcal{C}$ in four parts: $\alpha_1(t) = (t, 0)$, $0 \le t \le 2$; $\alpha_2(t) = (2, t)$, $0 \le t \le 2$; $\alpha_3(t) = (2 - t, 2)$, $0 \le t \le 2$; and $\alpha_4(t) = (0, 2 - t)$, $0 \le t \le 2$. Since $f(x, y) = 0$ on α_1 and α_4, the integral is $\int_0^2 2t\, dt + \int_0^2 (-2t + 4)\, dt = 8$.
(b) Parametrize $\mathcal{C}$ in two parts: $\alpha_1(t) = (2\cos t, 2\sin t)$, $0 \le t \le \pi$; and $\alpha_2(t) = (t, 0)$, $-2 \le t \le 2$. The integral is $\int_0^\pi 8 dt + \int_{-2}^2 t^2\, dt = 8\pi + 16/3 \approx 30.5$.

8. Independence of Parametrization I. (a) Change variables by substituting $u = h(t)$ and $du = h'(t)dt$. Then $\int_\alpha f\, ds = \int_{u=c}^{u=d} f(\alpha(u))\|\alpha'(u)\| du$
$= \int_{t=a}^{t=b} f(\alpha(h(t)))\|\alpha'(h(t))\|\|h'(t)\|\, dt = \int_{t=a}^{t=b} f(\alpha(h(t)))\|\alpha'(h(t))h'(t)\|\, dt$
$= \int_{t=a}^{t=b} f(\beta(t))\|\beta'(t)\|\, dt = \int_\beta f\, ds$.

■ 6.2 Line Integrals

1. Line Integrals. (a) $\int_0^1 (-e^{-t}, e^t) \cdot (e^t, -e^{-t})\, dt = -2$.
(b) $\int_0^{2\pi} (\cos^3 t, \sin^3 t) \cdot (-3\cos^2(t)\sin(t), 3\sin^2(t)\cos(t))\, dt = 0$.

2. Line Integrals over Curves. (a) $\alpha(t) = (2\sin t, 2\cos t)$, $0 \le t \le 2\pi$. $\int_0^2 (2\sin t, 2\cos t) \cdot (2\cos t, -2\sin t) dt = 0$.
(b) $\alpha(t) = (t, t^2)$, $0 \le t \le 2$. $\int_0^2 (t^4, t^2) \cdot (1, 2t)\, dt = 72/5$.

4. Path Independence. (a) Since $\mathbf{F}$ has the path independence property, $\int_{\mathcal{C}} \mathbf{F} \cdot \mathbf{T}\, ds$ can be evaluated by choosing any path with the same initial point and same endpoint as the original $\mathcal{C}$. Thus we may choose a C that goes from $\mathbf{x}_1$ to $\mathbf{x}_0$ to $\mathbf{x}_2$. Call the first portion of the curve $\mathcal{C}_1$ and the second $\mathcal{C}_2$. Since the initial point of $\mathcal{C}_1$ is $\mathbf{x}_1$ and the final point is $\mathbf{x}_0$, $\int_{\mathcal{C}_1} \mathbf{F} \cdot \mathbf{T}\, ds = -f(\mathbf{x}_1)$. Therefore, $\int_{\mathcal{C}} \mathbf{F} \cdot \mathbf{T}\, ds = \int_{\mathcal{C}_1} \mathbf{F} \cdot \mathbf{T}\, ds + \int_{\mathcal{C}_2} \mathbf{F} \cdot \mathbf{T}\, ds = -f(\mathbf{x}_1) + f(\mathbf{x}_2) = f(\mathbf{x}_2) - f(\mathbf{x}_1)$.

6. Potential Functions. (a) $\frac{\partial u}{\partial y}(x, y) = -1$ and $\frac{\partial v}{\partial y}(x, y) = 1$. Since these are not equal, $\mathbf{F}$ is not a gradient vector field.
(b) $\frac{\partial u}{\partial y}(x, y) = 1$ and $\frac{\partial v}{\partial y}(x, y) = 1$. Since these are equal, $\mathbf{F}$ is a gradient vector field. $f(x, y) = x^3 + xy + e^y$.

7. Path Independent Vector Fields. (a) $f(x, y) = \frac{1}{2}(x^2 + y^2)$. $\int_{\mathcal{C}} \mathbf{F} \cdot \mathbf{T} dx = f(-2, 0) - f(2, 0) = 0$.
(b) $f(x, y) = \sin(x)\cos(y)$. $\int_{\mathcal{C}} \mathbf{F} \cdot \mathbf{T} dx = f(1, 1) - f(-1, 1) = 2\sin(1)\cos(1) \approx 0.909$.

11. Work-Energy Theorem. (a) Since $\|\alpha'(t)\|$ is constant, $\|\alpha'(t)\|^2 = \alpha'(t) \cdot \alpha'(t)$ is also constant. It follows that $\alpha'(b) \cdot \alpha'(b) = \alpha'(a) \cdot \alpha'(a)$ for all a and b, so that $W_\alpha = \frac{m}{2}(\alpha'(b) \cdot \alpha'(b) - \alpha'(a) \cdot \alpha'(a)) = 0$.

(b) Since the argument in (a) only relied on the speeds being equal at the endpoints, it can be applied here to show the work is 0.

13. **Inverse Square Fields.** (a) From the calculation of Example 6.7, we see that $f(\mathbf{x}) = -K/\|\mathbf{x}\|$ is a potential function for $\mathbf{F}$.
    (b) Suppose $\mathbf{x} = \mathbf{x}_0 \neq \mathbf{0}$ and α is any path from $\mathbf{x}_0$ to $\mathbf{x} \neq \mathbf{0}$ that does not pass through the origin. Then $\int_\alpha \mathbf{F} \cdot \mathbf{T} \, dx = -\frac{K}{\|\mathbf{x}\|} + \frac{K}{\|\mathbf{x}_0\|}$. In the limit as $\|\mathbf{x}\| \to \infty$, the limit of the work is $\frac{K}{\|\mathbf{x}_0\|}$, the desired result.

■ 6.3 Integration over Closed Curves

1. **Positive Orientation.** (a) $\alpha(t) = (t, 0)$ for $0 \leq t \leq 1$, $\alpha(t) = (1, t - 1)$ for $1 \leq t \leq 2$, $\alpha(t) = (3 - t, 1)$ for $2 \leq t \leq 3$, and $\alpha(t) = (0, 4 - t)$ for $3 \leq t \leq 4$.
   (b) $\alpha(t) = (2\cos t, \sin t)$ for $0 \leq t \leq 2\pi$.

2. **Green's Theorem Calculations I.** (a) $\int_0^2 \int_0^{3x/2} y \, dy \, dx = 3$.
   (b) Using polar coordinates: $\int_0^2 \int_{\pi/4}^{3\pi/4} 3r^3 \, d\theta \, dr = 6\pi$.

3. **Green's Theorem Calculations II.** Integrate $\mathbf{F}(x, y) = \frac{1}{2}(-y, x)$ around the boundary of the given regions.
   (a) The boundary consists of the segment on the x-axis, $\alpha_1(t) = (t, 0)$, $0 \leq t \leq 2\pi$, and the arc of the cycloid, $\alpha_2(t) = (2t - 2\sin t, 2 - 2\cos t)$, $0 \leq t \leq 2\pi$. The area is given by $\int_C \mathbf{F} \cdot \mathbf{T} \, ds = \int_0^{4\pi} \frac{1}{2}(0, t) \cdot (1, 0) \, dt - \int_0^{2\pi} \frac{1}{2}(-2 + 2\cos t, 2t - 2\sin t) \cdot (2 - 2\cos(t), 2\sin(t)) \, dt = 0 - (-12\pi) = 12\pi$. Note that α_2 parametrizes the arc of the cycloid clockwise around C, so we must subtract the line integral over α_2 to obtain the correct value.

8. **Total Flux.** We use the divergence theorem to evaluate the integrals.
   (a) Using polar coordinates: $\int_0^1 \int_0^{2\pi} 3r^3 \, dr \, d\theta = 3\pi/2$.
   (b) $\int_{-1/2}^{1/2} \int_{-1/2}^{1/2} (2x + 2y) \, dy \, dx = 0$.

9. **The Plane Minus the Origin.** (a) Since C is a closed curve that misses the origin, there is a point on C that is closest to the origin. Let r_0 be half the distance of this point from the origin, and let C_1 be the circle of radius r_0 centered at the origin. Apply the argument of Example 6.10B to C, C_1, and the region between them to obtain the result.

11. **Normal Field to a Curve.** (a) $\mathbf{N}(t) \cdot \alpha(t) = \frac{1}{\|\alpha'(t)\|}(y'(t)x'(t) - x'(t)y'(t)) = 0$. Therefore, $\mathbf{N}$ and α' are normal.
    (b) Using the direction defined by α' (which is well-defined since $\alpha(t) \neq \mathbf{0}$), $\mathbf{N}$ always points to the right of and normal to α'. This can be seen by considering the different possibilities for the signs of $x'(t)$ and $y'(t)$.

(c) Since α parametrizes $\mathcal{C}$ in a counterclockwise direction, the region $\mathcal{R}$ must lie to the left of $\mathcal{C}$. Since $\mathbf{N}$ points to the right of α', it must point out of the interior of $\mathcal{C}$.

CHAPTER 7 SOLUTIONS

■ 7.1 Parametrization of Surfaces

1. **Parametrizing Planes.** (a) $\Phi_s(s_0, t_0) = \mathbf{u}$ and $\Phi_t(s_0, t_0) = \mathbf{v}$.
(b) No, they are independent of (s_0, t_0). This makes sense since s coordinate curves and t coordinate curves are families of lines parallel to $\mathbf{u}$ and $\mathbf{v}$, respectively.

2. **Parametrizing Polygons.** (a) Let $P = (1, 1, 2)$, $Q = (4, 1, 2)$, and $R = (1, 3, 2)$, so that $\mathbf{u} = (3, 0, 0)$ and $\mathbf{v} = (0, 2, 0)$. Use $\Phi(s, t) = (1, 1, 2) + s(3, 0, 0) + t(0, 2, 0) = (1 + 3s, 1 + 2t, 2)$ with domain $[0, 1] \times [0, 1]$.
(b) Use $\Phi(s, t) = (1 + 3s, 1 + 2t, 2)$ from (a) with domain $\{(s, t) : 0 \le s \le 1, 1 - s \le t \le 1\}$.

3. **Parametrizing Surfaces.** (a) $\Phi(s, t) = (3 \cos s, 3 \sin s, t)$, $\mathcal{D} = [0, \pi] \times [0, 2]$.
(b) $\Phi(s, t) = (2 \sin(s) \cos(t), 2 \sin(s) \sin(t), 2 \cos(s))$, $\mathcal{D} = [0, \pi/2] \times [0, \pi/2]$.

4. **Surfaces and Coordinate Curves.** (a) $\mathcal{S}$ is the parallelogram with a vertex at $(1, 0, 2)$ and sides emanating from this vertex given by the vectors $(1, 1, -1)$ and $(2, -1, 1)$. The coordinate curves are line segments parallel to these vectors.
(b) $\mathcal{S}$ is the quarter-cylinder of radius 2 and height 3 in the first octant, resting on the xy-plane with axis the z-axis. The coordinate curves are quarter-circles of constant height and vertical line segments.

5. **Tangent Planes.** (a) $\Phi_s(s, t) = (1, 1, -1)$, $\Phi_t(s, t) = (2, -1, 1)$, and $\Phi_s(s, t) \times \Phi_t(s, t) = (0, -3, -3)$. $\Phi(\frac{1}{2}, \frac{1}{2}) = (\frac{5}{2}, 0, 2)$, so the plane is given by $(0, -3, -3) \cdot ((x, y, z) - (\frac{5}{2}, 0, 2)) = 0$ or $-3y - 3(z - 2) = 0$.
(b) $\Phi_s(s, t) = (-2 \sin s, 2 \cos s, 0)$, $\Phi_t(s, t) = (0, 0, 1)$, and $\Phi_s(s, t) \times \Phi_t(s, t) = (2 \cos s, 2 \sin s, 0)$. $\Phi(\frac{\pi}{4}, 1) = (\sqrt{2}, \sqrt{2}, 1)$, so the plane is given by $(\sqrt{2}, \sqrt{2}, 0) \cdot ((x, y, z) - (\sqrt{2}, \sqrt{2}, 1)) = 0$ or $\sqrt{2}(x - \sqrt{2}) + \sqrt{2}(y - \sqrt{2}) = 0$.

6. **One-to-One and Regular I.** (a) Φ is one-to-one. Φ is regular since $\Phi_s(s, t) \times \Phi_t(s, t) = (0, -3, -3) \neq \mathbf{0}$.
(b) Φ is one-to-one. Φ is regular since $\Phi_s(s, t) \times \Phi_t(s, t) = (2 \cos s, -2 \sin s, 0) \neq \mathbf{0}$ for any s.

7. **Surfaces of Revolution I.** (a) The surface is a cylinder of radius 2 and height 6 extending from $z = -3$ to $z = 3$.
(b) The surface is the shape of an American football with tips located at $(0, 0, \pm \frac{\pi}{2})$.

9. Quadric Surfaces. (c) (i) This is a hyperboloid of one sheet. Rewrite the formula to obtain $x^2 + y^2 = z^2 + 4$. The horizontal slice at height z is a circle of radius $\sqrt{z^2 + 4}$; thus $\Phi(\theta, z) = (\cos\theta\sqrt{z^2 + 4}, \sin\theta\sqrt{z^2 + 4}, z)$ on the domain $[0, 2\pi] \times (-\infty, \infty)$.

10. One-to-One and Regular II. (c) (i) Φ is one-to-one on its interior, that is, for $0 < \theta < 2\pi$. $\Phi_\theta(\theta, z) \times \Phi_z(\theta, z) = (-\sin\theta\sqrt{z^2 + 4}, \cos\theta\sqrt{z^2 + 4}, 0) \times (\frac{z\cos\theta}{\sqrt{z^2+4}}, \frac{z\sin\theta}{\sqrt{z^2+4}}, 1) = (\cos\theta\sqrt{z^2 + 4}, \sin\theta\sqrt{z^2 + 4}, -z)$. This is nonzero on the domain, so Φ is also regular.

12. Surfaces as Boundaries. (a) $\mathcal{S} = \mathcal{S}_1 \cup \mathcal{S}_2 \cup \mathcal{S}_3$, where $\mathcal{S}_1$ is the lateral surface of the cylinder parametrized by $\Phi(s, t) = (\cos s, \sin s, t)$ with domain $\mathcal{D} = [0, 2\pi] \times [0, 3]$; $\mathcal{S}_2$ is the bottom face of the cylinder parametrized by $\Phi(s, t) = (t\cos(s), t\sin(s), 0)$ with domain $\mathcal{D} = [0, 2\pi] \times [0, 1]$; and $\mathcal{S}_3$ is the top face of the cylinder parametrized by $\Phi(s, t) = (t\cos s, t\sin s, 3)$ with domain $\mathcal{D} = [0, 2\pi] \times [0, 1]$.
(b) $\mathcal{S}$ is the union of the six faces of the box. The bottom and top faces can be parametrized by $\Phi(s, t) = (s, t, -2)$ and $\Phi(s, t) = (s, t, 2)$, each with domain $\mathcal{D} = [-1, 1] \times [-1, 1]$. The vertical faces parallel to the xz-plane can be parametrized by $\Phi(s, t) = (s, -1, t)$ and $\Phi(s, t) = (s, 1, t)$, each with domain $\mathcal{D} = [-1, 1] \times [-2, 2]$. The vertical faces parallel to the yz-plane can be parametrized by $\Phi(s, t) = (-1, s, t)$ and $\Phi(s, t) = (1, s, t)$, each with domain $\mathcal{D} = [-1, 1] \times [-2, 2]$.

◼ 7.2 Surface Integrals

1. Surface Area. (a) $\mathcal{S}$ is the parallelogram with vertices $(1, 1, 0)$, $(2, 0, 2)$, $(2, -1, 3)$, and $(3, -2, 5)$. $\int\int_{\mathcal{S}} dS = \int_0^1 \int_0^1 \sqrt{3}\, ds\, dt = \sqrt{3}$.
(b) $\mathcal{S}$ is a horizontal quarter-disk of radius 2 and height 1. $\int\int_{\mathcal{S}} dS = \int_0^{\pi/2} \int_0^2 r\, dr\, d\theta = \pi$.

2. Total Accumulation. (a) $\mathcal{S}$ is the parallelogram with vertices $(-1, 3, 1)$, $(0, 2, 2)$, $(1, 1, -1)$, and $(2, 0, 0)$. $\int\int_{\mathcal{S}} f\, dS = \int_0^1 \int_{-1}^1 (s - t + 2)2\sqrt{2}\, dt\, ds = 10\sqrt{2}$.

3. Surfaces of Revolution. (a) $\Phi_\theta(\theta, z) = (-\sin(\theta)f(z), \cos(\theta)f(z), 0)$, $\Phi_z(\theta, z) = (\cos(\theta)f'(z), \sin(\theta)f'(z), 1)$, and $\Phi_\theta(\theta, z) \times \Phi_z(\theta, z) = (\cos(\theta)f(z), \sin(\theta)f(z), -f(z)f'(z))$. $\int\int_{\mathcal{S}} dS = \int_0^{2\pi} \int_a^b |f(z)|\sqrt{1 + f'(z)^2}\, dz\, d\theta$.
(b) (ii) $\int\int_{\mathcal{S}} dS = \int_0^{2\pi} \int_{-\pi/2}^{\pi/2} |\cos z|\sqrt{1 + \sin^2(z)}\, dz\, d\theta = 2\pi(\sqrt{2} + \ln(1 + \sqrt{2})) \approx 14.92$. (*Hint:* Let $u = \sin z$.)

4. Total Absorption. The total absorption must be computed for the walls, floor, and ceiling. The calculation from Example 7.7 shows that the total absorption of the back wall is 98.85. For the wall lying in the xz-plane, $\Phi(s, t) = (s, 0, t)$ for $5 \le s \le 20$ and $0 \le t \le 3 + 0.5s$, and $\int\int_{\mathcal{S}} 0.2\, dS = \int_5^{20} \int_0^{3+0.5s} 0.2\, dt\, ds = 27.75$. The analogous calculation for the other wall also yields the value 27.75. For the front wall, $\Phi(s, t) = (s, 5 - s, t)$ for $0 \le s \le 5$ and $0 \le t \le 5.5$, and $\int\int_{\mathcal{S}} 0.2\, dS = \int_0^5 \int_0^{5.5} 0.2\sqrt{2}\, dt\, ds \approx 7.78$.

For the floor, $\mathbf{\Phi}(r, \theta) = (r \cos\theta, r \sin\theta, 0)$ for $\frac{5}{\cos\theta + \sin\theta} \le r \le 20$ and $0 \le \theta \le \pi/2$, and $\int \int_S 0.6 \ dS = \int_0^{\pi/2} \int_{5/(\cos\theta+\sin\theta)}^{20} 0.6r \ dr \ d\theta \approx 181.00$. For the ceiling, $\mathbf{\Phi}(r, \theta) = (r \cos\theta, r \sin\theta, 3 + 0.5r \cos\theta + 0.5r \sin\theta)$ for $\frac{5}{\cos\theta + \sin\theta} \le r \le 20$ and $0 \le \theta \le \pi/2$, and $\int \int_S .65 \ dS = \int_0^{\pi/2} \int_{5/(\cos\theta+\sin\theta)}^{20} 0.65\sqrt{1.5}r \ dr d\theta \approx 240.15$. The total absorption of the room is $\approx 98.85 + 2(27.75) + 7.78 + 181.00 + 240.15 = 583.28$.

■ 7.3 Flux Integrals

1. **Cylindrical Coordinate Parametrizations.** (a) S is a horizontal disk of radius 2 centered at the point $(0, 0, 3)$. $\mathbf{\Phi}$ is continuously differentiable, and $\mathbf{N} = \mathbf{\Phi}_r \times \mathbf{\Phi}_\theta = (0, 0, r)$, which is nonzero on the interior of the domain. Hence $\mathbf{\Phi}$ is regular. $\mathbf{N}$ is an upward pointing normal field.
(b) S is a vertical rectangle of base 3 units and height 1 unit with base the segment from the origin to $(3/\sqrt{2}, 3/\sqrt{2}, 0)$. $\mathbf{\Phi}$ is continuously differentiable, and $\mathbf{N} = \mathbf{\Phi}_r \times \mathbf{\Phi}_z = (\frac{\sqrt{2}}{2}, -\frac{\sqrt{2}}{2}, 0)$, which is nonzero on the interior of the domain. Hence $\mathbf{\Phi}$ is regular. $\mathbf{N}$ points clockwise around the z-axis.

2. **Spherical Coordinate Parametrizations.** (a) S is the portion of a sphere of radius 2 centered at the origin in the first octant. $\mathbf{\Phi}$ is continuously differentiable, and $\mathbf{N} = \mathbf{\Phi}_\theta \times \mathbf{\Phi}_\phi = (4 \cos\theta \sin^2\phi, 4 \sin\theta \sin^2\phi, -\sin\phi \cos\phi)$, which is nonzero on the interior of the domain. Hence $\mathbf{\Phi}$ is regular. $\mathbf{N}$ is an inward pointing normal field.
(b) S is a vertical quarter-disk of radius 2 with vertical side on the negative z-axis and horizontal side in the xy-plane and making an angle of $\pi/3$ with the positive x-axis. $\mathbf{\Phi}$ is continuously differentiable, and $\mathbf{N} = \mathbf{\Phi}_\rho \times \mathbf{\Phi}_\phi = (-\rho\frac{\sqrt{3}}{2}, \frac{\rho}{2}, 0)$, which is nonzero on the interior of the domain. Hence $\mathbf{\Phi}$ is regular. $\mathbf{N}$ points counterclockwise around the z-axis.

7. **Boundary Surfaces.** (a) (See Exercise 6.) $S = S_1 \cup S_2 \cup S_3$, the union of the lateral, top, and bottom faces of the region, respectively. S_1 can be parametrized by $\mathbf{\Phi}_1(\theta, z) = (\cos\theta, \sin\theta, z)$ with domain $\mathcal{D}_1 = [0, 2\pi] \times [0, 3]$ and outward pointing normal $\mathbf{\Phi}_{1_\theta} \times \mathbf{\Phi}_{1_z} = (\cos\theta, \sin\theta, 0)$. S_2 can be parametrized by $\mathbf{\Phi}_2(r, \theta) = (r \cos\theta, r \sin\theta, 3)$ with domain $\mathcal{D}_2 = [0, 1] \times [0, 2\pi]$ and upward pointing normal $\mathbf{\Phi}_{2_r} \times \mathbf{\Phi}_{2_\theta} = (0, 0, r)$. S_3 can be parametrized by $\mathbf{\Phi}_3(\theta, r) = (r \cos\theta, r \sin\theta, 0)$ with domain $\mathcal{D}_3 = [0, 2\pi] \times [0, 1]$ and downward pointing normal $\mathbf{\Phi}_{3_\theta} \times \mathbf{\Phi}_{2_r} = (0, 0, -r)$. $\int \int_S \mathbf{F} \cdot \mathbf{N} \ dS = \int_0^{2\pi} \int_0^3 1 \ dz \ d\theta + \int_0^{2\pi} \int_0^1 2r \ dr \ d\theta + \int_0^{2\pi} \int_0^1 r \ dr \ d\theta = 6\pi + 2\pi + \pi = 9\pi$.
(b) Intuitively, the total flux of a constant vector field out of a closed surface will be zero. (Why?). Here are the calculations: $S = S_1 \cup \cdots \cup S_6$, the union of the top, bottom, and lateral faces with $x = -1$, $x = 1$, $y = -1$, and $y = 1$, respectively. Letting $\mathbf{\Phi}_1(x, y) = (x, y, 2)$ with domain $\mathcal{D}_1 = [-1, 1] \times [-1, 1]$ and upward pointing normal $\mathbf{\Phi}_x \times \mathbf{\Phi}_y = (0, 0, 1)$ and parametrizing the other faces analogously, $\int \int_S \mathbf{F} \cdot \mathbf{N} \ dS = \int_{-1}^1 \int_{-1}^1 2 \ dy \ dx + \int_{-1}^1 \int_{-1}^1 -2 \ dy \ dx + \int_{-1}^1 \int_{-2}^2 -3 \ dz \ dy + \int_{-1}^1 \int_{-2}^2 3 \ dz \ dy + \int_{-1}^1 \int_{-2}^2 -1 \ dz \ dx + \int_{-1}^1 \int_{-2}^2 1 \ dz \ dx = 8 - 8 - 24 + 24 - 8 + 8 = 0$.

11. Total Acoustic Power. (a) Using the usual parametrization of the sphere with an outward pointing normal and the fact that $\|\mathbf{x}\| = R$ on the sphere, $\int\int_{\mathcal{S}} \mathbf{I} \cdot \mathbf{N} \, dS = \int_0^\pi \int_0^{2\pi} \frac{A^2}{\rho_0 c} \sin(\phi) \, d\theta \, d\phi = \frac{4\pi A^2}{\rho_0 c}$.

(b) It does not depend on the radius, which makes sense because the total amount of acoustic energy leaving any sphere containing the source should be the same. However, the larger the sphere, the larger the surface area and the smaller the flux per unit area.

■ 7.4 The Divergence Theorem

1. Gas Flow. (a) If the temperature is increasing while the pressure is constant, the volume occupied by a number of particles must increase. This means that there must be a net outflow of particles from a fixed volume of the space occupied by the gas.

(b) If the temperature remains constant while the pressure increases, the volume occupied by a number of particles must decrease. This means there must be a net inflow of particles into a fixed volume of the space occupied by the gas. Therefore, the divergence must be negative.

5. A Total Flux Calculation. Since div $\mathbf{F}(x, y, z) = 3$, the total flux is 3 times the volume of the region, which is $\frac{4}{3}\pi(R_2^3 - R_1^3)$. Therefore, the total flux is $4\pi(R_2^3 - R_1^3)$.

9. Total Acoustic Power. (a) By the divergence theorem, the total acoustic power is the integral of div $\mathbf{I}$ over the region enclosed by $\mathcal{S}$, which is zero.

(b) Suppose $\mathcal{S}_1$ and $\mathcal{S}_2$ are closed surfaces that enclose the source and with $\mathcal{S}_1$ contained in the interior of $\mathcal{S}_2$. Then Exercise 6 applies, and $\int\int_{\mathcal{S}_1} \mathbf{I} \cdot \mathbf{N} \, dS = \int\int_{\mathcal{S}_2} \mathbf{I} \cdot \mathbf{N} \, dS$. If neither $\mathcal{S}_1$ nor $\mathcal{S}_2$ is contained in the interior of the other, use a sphere $\mathcal{S}_3$ of larger radius containing both $\mathcal{S}_1$ and $\mathcal{S}_2$. Apply the argument to $\mathcal{S}_1$ and $\mathcal{S}_3$ and then to $\mathcal{S}_2$ and $\mathcal{S}_3$ to show that the flux integrals for $\mathcal{S}_1$ and $\mathcal{S}_2$ are equal. This proves the claim.

(c) $\mathbf{I}(\mathbf{x}) = \frac{A^2}{\rho_0 c} \frac{1}{\|\mathbf{x}\|^3} \mathbf{x}$; div $\mathbf{I} = \frac{A^2}{\rho_0 c} \left(\left(\frac{-3x^2}{\|\mathbf{x}\|^5} + \frac{1}{\|\mathbf{x}\|^3} \right) + \left(\frac{-3y^2}{\|\mathbf{x}\|^5} + \frac{1}{\|\mathbf{x}\|^3} \right) + \left(\frac{-3z^2}{\|\mathbf{x}\|^5} + \frac{1}{\|\mathbf{x}\|^3} \right) \right) = 0$; and integrating over a sphere of radius 1 centered at the origin with a unit outward normal, $\int\int_{\mathcal{S}} \mathbf{I} \cdot \mathbf{N} \, dS = \int\int_{\mathcal{S}} \frac{A^2}{\rho_0 c} \, dS = \frac{4\pi A^2}{\rho_0 c}$.

10. Gauss's Law. (a) This calculation is essentially identical to that of Exercise 9(c).

(b) See Exercise 9(a).

■ 7.5 Curl and Stokes' Theorem

1. Curl Calculations. (a) curl $\mathbf{F} = (-1, -1, 1)$. (b) curl $\mathbf{F} = (0, xy, -xz)$.

2. Right-Hand Rule Constructions. (a) $\mathcal{S}$ is the disk of radius 3 centered at the origin in the xz-plane oriented by the normal vector field $\mathbf{N} = (0, -1, 0)$.

(b) S is the elliptical disk of semi-major axis 3 in the y-direction and semi-minor axis 2 in the z-direction centered on the origin in the yz-plane oriented by the normal vector field $\mathbf{N} = (-1, 0, 0)$.

7. **Gradient and Curl.** (a) curl $(\nabla f) =$ curl $(\frac{\partial f}{\partial x}, \frac{\partial f}{\partial y}, \frac{\partial f}{\partial z}) = (\frac{\partial^2 f}{\partial y \partial z} - \frac{\partial^2 f}{\partial z \partial y}, -\frac{\partial^2 f}{\partial x \partial z} + \frac{\partial^2 f}{\partial z \partial x}, \frac{\partial^2 f}{\partial x \partial y} - \frac{\partial^2 f}{\partial y \partial x}) = (0, 0, 0)$.

8. **Potential Functions.** (a) $f(x, y, z) = \frac{1}{2}(x^2 + y^2 + z^2)$.
   (b) $f(x, y, z) = -\cos x - \cos y - \cos z$.

9. **Stokes' Theorem Calculations I.** (a) $\int \int$ curl $\mathbf{F} \cdot \mathbf{N} \, dS = \int_0^1 \int_0^{2\pi} (1, -1, 1) \cdot (0, 0, r) \, d\theta \, dr = \pi$.
   (b) $\int \int$ curl $\mathbf{F} \cdot \mathbf{N} \, dS = \int_0^1 \int_0^1 (-1, 1, 1) \cdot (0, -1, 0) \, dx \, dy = -1$.

10. **Stokes' Theorem Calculations II.** In each part, the choice of normal gives a counter-clockwise orientation of the boundary.
   (a) $\int_{\mathcal{C}} \mathbf{F} \cdot \mathbf{T} \, ds = \int_0^{2\pi} (\sin \theta, 0, 2\cos \theta) \cdot (-\sin \theta, \cos \theta, 0) \, d\theta = -\pi$.
   (b) $\int_{\mathcal{C}} \mathbf{F} \cdot \mathbf{T} \, ds = \int_0^{2\pi} (\cos^3 \theta, \sin^2 \theta, 3\cos \theta) \cdot (-\sin \theta, \cos \theta, 0) \, d\theta = 0$.

■ 7.6 Three Proofs

1. **Divergence Zero.** Let $\mathbf{F}$ denote the velocity field of the flow, and let $\mathbf{x}_0$ be a point in the domain of $\mathbf{F}$. If the amount of fluid entering a region of space is balanced by the fluid entering the region, then the flux integral of $\mathbf{F}$ over the boundary of the region will be zero. This means that the numerator in the definition of infinitesimal flux is always zero and, consequently, the infinitesimal flux will be zero. It follows from the proposition relating infinitesimal flux to divergence that div $\mathbf{F} = 0$.

3. **Stokes' and Green's Theorems.** Green's theorem states that $\int_{\tilde{\mathcal{C}}} \tilde{\mathbf{F}} \cdot \mathbf{T} \, ds = \int \int_{\tilde{\mathcal{R}}} (\frac{\partial F_2}{\partial x}(x, y, 0) - \frac{\partial F_1}{\partial y}(x, y, 0)) \, dA_{x,y}$. Applying Stokes' theorem to the integral of $\mathbf{F}$ around $\mathcal{C}$ yields $\int_{\mathcal{C}} \mathbf{F} \cdot \mathbf{T} \, ds = \int \int_{\mathcal{R}}$ curl $\mathbf{F} \cdot \mathbf{N} \, dS$. Parametrizing $\mathcal{R}$ by $\mathbf{\Phi}(x, y) = (x, y, 0)$ for $(x, y) \in \tilde{\mathcal{R}}$ produces the normal vector $\mathbf{N} = (0, 0, 1)$. Then curl $\mathbf{F} \cdot \mathbf{N} = \frac{\partial F_2}{\partial x}(x, y, 0) - \frac{\partial F_1}{\partial y}(x, y, 0)$ on $\mathcal{R}$. Since $dS = dy \, dx$ for $\mathbf{\Phi}$, we see that the double integrals are the same.

Index